AF797775

Positioning Playbook

The ultimate guide to narrative, positioning, & brand for B2B SaaS

Luke Stevens

First published by Positioning Science 2024

Copyright © 2024 Luke Stevens

All rights reserved. No portion of this book may be scanned, distributed, or reproduced in any form without written permission from the publisher, except as permitted by copyright law. No portion of this book may be used in training Large Language Models or any other form of artificial intelligence without the prior written consent of the author. The author/publisher is not associated with any individual, product, or company mentioned in this book (except where client relationships are disclosed), and no individuals or companies referenced in this book endorse the book.

Cover illustration by Milad Fakurian c/o Unsplash.

Cover design, book design, and ebook production by Luke Stevens.

Images, typefaces, and screenshots © their respective owners and are included under fair use and/or as permitted by their license.

Giving credit: Want to use any of these ideas in your social media think pieces? Credit is all I ask :) Otherwise, expect to be haunted through space and time by yours truly.

First edition v1.00

ISBN: 978-0-9756652-8-2

Contents

1

Welcome — start here

Welcome! This is a big book about a big idea that affects just about everything. It's written for B2B marketers in tech — and product marketers in particular — but the ideas we'll explore will be relevant to startup marketers, copywriters, sellers, founders, and just about everyone involved in the sales and marketing side of selling software or B2B tech more broadly.

Before we dive in, two things to note:

- I'd love to stay in touch via my newsletter — **https://lukestevens.co/newsletter/** or on LinkedIn **https://www.linkedin.com/in/ldstevens/**.
- This book is for practitioners and those who want to go deep. If you want something more accessible for your boss or colleagues, check out the science-based positioning articles on my site **https://lukestevens.co** and/or my book *Founder Vision*, which summarizes this book in ~160 pages and is available in audiobook form, too.

A book about brains

On a pragmatic level, this is a book about (almost) all things positioning and product marketing — narrative, positioning, brand, category creation, market strategy, and GTM leadership in B2B SaaS.

But it's also about brains.

Markets are made up of people; people have brains; and we're going to explore some profound insights into how those brains work.

And on that deeper, squishier level, the scientific theory that much of this book is based on — the hierarchy of attention — has the potential to profoundly change the way you think about brains, narrative and positioning, and indeed product marketing itself.

Let's think about where product marketing fits for a moment. On the one side, you've got a product, and the R&D organization that builds and supports that. On the other side, you've got a market and the sales and marketing organization that engages that market.

What we want, of course, is for those two sides to fit, hence the idea of product/market fit. And right in the middle of product/market fit sits product marketing.

The value of market/product fit

Startups and tech companies often obsess about building the right product, too often believing they're just one more feature away from the market beating down their door.

The reality is that startup success is just as much about market/product fit, that is, deeply understanding your market *and* competitive environment so you make something people actually want *and* that's sufficiently differentiated from the competition.

That is, of course, very hard to do, and that's why many startups fail. And because it's both hard to do and critical to the success of a startup, getting it right is very valuable.

It's so valuable, in fact, that when I snuck into a venture-backed unicorn as a copywriter-turned-product marketer, I saw them spend six figures — twice! — on six-week engagements on narrative work *alone.*

And that was work that was, dear reader, frankly *not very good.*

On the very good side, however, the value goes up even more. The world's leading B2B positioning consultant, for example, currently charges up to $80,000... for a *three-day workshop.*

The demand is there. It's *incredibly* valuable work... when done well. But back here on earth for us mere mortals, salaries and consulting rates tend to be more mid-range.

Why? Because the positioning work product marketers often do is more mid-range. I believe that is, in part, because product marketing is often seen as more low-level 'doing' work, i.e., creating the marketing assets for the product — the messaging, the decks, the competitor analysis, and the "positioning" in a very narrow, us-vs-them sense.

That stuff is all great — I've done it, enjoyed it, and helped drive messaging clarity through that kind of execution work.

But what's been missing are deep fundamentals that truly dedicated product marketers (and those in adjacent consulting roles) can bring to the table to do positioning work akin to those charging six-figures a *month.* And you, dear reader, may well be one of those dedicated product marketers. Let's face it, this is not a book written for the masses. It's written for those who, like me, to be totally candid, hope to deliver more value than typical mid-range work and make more money than typical mid-range rewards.

But to do that, we need to take positioning and product marketing further. Therefore, what would it mean for product marketing to:

- Truly own the narrative the startup runs on?
- Truly understand positioning and quarterback a GTM plan to match?
- Not just represent the voice of the customer, but the very way the customer attends to the world around them?

This is the opportunity for product marketing — and those that wear the product marketing hat — to own the *qualitative* side of a startup's entire

GTM operation, from helping articulate the vision, spot insights, and deliver messaging that truly does create a new position in the buyer's mind and in the market more broadly.

New fundamentals, new opportunities

This opportunity requires new fundamentals — it doesn't come from just doing the same things we've always done — and those new fundamentals come from our new theory: the hierarchy of attention.

The hierarchy of attention is a modern, breakthrough insight into the nature of left- and right-brain attention, and we're going to unpack what it means for the fundamental way we create narratives that sell. This is a profound new way for startups and tech companies to get clarity on their story and strategy: specifically, the two story types, your audience's two modes of attention, and how to address both in your sales narrative and go-to-market (GTM) strategy.

We'll also take a renewed look at positioning fundamentals, from creating categories to crossing chasms, diving into brand, growth flywheels, Jobs to be Done, SMB vs. enterprise, and competitive strategy while we're at it. It's all here, cast in a new light with a fresh, left-brain/right-brain twist.

These are the new and renewed fundamentals of narrative, positioning, and brand that, if harnessed fully, can change business trajectories, make marketing careers, and, to be blunt, help you sell a lot of software.

First principles for selling

We're going to spend a lot of time initially on how these new first principles of attention shape how we tell sales stories that connect with B2B buyers in particular, as B2B sales in the tech world often involve a lot of applied attention, especially in 'we not me' enterprise deals that may include everyone from end users to managers to the C-suite.

What drives these deals forward — from initial interest to overcoming internal organizational challenges so a deal gets done — is a strong, clear narrative that engages with both modes of attention so all involved — executives and folks in the trenches — know exactly why they're putting in the effort to make the deal happen, especially when the inevitable roadblocks appear.

Many deals fall over — or fail to get started — because of fundamental attention mismatches that, until now, have gone undiagnosed and unaddressed because we haven't had the language to describe them.

Science-based positioning

Now we do, and I call this approach science-based positioning.

Take this with a grain of salt — think of it as science-*ish* positioning if you like — but the fact is we now have new science-based primitives of attention, in terms of both mode and quantity, that allow us to take a far more science-informed approach to positioning, narrative, and brand.

Indeed, those terms — positioning, narrative, and brand — are ones I now just map to different modes and quantities of the brain's attention. As far as fundamentals go, it doesn't get much more fundamental (or customer-centric) than starting with how the brain works and connecting it to the fundamentals of your strategy as a startup or tech company.

This is how product marketing — and marketing more broadly — goes from being in-the-weeds 'doers' to strategic thinkers who can help conduct the GTM orchestra, making sure the pieces all fit.

> **Clarity & science-based positioning:** To get a little meta, the term "science-based positioning" is actually my attempt to use the clarity strategy techniques we're about to discuss and apply them to the contents of this book. In other words, it's a product of me taking my own medicine. The development of this book happened over three stages:
>
> 1. Research to understand the world I wanted to engage with.
> 2. World-building to create a new model of the world.
> 3. Distillation to create a gateway to this world.
>
> This book is largely about #2, with plenty of references to #1. This is the world-building I've done that I'll dine out on for years to come — and so can you, given you've made it this far. But I still needed #3 — a door to this world, and that's where science-based positioning (SBP) comes in, which, along with founder vision, is what I talk about publicly. But from here on out, while we're operating under the banner of science-based positioning, for the rest of the book we'll talk about your *clarity strategy,* which is the actual playbook you can put into action. Enjoy!

Haven't we been here before?

This new approach might sound a lot like an old approach. If you're thinking, "Right brain/left brain, split-brain stuff? I thought that was debunked years ago!" you're both right and wrong, and it kind of doesn't matter.

You're right that the simplistic pop-science takes on the two hemispheres — like "logic" *only* happens on the left side, or "creativity" *only* happens on the right — were wrong and were debunked. Both sides of the brain are involved in everything all the time.

But! The brain still has two hemispheres, and in recent years, a much more modern, thorough, and serious academic take on that hemispheric split has appeared. That theory says the hemispheric split exists for a reason and that each hemisphere has quite a different *take* on the world — a different *mode of attention*.

That's a big deal because it unlocks how we understand attention. And that understanding is so important — and describes such a common, universal experience we all have as humans — that it doesn't really matter how much you buy into *exactly* what's happening in each hemisphere to generate these two 'takes,' there's still a profound truth about our modes of attention that we can work with, and boy are we going to work with it. I'm going to stick with right-brain/left-brain labels to describe this world, but you do you.

We know what works

This is, therefore, an ambitious, science-based positioning book that goes from the micro of differences between the brain's two hemispheres to the macro of what we know about how tech markets fundamentally work.

It turns out we know how many of the ingredients of positioning works, including attention (sales and marketing), adoption (innovative technology), memory (brand), and competition (differentiation)... we know how these fundamentals work thanks to years of research and thinking on these topics by plenty of very smart people.

But many founders, operators, individual contributors, and consultants *don't know* that we know how these fundamentals work. My sincere hope is that this book changes that. This is a new era, with new fundamentals, and — in my not-so-humble view — a new default way to think about B2B narrative and positioning.

> **Sharing these insights:** These insights are here to be shared, too. If other folks you know could use these ideas, be they your leadership team, direct reports, management, or investors, be sure to pass on a copy of or link to this book, or the free content at lukestevens.co. I'm incredibly grateful to you for any help at all in spreading the word!

New narrative

This book wouldn't exist without the hard work of those folks who have studied, practiced, and shared their research and business experience over the years. In that sense, we'll be perched precariously on the shoulders of giants as we explore both theory and case studies, handy frameworks and hard-won lessons, with plenty of practical tips to get you started, too.

As I'll explain in the next chapter, we're going to start with a brief introduction to the core concepts, and then we'll get hands-on as soon as possible. I'm *super* keen to help you get started on clarifying your narrative first and foremost, and to that end, I've front-loaded the upcoming chapters with as many tools and insights as possible so that you can get hands-on with your sales narrative right away.

We'll start with your narrative because it's the fastest way to iterate on your ideas around your pitch, story, and positioning and test out directions before committing to them. Therefore, we'll get a snapshot of your narrative down first, and then the rest of this book is designed to help you stretch your thinking on what your narrative and positioning could and ultimately should be.

Your clarity project

Positioning, to me, is ultimately about clarity — clarity about what the market needs and clarity in how you, as a company, articulate how your product meets those needs in a unique way.

The whole point of this book is about getting clarity and indeed "Clarity" was the working title of this book. Therefore, we're going to talk about clarity, and every section in this book is about providing clarity on a critical area of positioning.

How do you put that clarity to work? I suggest you use the ideas in this book as the basis for a narrative and positioning *clarity project* you run internally (or with your clients) to help align the team and better engage your prospects and customers.

Have a read of the next few chapters, share your ideas with your colleagues, and then think about how you can develop your clarity strategy — the umbrella term we'll be using for this work — to unlock clarity in your company's product marketing in terms of story, pitch, positioning, and direction.

What's in it for you?

The result of reading this book should be an extremely clear narrative, but it can be much more than that, too, including:

- Strong messaging for recruiting, leadership, and sales.
- A clear, catchy micro-pitch that can meme a market.
- Targeted pitches to the enterprise C-suite, outbound folks, and/or folks

in the trenches.
- A deep understanding of how category, brand, and positioning really work.
- Clarity in product development and how you direct your product teams to solve customer problems.
- Greater sales and marketing efficiency that comes with having all your GTM ducks in a row.
- A more aligned leadership team with your R&D and GTM organizations pulling in the same direction.
- Clarity about how you fundamentally compete to win in your market.

These are the big juicy outcomes I hope this book pushes you towards, but they don't happen on their own. They happen because you take the lead and start a project that addresses the fundamentals that we will discuss in this book and make them happen.

So, step one: start thinking about your narrative and positioning project so you can bring folks with you.

Your new reference guide

Building enduring, high-growth companies requires more discipline, more insight, and, frankly, more clarity than most folks realize, and that's why I've packed this book with as much clarifying insight as I possibly can. Probably to a fault, but hey, this isn't some brain fart of a business book. This is world-building — I want a fully fleshed-out world to draw from, and I hope this approach inspires you in your own narrative world-building, too. Either way, I hope this book provides a handy positioning reference for months and years to come.

I'd love to hear what resonates with you: luke@lukestevens.co.

Next, let's take a moment to explore how this book works so you can chew on the chapters that are most relevant to you.

2

How this book works

The *Playbook* is a big, textbook-length deep dive that can deliver both immediate insights and serve as a reference for years to come, so it pays to take a moment to get oriented.

Two-in-one

First, it's worth understanding this is essentially two books in one:

1. **Sales narrative how-to:** The first third of this book is a practical guide to creating and applying your sales narrative for your B2B SaaS or tech solution. There's a deck outline, messaging tips, pitch outlines for different sales contexts, and plenty on the core idea that drives this book — the hierarchy of attention.
2. **Narrative & positioning textbook:** The rest of this book is stuffed to the gills with ideas on product marketing: concept, category, GTM strategy, and positioning — especially positioning. This is the more textbook-style half of the book that's intended to stretch your thinking — and your market strategy — so you can grow your business more strategically and less chaotically.

That is, we're aiming for both clarity in your narrative as a deck and clarity in the ideas that power your narrative conceptually. We'll go practical first so you have something to work with, and then we'll go deep so you can get your strategic ducks in a row.

Peeling the onion

Second, it's worth understanding the structure of this book.

It might be textbook length, but I've tried to keep it as accessible and practical as possible, with the core ideas up front in the first few chapters. If you want to go deeper and peel back more layers of the onion, then just keep reading, and we'll go deeper and deeper into the ideas until we arrive at the very foundation of strategy itself at the end of the book. But if those early outer layers of the proverbial onion are all you need, great! Grab what you need, try it out, and come back later when you face fresh challenges. The content will still be here!

But if you really want to go deep, the ideas are there. I've taken the kitchen-sink approach to this project, and if you enjoy nerding out on this stuff as much as I do, there's plenty of meat to sink your teeth into. Just don't let the book's length intimidate you!

> **Go down the rabbit hole:** I've also tried to be as thorough with my references as possible, and I hope the various experts, articles, and books I reference open more doors for further exploration of topics that interest you, too.

Clarity in 10 parts

This book is broken up into the following ten parts so that it is, I hope, easy to jump to whatever topic is of most interest to you. Consider this a table of *concepts* you'll learn. Here's a brief rundown of each part:

1. Clarity & developing a clarity strategy

You'll learn:

- The basics of the **hierarchy of attention**.
- The fundamental concepts of left-brain and right-brain attention (and resulting **'story 1'** and **'story 2'** stories).
- The importance of creating your own **clarity strategy.**
- The classic **narrative mismatches** that plague startups.

2. Clarity & your narrative snapshot

You'll learn:

- How **Segment** discovered that sales was vital for product/market fit in B2B.
- How to create a **narrative snapshot** in the form of a sales deck (with a ten-slide example).
- How to create a **micro-pitch** (your 'Hemingway story').
- How to take inspiration from your **origin story** (who, what, when, where, why, how).
- Why a **Slack**-inspired memo makes for a great long-form narrative.

3. Clarity & messaging

Now, with narrative in hand, you'll learn:

- The 'think, align, execute' framework for turning your narrative into **messaging.**
- What to look for with the **web messaging checklist.**
- The importance of **memes** and **metaphors** in your messaging.
- Where mission statements fit and how **Stripe** evolved their homepage

messaging as they matured.
- Don't miss the **naming cheatsheet**, either.

4. Clarity & pitching

You'll learn:

- Why pitching is like an arcade **claw game.**
- More about the two schools of clarity strategy, including.
    - The story 1 '**strategic narrative**' approach.
    - The story 2 **positioning approach** using **Snowflake** as an example.
- What happened in a **natural story 1 experiment** I saw play out.
- How to apply these approaches at different levels of customer awareness, with example deck outlines (**CTO** pitch, **Keap's** SMB pitch, and a mid-market **SEO tool** pitch).
- What the hierarchy of attention means for advanced **Challenger** selling.

5. Clarity & vision, alignment & insight

You'll learn:

- How the all-important **right/left/right-hemisphere jump** works, especially for vision and insight generation.
- The implications of the R/L/R jump for both sales and startup management, including the **vision → details → integration** alignment loop.
- How the R/L/R jump shows up in product development at **Figma.**
- The **reality layer vs. speculation layer** and the ever-present risk of **brain putty**.

6. Clarity & concept vs. category

You'll learn:

- Why right-brain, story 1 narratives are a function of secular change, with a look at **Gainsight** vs. **Hopin** and the **concept = change + playbook + tool** formula.
- How this played out in the CRM world of **Salesforce**, **HubSpot**, and **Drift**.
- What **unbundled category creation** looks like.
- How **Yammer** created an actual category.
- How the **category creation narrative** swept the VC world.
- Where the category creation narrative needs to be debunked, untangled (separating categories, concepts, and Aggregators like **Google**, **Amazon**, and **Facebook**), and better understood.

- How **Qualtrics** used a category-transcending playbook as a brand and narrative tool.
- How **Play Bigger** uses new names, theoretical new disciplines, and new playbooks in their own category creation playbook.

7. Clarity & JTBD

You'll learn:

- Why the **Jobs to be Done** theory is the linga franca of needs in the startup world.
- How the different **JTBD flavors** (Ulwick, Christensen, and Moesta) map to our right-brain/left-brain, story 1/story 2 modes of attention.
- What the **five 'fits' of JTBD** look like, including language/market fit, position/outcome fit, sales process/buyer jobs fit, innovation/user jobs fit, and analyzing competitor/market fit.
- *This is a handy primer for our positioning deep dive!*

8. Clarity & positioning

You'll learn... a lot:

- You'll learn about the different approaches to positioning, from **positioning-as-story** (language, story 2 pitching) to **positioning-as-GTM** (crossing the chasm, reach vs. niche).
- What the modern, left-brain, story 2, **April Dunford**-inspired approach to positioning looks like, especially when it comes to finding **user innovation.**
- How you can test your pitch with the **Eyes Light Up** approach.
- How early-stage **Dock** iterates through feedback on their pitch and the challenge of **vertical vs. horizontal** positioning.
- How **ConvertKit** used **niche positioning** and direct sales to build initial momentum.
- Why you need to consider **nerds vs. normies** when getting started and again when trying to 'cross the chasm' from early adopters to the mainstream.
- How distribution has changed from just **'segment & sell'** to now include **'cast & catch'**, and why distribution mismatches can bring your entire GTM strategy unstuck.
- How **Pebble** failed to cross the chasm, in part due to misplaced right- and left-brained attention.
- What **positioning for reach** looks like with **Squarespace** using B2C 'mental availability' brand tactics.
- What positioning across the **B2B spectrum** looks like from SMB (**Patreon**) to enterprise (**Whatfix** & **UiPath**).

- What positioning in the mid-market looks like with **wedge positioning** by attribute, with a case study of pitching the **LMS market**.
- Why B2B positioning is often an exercise in finding the **MAYA** (Most Advanced, Yet Acceptable) threshold.

9. Clarity & competitive strategy

You'll learn:

- Why **uniqueness** is the fundamental force of competition.
- How that plays out in two broad strategies: **competing to be unique** and **competing to be the same** (a.k.a. Porter vs. Hotelling).

10. Clarity in conclusion

You'll learn that:

- We now know how the fundamentals of the "market" in "product/market fit" work.
- We know how attention works; we know how adoption works; we know how brand works; and we know how positioning and strategy works.
- You can only put these ideas into practice if you start writing, **so get cracking on your narrative**.

Phew! They're the ten parts of the book — choose your fighter and dive in!

Jump around

You're welcome to read through those sections from start to finish (and a big high five if you do!), but again, I've included plenty of places where you can skip ahead to follow a particular theme, too. There's loads of value waiting for you in both the early, hands-on chapters and the later strategy-focused chapters, so in the immortal words of Kris Kross, jump around.

3

About Luke (shameless plug)

You know how some business books are thinly veiled advertisements for a consultancy?

Given the size and scope of this book (at ~225k words it's more than 3x the size of your average business book), and given I'm a consultant, I suppose this book is perhaps one of the most *thickly* veiled advertisements one could possibly write.

But fear not, dear reader, because there are no veiled ads here. We're all adults, so there's no wink-wink, nudge-nudge, hire-me stuff, apart from this very non-veiled page, in which I'll put it as simply and bluntly as possible: **You can hire me.**

- I'm a **solo, hands-on consultant** who works with B2B SaaS companies to help them clarify their narrative and positioning.
- I focus on **sales narratives & positioning deliverables,** and I've been doing it for a while now.
- I've got a **thorough framework** to take clients through, as this book demonstrates (!), and that can be applied in **short sprints** and **ongoing engagements**.
- If you've got some initial traction and need help nailing your positioning and narrative, **let's talk**.
- Email me: **luke@lukestevens.co** or visit **https://lukestevens.co**.

That's it; that's the plug!

I hope that wasn't too painful. I've poured everything I know into this book, and I've tried to give you all the tools and advice so you can DIY, but if you get stuck, you know where to find me.

— Luke

PART I

CLARITY & YOUR CLARITY STRATEGY

4

The hierarchy of attention

> [T]he left hemisphere yields narrow, focussed attention, mainly for the purpose of getting and feeding. The right hemisphere yields a broad, vigilant attention, the purpose of which appears to be awareness of signals from the surroundings, especially of other creatures, who are potential predators or potential mates, foes or friends.
>
> *- Dr Iain McGilchrist*[1]

The hierarchy of attention is the foundational idea of this book, popularized by one of the most important and, perhaps, underrated thinkers of our time: Dr. Iain McGilchrist, psychiatrist, neuroscience researcher, and philosopher.

We're not doing business-book neurobabble here, dear reader. McGilchrist's work is profound and far-reaching and it's what we're going to be dining out on all throughout the coming chapters. It's deep stuff, but let's start with a simple introduction.

Short version: Global radars, local laser beams

So, short version: brains. The big noodle between our ears. It's split in two: left and right hemispheres. For a while, as we touched on in the intro, it was popular to think each hemisphere did its own particular things (the left brain was 'logical', the right brain was 'emotional' or 'creative,' and so on). That was debunked, and the theory fell out of favor. That was the *old* hemisphere hypothesis.

Then along came McGilchrist.

McGilchrist has spent decades of his career researching and studying the split-brain phenomenon and, as far as your humble author can tell, he seems to have almost single-handedly put the new hemisphere hypothesis back on the map. That's what we're using here.

In McGilchrist's telling, both sides of the brain are involved in everything (contra the old hemisphere hypothesis), but *how* they're involved differs, and that is particularly true of attention, or how we 'attend' to the world around us. (It's important to note, therefore, that when I talk about "left-brain" or "right-brain," I mean that in terms of *dominant mode of attention,* not *exclusive neural activity* or anything like that. You'll have to forgive the shorthand.)

McGilchrist's approach is not without controversy[2], and it's obviously (and necessarily) reductive — the brain is vastly complex after all. Nevertheless, the hierarchy of attention works *beautifully* in a descriptive, metaphoric sense for us as we try and craft a strategy that engages with both modes of buyer attention.

Make it shorter

To make the theory even shorter, I've developed a bastardized version of McGilchrist's insight.[3] When it comes to our split brains, we have two types of attention:

- **Right-brain, global, outward attention is your *radar*.**
- **Left-brain, narrowly focused, downward attention is your *laser beam*.**

Radars and laser beams. Note that these are my terms and most certainly do not do justice to the depth and nuance McGilchrist brings to the topic. (Sorry, Iain!) McGilchrist makes his case so rigorously his book *The Master and His Emissary* has *thousands* of footnotes and was written over *two decades*. In this little book we're just trying to extract his powerful mental model of attention for our own greedy little ends. (Mwahaha!)

The mental model

While you could conceive of these two modes purely as two ends of a spectrum (from broad to narrow), and while there's some truth to that, the idea of a spectrum doesn't quite capture the essence of the insight. It's more about having two opposite modes of attention that we engage to varying degrees *at the same time*, perhaps with one or the other being dominant.

Therefore, the way I think about it (and, again, this is but a humble business book) is that these modes of attention are like tools we wield — broad radars and narrow laser beams of attention — operating both simultaneously, but usually giving priority to one or the other.

This is a surprisingly powerful mental model because it clarifies the key ways we generally *try* and do discovery (dig into the specifics) and *try* to sell our solution (put it on folks' radar) without really knowing *why*. But if we understand the mechanics of attention, we get insight into why.

For example, we — and our prospects — might be directing attention:

- **Down in the details:** We might try and engage our left-brain, narrow-focus attention to understand the details of what customers do. But do we use our narrow-focus attention enough? Do we really scrutinize our customers' problems? Are we prepared to engage that laser beam of attention to drill right down, not lazily pattern match, and then connect that all the way back up to what our product is about?

- **Up in the clouds:** We might try and tell big, right-brain, we're-the-new-hotness stories but fail to understand what drives these narratives, and have our attempts at 'creating a category' fall flat. Or we might paint a picture of a glorious future, but fail to join the dots between our vision and the left-brain analysis of customer needs.

It's *really* hard to use both modes of attention effectively and connect the two. It's easy to get stuck in one mode or the other, but ultimately they need to connect, and this book will give you the tools to help deploy both kinds of attention in appropriate and, I hope, growth-accelerating ways.

The two stories

If there's two distinct modes of attention, then it stands to reason that we need two types of stories to match. As we'll see in the chapters ahead, the two modes of attention roughly correspond to the two schools of narrative that we see in B2B SaaS: 'change stories' on the one hand, and 'positioning' on the other.

I like to simplify this as 'story 1' and 'story 2.' Here's how each story *type* matches a particular mode of attention:

- **Story 1 — Change stories (a.k.a. strategic narratives, category creation, or vision-driven stories):** Right-brain radar attention maps to big-picture change stories — megatrends, hype cycles, tech waves, secular change — it's about what's happening 'out there' in the world.
- **Story 2 — Positioning:** Left-brain laser beam attention maps to the narrow, job-to-be-done, hands-on-the-tools type of stories that often appear late in the sales journey, especially when folks are trying to figure out where you fit in their mental map of the market.

I like to think of these two stories as *out* vs. *down*, *observation* vs. *analysis*, and even *x-axis* vs. *y-axis* (as we'll see), but these are all just my descriptive labels that act as a shorthand for how we address the two fundamental modes of attention that McGilchrist describes in the opening quote.

We're going to explore these approaches in a probably ill-advised amount of depth in the rest of this book, but you can start thinking about them right now, given we'll soon get hands-on with a basic narrative sales deck in **chapter 9**.

> **Lean vs. 'fat' startups and analysis vs. observation:** The hierarchy of attention helps illuminate a fundamental debate about what startups should be — should they be 'lean,' experiment-driven operations, or

> 'fat,' vision-driven ones?[4]. Some folks gravitate to left-brain, narrow-focus attention, and that suits the classic lean, build-measure-learn loop where *analysis* is king. Others prefer the right-brain, vision-driven approach where founders bet their company on a future only they see — the 'fat' approach where future-looking *observation* is king. Neither is the 'right' way to go of course; both modes of attention are just tools to be used in line with your ambition and resources. But if we understand these modes of attention for what they are, we can then deploy them together to greater effect, wielding both radar and laser beam, both vision-driven observation and analysis, without being constrained to one or the other.

Hierarchy of attention and product/market fit

Fundamentally, the hierarchy of attention is a lens through which we can explore and develop product/market fit. That is, if we understand how attention works, particularly the two modes of attention, then we can be far more deliberate in how we develop our product and meet our market.

- **Product fit:** We can *direct* our company or startups' collective left- and right-brain attention, ranging from how we articulate our vision (right-brain), discover opportunities (left-brain), and solve problems (connect both), as we'll see with Figma in **chapter 27**.
- **Market fit:** We can *meet* our audience where *their* attention is at to help them discover us (putting our product on their radar) and showing how we solve their problems (working with their laser-beam attention) while avoiding the all-too-common story mismatches we'll discuss in **chapter 6**. This is the primary focus of this book.

The hierarchy of attention, then, cuts to the heart of product/market fit.

If you can better direct your internal attention to meet the attention of your customers, then it stands to reason you'll have a much greater chance of finding and strengthening product-market fit. This approach will also save you from wasting years in the market spinning your wheels, for example:

- **Founders** won't waste years stuck in one mode of attention or the other, unable to fully connect with their market.
- **Sales and marketing** won't waste their efforts failing to connect with prospects because they misjudge where their prospects' attention is at (e.g. rank-and-file folks vs. the C-suite in **chapter 22**, or awareness in **chapter 24**, or shifting prospect attention dynamically as we'll discuss in **chapter 25**).
- **Product folks** won't build pointless things because they failed to connect

the specifics of needs and features to the big picture of vision and mission.

Our simple hierarchy of attention — radars and laser beams included — is a deceptively simple mental model that can have very deep, very profound effects across your entire organization. But only if we have the appropriate framework to put it to use, and that framework will look different for different companies at different stages. Therefore, you need to craft your own clarity strategy, and that's what we'll explore next.

Notes

1. McGilchrist, Iain. *The Master and His Emissary* (p. 27). Yale University Press. Kindle Edition.
2. For a critique of McGilchrist's thesis see, for example, "McGilchrist and hemisphere lateralization: a neuroscientific and metaanalytic assessment" by Dr. Michael Spezio (**https://psycnet.apa.org/record/2019-63760-008**), and McGilchrist's response to Spezio and others (**https://www.tandfonline.com/doi/full/10.1080/2153599X.2019.1604418**). For my part, I obviously believe McGilchrist's work is worth engaging with for its insights into attention, the nature of the human experience, and just to enjoy the utterly expansive scope of his thinking. He's also a prophetic voice regarding the direction our civilization is heading in, and even if you don't buy into the cultural/philosophical side of his arguments, one of life's gifts is to be able to appreciate a genius mind at work, and his is surely that and more.
3. I'm well aware of the irony of trying to make a popular version of a sophisticated theory that had to replace a bad-but-popular theory before it!
4. Hal Koss's Aug. 10, 2021 article "The Tech World Fell for Lean Startup. Was That a Mistake?" for *Built In* has a nice overview of the debate: **https://builtin.com/founders-entrepreneurship/lean-startup**

5

Creating a unique clarity strategy

> If I'm right, then [the new hemisphere theory] should apply to everything, because we bring these two hemispheres to everything we do.
>
> *– Dr Iain McGilchrist on the Uncensored CMO podcast*[1]

In the previous chapter, we introduced the hierarchy of attention: right-brain 'radar' attention vs. left-brain 'laser beam' attention. That's the core theory behind this book. We also touched on how these two modes of attention tend to map to two broad types of stories — change stories and positioning stories. In the chapters ahead we're going to explore each approach in depth, and we're going to look at specific narrative examples so you can get hands-on and put these ideas into practice.

But we still need a framework for this theory — what you actually *do* or *implement* — for a couple of reasons.

First, there's an extra dimension to attention that I want to throw in the mix here, too. We've discussed the two *modes* of attention (the right-brain radar and left-brain laser beam), but as we'll see much later in the book, there's also the question of *quantity* of attention too, i.e., what folks are passively aware of what's 'out there' on their radar. This is the world of brand, specifically memories and memory associations. (We'll only touch on brand here before revisiting it later in the book.)

This gives us three tools for clarity — **narrative**, **positioning**, and **brand**. (That's NPB if you really want a TLA.) These are the core tools you can use to diagnose and solve your specific clarity challenges. We'll call this approach your **clarity strategy**, and it's going to be custom to you.

The second reason for a framework has to do with the reason we use frameworks in business at all: confidence. Confidence that we have a structure in place, confidence that we have somewhere to start, and confidence that this all hangs together in a coherent way that will lead to a nice outcome. In that sense, in this chapter, I want to give you the confidence to get started and to go as broad or as narrow as you need to with your clarity strategy. You can do it!

> **Why "clarity strategy"?** I particularly like the *clarity* in clarity strategy because it sums up what we want out of our GTM organization: clarity.

> Clarity on who we reach, why we reach them, how we reach them, how we communicate our value, and how we connect with their needs and pains. It's ~~turtles~~ clarity all the way down. And when so much of B2B tech is new, strange, complicated (hello enterprise products), or just plain confusing to your average buyer, it's hard to imagine any context where more clarity wasn't a good thing.

Your clarity strategy

We now have:

- **Theory:** The hierarchy of attention.
- **Framework & tools:** Your clarity strategy built on our three tools — narrative, positioning, and brand (NPB).
- **Outcome:** An owned narrative that gives you clarity in communication and leadership.

Note the "your" in "your clarity strategy." There *isn't* some universal thing called "clarity strategy," but there is very much *your* clarity strategy, which is a function of how you use the three tools of narrative, positioning, and brand as they relate to your specific context. Your clarity strategy is, therefore, how *you* apply those things in *your* situation at *your* company's level of ambition and maturity.

"Every company is different" is a cliche for a reason, but even within the niche of B2B SaaS, it's amazing to think about just how vast the difference is between, say, an early, SMB-focused product-led company using B2C-style tactics and a mature enterprise-focused SaaS company that's 15 or 20+ years old. They're really nothing alike, as we'll see, so it doesn't make sense to talk about a top-down universal framework that applies equally in all contexts.

Instead, it's better to talk about the bottom-up fundamentals of attention we're dealing with, the tools to address those contexts of attention, and DIY'ing your own framework on top of those fundamentals so you can create clarity for your team and your customers.

The attention 2x2

I wouldn't be much of a consultant if I didn't try and summarize my approach in a 2x2, and I have not one but two 2x2s for you.

The first captures the science-based positioning tools of your clarity strategy. Each tool (or lens, if you like) addresses one of the different combinations of *mode* and *quantity* of attention that we commonly see in B2B.

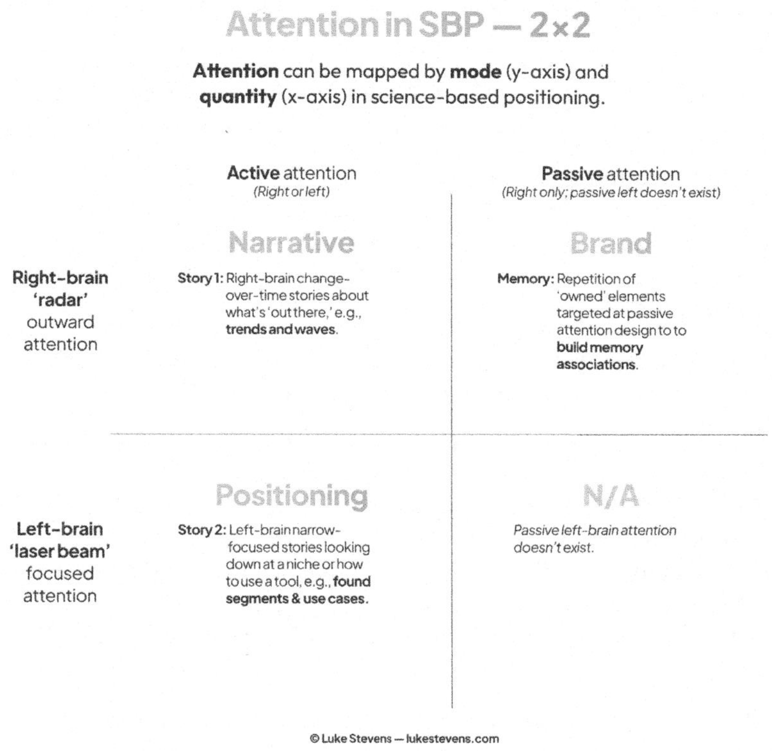

Figure 5.1: The SBP Attention 2x2.

I find this graphic is a helpful starting point for clients and leaders who would otherwise start nodding off if I was to walk them through all the theory first, and perhaps you'll find it helpful in similar situations, too. 'Bottom line up front' really helps. And this really is the bottom line — in some ways, it's the entire summation of this book — and a quick visual snapshot like this can help set the context for the work and ideas to come.

The science-based positioning 2x2

The attention 2x2 is still quite abstract, though, so to make things more concrete, let's talk about specific *strategic choices* we can derive from our understanding of attention. Which approach most resonates with you?

- **Wave riders (narrative):** You are, for example, riding the generative AI wave and building the new AI-first solution to a broad business problem.
- **Problem finders (niche positioning):** You've zeroed in on a specific need, you have a solution, and you know exactly who needs it and why.

- **Memory builders (brand):** You've found the sweet spot of a simple approach to a common need, and you want to be the first company that comes to mind when people have that need.

These aren't mutually exclusive, and you may well be trying to thread the needle through all three. When they all come together, we get a fourth option:

- **Riding a winning position:** This is when things click and you just need to execute like crazy. It's worth remembering we're not looking for a position in some abstract, intellectual, 'correct' sense; we're looking for a position where you win — win in deals, win in your niche, and win in your market and adjacencies.

We're getting closer to strategic and positioning clarity here, but we can go further. I like to boil these four choices down to:

- **Prove it:** Catching a wave is easy, *proving* the value of your solution is the hard part.
- **Find it:** Picking a segment is great, but you've still got to drill down to *find* the unique value you can offer.
- **Own it:** If you want to build a brand — that is, build a tight need/name memory association — you've got to *own* your message.
- **Ride it:** Then, when you've found that winning position, you need to *ride it* and hold on tight.

That brings us to the 2x2 *pièce de résistance* that summarizes the core ideas in this book (and that of my framework on **lukestevens.co**):

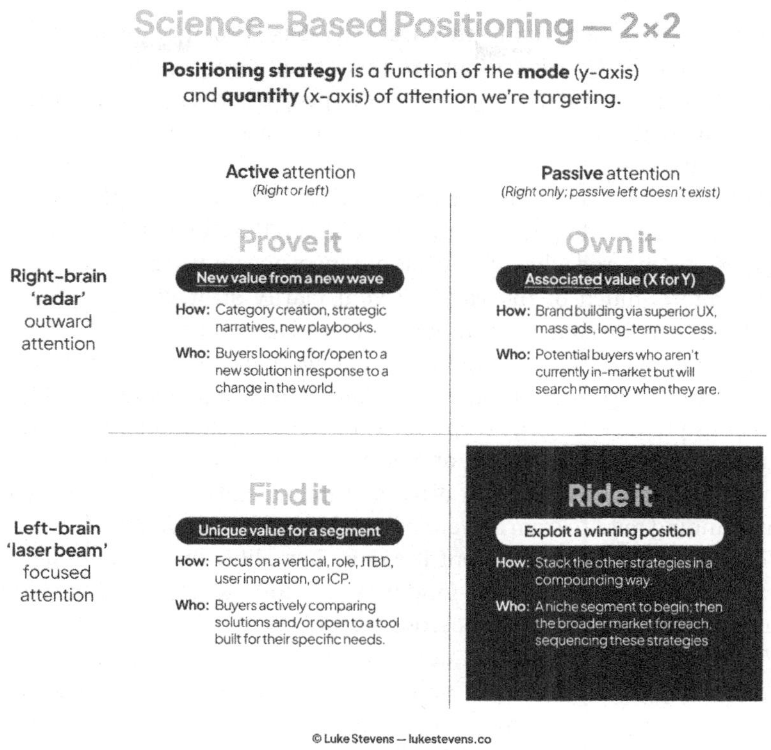

Figure 5.2: The science-based positioning 2x2.

There's a lot going on there! But this is what it means to get started with science-based positioning and building out your clarity strategy. We've boiled things right down to clear choices and actions — down to the literal verbs (prove, find, own, ride) — and simply knowing that these choices *even exist* puts you well ahead of the game.

> **Bookmark and return:** These 2x2s will make much more sense as we work through these techniques in the chapters ahead, so consider bookmarking this page to return to later. We're going to cover a lot of story 1/story 2 sales-focused narrative and positioning before we hit brand in chapter 43, for example, but we'll get there. There's also a lot of overlap between these quadrants — your narrative might be part of your brand, or your brand might be shaped by your positioning choices, and so on. Don't get hung up about the labels; clarity comes from focusing on the type of attention you're addressing.

Clarity and coordination

The point of these 2x2s and, indeed, science-based positioning more broadly is to help you find clarity. When we talk about clarity, we usually think in terms of saying the right things to the right people at the right time, and often that points to execution — the right words on the website, the right slides in a deck, and so on.

But when it comes to startups and technology companies, there's a lot of work that needs to happen in terms of aligning folks on who, exactly, those 'right people' are and what, exactly, is our message for them.

Therefore, much of the value of your clarity strategy is in creating a coordination point that can help you transcend functional silos. That is, your clarity strategy can be something you can work on that's bigger than your sales strategy, bigger than your marketing strategy, and perhaps even bigger than your product strategy (especially if you've reached product maturity). Your clarity strategy gives you, as a founder, functional leader, or gets-stuff-done contributor, a chance to drive alignment and ensure you're getting compounding results from all your efforts across all your functions.

It's not the sexiest work, but it can be incredibly impactful to get folks on the same page about your fundamental story, whether you tackle that through the lens of narrative, positioning, or brand.

That is, these approaches don't just help you communicate with the market, they clarify how you and your company see everything you do, including your product, direction, and place in the world. That makes your clarity strategy vital, high-leverage, high-value work.

Starting small or going big

We've got a theory. We've got our three main tools. We've got the bones of your DIY clarity framework. Hopefully, you're getting a bit of confidence in the fact you can do this. (You can!)

Now, where should you apply those techniques in your business? How ambitious should you be?

These questions will be highly specific to your context, dear reader, but I want to give you a few ways of thinking about it so that (a) it's easy to get started and (b) you can go big if you really want to.

Firstly, let's think about a project-based approach.

This might consist of:

- Workshops with the leadership team to get everyone on the same page.
- Research or analysis to bring customer insights and fresh trends to the table.
- A deliverable as the project's outcome — perhaps a new sales narrative

that then flows through your customer-facing messaging.

This approach focuses more on the "front end" of your clarity strategy, if I can put it that way, i.e., the execution. It's usually aimed at improving customer-facing assets and messaging, though a byproduct of the process might be a stronger internal narrative for leadership, team coordination, and investor relations, too.

In terms of your framing for this kind of project, you might pick one of the three NPB tools, depending on your needs. Perhaps that's:

- **Narrative** for your sales efforts to create a stronger, clearer 'why' for your product that crystalizes what's happening in the environment around you.
- **Positioning** for folks drilling down to find their best customers, who need to think harder about who they actually want to sell to in the first place.
- **Brand** for companies with a clear need/solution pair that they want to become known for (where the company isn't going to pivot or change their ICP anytime soon).

The deliverables here might be a stronger sales narrative, a refined ICP and messaging, and/or a new visual identity and brand association you can drum into people's heads.

That's all very practical and execution-focused, and we'll pick up more practical ways to do this work with a simple 'think, align, execute' framework in **chapter 13**, too. But what if you want to go bigger? What if you want to tackle some of the fundamentals of the business itself?

Clarity and your company

If the "front end" of your clarity strategy leans more towards execution and projects, clarity on the "back end" leans more towards an ongoing strategy that you live and breathe, day in and day out, where the two modes of attention are never far from what you do.

This is where you can go big. Part of the challenge that the hierarchy of attention presents is that, as McGilchrist says, it touches everything we do. Here's the opening quote again:

> *If I'm right, then [the new hemisphere theory] should apply to everything, because we bring these two hemispheres to everything we do.*

The theory starts from such a foundational place — the functioning of our brains — that's so far upstream of everything else that it has implications for how we think about, well, everything else, including the business itself.

This raises big questions about how we innovate and do product discovery work, how we organize our GTM teams, and how we think about the right-brain wave we are (or aren't) riding and the way we use our left-brain laser beam to repeatedly bring the best of what our customers do back to the whole of what the company does.

The answers to these questions are (thankfully, given the size of this book) beyond the scope of what this *Playbook* hopes to tackle. But I do want to tease the possibilities here because the implications are so vast. Thinking about the application of attention alone, we can consider:

- **Sales narratives** (the main focus of this book) tailored to the *mode* of attention.
- **Marketing strategies** tailored to the *quantity* of attention.
- **Product development strategies** driven by your *application* of attention.
- **Positioning** based on adoption stage and the segment's *focal point* of attention.
- **Competitive strategies** when we bring all these things together as a compounding whole.

There's so much potential here to take this theory and run with it. I've taken it as far as I can on the product marketing and GTM side, but I still feel like I've only just scratched the surface of what the theory can offer. There's a management strategy waiting to be found that's much bigger than this book, so when I say you can go broad with your clarity strategy, that's what I mean. See just how far you can take it!

DIY frameworks & existing contexts

In the spirit of DIY frameworks[2], let me stress that how you formulate your clarity strategy is entirely up to you. You can go narrow and focus on only *story 1*, or *only* story 2, or you can go broad and push for attention-driven clarity across your entire organization.

If you've got the power to push for a new approach — i.e., you're the founder or functional leader — then go for it. If, however, you're trying to make an impact within a particular function then let me encourage you to try and work within whatever existing context you find yourself in so your ideas don't get instantly shot down. Consider taking an improv-style "Yes, and..." approach to get the ball rolling.[3]

If, for example, brand, category creation, or some other approach is the overarching mental model leadership uses, great, run with that. Call it your "brand clarity strategy" and sneak the NPB model in under that umbrella because it's 'good for the brand.' At the end of the day, what matters is meeting the different modes of attention in a clarity-creating way, not the labels we use. If you can have an impact by riffing on what folks are already familiar with, all the better. I'd rather see you succeed (I believe in you!) under the existing context rather than fail trying to change your leadership's entire mental model, for example, as worthy a goal as that might be.

> **Stealing customer language:** Working with existing points of reference is, in part, how the story 1/story 2 idea came to be. It's a riff on the System 1 and System 2 insight of Daniel Kahneman's famous book on how we think, *Thinking, Fast and Slow*[4]. That insight actually came from a client of mine — the best lines and ideas often do — so I borrowed the concept and ran with it, and I'm sure it will resonate with other folks who are familiar with the concept. It also provides a simple way to get started, as the two stories concept (radars and laser beams included) is an easy way to teach people the dual modes of attention and how they can be engaged. A simple binary can go a long way!

Selling clarity strategy — old way/new way

The weird part about trying to sell *with* clarity is that sometimes you have to sell new ideas *about* clarity first. To do that, you can use some of the right-brain tricks of the narrative trade, like a good, strong 'old way/new way' or 'from/to' story.

If you're pitching folks on the idea of a clarity strategy (and perhaps a project to kick it off), you'll need a 'why.' In other words, why do you really *need* a clarity strategy, anyway? What was wrong with the old approach?

I'm sure you'll have plenty of evidence from what you're seeing in your company — prospects that don't quite get it, a product that's not quite catching on, growth that's slowing down, competitors that are passing you by, a team that lacks alignment and direction, investors that are pushing one way but customers seem to be pulling another, and so on. (Or it might just be the fact you have something new and cool, but B2B buyers think it's weird and unknown. That's where even some of the most ubiquitous tools like Slack started, as we'll see in **chapter 12**.)

But what can really help sell the strategy internally is not just the 'old way' pains but your diagnosis about *why* those pain points exist and, therefore, what you can do about them. For example, in terms of the old way you've been doing things, it may have been a case of:

1. **Random walk:** A lack of any framework at all for developing clarity in your market, and you've just been on a random walk, grinding out sales.
2. **Hammer only:** The old when-all-you've-got-is-a-hammer approach

where a simple framework has become all-consuming — category creation obsession has taken hold, for example, or you're wedded to a vision/narrative that needs revisiting — then it's time to introduce some balance with the other NPB tools.
3. **Poor execution:** You might have good ideas, but the resulting writing stinks. It happens. Execution matters!

For a more precise diagnosis, we'll look at diagnosing missing, mismatched, or disconnected story 1 or story 2 narratives next in **chapter 6**. There, the old way is a lack of a modern, scientific understanding of how attention works and how we can address it, and the new way is applying the right tools in the right way to hit the right mode of attention.

A strong, credible diagnosis creates trust that your suggested course of action is plausible in achieving the outcomes you're after. It builds buy-in and gets everyone on the same page so that you can *get everyone on the same page,* if that makes sense.

Clarity outcomes

In terms of those outcomes, they again will be specific to your context, but it's worth articulating how you think your diagnosis and chosen course of action will lead to greater:

- **Clarity for you and your team** in terms of how you understand your market, your strategy, and the individual buyers and dominant types of attention in that market.
- **Clarity for your buyer** in their mind as they search for a solution to their problem and/or ride the waves of change in their industry.

In broad strokes, the course of action I'd suggest in getting to those outcomes consists of two steps — snapshot & stretch:

1. **Snapshot:** The first step in moving from the old haphazard way to the new way is to use what we know about right-brain narrative and left-brain positioning to your current sales narrative and ensure it's tight, punchy, and not making the mistakes we'll discuss in the next chapter.
2. **Stretch:** The second step is to use all three tools — narrative, positioning, and brand — as tools of innovation in the way you tell your story, address your market, and ultimately unlock new growth.

We're going to cover both at length. That's why you're here and that's what this book is all about — getting the clarity outcomes you need using the clarity strategy appropriate to your business.

The fastest way to get started is to work on your snapshot and refine what you've got. So that's where we'll start in the coming chapters. But if you're swinging for the fences — and if you've raised capital, you *have* to swing for the fences — you need to know how to stretch your narrative and positioning, too. Most folks think that means coming up with the most grandiose story possible, but that's wrong. It may be thinking big, but it may also be thinking very, very specific, too. That's all ahead of us; what matters now is that you have a theory, can diagnose your situation, and suggest a course of action that will lead to greater clarity for your buyers and your team.

Clarity starts with you

Working on your clarity strategy can be exciting, daunting, and, I hope, transformative. This is a chance to think through the fundamentals that drive your business and how you connect those to the attention of the market — the very heart of product/market fit.

In more practical terms, it's a chance to translate the vision and insight at the heart of your company into a narrative that sells — both internally for your team and externally for your customers. But again, all good strategy work must start with a diagnosis, so in the next chapter, let's dig in and diagnose the common clarity issues and narrative mismatches in B2B startups.

Notes

1. McGilchrist, Iain. Interview with Jon Evans, "The divided brain, attention and how we see the world - Dr Iain McGilchrist," *Uncensored CMO*, November 15, 2023. **https://youtu.be/tY89D1UC9Dw?si=XRGNCPhZcb5iUGiR&t=2605**
2. For more, see Tom Critchlow's essay, "How (and why) to roll your own frameworks in consulting engagements," *The Strategic Independent*, June 27, 2019. **https://tomcritchlow.com/2019/06/27/frameworks/**
3. Another tip o' the hat to Tom Critchlow: "Yes! And... How to be effective in the theatre of work," *The Strategic Independent*, November 18, 2019. **https://tomcritchlow.com/2019/11/18/yes-and/**
4. If you're not familiar with the book, Wikipedia has a decent overview (**https://en.wikipedia.org/wiki/Thinking,_Fast_and_Slow**), describing the book's main thesis as "a differentiation between two modes of thought: "System 1" is fast, instinctive and emotional; "System 2" is slower, more deliberative, and more logical." Note Wikipedia's section on the replication crisis, too. For more on the running debate on the replication crisis and cognitive biases, see, for example, Scott Alexander's August 31, 2023 article "Here's Why Automaticity Is Real Actually": **https://www.astralcodexten.com/p/heres-why-automaticity-is-real-actually**.

6

Diagnosing clarity challenges

The left hemisphere is better at "how" and "what," while the right is better at "why"—though McGilchrist points out that both are capable of addressing either set of inquiries depending on context. The left sees constituent pieces, whereas the right sees the whole. The left explains, the right understands. The left apprehends, the right comprehends. Perhaps most importantly, the right hemisphere is present in the world—acutely aware of its surroundings—whereas the left is prone to represent (literally re-present) it—preferring, as it were, the map to the territory.

- Nick Spencer, reviewing McGilchrist's work in Prospect magazine[1]

Every B2B tech company — especially ones with a sales motion — already has some form of clarity strategy in play.

It might be the vision of category creation the founder pitches, the nuts-and-bolts narrative that your sellers use, or maybe there's a positioning statement buried somewhere in your shared files courtesy of the marketing team.

But often the current approach is heavily siloed, lacking any real underlying theory, and failing to deliver the desired clarity the team and your prospective customers need.

Sometimes, in these cases, it's easy to diagnose why things aren't clicking: Everyone might be doing their best, but they're doing it silos, and Sales and Marketing aren't talking to each other, for example. Or perhaps leadership hasn't had the tools or time to get together and hash it out and truly find clarity across the team.

Your clarity strategy might start there, bringing everyone together, hashing it out, and seeing where you all land. Perhaps it doesn't need to be more complicated than that, and a few workshops over a few weeks with a deliverable in mind is all you need to get that much needed clarity.

If so, great! However, I want to suggest that there may be deeper, harder-to-diagnose issues going on that result in this disconnect between departments, or between investors and the founder, or between your company and the market itself.

These issues can be much more insidious and harder to detect because, until now, we haven't had the language to put a finger on exactly what's going on. And what's going on, in these cases, is mismatched attention. Below, we'll examine these forms of mismatched attention as they apply to your sales narrative.

> **"Sales narratives" include positioning:** In the previous chapter, we discussed NPB — narrative, positioning, and brand. That is, narrative and positioning were distinct. In this chapter, I'll use "sales narrative" and "narrative" as a shorthand that *includes* positioning and spans both story 1/story 2, i.e., both your change-over-time narrative your your value-here-and-now positioning. Apologies for the dual use of 'narrative' — if you have any better suggestions, I'm all ears!

Your current sales narrative

To diagnose your clarity challenges in general and your sales narrative in particular, first, let's think about your two primary 'story' decks: your investor pitch deck and your sales deck (or website messaging). Consider the job each deck has to do:

- **Next Big Thing:** Pitch decks often tell the big, game-changing story about how all the stars have aligned for you to be The Next Big Thing.
- **Do The Thing:** Sales decks (or demos and websites), on the other hand, tend to get into features and use cases pretty quickly, with a focus on how to Do The Thing instead.

One tells a vision-driven, change-over-time story. The other tells a narrow, here's-how-you-use-the-tool story. This is the narrative baseline for most B2B tech companies and startups, and you can already see shades of our two story types at work.

On one level, this is fine. You have to start somewhere, and if this approach is working for you, great. But on another, there's often not a lot of actual narrative or positioning *strategy* there to deliver market-shaping clarity. Usually, sales narratives are cobbled together either based on 'the way we've always done it' or the last decent deck someone saw and decided to copy. There's no deeper theory at work, and these naive approaches tend to be more superficial than they are strategic and often miss what makes successful narratives stick.

Diagnosing narrative mismatches

Now, let's diagnose three common narrative problems that play out in reality, specifically:

1. **Weak or missing story 1 or story 2.**
2. **Mismatched story 1 and story 2.**

3. **Disconnected story 1 and story 2.**

Then, we'll run through a bunch of more in-depth diagnostic questions that can help you further locate specific narrative issues that might be holding your story back.

First, the common issues.

1. Weak story 1 or story 2

Let's recall that story 1 is your story for right-brain radar attention, and story 2 is your story for left-brain laser beam attention. For now, consider story 1 as more the why, what's out there, the change in the environment, and story 2 is more the how or what (to borrow from the opening quote above), the narrow-focused analysis, the tool itself and how you use it.

Perhaps either of those aspects of your story are weak or even non-existent. You know you need to target folks with a sharp messaging arrow, but it feels like you're firing squishy nerf darts instead and the message just isn't sticking. Perhaps that's because:

- **Strong vision, weak positioning:** Maybe you have a solid story 1 (a vision-driven change narrative), but your story 2 (perhaps a positioning wedge) is a little flimsy. Investors love your big-picture deck, but how do you translate that to something customers care about?
- **Weak vision, missing positioning:** Or perhaps your story 1 isn't that strong in the first place. Folks can get a little carried away with their story 1 — every startup is going to take over the world, create that category from scratch, or deliver the AI god we've been waiting for — when really you need to start with story 2 and nail the narrow problem you solve better than anyone else first. Your story 1 can come later.
- **Strong positioning, missing vision:** Perhaps you *do* have an excellent, narrow-focus technical story of the problem you solve. Your tech *really is* harder/better/stronger/faster, and technical customers love it. But how do you turn that into something investors, less technical buyers, executives, or even the unaware market at large should care about? That is, how do you put it on their radar in story 1 form? This can be a challenge for technical founders.

These problems of weak or missing stories 1 or 2 can manifest in a variety of ways. Consider the front and back end of your clarity strategy, for example:

- **Missing an edge on the front end:** You might be a leader with a strong intuitive vision for what the market needs — you *know* how the future should look — but you haven't found a catchy way to communicate that. (It can be hard to sell — internally and externally — what you can't fully describe.)

- **Missing the substance on the back end:** Maybe your investors are on your back telling you that you should be 'creating a category,' but what does that even mean? And why do your attempts to do so seem to fall flat? (We'll find out.)

Therefore, in terms of a starting point when trying to diagnose clarity strategy issues, we want to focus on whether story 1 and story 2 exist in the first place and then whether they have sharpness and depth.

> **Your story 1/story 2 split:** Take a moment to think about your current investor pitch deck and sales deck or demo script for prospects. How much are they, respectively, calibrated to the broad story 1 change narratives or the here's-how-the-tool-works narrow story 2 pitch? If you had to allocate a percentage for each, what would it be? 90/10? 20/80? 50/50?

2. Mismatched story 1 and story 2

Two types of stories × two modes of attention = four possible outcomes.

But two of those outcomes are mismatches. In slightly abstract terms, the problem looks like this:

- Broad story 1 → broad, right-brain attention = match.
- Narrow story 2 → narrow, left-brain attention = match.
- Broad story 1 → narrow, left-brain attention = **mismatch**.
- Narrow story 2 → broad, right-brain attention = **mismatch**.

That is, you can have strong stories but deliver them to the wrong people or in the wrong way.

Let's say you've got a killer story 1 change narrative *and* a compelling story 2 product wedge. You're just misreading when to use which story and with whom.

Maybe you're trying to tell prospects in demos they're doing it all wrong — *and they are* — and your way is better — *and it is* — but the message just doesn't land. I've had a founder (and great seller) run into that wall before due to mismatched story 1 and story 2 (we'll pick this up in **chapter 46**).

I've also seen folks try and sell customers with specific needs on a big change narrative that's well above their pay grade... and have it fall flat, for obvious reasons. Again, mismatched stories.

And sometimes startups hit "second meeting syndrome," where they successfully sell to the potential *user* of their solution but not the *economic buyer* (the person with final buying authority) because they didn't know when to use story 1 or story 2.

Mismatched stories can sink deals. All because the story being told didn't match the mode of attention the buyer was using.

Getting your story × attention match wrong — especially at the exec/enterprise level — can kill your chances of moving upmarket, too. The problem is you're addressing folks with a completely different mode of attention, and you end up saying one thing while the folks you're pitching are listening for another. The worst part is founders (and marketers and sellers, too) often *don't know that they don't know this.* It's an unknown unknown. They think they've got a good story, and they do! It's just the wrong one for the kind of attention they're targeting.

3. Disconnected story 1 and story 2

Or, finally, maybe you have a strong version of both story 1 and story 2... the only problem is they don't connect. A isn't leading to B. This happens when folks come up with naive category stories without knowing what actually drives category creation. They end up wasting an enormous amount of time and resources on something that was never going to work.

That's what happens when there's no bridge between story 1 and story 2. There's a glorious story 1 vision for a new future... and a helpful, precise, story 2 pitch on how the tool works... but one simply doesn't lead to the other, no matter how glorious the vision may seem.

You don't want to feel like you're pushing two startup stories uphill — one is hard enough. That's why it's so important to have a coherent whole — your clarity strategy — driving your story.

> **Narratives disconnected from markets:** There's another kind of narrative disconnect that became disturbingly common in the ZIRP era, and that's story 1 narratives disconnected from any kind of market reality. The startup ecosystem works when the foundational narrative — capital today provides returns tomorrow — accords with reality. When startup narratives get disconnected from any kind of market discipline, then all that's left is narrative, and that quickly descends into zero-sum cynicism and speculation with folks trying to line their pockets with ridiculous put-jpeg-urls-on-the-blockchain narratives (or whatever). This causes enormous damage and puts the whole ecosystem at risk. Venture works when capital invested generates returns that then gets invested in the next round of founding teams so they can take their shot. That's the venture flywheel. But if the LPs that supply many VCs with capital don't get any returns because all the

> capital was burnt on insane narratives disconnected from the market (or reality itself, as we'll discuss in chapter 28), then when the next round of genuinely interesting tech comes around, the cupboard is, in some cases, bare. And that's a great shame. Narratives are a powerful tool, but they need to be in service of the ecosystem as a whole. If that doesn't make sense to you, consider a career in finance instead.

The ETC test

The real test for your clarity strategy is whether your prospect's eyes light up (**chapter 38**) and deals get done. But as a quick little litmus test using the above criteria, consider whether your current clarity strategy is:

- **Existent:** Having the right story types in the first place, with strong, clear, insight-rich versions of your right-brain change story (if appropriate) and left-brain positioning story. These should have a strong creative and strategic foundation and have survived contact with customers.
- **Tailored:** Having the right stories used in the right context, and knowing how to dial up and down your two stories to meet your audience where their attention is at.
- **Connected:** Having a connected story 1 and star 2 that are mutually reinforcing — i.e., they're a product of your greater sense of a cohesive whole.

If your current clarity strategy and narrative passed the test, congrats! You can stop reading here. It's been real.

But I suspect you're here to improve your narrative and positioning, not just diagnose problems. With that in mind, and noting that your clarity strategy can be as simple or complex as you need it to be, below are some critical diagnostic questions that make excellent jumping-off points into the deeper recesses of this book.

Always Be Diagnosing

- **Have you neatly captured your idea?** A few well-chosen words (**chapter 10**) or a catchy metaphor (**chapter 15**) can be very powerful indeed, as can building context through your origin story (**chapter 11**) and, most fundamentally, nailing your narrative deck (**chapter 9**) and ensuring it gets applied across your website (**chapter 15**).
- **Have you understood the two schools of narrative?** Much of this book is about understanding the two schools of narrative, i.e., change stories

(**chapter 20**) and positioning (**chapter 21**), and how to use them in the 'claw game' of attention (**chapter 18**) to build a pitch tailored to your audience's level of awareness (**chapter 24**).

- **Do you have the necessary insights?** You might reflect on how you direct attention internally (**chapter 26**) to find the insights and greater whole that can drive your narrative and your pitch (**chapter 27**).
- **Are you trying to create a category?** "Category" is a thorny one, as it gets used to mean everything from secular trends (**chapter 30** and **chapter 31**), literal categories of software (**chapter 32**), and C-suite targeting playbooks baked into your brand (**chapter 35**). 'Category creation' can be quicksand if you're not careful though, and in **chapter 33** we'll see why it's such a slippery concept.
- **Do you know your customers' jobs-to-be-done?** Jobs is the *lingua franca* of customer needs in the startup world, and JTBD — in its variety of flavors — sets up the positioning prong of your clarity strategy. If you've been looking for better language to capture *how* your customers are trying to succeed, **chapter 36** is for you.
- **Do you need to build momentum in a niche?** Everyone loves positioning, but not many folks use it to maximum, momentum-building strategic effect. We'll explore how positioning can drive your clarity strategy (**chapter 39**) and your niche-to-mainstream GTM strategy (**chapter 44**). Or maybe you need to find a wedge by *attribute* (**chapter 46**), not segment, in a crowded market?
- **Are you trying to cross the chasm?** Had success with the "nerds" (innovators and early adopters) but struggling with the "normies," mainstream folks who seem weirdly cautious and dismiss you as too new or too early? You need to understand the different stories that nerds vs. normies want to hear and the chasm that can exist between them (**chapter 40**).
- **Are you trying to move up or down market?** "B2B" is a pretty meaningless term — you need to tailor your clarity strategy to your segment, be that B2SMB (**chapter 43**), B2MM (mid-market, **chapter 46**), or B2Enterprise (**chapter 47**), as those strategies look very different.
- **Are you trying to build a brand?** For folks down the SMB end of the B2B spectrum, we'll look at what the B2C brand experts think brand actually means, with a quick tour through one of the most influential books on B2C marketing to see what it can teach us in B2B (**chapter 45**).
- **Are you trying to engage conservative buyers?** Back at the enterprise end, there's always a tension between the *most advanced* and what's *yet acceptable,* a.k.a the MAYA threshold (**chapter 48**). This is a classic positioning problem that's existed for as long as companies have been selling products, and you may need to finesse your positioning to get

folks over their *fear of the new* in more conservative markets.

- **Are you trying to compete in a crowded market?** Everyone loves to talk differentiation — uniqueness is at the heart of most popular business books, after all, and we'll look at how uniqueness is at the heart of competitive strategy, both in terms of competing to be unique and its evil twin, competing to be the same (**chapter 49**).
- **Do you need to bring it all together?** Finally, it's unlikely you'll have one decisive advantage that helps you win your market. Instead, you'll stack advantages, as we'll finish on (also in **chapter 49**).

There are plenty of places to start investigating! Pick one, any one, and dive in. This is the final boarding call.

You certainly don't need to do *all* of these things, either. If you told me you were trying to *define your brand* while *creating a category* so you could *move up market* and *sell to conservative buyers*... I'd suggest you take a deep breath and just pick one to start.

These things take time, and *making time* is the first step in going from diagnosis to a vastly improved clarity strategy. In the next chapter, we'll briefly run through what it means to invest in your clarity strategy.

Notes

1. Quote taken from Spencer's 2022 review of McGilchrist's *The Matter with Things*, **https://www.prospectmagazine.co.uk/culture/38439/iain-mcgilchrist-and-the-battle-over-the-left-brain-right-brain-theory**

7

Investing in your clarity strategy

Clarity strategy work is incredibly high leverage — if you invest in it.

We'll talk about financial investment in a moment, but first let's discuss time and attention. To get the most out of your clarity strategy, it pays to create the space to think — to really think — about the whole of what you're doing with your startup, product *and* market, both broadly and in the specifics.

Executive team involvement

Your clarity strategy will help *clarify* perceptions internally from the founder down, and *shape* perceptions externally from the founder out. It's foundational work, ideally involving the founder/s and your leadership team. Teamwork makes the dream work, and that has to come from the top. Pushing clarity work down the org and crossing your fingers doesn't cut it. Alignment starts with you.

If you're going to quarterback your startup's narrative and positioning across product, sales, marketing, success, and other teams, and have them all playing in harmony with each other and with your customers, you're going to need an impeccably clear narrative.

That narrative is the fundamental message that will shape what your company does, what customers hear, and how *you* think about your — and their — success, so be sure to give it the time and attention it deserves. Block time out in your calendar, book time in for workshops, whatever it takes. If there's no time, there's no thinking, and if there's no thinking, there's no clarity.

Where's your ~~head~~ attention at?

Your clarity strategy can be a challenging intellectual exercise because it requires shifting in and out of our two fundamental modes of attention when we may be much more comfortable with one or the other.

For example, I've seen:

- **Too narrow:** Founders struggle to engage with the strategic implications of their story while fixating on individual words on their homepage.
- **Too broad:** Founders get completely obsessed with their big strategic message but fail to meaningfully apply it to their customer-facing assets… like their homepage.

These are classic examples of getting stuck in one mode of attention, narrow or broad, and struggling to switch to the other. It can be hard to grapple with the big, right-brain picture when you're in a left-brain mode of attention — narrowly focused on the words and not the implications. Similarly, it's easy to get lost in your glorious story of industry change and forget to connect that to what customers actually see.

That is, it's hard to focus on the whole. But being able to see the whole — and switch from that broad to narrow focus and back again — is essential for leading with a coherent clarity strategy. (This is the right/left/right hemisphere jump we'll discuss in **chapter 26**.)

How much should your clarity strategy cost?

How much should you invest financially in your clarity strategy?

I've seen startups spend *hundreds of thousands of dollars* on agencies and consultants (well, literally *an* agency and *a* consultant, each charging six figures a pop) for narrative engagements where the work appeared to be:

1. The product of a month-long acid trip that produced work that was immediately forgotten, or
2. Based on a copy-and-paste formula that ended up telling porkies to prospects.

Dear reader, I thought that was *insane*.

Nothing against consultants and advisors (I am one) or charging what you're worth, but I thought there *had* to be a better way — hence this book and custom clarity strategies as a concept. This work just seemed far too important to devolve into BSing for a buck or two (or many thousands of them, as the case may be).

> **Consultants and sacrificial solutions:** To be fair, I get that it's often the job of the outside expert to generate 'sacrificial solutions' that help the group arrive at a solution and that not every attempt will be a home run on the first swing. Sometimes consultants can do everyone a favor by putting a rough first draft up for people to react to and tear to pieces, taking the risks internal folks won't or can't. I appreciate that. But neither acid trips in deck form nor a single formula for all occasions made much sense to me, so here we are.

How much can you learn, and how quickly?

Your clarity strategy — and the narrative, positioning, and brand work it entails — will be an iterative process. Don't get stuck on whatever your first attempt may have been — significant outcomes can require significant investment. But that should be tailored to the stage you're at. By definition, you're not going to know everything upfront. Insight comes from learning, and learning is a function of time in market.

For a small, scrappy startup, your initial clarity strategy can be as simple as getting a rough sales deck in the founder's hands and a reworked home and product page up. Start learning. Don't overthink it.

For later-stage companies, coordination becomes a much big challenge. There's the leadership team to collaborate with and an execution process that might unfold over months with customer testing and feedback, internal presentations, asset updates, and new training for reps. Don't underestimate the effort to roll it out.

How much does *not* having a clarity strategy cost?

Not to lean too hard on the old 'If you think it's expensive to hire a professional to do the job, wait until you hire an amateur' cliche, but the cost of not understanding your clarity strategy early on is, shall we say, considerable, especially in the form of what I call *brand debt*.

Your clarity strategy, in whatever form it takes (be that focusing on a very specific position, or trying to create a category, or launch with a very specific vision), is so fundamental that it gets baked in at a very deep level. Product direction, marketing material, sales collateral... it's all too easy to pile up the brand debt if you run as fast as you can in the wrong direction.

Sooner or later, however, you'll find your groove — or at least a groove — and start building memories in the market (that's what brand is, as we'll see in **chapter 44**), and that can take years. But you want those to be the right memories! This is what you become famous for, after all. If you're going to work your collective butts off getting a story out there — and making your company famous for something — then it pays to have a clear idea about what that something is. Otherwise, you might end up in one of the situations I've seen over the years, and I'm sure you have too (I'll use "narrative" here, but "positioning" or "brand" would work just as well):

- **Narrative mystery:** What even is your product? What does it do? No one can tell, and all that marketing fluff isn't helping.
- **Narrative decay:** The homepage hasn't been updated in years. The team — and founder in particular — is wedded to what worked a decade ago, and they're afraid if they change it even a little bit it'll all go up in a puff of smoke.
- **Narrative thrashing:** One day the startup is <X> Intelligence. No, wait, now we're an Operating System for X. No, no, we're... <insert the hype-y term of the month here>. The website keeps changing, sales decks come

and go, but nothing seems to stick.

- **Narrative weirdness:** There's been so many cooks in the kitchen your sellers end up trying to sell from dog's breakfast decks and ill-advised slides. Do you know who wants to see a giant egg and sperm as an illustration in a martech pitch? No one, that's who. (That is, dear reader, an all-too-real example.)
- **Narrative hamster wheel:** You can't quantitatively A/B test your core narrative, but that doesn't stop folks trying. You can — and absolutely should — test it qualitatively with real customers to see if it lands (as we'll explore in **chapter 38**), but if you operate without a coherent narrative, you'll have marketers doing endless, meaningless A/B tests of messaging minutiae for absolutely no material gain.

All of this can go away when you and your team lock in a clarity strategy that works and embrace sales as a tool for finding and strengthening product/market fit. Embracing sales is what we'll explore next, touching on Segment's experience discovering that sales — and uncomfortable discovery questions — can be a profound source of insight into product/market fit.

PART II

CLARITY & YOUR NARRATIVE SNAPSHOT

8

Sales as a PMF superpower & Segment

Most clarity strategy work results in a sales narrative, and in the next chapter, that's exactly what we're going to dig into. Before we do that though, let's set the scene for why a *sales* narrative, in particular, is so important, by making sure we're all on the same page in terms of using B2B sales as a means of finding product/market fit.

Some founders take to sales like a duck to water, and if that's you, this might seem like a weird introduction to a chapter. But the reality is not everyone appreciates sales for what it can do — some folks hold their nose and treat it as an unfortunate necessity, while others have an allergic reaction to having to sell at all. This chapter is particularly for those founders who feel they are allergic to sales or are bored to death by it. You need to know that sales can be a superpower for nailing your narrative. There is, however, an art to selling that goes far beyond what all too often passes for sales in most SaaS companies (i.e., asking a bunch of rote questions and giving a rote presentation clicking through the product).

If that *isn't* news to you and doing sales is The Most Obvious Thing In The World, then feel free to skip ahead to the next chapter. There, we'll create a sales narrative deck as a starting point for your clarity strategy, and in later chapters, we're going to tease out some of the finer points of B2B selling, too. Don't miss 'em.

Again, if you're not sure if this is applicable to you because you don't do sales, or you have someone doing it for you, or you do it but you don't think there's much leverage there, I get it. No one really launches a startup because they think, "Yay, I get to sell all day long!" But leading is selling, and if you're a leader, I bet you are selling all day long: to new hires, to investors, and to the market more broadly (and perhaps to customers, obviously, too). That means there's an incredible amount of leverage to use sales as a tool for finding product/market fit and, ultimately, your narrative. And that includes technical founders selling technical products to developers, too.

Here's Peter Reinhardt, former CEO of Segment (acquired by Twilio for $3.2B) in an excellent interview with Logan Bartlett on *The Logan Bartlett Show*[1] talking about what he learned through the Segment journey (my bold, lightly edited for clarity):

> *What I've learned is that [...] If you can test that value [of your B2B*

> *product],* ***the process of discovery and sales qualification is effectively the same thing as testing for product/market fit*** *[...] early sales and the search for product/market fit are actually very tightly intertwined, and [sales] is a much more repeatable way of actually finding interesting opportunities.*

Segment is a great case study in how powerful sales can be for a technical, originally product-led product with technical customers, and we'll return to their story in a moment. For now, the point is that the early sales process itself *is* customer development.

To do this kind of selling — to do this exploration of product/market fit — you're going to need a narrative sales deck, and I've got a nice outline for you in **chapter 9**. The point of the deck isn't to speak to it robotically (we'll see what a more natural approach looks like with Dock in **chapter 38**), but to know what it is you're testing.

A clear narrative sales deck helps align your team at later stages too. Once you've got traction in the market, you're going to find yourself with a lot of new faces looking for guidance. To keep your collective foot on the gas, it helps to have everyone telling the right story for the right context in a highly coordinated way, but the team can't be clear if you're not clear.

All that starts with a clear narrative sales deck. From there, you can derive messaging and copy to apply to all your customer-facing assets and communication — from keynotes to homepages, from one-pagers to whole content strategies.

So whether you're looking for initial product/market fit or you've powered well beyond it, a tight sales narrative is essential. But that can be easier said than done.

Narrative is a team sport

You need to smash your silos when developing your narrative. Clarity strategy work often gets chopped up arbitrarily — leadership sets the "vision," Marketing handles "messaging," Sales does the "pitching."

It's hard to write a coherent story when you have three different authors — or groups of authors — all writing their own chunk. Imagine reading that story as a customer — it's unlikely to make much sense.

In reality, this is all just selling, and if you want a tightly aligned ship with everyone is rowing in the same direction, then you need some ground truth that's anchored in customer reality that everyone can get behind.

The ideal is:

- Your narrative...
- Is what the CEO is talking about

- Is what Marketing is publishing
- Is what Sales is pitching
- Is what Product is building
- Is what Success is teaching
- ...with a customer success feedback loop going back to the top — the CEO needs to be the #1 listener, not the #1 talker — so everything is in sync.

That is alignment, and that is hard. And that's all the more reason to have one simple starting place — one clear foundational narrative — so everyone is singing the same tune.

One deck to rule them all?

Having one sales narrative is a fantastic exercise in clarity — clarifying who you're serving, the value you offer, and the way you frame that offer. It gives you consistent messaging, themes, and talking points.

But it doesn't mean everyone has to robotically and monotonically parrot the same lines. Nor does it mean you have to plaster the exact same story beats on every surface in the exact same way.

For example, I've seen folks attempt to plaster a sales deck 1:1 on their homepage, and it just doesn't work. Buyers are at very different stages of their journey when they first hit your website vs. when they take a demo call. They have different expectations of the medium and vastly different amounts of attention that they're willing to dedicate to understanding what you're about — maybe 30 seconds for a homepage vs. 30 minutes on a sales call. (That's why I've got messaging tips for you in **chapter 17**, as well.)

What matters is:

- Having a ground truth for your narrative.
- Tailoring that narrative for different contexts based on the job the narrative has to do.

Narrative JTBD

How do you know when and how you should adapt your core narrative to a specific situation?

How you *apply* your narrative will vary significantly based on the communication job-to-be-done you're solving for (we'll explore JTBD in **chapter 36**).

- Your investor pitch doesn't (or shouldn't) look like your customer pitch, as we've discussed.
- Your homepage and sales deck will be tailored to the vastly different amount of attention paid to each.

- Your keynote talking points will look different from, say, your pitch to new hires.
- Your sales pitches themselves might look different — high-velocity sales will look different to high-stakes enterprise deals; inbound and outbound might start at different places.

That's a lot of variation in your story.

But it's precisely *because* there's so much storytelling going on that it helps to have some ground truth, some common thread, to draw all these stories together.

While we don't want robotic repetition, we do want *thematic* repetition — you and your team should be hitting the same themes and core ideas consistently in your communication so customers (and employees) get a clear picture of what you're about, but you can and should tailor your pitch to the JTBD you're addressing.

What about the actual selling?

A good narrative helps you actually sell, too. (You're allowed to sell in demos, folks!)

Sales is fundamentally about the intersection of your story and your customer's story. You have your take on change, pains, solutions (yours), and outcomes.

Your prospect has their own story — an instigating incident perhaps, with pain, the search for a solution, and some potential outcome in mind.

"Discovery" in sales is essentially digging up the relevant parts of your prospect's story so you can draw the dots between their story and yours. If you can find some happy overlap, great, you'll probably get a sale. If there's no overlap, you've got qualification issues — your story wasn't clear enough to *turn them off*. Sometimes your narrative is about making clear what good fit/bad fit looks like as soon as possible so you can sell as efficiently as possible.

The great thing about the discovery process is that you're not reading the tea leaves and trying to infer what the customer is after or why they're struggling — you can ask them point-blank. You can ask, directly or indirectly, if they're an early adopter (or if they want to be a "design partner" in the very early MVP days)[2]. You can ask — and dig for — the inciting incident that sent them on this journey.[3] Or you can ask them what's stopping them from building a solution themselves, as we'll see in a moment.

B2B selling can, therefore, be value-creating in that it's valuable for prospects to clarify their own story — what it is they want to achieve and what their real jobs-to-be-done are. Those discussions alone can paint a picture of an end to their story — in solution and potential outcomes — they may have never otherwise considered.

If you're skeptical of sales, you may be thinking about sales as high-pressure, slick-but-fake, monorail-for-Springfield type of selling as "persuasion." But B2B selling, when done well, is a million miles from that — it can be both inherently value-creating and highly profitable for both you and your customer.

High-performance sales

High performance exists in B2 sales like anything else. And high performers aren't ones who bend over backward to meet the customer's every whim, either.

High-performing B2B sellers tend to (very respectfully!) challenge folks around what they want to achieve and walk side-by-side with prospects to try and help them make sense of an often very complex landscape. There's research into what tends to characterize a high-performance B2B seller; it's not a secret.[4]

But many founders (and marketers too, I should add) have never been exposed to B2B sales, let alone high-performance B2B sales. They've never bought a high-ticket B2B product, they've never encountered the B2B research literature, and they've never been responsible for selling B2B products before. Therefore, sales is a bit of a mystery.

Reality tester vs. human tape recorder

Sales, to these folks, can seem like a slightly icky, necessary evil that must be done while holding your nose. Or maybe it's a temporary inconvenience that would be better left to someone who actually *enjoys* this kind of thing (weird), so they can get back to more meaningful, 'high leverage' work.

And I get it — being the leader and visionary for your startup *and* feeling like a human tape recorder that hits play at the start of the same sales call day after day is a weird combination.

But if you start treating sales as customer discovery and a tool for developing product/market fit — a tool for testing reality — it becomes much more interesting. It doesn't have to feel like a random lottery or a repetitive grind... or at least it can be less of one.

It's a skill like anything else, and in this book, you'll get a skill few folks have — the ability to tailor your message to meet your prospect's attention where it's at, to tell your story in a way that intersects with theirs.

> **"Ugh, they're just good at sales":** Effective sales is so powerful some tech companies succeed on the back of their sales force alone. Or so some folks would have you think. And that's kind of true! Sales *is* how

> business gets done in the bigger B2B world. But rather than scoff at it, why not see it as a force multiplier for whatever it is you're doing, for whatever great product or technology you're creating? It's like being a skillful tennis player and watching someone really strong but less technically skilled win a game and then rolling your eyes, saying "Pfft, they're just brute force, not skills!" But that's the game! If you're in it to win it, why not both?

Segment — early-stage sales and discovery

You might be familiar with the S-curve of product adoption — a slow start, an acceleration in the middle, an then a flattening out as the point of saturation is reached and the market reaches maturity.

Some folks see sales like that, too, with an S-shaped learning curve[5] that takes you to, and beyond, the threshold of product/market fit[6].

At the mature, stable end of the spectrum folks have usually iterated their way to a pitch that works, perhaps backed by a sophisticated sales machine they've spent years building.

But that's the end of the curve. What happens at the start? Early on, sales is about learning (we'll explore this more in **chapter 38** when we look at Dock), and that means a lot of discovery, sometimes asking uncomfortably direct questions.

This is something Segment found out through lived experience. Segment, a customer data platform, was founded in 2011 and acquired in 2020. I love Segment co-founder Peter Reinhardt's Twitter thread[7] (and podcast interview mentioned earlier[8]) about discovering sales in their early days, which goes something like this:

- **Uncomfortable questions:** Reinhardt found that, to his surprise as a product person, "our biggest learnings in PMF were from wildly uncomfortable questions from our sales team".
- **Directed at engineers:** That discovery came from their second sales hire who crushed his numbers, was loved by customers, and had a knack for asking engineering buyers very direct questions like, "Why don't you just build Segment yourself?"
- **Who would sell themselves:** The customer, surprised and perhaps a little discombobulated, would explain their thinking, and then they were easy to sell to. "It wasn't really even selling" at that point, as Reinhardt describes, "We just explained how Segment [...] fit into their problem and their approach."

As their sales process matured, Segment got far more clarity on customer problems, eventually implemented the sales MEDDPICC framework, and leveled up again. According to Reinhardt, "it was like a new product/market fit superpower."

All from getting serious about sales.

That's how you accelerate up the sales S-curve, and it's a fantastic little case study in unpacking the "market" side of product/market fit *with your customers, as you sell to them* by having the guts to ask the hard questions.

But that starts with having a sales narrative, and sales narratives start from understanding the hierarchy of attention.

Notes

1. "Founder of Company Sold for $3.2B (Segment) on Quitting Software Industry to Save the Planet," *The Logan Bartlett Show*, January 21, 2023. **https://youtu.be/gPGVuL1KLhg?t=2092**
2. Early-stage startup sales consultant Jen Abel has tweeted about this, including asking your prospects point-blank if they like being an early adopter/design partner (**https://twitter.com/jjen_abel/status/1433028276331716611**) and looking for indirect signals ("In most cases, historical intent is the leading indicator of an early adopter. "Have they tried to solve this unsuccessfully in [the] past"?" **https://twitter.com/jjen_abel/status/1567803434434281472**). Don't overlook the language here either — "design partner" is the term of art and, I imagine, acts as something of a secret handshake for those who understand what kind of relationship the term implies. See **https://www.unusual.vc/articles/what-are-design-partners** for more.
3. The 'inciting incident' is storytelling jargon for the event that sent the person on their journey. That may be as simple as "Our contract is up with our current provider, and we want to try something new," or it might be more fundamental and help you uncover the deeper job-to-be-done. The more pain in that incident, the more motivation they'll have. If they just want to try a new flavor of proverbial ice cream for their next contract cycle, fine, but that's not a sign of great pain you can tap into. If they really are dealing with an acute problem, however, and you can uncover exactly what's driving that problem, then that can be insight gold that gets you closer to product/market fit.
4. *The Challenger Sale* series is a case in point, and the most recent example of that style of sales research (as of writing) comes via *The JOLT Effect*, which we'll touch on in **chapter 23**.
5. As described in the classic *HBR* article "The Sales Learning Curve" by Mark Leslie and Charles A. Holloway from 2006: **https://hbr.org/2006/07/the-sales-learning-curve**. Leslie also wrote about it for Sequoia in 2016: **https://www.sequoiacap.com/article/the-sales-learning-curve/**.
6. There *is* a measurable-ish definition of product/market fit for enterprise software, by the way, and it's based on the Sales Learning Curve. When Marc Andreessen wrote his famous 2007 essay on product/market fit ("The only thing that matters,"

https://pmarchive.com/guide_to_startups_part4.html) he was riffing on Andy Rachleff's ideas, which Andreessen cites in his essay. Rachleff gets credit for originally popularizing "product/market fit," which is a neat addition to a resume that includes co-founding Benchmark Capital, co-founding and being CEO of Wealthfront, and teaching classes on entrepreneurship and product/market fit at the Stanford Graduate School of Business. So what's the definition? In a podcast interview with Mike Maples, jr (**https://greatness.floodgate.com/episodes/andy-rachleff-on-how-to-know-if-youve-got-product-market-fit-XxGvX8DH/transcript**) Rachleff actually credits Mark Leslie (of Sales Learning Curve fame above) with the idea of enterprise software product/market fit being the point when sales yield ("contribution margin of a sales team divided by the total cost to field the sales team") is greater than one. That's not his only definition — the other heuristic Rachleff offers comes, he says, from Sequoia's Doug Leone, which involves (supposedly) pulling a trial, no matter what, when a prospect's 30 days are up. Then, Rachleff says, "If the customer doesn't scream, you don't have product-market fit." I'd take that with a grain of salt, but it's fascinating to see how folks who have been in the venture game for a long time talk about product/market fit.

7. Reinhardt, Peter, January 19, 2022. **https://twitter.com/reinpk/status/1483476822897668099**
8. "Founder of Company Sold for $3.2B (Segment) on Quitting Software Industry to Save the Planet," *The Logan Bartlett Show*, January 21, 2023. **https://youtu.be/gPGVuL1KLhg?t=2092**

9

10 slide narrative deck

> There will be a circular feedback loop between the product narrative and your sales narrative as it meets the market and either fails, succeeds, or does a little of both. So this shouldn't be looked at as something set in stone, but rather as a hypothesis that will change over time. But you still have to have a coherent rough draft to start.
>
> *- Pete Kazanjy, author of Founding Sales*[1]

In this chapter, I'm going to walk you through how I might construct a simple narrative sales deck for a B2B SaaS company. This is based on my real-life client work, and I suggest you follow along to either sketch out a first draft (if you don't already have a deck) or compare it to your existing deck if you do.

Here, we're just setting up the basic foundation for the clarity strategy work to come, but I want to stress that it's worth thinking of your sales narrative, including your pitch, as a product that you need to develop with your actual product. The same rules apply — prototyping, testing, iterating, tailoring it for specific sales jobs-to-be-done, and then polishing it as you scale. This is about the prototyping stage.

Before we dive in, a few caveats:

- We're just focusing on the narrative half of your pitch to set the broader context. We're not focusing on your product or your customers in any particular depth, as they are obviously unique to your situation, and you can segue into them from your narrative.
- This is intended for folks who *don't* have a strong sales narrative. If that's not you — i.e., you think the deck you've got is pretty decent — that's great, don't throw it out for the sake of my basic outline below. The rest of the book exists to help you improve what you've got.
- A strong narrative isn't a substitute for a strong sales process with appropriate qualification and discovery, whether you're getting started with founder sales[1], using something like BANT[2], or running a heavy, enterprise-oriented methodology like MEDDPICC[3], for example.
- I'm throwing you in the deep end here — if you want a simple framework to help you work through your messaging more methodically, check out **chapter 13**, which also includes a mini Design Thinking primer.

Next, let's look at how you can assemble a basic sales narrative as a deck.

> **Sales is contextual — different decks for different contexts:** This chapter isn't a prescription for The One True Magic Deck Formula — that doesn't exist because sales is highly contextual, depending on the amount of story you have to tell, who (and what) you're pitching, and what stage you're at. The *amount* of story you have to tell radially shapes what your deck looks like. What we spend 6 or 7 slides on here might only need 1 or 2 slides depending on what you have to say. Sales is also highly dependent on *who* you're pitching — we'll explore three very different deck outlines in chapter 23, for example, exploring the difference between a CxO pitch, a high-velocity SMB pitch, and a positioning-driven pitch that sits in between. Your pitch will evolve based on your *stage*, too, and we'll see what very early stage, discovery-driven sales looks like in chapter 38.

Deck basics — wireframing & outlining

Anyone who has ever done any exploratory creative work, be that design or code or writing, will tell you it's usually a mistake to jump in and start working on The Final Thing. You need to start with something low fidelity first, so here we're going to start with:

1. A place to sketch out or wireframe your deck.
2. A simple slide template you can use to keep your messaging tight.
3. An outline you can flesh out.

Step 1: Set up your doc/slides with no or minimal styling

On the first point, I recommend starting with a scratch deck — the point here is rapid iteration and the chance to explore directions that may turn out to be dead ends. There's no need to design anything beyond the absolute bare minimum for layout. All custom style is distraction. This is words, words, words.

For that reason, I suggest starting with a doc-based tool because this is a *thinking* exercise, not a production one. I personally find Notion a great place to do this thinking and writing because there's *just* enough formatting for structure and nothing else to fuss with when writing, but any other similar tool would work. You can use Google Docs, of course, or if you're more comfortable in Powerpoint, Google Slides, or your app of choice (I use Figma for decks occasionally, too), that's fine. Sketch it out in a notes app or a physical notebook if you want to. The tool doesn't matter so long as it stays out of your way.

Putting together 'slides' in a document editor may sound weird, but that's how I like to do it, given that's what 'slides' at this stage are: just text broken up over a doc, perhaps with 'speaker notes' in callouts or comments. This makes it easy to scroll through, share, and otherwise take in as one end-to-end whole, to understand it as raw story — raw *thinking* — not a pretty presentation. Keep it rough and keep iterating.

Step 2: Constructing a basic slide template

Next, create some dummy slides you can copy & paste. You'll replace the dummy content with the content we'll discuss below. Here's how I construct these rough slides (in awkward text format):

Slide title

Slide header

A concise sentence or two that captures the idea and states your key value prop.

1. **Bullet one:** You can also use two or three bullets to make your point.
2. **Bullet two:** Consider arranging them side by side if that helps.
3. **Bullet three:** You want to avoid slabs of text — keep refining until it's clear.

> *I like to include speaker notes here in a callout box like this. Here, you might add more context about how you plan to engage with your conversation partner in the sales discussion or how you might segue from one slide to the next. This box helps prevent you from trying to jam everything you want to say into the slide itself.*

That's it — dead simple.

Step 3: The initial outline

Next, write an ultra-simple outline for your deck, especially if you know there are key points you want to hit or you have a rough idea of the flow in your head. If you're not sure how to start, review the flow of the slides below and then come back to it. Prompt ChatGPT to get a bad example of an outline if it helps.

Some folks like writing strictly to an outline; some folks like writing to *find* an outline. Whatever your approach, this is where you can dump the rough, semi-structured outline of your core value proposition in your head onto the page so you can then flesh it out from there. This is about getting over the intimidating blank page. You don't have to write a whole deck, start to finish, in one sitting, starting with a blank first slide. Start experimenting with a simple outline and see where you land.

Here's an example (not a prescription!) of a fairly typical insight-driven B2B narrative deck outline I use with my clients for a basic pitch that balances story 1 & story 2 in their narrative:

1. **BLUF:** Bottom line up front to introduce your story and product.
2. **Context (macro/micro):** Set up your story 1 — the broader context and/or your perspective on the market. What's your unique insight into their broader (or specific) context?
3. **Tension:** What's the tension that brought the prospect here? What's stopping the prospect from moving forward? What's wrong with the alternatives? And what does the *positive* opportunity for them to keep succeeding in their role look like?
4. **Solution:** Switch to story 2 — the product reveal that addresses the tension we've set up from the story 1 change in the environment, perhaps with a look at major product pillars.
5. **Resolution:** The impact and results the solution delivers, and what the solution means for the broader team and organization. This is the outcome — the integrated whole — they can expect and others have achieved.

> **Don't forget the prospect's story:** Note that these slides are purely about *your* story, and an effective sales conversation would spend a considerable amount of time uncovering your prospect's story (a.k.a. discovery) so you can find the intersection between the two stories. You'd then follow these slides with a call to action and next steps, perhaps outlining the 'time to value' for your product. We're just focusing on the narrative slice of your pitch here, however.

The deck

Here we go, with a slide-by-slide rundown of ideas to flesh out your basic sales deck. Don't just read; get hands-on. Open your editor of choice, create some dummy slides, and then follow the bullet-point recommendations for each suggested slide below. Feel free to riff, remix, rework, and/or simplify however you see fit!

Remember, we're following the slide format outlined above (title, header, contents). I'll suggest a slide title, leave the slide header to you, and suggest the points you might hit in the slide's bullets.

Slide 1: Intro

- **Title:** {Product name}, Introducing (or meet) {product}, or leave your opening slide without a title.
- **BLUF:** This is the BLUF (Bottom Line Up Front) slide, perhaps with your three-or-so-word value prop (a.k.a. your Hemingway Story, which we'll explore next in **chapter 10**) as your header.
- **TL;DR summary:** Add a sentence or two unpacking what you do and who you do it for, perhaps hitting max 3 key pain points or jobs-to-be-done. Or if the product is novel, explain with "{Product} is an X that does Y (value added, pain point solved) by Z (unique approach)." Very direct and to the point.
- **No backstory, no features:** As a rule of thumb, don't start by walking the prospect through your journey (unless it's super relevant) or by jumping straight into the product. A demo includes demonstrating your thinking, not just your product, so that's what we're focused on first.

By framing the conversation with a brief tease of the solution first, you're scratching the prospect's main itch — whether you have a solution that's relevant to their problem *at all*. Everything that follows is there to help them judge if this is the *right* solution. We'll build some tension along the way around that question, but folks want you to get to the point, so I like to start with just that: the point, the bottom line, right up front.

Slide 2: Macro trends/specific pains

- **Title:** Why {product}, Megatrends, Trends, Why now, What we see, {The name of your category concept}, etc.
- **Context:** You've planted your flag. Now, let's zoom out to provide some story 1 context before we zoom in on the product. We want to set up why your solution matters; why *now*.
- **"What we see":** This is a good place to introduce 'what we see' in the market. Your slide header, for example, might name a change in the world or a macro trend that's relevant to the prospect (if one exists) or the broad challenge that's causing pain or preventing prospects like them from succeeding in particular ways.
- **Be specific:** Don't rely on generic trends here — you need to have a specific insight or observation. You can go big picture if a trend really is moving the market (we'll look at trend narratives in **chapter 20**), or you can zero

in on a typical 'inciting incident' that acts as a trigger for the prospect's buying journey (e.g., out-growing spreadsheets, drowning in some kind of complexity, needing to collaborate in a specific way, and so on).
- **Support and elaborate:** Sum up the trend or pain points in three bullets max, and then sum up or title each bullet in a word or two. (A bit like I'm doing here.)

Whether you go big picture with a change or trend narrative or specific with a problem and a tightly positioned point solution is actually a profound question around right- and left-brain attention that we'll spend much of the rest of this book exploring, so if you're not sure how macro or how specific to get here, don't worry, there's plenty of advice to come.

Slide 3: The need/the insight

- **Title:** The need, The problem, The insight.
- **Build a bridge:** We've set the broad context. Now, let's connect to the prospect's specific situation. Here, we want to build bridges to the prospect's reality — their pains and challenges. This will usually take 2-3 slides to unpack, maybe starting with an "Insight" or "Need." This is where we start to shift attention from story 1 to story 2.
- **Locate the prospect:** For the next two slides, we're trying to establish our unique insight in the market (be it a market, product, or technology insight) and locate where the prospect exists in that context. *The insight must be relatable and familiar if it's going to connect.*
- **Finger on the need:** We're looking for the Venn diagram overlap between our insight and their acutely felt need. For the prospect, they might not have connected their pain to our bigger picture — they just know that metric X is going down, or the volume of Y is too high, or they're fed up with their current way of doing things. It's up to you to help them join the dots.
- **Read their mind:** What are the 1-3 things they see that make things particularly challenging in their role? We're looking for a "You got me," "You read my mind," or "You just described exactly what we're going through" reaction.

Your ability to connect here is only as strong as your discovery. How folks do discovery varies, and it can happen in an initial call (literally a 'discovery call'), before you present, or as you present. Either way, in complex sales, the magic words are "Tell me more about that" and "Why is that a problem for you?". Buyers want to feel heard, so while we're focusing on the stuff you say here, listening carefully is just as important.

Slide 4: The challenge

- **Title:** The challenge, The reality.
- **Name the challenge:** We've set the context. We've located the prospect in it. But what are the specific reasons that *this person* in *this role* finds it such a challenge? We've established the problem is there, and it's there for them, but *why* is it there for them? Consider naming the challenge if you can (if it doesn't sound too corny). Show you've thought hard about it. "We call this challenge the _____." or "We often find that the reason that drives this is _____."
- **Pain and tension:** Can you provide three reasons that flesh out why this challenge is so difficult, frequent, or otherwise painful for the prospect? We're trying to build some tension before the full reveal of our solution — the more we can get them invested in and relating to the problem, the more of a relief or revelation our solution will be.
- **Keep succeeding:** Your prospect is likely doing a lot with a little. They're doing their best in (usually) somewhat adverse and dysfunctional circumstances — i.e., those of just about any company, anywhere. Consider framing this slide around how you can help them *keep* succeeding. Think more "You've done great to make it this far, you beautiful human," not "You need to stop punching yourself in the face, you big dummy."

There's a big emotional component to selling, and there's a big difference between making someone feel doomed vs. giving them confidence in their own potential for success. Some folks like to hammer their prospect with scary YOU'RE DOING IT WRONG AND YOU'RE GOING TO BE A BIG LOSER stuff here, but that's both a bad experience for the buyer and ineffective in helping them take confident action. Recent B2B sales analysis of 2.5 million sales conversations suggests buyers are more scared of *messing up* (and buying the wrong thing) than *missing out* on your solution (to quote the book that presents the research, *The JOLT Effect*, which we'll discuss in **chapter 23**), and building buyer confidence *in themselves* is crucial for the sale.

Slide 5: The status quo

- **Title:** Status quo, Alternatives, What doesn't work.
- **Goldilocks it:** I love using a Goldilocks slide here if I can, first framing alternate solutions as too hot, too cold, or too whatever (simplistic, expensive, time consuming, cumbersome, etc.).
- **Why alternatives fail:** The classic B2B approach might be to highlight that spreadsheets are too brittle, manual processes aren't good enough,

generic solutions weren't built for your specific needs, and so on. Too hot, too cold, in preparation for what's *just right*.

Again, your insight here is only as good as your discovery, both in the market more broadly and for this prospect in particular. The more you can draw from reality — especially your prospect's reality — the better. That sounds obvious, but startups often index off peer competitors that the prospect may never have heard of. The real 'competition' is the customer's status quo alternatives for solving their job-to-be-done, which may well be spreadsheets, manual processes, or an antiquated legacy system. How do you know what these are? Ask! Ask every prospect you can, "How are you currently trying to solve this problem? What else have you tried?" The patterns will quickly emerge.

Slide 6: The opportunity

- **Title:** The opportunity, What if..., The new way, Succeeding with/through {Category}
- **Wouldn't it be great if...:** We set up our big reveal by first teasing out the logic of the *kind* of solution that would be great... if it existed. Making this too obvious ('Wouldn't it be great if a product existed with feature X AND feature Y?!?') is cheesy as, so tread carefully. Keep the framing around *their* success, their job-to-be-done, and the relief of their pain — wouldn't it be great if you could do A and B without worrying about C? Wouldn't it be great if you never had to deal with D or E ever again?
- **Tease success:** Again, we want to make it clear we understand what their success is *not* merely having our product or using a specific feature, it's the outcome they get, and that we understand that success outcome as much as we've understood the pain that brought them here.
- **Paint the picture:** The rule of threes comes in handy here — consider highlighting the top 1-3 outcomes that prospects might care about here as concise bullets, perhaps ones that connect to your 3-word 'big idea' value prop/Hemingway story from the very first side. Resist the urge to cram in everything. Curation keeps folks curious. And make sure you hear their side of the story by asking if these outcomes would hit the mark for them.

Proof makes claims real, and as a general note, you want to anchor your claims in some proof that sits outside your company. What is it that makes your solution possible now in a way that's unique to you? Maybe it's an underlying technology breakthrough, maybe it's just that folks in some other function have tools that do this (and now it's your prospect's turn), and/or maybe you have compelling case studies and references from other customers. This gives you the ability to go from 'Wouldn't it be great if...' to

'And that's what we see, with X of our customers achieving Y in Z time.' And remember, the trick to good proof is to make sure it's *actually proof*.

Slide 7: The product

- **Title:** Introducing X, Meet Y (if you didn't use that on slide 1), The Z difference
- **The big reveal:** Now we finally pull back the curtains and reveal our solution. We've hopefully established common ground and agreement with the prospect about what their pain and potential success looks like. If they've bought into our perspective and insight, they should now be invested in what our solution might mean for them.
- **Answer the tension:** This is where we relieve the tension we've been building in the narrative thus far — we've established we know their problems, we know their real challenges, we know their alternatives don't work, we know 'it would be great if...', and now **drum roll** we've got the solution just for them. We can now get into the story 2, narrow-focus specifics.
- **Let the prospect sell themselves:** As you introduce the product and its capabilities, give the prospect the chance and space to think things through and sell themselves on it. Prompt the prospect to see how it could help with their specific challenge. Ideally, they should be selling themselves at this point. Complex B2B sales are more rational purchases, and folks need space to verbally process what they're hearing, so be sure to provide it.

If your product is complex, consider organizing its functionality into key pillars for the sake of your sales and marketing efforts if you haven't already. Then you can hopefully connect those pillars to the problems your prospect has been discussing and the success they want to achieve — how it works end to end, how it's a unique combination of tech & UX, how it addresses their unique situation, and so on.

Slide 8: The impact

- **Title:** Team impact, Levelling up.
- **What it means:** Your product isn't an end in itself; it's the outcome that matters, and complex B2B products often have different outcomes for folks up and down the org. This slide is a great place to neatly articulate the benefits of your solution for **end users**, **the team/function** more broadly, and **the company on the whole**.
- **Ladder up benefits:** These benefits can go micro to macro, from workflow improvements from those using the tool, to net business outcomes for

management.

A handy device consultants use in framing these changes is from → to, i.e., then → now, or current state → future state. Here, this can be a useful way of succinctly describing what happens with and without your product for different roles or departments, e.g., front-line workers may go *from* 'struggling with X' *to* 'confident with Y,' and management might go from 'lacking insight about A' → 'full visibility due thanks to B' and so on.

Remember, in complex sales it's not just the end-user buying the product. It's the team, the department and management, and the company or enterprise overall. The more credibly you can speak to the value and impact of your solution across those roles, the less resistance you're likely to have from the buying committee.

Slide 9: The product pillars

- **Title:** The name for the feature, e.g. (to use generic examples) Dashboards, Workflows, AI, etc.
- **The how:** We've put in a lot of effort to set the context for the product and its impact. Now it's time to show *how* it does all these great things we've been discussing. Time to walk the talk.
- **Features, finally:** This means features, ideally organized into key themes or product areas. This is where you can get as tailored as you like. Maybe you dive into the product itself with a live demo, or maybe you work through the prospect's particular requirements — whatever best tells your product story *as it relates to the prospect's pain.*

We want to keep drawing connections between our key features and the big-picture value we've been describing throughout the deck. Be sure to hit objections as you go, too. Maybe folks always ask about a Foobot feature when you know they need your Whatsit automation. Knock objections on their head, and pre-emptively if you can, as that's what high-performing sellers do.[4]

Slide 10: The result

- **Title:** The outcome
- **Full circle:** This is where we bring things full circle and tie the narrative back to that initial three-word value prop we opened with, elaborating on the glorious future that awaits the prospect. This is the big idea we want to leave them with, having made a strong, compelling case both in terms of insight, discovery, and solution by way of your product.
- **Value, value, value:** Consider using very succinct bullet points to round

out this slide that describe the new successful future that awaits your prospect enabled by your solution, hitting your major value themes and introducing them with a single word (productivity, efficiency, safety, speed, etc.).

Really spell things out for folks here. You could also consider using an A → B → C sequence that recaps (a) how your product solves the challenge we raised, (b) delivers meaningful quantitative or qualitative outcomes, and thus (c) achieves the original value proposition (from the very first slide) to sum up this part of your narrative.

Slides >10: Product, proof, and next steps

- **Title(s):** Features, Testimonials, Customer stories, Customer success, Results.
- **Product:** So far, I've kept this deck outline relatively concise and focused on the value *around* your product, but if your particular product needs deeper explanation — perhaps you're selling something very technical to technically sophisticated buyers, then by all means elaborate where appropriate.
- **Social proof:** A great narrative is nice, but talk is cheap — where's the proof? People want evidence you can deliver, so consider introducing some compelling social proof or references here. You can drop in logos or customer outcome stats earlier, too, if it helps. (Remember, I've ignored the contextual stuff to focus on the narrative here.) Either way, the more you know your customers *and can capture their success*, the more you'll have to work with.
- **Next steps:** Your deck will naturally end with a focus on next steps for qualified buyers — that might mean further demos for other stakeholders, pilots, trials, contracts, etc. That's outside the scope of our narrative, but (for those new to sales) it's incredibly important to end actual sales calls with clear next steps so the prospects can either evaluate further (e.g., getting hands-on) or otherwise make progress towards a decision. No "So... whatcha think?"

Remember, a successful narrative isn't one you've delivered; it's one the prospect has truly heard. Sales is a conversation, not a monologue. How do you know the prospect has really heard and internalized what you've discussed? It's often in their reaction. For early-stage folks, anything less than a 'YES!' is probably a 'no', *even if they sound interested*. Early-stage founders can struggle with this. You'll get a lot of polite agreement and even requests for getting hands-on (as sales pro Jen Abel notes[5]) from folks who are actually unqualified to buy from you. Ruthless qualification is necessary to separate the tire kickers from the genuine buyers, but, like everything, the only way you get there is through putting in the work, knowing what to look for, spotting the patterns, and iterating accordingly.

That's it. That's your basic narrative

Phew! That was a lot to take in if you've never created a sales narrative before, or it might be old hat if you've been polishing your deck over years of selling your software.

Either way, you've now got a narrative to work with! You can test it on the team and on prospects, see what lands (and what doesn't), turn it into real slides, and roll it out for further testing... but I suggest making your way through the rest of this book first. The big insights are yet to come.

'Is that it?' vs. 'That's too much!'

Now, here's where it gets tricky, because I've tried to please everyone with a single deck format. There's two possible reactions at the extremes of the spectrum here: you may either underwhelmed or overwhelmed. (If you're just pleasantly whelmed, then great, keep powering ahead.)

If you're a startup veteran or sales pro and feel **underwhelmed**, then fear not. What we've fleshed out here is just the basic *format* of a deck, but it's the *content* that really matters, especially the nature of your story 1/story 2 insights and how they engage with a prospect's attention. That's what we're going to explore in *quite* some depth in the coming chapters, and feel free to skip ahead if that's what you're here for.

If you're new to this and feeling **overwhelmed**, or that things are still too fuzzy and you're not sure what to say for each slide, that's totally understandable. I really did throw you in the deep end, dear reader, and I get wanting more explicit recommendations about what to say when. However, keep in mind that (a) that's where the magic of creativity comes in where you tell your story in a way that's unique to you, and (b) the next few chapters will help you flesh out the fundamentals a little more, as we look at expressing your core idea next in **chapter 10**, then your origin story and 5Ws in **chapter 11**, a messaging framework in **chapter 13**, your memes and messaging in **chapter 15**, and alternate, more contextual deck formats in **chapter 23**.

That is, there's plenty more to come in both figuring out the particulars of your message as well as the deep insights that power your narrative and positioning strategy, so hang in there.

Narrative as product

Let's also remember that we're trying to seed the idea that your sales narrative is a product here, one with its own virtuous loop of explore/exploit, where you learn what you can, create a narrative snapshot, test that out and learn some more, create a new snapshot, and repeat.

1. **Snapshot:** Start with a narrative that captures where you're at now ← **you are here**
2. **Explore:** Do the deep work on stretching your narrative *out* and/or *down* ← **where we'll go next**
3. **Snapshot:** Come back to your narrative to incorporate the fresh insights you've developed and create a new snapshot.
4. **Explore:** Rinse and repeat.

That is, your narrative is a tool for **capturing** and **generating** insights that inform every part of your operation, not just a deck you slap together and call it a day. It's a profoundly important set of ideas that expresses your working theory of product/market fit and creates a two-way bridge between you and your market so you can test what works.

Next, let's explore your core idea and see if we can turn it into a 'Hemingway story.'

Notes

1. Kazanjy, Peter. *Founding Sales,* 2020. You can read the book free online after signing up: **https://www.foundingsales.com/2-product-marketing**
2. BANT is a simple qualification framework focused on Budget, Authority, Need and Timeframe. Sales folks love making up their own variations and frameworks, and many riffs on this exist. What matters is matching the complexity of how your customers buy to the complexity of the framework, and we'll look at the vast differences between B2SMB vs. B2Enterprise buying, for example, as we go.
3. MEDDPICC is both an absolute mouthful of an acronym and a sophisticated qualification framework that stands for — *deep breath* — Metrics, Economic Buyer, Decision Process, Decision Criteria, Partner, Paper Process, Identify Pain, Champion, and Competition. Phew. Gitlab has a nice explainer that also provides a bit of insight into why they're called *complex* sales: **https://about.gitlab.com/handbook/sales/meddppicc/**.
4. Dixon and McKenna write in the *The JOLT Effect* that "[O]ur research shows that high performers will sometimes anticipate an objection to be coming and preemptively offer a rebuttal—a "pre-buttal," if you will. High performers are always on the hunt for signs of "implicit non-acceptance"—that is, when a change in tone or a slight pause on the part of the customer tells them that something is amiss and the customer's not buying it." They also write that "the average win rate for sales interactions in [their] study was 26 percent," but if objections that go unanswered appear, that drops to 17 percent (a 35 percent relative decrease), but if "just one" rebuttal is offered, win rates go up to 31

percent (a relative jump of almost 20 percent). It's hard to know with these observational studies what's cause and what's correlation, but it at least goes to show that listening carefully for and responding to objections — perhaps preemptively — is a sign of a strong seller, and that's something you may consider building into your pitch as a talking point if a particular objection comes up frequently enough.

5. Abel has a great tweet about this (**https://twitter.com/jjen_abel/status/1633423561439281152**), writing (with slight reformatting) "Startups: If any of these are said on a first call, here's the proper interpretation of the market feedback: "This is really cool. I'll reach back out shortly" = unqualified. "I have limited time, but send me a link/login so I can explore for myself" = unqualified. "Need to jump. Shoot me a proposal, and I'll circulate it" = unqualified. Qualified customers don't say these lines. These are parting lines used to quickly jump off the call." That is, when we're excited about our product, it's easy to assume that *any* hint of interest from a prospect indicates a potential sale. And isn't that what we're trying to do — sell to as many people as possible? But, counterintuitively enough, having a high bar for real interest forces you to generate a solution — and pitch — that really does generate a 'Heck yeah!' from prospects, which puts you on a much more plausible path to success due to much greater market pull.

10

Your 'Hemingway story' micro-pitch

Most B2B narratives can be summed up in six words: so many words, so little said.

And that creates an opportunity for you, dear reader, to stand out in the sea of meaningless dross. When it comes to your narrative, a simple, punchy message acts as the perfect tip of your narrative spear.

In my work with clients and from what I've observed studying other practitioners, sometimes the real power of a narrative is in the three-or-so words that *finally* captures what a founder has been hammering away at for years.

In the previous chapter we covered your narrative deck end-to-end, and in this chapter we're going to go back to that first slide, and think about the sharp, concise, hit-the-nail-on-the-head micro-pitch that frames your narrative.

Your Hemingway story

I call this pithy little expression of your idea — ideally about what your product *does* for your customer — your Hemingway story.

Ernest Hemingway, legend has it, once crafted a short story to win a bet with his drinking buddies. The bet? That he could make them cry with a story of just six words. The story?

"For sale: baby shoes, never worn."[1]

Your Hemingway story doesn't need to make anyone cry (or roll their eyes for that matter) but it does need to explain what you do, that is, what your core idea is for your customer, with absolute clarity.

The idea

This is the idea at the heart of your clarity strategy. We're not talking about your product — your product *expresses* your idea (and might require a longer description), but it isn't your idea. We're talking about the outcome, behavior change, or conceptual revolution your product brings about.

This is the core concept that's strong enough to carry you through the early, scrappy days with big dreams and a barely functional product to something much more consequential.

This core idea is your conceptual North Star — and may well inform your North Star metric too.

Rescuing revenue

Here's an example from my own work that illustrates the point. One client I worked with had a chatbot that acted like a sales assistant for ecommerce stores. There was a nice Goldilocks aspect to it — it didn't pop up for everyone (too annoying), and it didn't make lame attempts to capture you on your way out (too little, too late). Not too hot, not too cold, just right.

The client's messaging, however, piled on the B2B-style value props about all the ways the tool could help, but the result was value prop soup — a big mix of everything, but it wasn't *about* anything.

To the extent it was about anything — improving conversion rates — buyers had heard it all before on the one hand, and couldn't accurately gauge meaningful conversion rate improvements on the other. That's a difficult value story to sell.

What's more, the client had a mix of hype-y, story 1, future-of-ecommerce, 'category creation' stuff mixed in with their product pitch. This was a classic two-story problem: they needed to better distinguish between their big-picture story 1 stuff (and ideally save it for new hires and investors) while sharpening their customer-facing, what's-in-it-for-me story 2.

Their story 1 pitch could be summed up as 'the future of e-commerce is chat.' You could debate whether that's true or not, but either way it's not what buyers care about. They're not buying the speculative future, they're buying the a better customer experience and resulting revenue bump they can get started with *today.* And that laser-focused story 2 pitch needs its own micro-pitch too.

Long story short, I landed on a simple three-word line for this part of their narrative: "{Product} rescues revenue."

I used that as a hero headline at the start of a narrative deck, and it stuck. In fact, for all the other narrative work I did, this idea was the most consequential. (Though sadly, it never saw the light of day — such is life with client work!)

You might think that is... underwhelming. I mean, it's just a short slogan, right? Actually, those three words were immensely helpful in thinking through their story, not because the words themselves are magic, but because the process *captured what their product actually did for buyers*, including *how* it did it.

From there, the startup could have:

- **Shift the sales conversation** from fuzzy, fluctuating conversion rates (where going from 1.0% to 1.1% is a significant win, but hard for buyers to see) to absolute, concrete dollars-and-sales rescued ($15,000 in June, for example).

- **Build a brand with a theme** — 'rescuing' is heroic and suggests the character of the brand that they can run with.
- **Reframe their North Star metric**, focusing more on absolute gains and less on incremental improvements.
- **Build an end-to-end narrative** that neatly articulates their value, from first touch, to purchase, to the product itself, with the same language potentially being used in reports and dashboards.

All from three well-chosen words. Again, it's not that the words are magic. But the distillation of an idea kind of is.

> **What's your Hemingway story?** How many words does it take to get down your core idea? What did you use for the opening slide in your narrative deck in the previous chapter? Try making different variations to see what sticks. Also, consider asking:
>
> - Does your Hemingway story tie into your north star metric?
> - Does it reflect a behavior change or a from/to state in some way?
> - Does it appear (at least thematically) in your campaigns, product, onboarding, and visuals?
> - Does it distinguish you from competitors?

Radical clarity

That's just a simple example from your humble author's client work. But we see this same kind of radical clarity in big SaaS success stories *all the time.* I was often struck by this when thinking of examples for this book, especially how easy it was for me, an interested outsider, to express their core idea in just a few words.

It's easier to show rather than tell. In that spirit, here are a bunch of short SaaS examples off the top of my head:

- Use cloud software not on-prem. *(Salesforce's "End of Software" campaign at the turn of the millennium, before either 'the cloud' or 'SaaS' had been coined.)*
- Use chat not email. *(Slack, as we'll discuss in **chapter 12**.)*
- Use chat not forms. *(Drift and "conversational marketing", as we'll discuss in **chapter 31**.)*
- Do 'inbound' marketing. *(HubSpot, riding the digital marketing wave, as we'll also discuss in **chapter 31**.)*

- Use agile not waterfall. *(Jira riding the agile wave.)*
- Use Linear not Jira. *(Linear riding a classic anti-incumbent wave.)*
- Mine sales calls for insights. *(Gong, riding the Zoom call recording wave.)*
- Design collaboratively in the browser. *(Figma, initially dismissed as an unserious tool that, oops, won the market and Adobe almost bought for $20B.)*
- Build your own workspace. *(Notion, with its B2C2B strategy that we'll touch on in* ***chapter 45****.)*
- Payments for developers. *(Stripe, as we'll briefly look at in* ***chapter 15****.)*

And on and on it goes.

These are just a few famous examples, to be sure. Some are about the product; some are about a concept or new behavior; some are a mix of both. You might express their core idea differently. But fundamentally, *there's a core idea to be expressed.* (I'm sure there are many more niche examples out there too — if you've got a good one, let me know: **luke@lukestevens.co**.)

> **Gall's Law in action:** Gall's Law says "A complex system that works is invariably found to have evolved from a simple system that worked." These startups are a demonstration of how that can play out in reality — complex products born out of a simplicity in concept. This is a topic we'll revisit again at the very end of the book in chapter 49, when we discuss strategy, complexification, and value.

Do this not that

We'll look at some more tips on refining your Hemingway story next, but here's an easy way to start thinking about your story: think about the *verb* or the *change*.

Typically, SaaS one-liners are often very declarative — they just describe the product and its position: We do X for Y; we solve Y niche problem for Z specific vertical; we're the {Uber/Stripe/Linear/buzzy company} for {different market}.

These approaches can be helpful as pure description, but they're not really the heart of a great narrative. Some of the most memorable SaaS Hemingway stories are, instead, much more about the *change*. The 'do this not that' pitch; the fundamentally new approach.

So consider, as a start, what your Hemingway story might look like if you had to articulate the behavior change, the verb, the *do this not that* that you expect customers to achieve with your product. Can you name it?

Loom and the "perfect" idea

Here's a neat Hemingway story: Send videos, skip meetings. But what's less neat is how startups arrive on those ideas — often through *user* innovation and messy serendipity, not pure vision.

That was certainly true for Loom and Shahed Khan, co-founder of the video messaging SaaS app that was acquired by Atlassian in 2023 for $975 million. Khan described on Twitter[2] both how Loom found their idea and the paradox of *not* starting with a perfect idea so that you *do* find your winner.

Here's the story in a nutshell.

Loom began as a user testing marketplace that failed. They pivoted, rebuilt the product as an internal testing tool, and stumbled across *one single use case* where someone used the tool in an unexpected way: not to share the user tests, but to share a one minute video summary with their team.

The team pivots again, relaunches as a video communication tool on Product Hunt, and instantly gets 3,000 sign-ups, "2,990 more than we'd had in the 9 months prior" according to Khan in his thread. They would go on to raise over $200M, gain 25 million customers, and be acquired for just shy of $1 billion.

That is, as Khan shares, "the story of how we came up with Loom: An idea that worked, after a couple that didn't." They didn't start with a perfect idea, but by *doing* and paying close attention they managed to find one. Their homepage, as of writing, captures it nicely: "Loom on. Meetings off."

Category creation vs. user innovation: I've ironically seen this same story told another way by someone not involved with the company to make the opposite point: that Loom *did* have the perfect idea. Loom, they say, was just another screen recording tool. Then, out of thin air, the company developed a *point of view* and *et voilà*, a new category was born: video messaging for work. But that skips the messy reality of innovation, with Khan saying for Loom it was a product of focus, timing, and "pure luck." This is narrative fallacy — glossing over the facts to make a just-so story — and it's endemic in startup stories, especially 'category creation' stories, which often lionize founders as visionary product savants. Sometimes, that description is deserved, but often, startups succeed when they double down on *user* innovation, not the purity of their original vision. We'll see this again and again in this book — sometimes the 'user innovator' is a team (Slack), entrepreneur (Nathan Barry and ConvertKit), or just a few needle-in-a-haystack regular users (as April Dunford spotted in one of her famous positioning stories[3]). This is actually a well-theorized phenomenon,[4] and it's vital to understand for narrative and

> positioning work. And if you can find that nugget of user-innovator gold, like Loom did, then who knows, maybe there's a well-deserved, near-billion dollar exist waiting ahead for you, too.

Notes

1. The story even has its own Wikipedia entry discussing its origins, including the fact it was already being parodied in 1921: **https://en.wikipedia.org/wiki/For_sale:_baby_shoes,_never_worn**
2. Khan, Shahed. September 14, 2022. **https://twitter.com/_shahedk/status/1570050492159627265**
3. We'll discuss this story in **chapter 37**, but the gist is Dunford calls the entire user base of a seemingly dead-on-arrival database product and finds a few folks who were using the product in an innovative, unexpected way. That discovery creates the impetus for the team to reposition the product, which goes on to be very successful. But it wouldn't have happened without the discovery of those 'user innovators.'
4. Eric von Hippel has researched and written extensively about user innovation from 'lead users,' including in his 2005 book *Democratizing Innovation,* which is available as a free PDF: **https://direct.mit.edu/books/oa-monograph/2821/Democratizing-Innovation**.

11

5W1H origin story

Let's stick with sharpening your idea.

If you're a founder, or if you can put yourself in your founder's shoes, this chapter is a chance to reflect on where you've been and where you're heading. I'm going to ask you a bunch of questions, and you can reflect and write the answers down in a doc if you like. (Or have someone on your team ask the questions if that helps you dig into the back story.)

One of the principles of this book is that your clarity strategy is only as strong as your discovery. And our starting point for discovery is, well, *your* starting point: the origin story for your startup.

Your origin story is a great place to start applying our hierarchy of attention, too. I want to know where your attention was when you had that initial spark of insight that led to the creation of your company. Was it:

- **Broad attention** looking *out* at a trend?
- **Narrow attention** looking *down* at a specific pain or problem?

Once you had that epiphany, what did you do? How did you explore it? Did you start building, start testing marketing channels, and/or start talking to customers? All of the above? None of the above?

Let's find out.

The original ingredient

The stories that make up your clarity strategy *aren't* a rehashing of your origin story, particularly beyond seed stage. People don't need all the gory details that led you to where you are today. However, origin stories are hugely important for context that *shapes* your clarity strategy.

Founders often have an idea, decide to pursue it, and spend years hammering away at it. That's a lot of time and effort spent on one initial insight, and it pays to make sure we have clarity about that insight. Especially as, over time, that initial clarity can get lost in the noise.

Clarity around your origin story provides insight into:

- **Your GTM preferences**. (Did you jump to sales, marketing, or product-led growth?)
- **How you learn and lead.** (Do you lean towards intuition and feel or particular frameworks and methodologies?)
- **What's changed since founding?** (Is the initial insight still valid? Has it

evolved, and if so, how?)

Intuition talks

For seed-stage folks, the origin story is still fresh, making it easy to apply to your clarity strategy. For later-stage companies, though, with years of experiments and feedback under your belt, it may take some digging to rediscover that initial spark.

For example, in one later-stage startup I observed, the founder had been pushed to tick all the boxes by their investors. They were:

- Product-led but sales-driven.
- Bottoms-up but chasing big enterprise accounts.
- Pushing a big-picture vision but using me-too performance advertising.

All of that was very confusing. What were they actually trying to do? What did they actually believe beyond the box-ticking? It wasn't until I heard their origin story that I could answer those questions. In their origin story, the founder's true *implicit* preferences around product and GTM became abundantly clear. I could better understand what the founder was trying to do and the story they were trying to tell.

Clarity strategy work is your chance to revisit, refine, and refresh your thinking in terms of your original insight and the assumptions behind it, too.

And while, as a founder, the original insight may be old hat to you, it's still an epiphany you need to share. The more your team understands that initial spark and has the same epiphany, the more you all share context and can collaborate intuitively. But only if you share that initial spark.

How do you capture that?

5W1H and your origin story

In journalism, people talk about the "5 Ws" or the "5 Ws + 1 H". They are, of course, *who, what, when, where, why*, and *how*. That's a good place to start for capturing your origin story, too. Consider:

- **Who** was involved at the start, both as the founding team and initial customers or your target ideal customer profile (ICP)?
- **What** did you observe in the market, tech, or product world that suggested there was an opportunity? What was the core insight?
- **When** did this all happen? What was happening around the same time? What does that opportunity look like today?
- **Where** was your attention at? Were you looking outward at the environment and saw a change or trend, or downward at a specific challenge you experienced up close?

- **Why** did you decide to tackle it? Why was it important to you, specifically?
- **How** did you solve it in terms of the initial product and the initial go-to-market strategy? Did you build momentum through outreach, organic channels, advertising, word of mouth, or some other approach?

What's changed?

With that snapshot of your origin story in place, we want a sense of then and now. What's changed?

For example:

- **Insight then → insight now:** Is that original insight what powers your startup today, or have you found something else? Are you attacking the same market or adjacent ones? Are the job(s)-to-be-done you serve the same then as now?
- **ICP then → ICP now:** How has your customer base changed? (Bonus points if you've analyzed your recent deals and know exactly who you're selling to and, more importantly, closing. Minus points if you're using made-up personas.)
- **Product then → product now:** What significant product developments have moved the needle? What's the #1 feature that's resonated most with customers?
- **GTM then → GTM now:** Are the original GTM strategies (advertising, sales, PLG, etc.) still in place, or have you iterated on your GTM approach? What's your current #1 distribution channel?
- **Explore then → exploit now:** Where are you in the explore/exploit loop? Have you found an opportunity you're happily exploiting, or are you still looking for that spark that lights up your growth?

These basic questions help us tease out both your origin story and several secondary insights, including:

1. How you currently pitch.
2. How you're positioned.
3. Your distribution preferences.
4. How you learn.

Let's briefly touch on each.

1. How you pitch

Does your current pitch accurately reflect where the company is at?

By exploring the 'then → now' of your origin story and comparing that to where you're at today, we have a brief outline of what's current today, specifically:

- *Who* you serve (ICP).
- *What* you've built.
- *How* it has an impact.
- *Why* it matters.

The question for you is: does your current pitch accurately reflect those fundamentals? What's your current process for improving your pitch and rolling out updates — is it working? Are your sellers happy with it? And do you evolve your pitch methodically (perhaps in the context of a sales framework), or does it evolve more organically?

2. How you're positioned

How much work has gone into your current positioning?

It's not uncommon for SaaS companies, big or small, to have been through some kind of traditional branding or positioning exercise that resulted in some descriptive statement like:

> *We make X product for Y customers and, unlike competitors A, B, and C, we also do Z.*

That may have been fine as a starting point, but positioning is a case where you get out what you put in, and most folks put very little effort into their positioning beyond those fill-in-the-blanks exercises. (Everybody's favourite B2B positioning pro happens to *hate* those, by the way[1]).

Or alternatively, there may have been some effort towards 'category creation':

> *We're creating a new category of X for Y.*

This tends to be equally superficial — a fine first attempt and admirable in its ambition, but pretty light on the mechanics of how category creation actually works.

I'm guessing, by virtue of the fact you're reading this book, that either statement didn't exactly capture the team's — or your customers' — imagination. It didn't set the world on fire, become a rallying cry, or catch on as a market-defining concept.

I'm not suggesting you have to throw everything out and start again. You may well be on the right track. Regardless of your approach, it pays to understand the logic of selling on concept or position — story 1 or story 2 —

and what that means for your narrative. Our narrative development journey in the back half of this book is about exploring exactly that. And by the end of that process, dear reader, we're going to end up several galaxies away from those kinds of superficial treatments.

3. Distribution preferences

How do you sell, and does that match your positioning?

Positioning and distribution work hand in hand, as we'll see in later chapters, and that means how you sell, i.e., how you 'go to market,' is a critical ingredient for your clarity strategy.

Are your positioning and distribution strategies working together, especially in terms of the market segment you're chasing and the way you're selling to them? Is there compounding momentum there, or have you simply not given it much thought? (That's fine if not — the back half of this book is for you.)

Origin stories are great for teasing out founder GTM preferences here, especially as, I imagine, they still shape your GTM strategy today. Earlier, we asked *how* you went to market — did you start with sales, organic growth, advertising, or something else? Here, I'm curious about what that has meant for your positioning/distribution feedback loop. Are you:

- **Position-driven:** You find distribution channels and tactics to make your positioning work.
- **Distribution-driven:** Your success with a particular channel (e.g., paid ads) or expertise in a particular approach (e.g., sales or, say, SEO) makes that a fixed variable, and you need to develop a position to match.

There's no right or wrong here, only that your positioning and distribution must be coherent.

4. How you learn

What have you learned in market?

This is one of the most important questions, simply because it can turn up the most important insights: what you've learned from your time in market. Insights that drive clarity strategy are often emergent, appearing through sales calls, user behavior, or pure serendipity. (Slack started off as an internal chat tool for a video game company; Loom, as we discussed in the previous chapter, happened to spot an unexpected usage of their tool and pivoted around it.)

Reflecting on your origin story and contrasting it to where you're at today helps shed light on those insights. And if you need fresh insights, where will you find them? For example:

- What fundamental truths have you discovered about how your customers succeed? How did you get there — through meticulous customer study, ad hoc learning, or your initial vision?
- To what extent do you use *quantitative* data and run rigorous experiments to learn?
- To what extent do you rely on *qualitative* research and customer feedback to learn?
- Do you follow a strict approach to customer development or other methodology, or tend to feel your way through sales calls and customer requests?
- Do you have any other sources of organizational learning? (In-house subject matter experts, market research, competitive analysis, etc.)

No wrong answers, other than to say an ability to learn *why* people are doing what they're doing can go a long way. People like the idea of the qualitative, data-driven approach, and that's great if you can do it, but it's the qualitative insights that help you learn the 'why' behind customer behavior.

Learning — parts vs. whole

I like to think of startups as, in part, a game of 'who learns fastest wins.' Building product and closing deals is obviously fundamental to your success, but what fuels your product direction and market insights? It's what you learn.

There are an enormous number of techniques startups use to drive that learning, and you might be using any or all of customer development and MVPs, product analytics and A/B tests, NPS and surveying for product/market fit[2], and/or product management methodologies like continuous discovery[3].

Lots of learning, but learning about *what?* Usually, it's about the parts and pieces: this new feature, that new idea, this particular cohort. And all that's great. But there's rarely the chance to step back and think about the whole. Clarity strategy — what we're doing right here — is your chance to think about the whole. It's a chance to think about your core ideas — what you've *really* learned, what's *really* working — in a way that can unlock alignment, growth, and new momentum in your company.

And if these ideas come from your customers, from what you've learned in the market, all the better. This is your chance to take a moment and assess those emergent ideas, especially in the context of your origin story and where you've come from, teasing out those implicit preferences and lessons learned and giving us that all-important context for how your narrative fits together.

Next, we're going to look at the origins of one of the fastest-growing startups of all time and the narrative memo that the founder penned at the very beginning — or was it? — of their journey.

Notes

1. I'm referring to April Dunford, who we'll meet in a few chapters' time. Dunford writes, for example, that she believes "this ['mad libs' style positioning statement] exercise is not only pointless but potentially dangerous. The exercise assumes that there is only one answer for each of the blanks, and you simply "know" what it is. However, most products could easily be positioned in multiple different market categories, with different competitors, providing different value for different kinds of customers." **https://www.aprildunford.com/post/a-quickstart-guide-to-positioning**
2. For example, in the prosumer world, see the classic Superhuman product/market fit survey approach: **https://review.firstround.com/how-superhuman-built-an-engine-to-find-product-market-fit**
3. As per Teresa Torres's popular product management book, *Continuous Discovery Habits*, published in 2021.

12

Slack's memo & implicit ideas

> [W]hat we're selling is organizational transformation. The software just happens to be the part we're able to build & ship.
>
> - *Stewart Butterfield*[1]

We've been attempting to take a snapshot of your narrative as it currently exists — as a deck, as a tiny Hemingway story, through your origin story, or perhaps, as we'll see in this chapter, a memo.

Below we're going to explore one of the greatest narrative memos of all time.

This was for a PLG B2B SaaS ultimately gunning for bottoms-up team and enterprise adoption. What does narrative look like in that context? And what does it look like when the founder, Babe Ruth-style, points to the stands before hitting one of the biggest home runs of all time?

In this all-time classic example, the narrative was, per the quote above, focused on an outcome: organizational transformation.

The author of the memo was Stewart Butterfield.

The product was Slack.

Peek inside a founder's mind

Some founders have been on the startup merry-go-round before, and the richness of insight they bring to their later startups is something we can all learn from, and that includes Slack co-founder Stewart Butterfield's famous *We Don't Sell Saddles Here* memo[2].

The key to understanding this memo is that the ideas Butterfield is trying to unpack are implicit. They can't be directly explained to the market because the market has no concept of what's about to be unleashed.

Butterfield appears to have great clarity about Slack's strategic direction. He *knows* the value is there — they'd been using the tool internally, after all. But the great challenge was communicating that value to folks with no concept of Slack whatsoever.

Let's dive into the memo and peek inside the mind of a founder who built one of the fastest-growing B2B startups ever.

Squad goals

> *Just as much as our job is to build something genuinely useful, something which really does make people's working lives simpler, more pleasant and more productive, our job is also to understand what people think they want and then translate the value of Slack into their terms. […] Therefore [that]* is something we all work on. *It is the sum of the exercise of all our crafts. We do it with copy accompanying signup forms, with fast-loading pages, with good welcome emails, with comprehensive and accurate search, with purposeful loading screens, and with thoughtfully implemented and well-functioning features of all kinds.*

Butterfield wanted folks to live the Slack narrative — it was the whole team's job, per his emphasis above, right down to the nitty-gritty details.

But what was the actual message? What was it they were trying to communicate? A header in the memo makes it clear: "Sell the innovation, not the product." That helps… but does that mean category creation? Butterfield *does* describe their path as defining a new market:

> *[W]e are not battling it out in a large, well-defined market with clear incumbents […] we are setting out to define a new market. And that means we can't limit ourselves to tweaking the product; we need to tweak the market too.*

But what happens when no one knows about your supposedly new category? (We will explore category creation in *quite* some depth in **chapter 32** and beyond.)

> *We are unlikely to be able to sell "a group chat system" very well: there are just not enough people shopping for group chat system.*

So Slack's clarity strategy was *not* about 'creating a category' in so far as they would name and claim 'group chat systems,' for example.

(This highlights yet another of the pitfalls of naively attempting to 'create a category,' by the way. Category labels work when people are shopping *by* category — think enterprise buyers asking analysts what they should buy. But for end users dipping their toe into a new tool, the category name is irrelevant — no one is shopping for that.)

> **Not 'just' chat:** The category question is interesting because Slack wasn't *just* chat in the ephemeral sense; it was chat implemented in

> a very specific way, which is captured in 'Slack' as an acronym (or backronym) of "Searchable Log of All Conversation and Knowledge."[3] Slack, in that sense, did what many great startups do — take some hidden data (communication through email buried in inboxes) and make it legible and useful, or visible and searchable in Slack's case.

TTT and selling transformation

> *However, if we are selling "a reduction in the cost of communication" or "zero effort knowledge management" or "making better decisions, faster" or "all your team communication, instantly searchable, available wherever you go" or "75% less email" or some other valuable* result *of adopting Slack, we will find many more buyers. That's why what we're selling is organizational transformation.*

It's not a category they're creating; it's an outcome they're selling — "organizational transformation." Butterfield rattles off the value props describing that outcome, from less stress to better search, organizations, and teams.

To get to the core of their message, Butterfield works through what I like to call the TTT test — tech, tool, or transformation. In Slack's case, that meant:

- **Tech:** Chat, in Slack's case, wasn't a fundamental technological breakthrough. As Butterfield says, Slack wasn't "as eye-catching as self-driving cars or implantable chips — it is not basic research-y kind of stuff."
- **Tool:** It wasn't a breakthrough as a tool either, at least not superficially. It's not like Slack invented workplace chat. However, it *was* vastly superior to the alternatives at the time (e.g., HipChat, which one HN user described as "a pain to use and every employee I talked to hated it"[4]).
- **Transformation:** Slack was, however, a dramatic transformation in workplace communication. Positioned against email as a new way to work, Slack offered a new "Digital HQ" (as they describe it today). Slack proposed — and largely achieved — a fundamental change in the way teams work which, crucially, *the Slack team had experienced firsthand.* (The team famously built Slack while working on Glitch, a social MMORPG, acting as their own 'user innovators.'[5] That was the second time (!) Butterfield's attempt at a game resulted in a wildly successful incidental product, with Flickr being the first.)

Pitching the transformation — the outcome — is a much more compelling clarity strategy than pitching just another chat tool. And while that pitch was obviously aspirational, it wasn't — and I can't stress this enough — just made up. They *knew* it worked. They *knew* they, as a team, needed it. They *knew* they were better off with it and others would be too, with Butterfield saying "[A]lmost any team which adopts Slack [...] would be significantly better off than they were before."

There's a lot to be said for being your own accidental product lab.

Evoking the right brain

With the focus on transformation, Slack's clarity strategy had a strong, foundational, right-brain radar hook. The point was to evoke universal themes of productivity improvement. It wasn't about a niche, vertical positioning play or a battle with superficial competitors over features. It was a big 'story 1' style transformation narrative.

Butterfield wanted his team to evoke something more. That's the core idea of the memo. "We don't sell saddles here" alludes to his imagined Acme Saddle Company, a company that *doesn't* sell on left-brain specifics of features and price but on the evocative right-brain appeal of horse riding itself — the real outcome, the real transformation.

In Butterfield's view, it's that broader context of outcome and transformation that enables his fictional Acme Saddle Company to talk about saddles while growing a new market, establishing leadership of that market, and enabling the transformation to this new approach to horseback riding (Butterfield takes the analogy back ~4,000 years) or, in Slack's case, team communication.

Implicit > explicit

Butterfield knows this approach to narrative — selling the evocative outcome — isn't a novel concept, citing Harley Davidson selling motorcycle riding and Lululemon selling yoga. It's a tried and true technique... if you're capable of (and sufficiently resourced to) pull it off.

And that implicit message in the pitch includes how Slack's customers should feel, with Butterfield detailing what the reward should *feel like* for using Slack so that customers "understand what's at the end of the rainbow."

> *We want them to become relaxed, productive workers [...] We want them to become masters of their own information [...] We want them to feel less frustrated by a lack of visibility [...] We want them to become people who communicate purposively, knowing that each question they ask is actually building value for the whole team.*

But there's one more thing that sticks out in the memo, and it took me a long time to understand why.

Deep meaning vs. generic language

What stood out to me is the use of "organizational transformation." As a descriptor that's pretty... generic, right? I mean, it's true, of course. Butterfield called it and they hit the home run, and it does means something specific in Slack's context, but still... is that it?

This is the challenge Butterfield and just about everyone with a strong product vision faces when trying to communicate a product's value — it can be genuinely difficult to put it into words. You can talk about value props like 'less stress' and 'more productivity' and they can be true, but again, is that it?

Software as art

Let's go deep for a moment. I want to try and capture something that I think is pretty profound by returning to the work of ol' mate Dr Iain McGilchrist, who writes[6]:

> *In other words, knowing a piece of music, like knowing other works of art, is a matter of* kennenlernen. *Coming to us through the right hemisphere, such living creations are seen as being essentially human in nature. In an earlier book I argued that works of art – music, poems, paintings, great buildings – can be understood only if we appreciate that they are more like people than texts, concepts or things.*

And:

> *As long as [words in a poem] remain implicit, they cannot be hijacked by the conscious mind and turned into just another series of words, a paraphrase. If this should happen, the power is lost, much like a joke that has to be explained (humour is a right-hemisphere faculty).*

McGilchrist likes to draw out the differences between implicit and explicit meaning and how turning implicit meaning — like that in a poem, a song, or a work of art — into explicit descriptions destroys much of its meaning. Like, as he says, a joke explained.

For McGilchrist, the left brain is where things get turned into parts and pieces, while the right brain is the domain of the whole — the whole meaning, the whole person, and in fact, the right brain understands those things *more like people*.

Great software, I want to suggest — and granted this may be a stretch — can be something you 'know' like you know a person or perhaps your favorite artist or band. There's something irreducible about the meaning it evokes — and evokes consistently — that makes it work as a whole.

We talk *so much* in software about the parts — the GTM tactics, the tech stack, the feature lists, the market potential, the revenue numbers, the big exits, the money money money, but we don't talk very much about software as art. Even then, "art" is just a metaphor that doesn't fully capture what a brilliant and coherently realized product truly *is*.

Perhaps that's because, like a joke explained, when we switch to left-brain analysis, the humor, or meaning, is lost. Perhaps that's why founders have to be so oddly insistent on *their* way at times, too — someone has to hold the implicit vision of what the thing can be, at least until it's big enough to walk and take on a life of its own (which are, again, metaphors for personhood, to echo McGilchrist's observation).

And when it comes to Slack, it's fascinating to me just how much of "Slack," the metaphorical person or whole, the concept they were trying to evoke, managed to ooze through every part of their execution, including Butterfield's memo.

> **Linear and 'quality without a name':** If Slack was one of the 2010s' best examples of software-as-art capturing the imagination of users, the 2020s' standard-bearer for this obsession with complete user experience in business software is probably Linear, a project management tool for developers. Linear co-founder and designer Karri Saarinen writes about many of the themes we're discussing here, including the implicit nature of quality (and brand) that can't be put into words versus the explicit quantification of just about everything that tends to happen in tech.[7] The beauty of a tool like Linear, in my view, is in its ability to capture *both* the implicit and the explicit, and apply both modes of attention in a profound way, having a deep appreciation for taste and 'quality without a name' on the one hand, in a design-is-how-it-works sense, while also having a precise understanding of the explicit developer workflow they're building for as well. That is a formidable combination. It's easy to do pretty but empty, or ugly but precise; it's hard to marry both in a unified way. We'll return to this when we look at Figma in chapter 27.

Apple and the liberal arts

Let's stick with this concept of the whole and right-brain implicit meaning.

At the launch of the iPad 2, Steve Jobs said:

> *It is in Apple's DNA that technology alone is not enough—it's technology married with liberal arts, married with the humanities, that yields us the results that make our heart sing.*

If we think about the arts more broadly — including music, poems, paintings, the things McGilchrist talks about above — we can see the overlap between where Apple has excelled (creative industries, music, user experience) and where implicit meaning lies.

On the other hand, if we think about the more left-brain, parts-and-pieces focus, then we can see traditionally where companies like Microsoft (and Google with Android) excelled. They are obviously fantastically successful too, but they've never embraced the *art* of software like Apple[8], and its ecosystem, have.

John Gruber, longtime Apple watcher, picks up this thread on software and art and the 'cultural chasm' between platforms[9], putting the boot into the Android ecosystem, writing:

> *Android enthusiasts don't want to hear it, but from a design perspective, the apps on Android suck. They may not suck from a feature perspective (but they often do), but they're aesthetically unpolished and poorly designed [...]*

And:

> *I will offer another quote from Kubrick: "The test of a work of art is, in the end, our affection for it, not our ability to explain why it is good."*
>
> *Art is the operative word. Either you know that software can be art, and often should be, or you think what I'm talking about here is akin to astrology.*

Some folks don't get this, as Gruber goes on to discuss. In my view, some folks want a left-brain explanation for what exactly it is you want from an 'experience,' but that's impossible. Your right brain gets it, or it doesn't. And if it doesn't, you might think this extended riff about Slack, art, and right-brain understanding of implicit meaning is akin to astrology, too.

But I don't think it is. I think it's a natural product of our natural right-brain attention. You can certainly succeed without it, but you can succeed with it, too. And for what it's worth, I much prefer living in a world where people are succeeding with it.

Apple Vision Pro & the right/left/right jump: Later in chapter 26 we will explore McGilchrist's description of the way we process information from the whole, to the parts, and back to the whole (a.k.a. the right, left, right hemisphere jump). That is, at their best both hemispheres, and their respective modes of attention, work in complementary ways. Apple's Vision Pro, announced in 2023, was a great example of Apple flexing both modes of attention — the far-reaching, right-brain view of what could be possible, the intense left-brain dive down into the details to make that a reality, and the right-brain reintegration back into a cohesive experience. This is the Steve Jobs model of starting with the customer experience and working backward to the technology. And while the right-brain starting point is fundamental, you still need that laser-focused attention to detail, as documented for the Vision Pro in a Twitter thread from Jukka Ursin, who briefly worked on the project[10]. Audio ray tracing of the room so the sound is exactly right; getting lighting and shadows perfect; making sure the visuals are absolutely stable — it's this unified approach in applied attention to a unified experience *in all its minute detail* that very few companies are capable of.

The Cheesecake Factory

I might enjoy living in a world where companies succeed by using their right-brain attention but I am not, dear reader, an enterprise software buyer.

> *Enterprise products are much more unwieldy than jewel-box consumer products, as they have to support many more users and workflows. Consumer products are like Chipotle: a small, carefully curated set of menu items, built for elegance and efficiency. Enterprise products are like the Cheesecake Factory: you can get steak, pasta, a milkshake and 4 kinds of margarita in the same meal.*
>
> – Stay SaaSy[11]

Slack, at least in its original incarnation, was more at the Chipotle end of the spectrum — 'built for elegance and efficiency'.[12]

Enterprise products are... not like that. We'll explore why that is a little more in **chapter 43**, but the short answer is: more growth requires more sales, and more sales requires serving more of the market. And when that market is made up of enterprises with unique ~~needs~~ demands, then your product inevitably becomes a 1:1 reflection of the diverse needs it serves. Throw in a strong commitment to backwards compatibility — enterprises do not want to retrain thousands of staff every software update — and you get 'enterprise software.'

And, of course, the enterprise buyer is not the user and vice versa, and it was ever thus. Here's Steve Jobs quoted in the *WSJ*[13] back in 2010:

> *What I love about the consumer market, that I always hated about the enterprise market, is that we come up with a product, we try to tell everybody about it, and every person votes for themselves. They go 'yes' or 'no,' and if enough of them say 'yes,' we get to come to work tomorrow. That's how it works. It's really simple. With the enterprise market, it's not so simple. The people that use the products don't decide for themselves, and the people that make those decisions sometimes are confused. We love just trying to make the best products in the world for people and having them tell us by how they vote with their wallets whether we're on track or not.*

Users — at least some of them — appreciate the art of fine software, and if you're going bottoms-up, that value may be reflected in folks voting with their wallets (or at least telling their friends). Enterprise *buyers*, on the other hand, are in Jobs' world — and I love the euphemism here — "confused."

But they're not so much confused as just using a different mode of attention to Jobs' more right-brain product perspective. They're often in left-brain laser-beam mode. They're narrowly focused on lists of features and requirements. They're not buying on *feel;* they're buying in a very considered way; and (sadly) a great UX has not, historically, been as big a part of that consideration set.

That is, thankfully, changing. A rising tide lifts all boats, and that's certainly true of UX in the enterprise as well. The 'consumerization' of the enterprise is a thing, and we have the fantastic work of user-focus bottoms-up tools to thank for that.

If you embrace the art of software as part of your enterprise offering, fantastic. Just know that "easy to use" and a "great UX" is what pretty much everyone else says too, and given the buyer has no way of *feeling* that, it's unlikely to be a growth lever — it is, sadly, just another box to tick on the checklist.

> **Right-brain attention for concept design vs. product design:** If you're selling in an environment where "great UX" is more of a nice-to-have and not a PLG growth lever, does that mean right-brain attention is less relevant? Not at all, it just moves up from the product to the concept level, because chances are you're selling something more

> conceptual than just a nice bundle of features. That concept might be your change story, new playbook, fresh perspective... all things we're going to cover in the coming chapters. This is where you compete on clarity strategy as much as product, and calibrating your story to win the claw game of attention (chapter 18) is exactly what this book is about.

The embodied meaning

Returning to Slack and its initial narrative, the conundrum we were exploring was the deep implicit meaning behind Slack, and the seemingly generic label of "organizational transformation."

The reality is that left-brain labels will always be somewhat reductive when describing the right-brain art of our products.

That doesn't mean the label is useless — far from it. Butterfield still had to communicate his vision to the team and turn the implicit explicit and does as well as anyone could, elaborating on "organizational transformation" in a variety of ways, from value propositions to how the *customer* would be transformed from using Slack.

And to be fair, there was plenty of implicit meaning in "organizational transformation" that would have been apparent to the Slack team — the audience of the memo. They were using the software and would have understood the 'from/to' of email to chat for team communication.

But it was the product itself that embodied that deeper meaning — the new way of working — and that's the great challenge of clarity strategy: expressing that deeper meaning through the limited, explicit vocabulary of business and marketing.

The details

That big idea doesn't exempt you from getting the details right, either.

If you want to live and die on the strength of your product experience — if your product experience is what drives that organizational transformation — then it better be an incredible one. To that end, Butterfield drives home the need to execute over and over with comments like:

- They need to "do an exceptional, near-perfect job of execution" across things like sign-up forms, fast-loading pages, good welcome emails, and accurate search.
- "Going big also means that it will have to be really, really good. But that's convenient, since there's also no point doing it if it is not really, really good."

- "There is a great nobility in being of service to others, and well-run restaurants (or hotels, or software companies) serve with a quality that is measured by its attention to detail."

You get the picture.

But again, don't be fooled by the left-brain approach of 'Aha! It was about the details!'. It *was* about the details. And it was about the value. And it was about the transformation. And it was about the opportunity. It was about everything coming together — it was about the whole. It was that implicit insight into what the finished product should be — not 'product' as in 'a dish at a restaurant', but the entire restaurant experience itself, to use Butterfield's metaphor — and being able to articulate what that should look like for the team.

Let's do this

I love Butterfield's memo for its clarity — clarity about the opportunity in front of them, clarity about their purpose, and clarity about how they'll get there. To me, and many others, that's what makes the memo a classic. (Keep in mind it was sent two weeks before Slack's preview release in 2013.)

They had the transformation story, the ambition to go big, the focus to get all the details damn near perfect, and the dedication to their customers to make it happen.

And as for why, when the outcome was entirely unknown? Well...

> *The answer to "Why?" is "because why the fuck else would you even want to be alive but to do things as well as you can?". Now: let's do this.*

Slack was officially released on February 12, 2014.

On December 2, 2020, Salesforce announced it would acquire Slack for $27.7B.

Notes

1. Butterfield, Stewart. "We Don't Sell Saddles Here," *Medium*, February 18, 2014. **https://medium.com/@stewart/we-dont-sell-saddles-here-4c59524d650d**
2. ibid.
3. Griswold, Alison. "Slack is secretly an acronym," *Quartz*, September 28, 2016. **https://qz.com/794524/messaging-startup-slack-is-secretly-an-acronym-for-searchable-log-of-all-communication-and-knowledge/**

4. lustig, commenting on "Ask HN: How did slack beat hipchat?," *Hacker News*, November 10, 2017. **https://news.ycombinator.com/item?id=15670275**
5. We discussed 'user innovators' in the context of Loom at the end of chapter 10.
6. McGilchrist, Iain. *The Master and His Emissary* (p. 116). Yale University Press. Kindle Edition.
7. In October 2023, Saarinen posted on Twitter about "Quality without a name," citing architect and design theorist Christoper Alexander, who coined the phrase. Saarinen writes "What [Alexander] is saying is that quality exists, it can be experienced and perceived, but it cannot be described by words. A town, place, or building that just feels good and natural. [...] This throws off many people in the tech world, where often things are seen as binary and belief that the world or anything can be reduced to mathematical formulas. It's almost as if you have to stop thinking for a minute, and just focus on the experience. [...] Quality, brand and culture are some of those intangible things you need to create as a company founder." **https://twitter.com/karrisaarinen/status/1710331542369914903** I'm sure you can see the very strong overlaps here with the McGilchristian, implicit/explicit, right-brain/left-brain view of the world that we're exploring in this book.
8. Jibes here about iTunes/Music app being a dumpster fire are entirely justified.
9. Gruber, John. "Making Our Hearts Sing," *Daring Fireball*, February 1, 2023. **https://daringfireball.net/2023/02/making_our_hearts_sing**
10. Alas, Ursin's thread has since been deleted on Twitter — Apple, one imagines, probably isn't super fond of interns giving the world a deep inside look into its process — but the content is still out there if you look for it. **https://twitter.com/JukkaUrsin/status/1665973161295265797**
11. Anon., "The Rogue Wave of Enterprise SaaS," *Stay SaaSy*, July 29, 2020. **https://staysaasy.com/scaling/2020/07/29/the-rogue-wave-of-enterprise-saas.html**
12. I've never eaten at Chipotle, so I'm trusting the instincts of the person I'm quoting here!
13. Steve Jobs in an interview with Walt Mossberg and Kara Swisher of *The Wall Street Journal*, quoted in "What Steve Jobs hates about the enterprise," *ZDNET*, June 29, 2010. **https://www.zdnet.com/article/what-steve-jobs-hates-about-the-enterprise/** (I'd cite the *WSJ* but link rot has taken its course.)

PART III

CLARITY & **MESSAGING**

13

The 'think, align, execute' framework

After seeing hundreds of message testing results for SaaS home pages, one thing is becoming clear: people really want you to get to the point. If you take your time telling your story, people get impatient and lose interest. Don't mince words.

– Peep Laja[1]

We've looked at a handful of ways to create a snapshot of your narrative as it currently stands:

- As a deck.
- As a micro pitch.
- Through the lens of your origin story.
- As a memo, if you were to riff on Slack's example.

It's tempting — as the rest of this book demonstrates (!) — to try and construct an internal narrative and positioning of exquisite intellectual detail and creative perfection and assume that's job done. Sadly, alas, that's not the case. Your narrative is ultimately what you ship, because that's what customers see.

Therefore, over the next few chapters we're going to briefly touch on messaging and copy to understand where they fit in your narrative and positioning strategy.

> **Choose your own adventure — tactics vs. strategy:** In the spirit of trying to include as much practical stuff upfront in this book, this section, Part III, is more tactical and hands-on. If you're a founder or operator and want to skip to the meatier strategic stuff, be my guest and head straight to chapter 18.

Execution matters

I've seen folks put a lot of time, money, and brain sweat into their narrative and positioning strategy only to have it fall flat in execution. If that happens, what was the point of all that upfront work? You don't get a gold star for trying. It's what you ship that counts.

Why does this happen? In my experience, execution is often treated as a bit of an afterthought. Leaders do the thinking; doers turn the crank to execute. But that waterfall-style process creates a big attention disconnect when the right-brain big-picture work gets 'handed over' to someone doing the left-brain, narrow-focus execution. That's a chasm where a lot of good strategy and execution work goes to die, so I've created a dead simple collaborative framework to help.

> **Your narrative and positioning is not your homepage:** Often, the desire to improve your narrative or positioning stems from a very specific pain point, like *"We need to improve our homepage"* or *"We need a stronger pitch."* Usually, the asset then becomes a proxy for bigger strategic questions lurking underneath. And that's fine — there's nothing wrong with taking action because your current homepage, deck, or C-suite pitch no longer captures what you're about. You just need to decide whether you want a new lick of paint or if there are deeper structural questions that need to be addressed. If it's the latter, embrace it, and resist the urge to skip straight to execution. You're tackling a much higher-leverage problem than the wording in your hero section or opening slide, after all, and it's worth creating space for that. Moats are built by doing the thinking and discovery other folks won't, and this is your chance to think deeply about the exact value you offer and how well you know your customers. That can have profound effects on your company's direction and chances for success. Don't shortcut the process. There's so much more to be gained by doing the strategic work first. Just don't forget the execution side, either!

Think, align, execute

Here's the braindead-simple framework you can use from concept to execution:

- **Thinking — narrative & positioning:** Your high-level sales narrative and positioning in the form of a concise deck or doc that you've thought deeply about. (This is the kind of thing we fleshed out earlier in **chapter 9**, and will go very, very deep on in the latter half of this book.)
- **Alignment — messaging:** A shared doc that captures your key value props, themes, one-liners, longer descriptions, and other relevant messages. This doc helps your team stay aligned when executing across your website, outreach, ads, social media posts, decks, email, one-pagers, and other customer-facing assets. Your messaging doc might include breakdowns by ICP or vertical too. Consistency in messaging

matters. Message discipline is the name of the game.

- **Execution — copy:** The actual execution, in the form of the final words you see on a landing page or an ad, for example. This may include tight, concise explanations of your concept, how your product works, benefits, features, contrasts (with/without, before/after), and ideally draws from language your customers use (we'll explore "language/market fit" when looking at Jobs to be Done in **chapter 36**).

This book is largely focused on the first step — getting the initial narrative and positioning right — as everything else flows from that. But having worked across all three levels, let me give you a quick tour of the messaging world.

Diagnosis comes first

First, this framework makes for a simple diagnostic checklist. Specifically:

- **Have we done the thinking?** Have we gotten the critical people together to work through a narrative and positioning process? (We'll touch on such a process in **chapter 37**).
- **Have we got alignment?** Have the heads of relevant functions signed off on a set of core messages that prospects can expect to hear from the founder down? Founders may have neat ways of expressing the value of the product, but is that what your AEs use in deals? Is alignment obvious by virtue of what appears on your homepage, decks, ads, landing pages, success QBRs, and other touchpoints?
- **Have we executed in a clear, catchy, customer-validated way?** By "customer validated," I mean, can you point to a specific customer (or several) that has seen it and gets it? Are you using their language? Have you found a way to sum up your message that you can use across your customer-facing assets? If you pulled all your assets together, would they be hitting the same theme?

Diagnosis always comes first, and accurately diagnosing your existing situation can help you determine where you need to get to work. Maybe you already have a strong narrative, it's just not reflected consistently in your messaging. Or maybe there are creative ideas floating around, you're just not aligned across functions. Or maybe the insight is lacking at the top, and if so, you're in the right place: the rest of this book is for you.

Keep in mind that this 'think, align, execute' framework is intended to be collaborative, not a classic chuck-it-over-the-wall waterfall process where the executives do the thinking, the leaders agree on messaging, and some poor copywriter or product marketer is left trying to work with some decidedly bland ideas.

Think about it more as an iterative, collaborative loop. The stages inform each other, after all, and you need folks in the room who can write *and* who have contact with customers.

Speaking of creative, collaborative loops...

Design Thinking & narrative design

If you try and bite off everything in one go with no steps or process in place, you're in for a rough time. Fortunately, there's a battle-tested process we can rely on instead, a process that's particularly useful for internal teams working collaboratively.

Let's say you're starting with the thinking stage, focusing on your narrative and positioning deck, and aiming to get clarity there. Like any recipe, broadly speaking, you need two things:

1. Ingredients that go in the mix.
2. Steps to get to a result.

On the ingredients side, the rest of this book has you covered there. (Take your pick from creating right-brain concept stories in **chapter 20**, understanding Jobs to be Done in **chapter 36**, doing left-brain positioning in **chapter 37**, or picking a wedge attribute in **chapter 46**, for example.)

On the steps side, if you need a more concrete process to follow, I suggest borrowing from the Design Thinking world for reasons that the excellent Christina Wodtke[2] outlines, a few of which are:[3]

- *Move memory out of your head into the world, freeing up processing power.*
- *Allow you to see connections and relationships you can't easily visualize in your head.*
- *Create a shared understanding for teams. Now your team can be part of that distributed world you think with.*

That's great, but what's the actual process for this work?

The Design Thinking process

I'll keep this very brief as it's covered by Wodtke and others at length elsewhere, but here are the basic Design Thinking steps from the Standford d.school, with my summaries:

1. **Empathize** (Wodtke prefers "context," and I'd suggest "discover" here, i.e., drawing on customer interviews, insights, and observations in a right-brain out, left-brain down sense).
2. **Define** (define what problem you're solving and for whom, perhaps

drilling down with narrow focus).
3. **Ideate** (generate multiple ideas, diverge/converge, go for quantity then prioritize for quality, and you might start with the simplest story possible, like your Hemingway story in **chapter 10**).
4. **Prototype** (create drafts and mockups — might be simple deck outlines, sketched-out decks, or wireframes, which is what I outlined in **chapter 9**).
5. **Test** (try it out on actual humans — the team, best-fit customers, and ultimately prospects, as we'll discuss in **chapter 38**).

That gives us a simple tried-and-true set of steps to follow when cooking up your narrative, positioning, or messaging. Feel free to take the steps and make them your own. The God of Process isn't going to smite you for doing your own thing and making it fit your specific context.

Why use a design process for this work? Well, it's a great fit. Your narrative is a product and it can help to design it in the same way. But why *design?* Here's Wodtke's take:

> *A long time ago, when I was a manager of a team of 80 designers at Yahoo!, I wondered, "what is design's unique capability?" What did design do for the business that product management, research, marketing and engineering couldn't? I decided it was:* "Design makes ideas tangible so that they can be understood and evaluated before committing them to reality."
>
> *Design work is two kinds of world modeling work:*
>
> 1. *Modeling the world as it is.*
> 2. *Modeling the world as it could be.*

This is what narrative and positioning work — especially in the form of a simple deck — can do for the business in a strategic sense that very few other tools can. We can make these ideas tangible in a deck (and as messaging) *before* leaping to execution where we commit them to reality. And doing this in raw words on a page, deck, or memo can help us both model the world as it is *and* model the world as it could be. That's very powerful for understanding both where you're at, where you're going, and where you *can* go.

Writing and thinking vs. doing

The ideas do, however, have to come from somewhere. There needs to be fresh insight that can drive the whole process. And that insight needs to come from the leadership in collaboration with folks on the front lines — those who are looking *out* at what can be (leadership) and those who are looking *down* at what is (front-line folks).

This kind of collaboration takes time and needs to be treated with appropriate care. The implications of the "thinking" side are so far-reaching — touching product, GTM, why you exist, where you're headed, and what (crucially) your differentiated value is — that you really do need to give yourself and the team the time and space to think it through and *keep* thinking it through.

I've been on the other side of the process, trying to push the ideas up from the bottom on the execution side, and it rarely works, because leadership isn't in radar mode and hasn't created any space to think.

Get folks together. Get folks who can write in the room. Get someone on the frontline with customers in there. Hash it out. It doesn't have to be perfect instantly, and it won't be. Write bad drafts. Write better ones. Test them out. See what actually excites customers. But make sure you've done the thinking upfront to have the ideas to drive the whole process. Insight trumps all.

> **Investors don't write homepages:** As an illustration of how tightly message execution and strategy are linked, I recall hearing one product marketer talking about their messaging work on a new homepage for a very unicorn-y PLG B2B SaaS. Not only did they have to coordinate with all their internal stakeholders, they were getting input from their investors, too. On their *homepage.* Can you imagine great messaging or copy coming out of that process? Neither can I, and the homepage headline for this company is, predictably, as bland as a manilla envelope stapled to a beige wall. But it does illustrate how tightly interwoven your message (down to your homepage copy) is with your strategy (up to your big-name investors wanting input on direction). And while I'm a big believer in collaboration and closing the gap between strategy and execution, external folks need to know when to let the team do their thing. Having proxy discussions around marketing assets is a poor way to do strategy — better to start at the narrative level and hash things out there.

Implicit → explicit

So you've got the ingredients, you've got a bit of a framework and process, now what?

Now you've got to find the right words. This is hard stuff. We talked about turning the implicit explicit when unpacking Stuart Butterfield's famous memo in the previous chapter. But I want to highlight how difficult it is for everyone — even journalists writing about technology, and they're professional technology describers! This is why getting your narrative and messaging down needs to be far more iterative and collaborative than any waterfall process — it's *never* a matter of just turning the execution crank.

Let's return to Slack as an example. A product rich with vision and interesting, implicit ideas is a compelling one, but when it comes to messaging, it's *really hard* to capture those ideas in just a few words. But it's not impossible.

Understanding Slack

Everyone gets what Slack is, don't they? Well, maybe now it's ubiquitous, and team chat is passé, but what about when it was new?

Let's cast our minds back a decade or so to 2014. In a Wired profile of Slack's founder Stewart Butterfield[4], written a year after Slack launched to the public, journalist Mat Honan attempted to describe Slack in multiple ways, and it's striking to see just how much of a challenge it was to describe the product and what it could be. Honan comes at it from a variety of angles, writing in terms of (my labels, his quotes):

- **Novelty:** "On first blush, [Slack] sounds boring. Worse, it's a bit hard to explain because you haven't used anything like it before."
- **Category and analogy:** "It's a communications application" with "a supercharged group chat function, a modern take on the AOL rooms of yore."
- **Ambition and narrative:** But chat is just "a trojan horse for bigger ideas," including the ambition "to become the hub at the center of all your other business software. [...] Once they're all connected, it can keep track of most everything you do with them. [...] Slack is the one mutual platform where all those things come together. That's the longer-term thinking."
- **Behavior change:** Slack wants to "end interoffice email."
- **Company ambition:** "It wants to be the next Microsoft—not the current, cowed, pathetic company looking for a comeback." (How things have changed since 2014!)
- **Difficult to describe:** "And maybe the biggest challenge of all—if they want to get not just big but huge—will be explaining what the hell Slack is and who might want it in a way that mom and pop small businesses across this great nation can easily understand."

Explaining unfamiliar things is a challenge — a fun one for some of us, and a not-so-fun one for others. And that's especially true for products with a lot of right-brain, high-concept potential. (Try neatly explaining Notion, for example.) That very much included Slack, with Honan describing their biggest challenge in 2014 as *'explaining what the hell Slack is.'*

"Where work happens"

Eventually, the team did find a catchy way of explaining what Slack was: it was where work happens.

That line, one of Slack's strongest, came from Kelly Watkins, Slack's former VP of Global Marketing. Watkins mentions coming up with the story in passing in her newsletter[5], writing:

> *For the most part though, stories feel like they just come to me almost as if out of thin air. Slack's story about being the place where work happens woke me up in the middle of the night. Understanding the science behind stories helped me know it was a good one, but I didn't build that narrative from component parts and a checklist. It just arrived.*

It just arrived! We've discussed a lot of left-brain process in this chapter — breaking things down to their parts so we can manipulate them — but I don't want to discount the role that your subconscious plays in creating great stories, either.

In fact, part of the point of all this process — and indeed all the ideas in this book — is to prime your subconscious so it can get to work in the background and slowly cook up some ideas while life happens. Maybe the timer will go off — ding! — when the ideas are ready and you should be asleep. Maybe it'll be a classic 'shower thoughts' moment. Or maybe it will come out through discussions with your team or while you're reading this very book.

Whatever it is, it's ultimately the idea and insight that matters, wherever (and whenever) it may strike, with process or without. But for that to happen you've got to start stoking the fire; you've got to give your subconscious the ideas to chew on.

Value propositions

Once you've got the big idea, it's time to switch to specifics — specific *value propositions* that concisely articulate exactly what benefit or outcome a customer can expect with your product.

Value props are usually the domain of marketing, given most marketing material is made up of value props. The bullets on your homepage? Value props. The bullets on your sales deck? Value props. The message in your ads? One key value prop. The key to most high-quality B2B messaging and copywriting is the ability to write catchy value props.

The best teams I've worked with have a forensic approach to their value props. They can trace them back to the specific pain of a specific customer they interviewed, they know how they ladder up to serve that kind of customer's broader challenges, and they know which value props *don't* apply in specific industry contexts as much as they know which ones do. This approach to value props is a very B2B thing, especially as we're often dealing with complex products that can be positioned in multiple industries and segments.

Specifics as a superpower

The superpower of great messaging in B2B, then, especially when it comes to your value props, is making the implicit explicit in a way that's relevant to the market, to your specific ICP. It's not just that you made a thing; it's that you made a thing *and it helps in these very specific ways we know will be transformational for you, and we can express it very neatly in this way*.

You see why this is a superpower when folks who aren't strong writers try their hand at messaging, and the results are just... super generic. It's not specific and it doesn't sell. It's not that they don't have good ideas — they may have great ideas — but they lack the ability to make them explicit, to find the words to bring them to life. So they pile on the superlatives instead — it's not just a BizThing platform; it's the *first ever ultra mega super AI-powered BizThing platform! It's changing tomorrow, today!*

But adding fluff doesn't cut it. More fluff doesn't suddenly make a weak value proposition compelling.

There's an 80/20 rule here where you've got to have the discovery driving the insight. Writing value props is, like anything else, garbage in, garbage out or, preferably, insight in, insight out. And the ratio to generating insight to actual writing is weighted very heavily towards the insight side. It's a common refrain in messaging circles to say the work is 80% research and 20% writing — you've got to do the legwork to find the insights that you can bring to life in words.

That legwork means talking to customers and being very analytical in unpacking what they say. There's no shortcut: if you're going to do something unique for a set of buyers, you need to have a unique quantity and quality of insight into their world and their problems. If you take this seriously not just as a marketing discipline but as a company discipline, you end up with something like an auditable startup — you've got value props that can be traced back to specific customer evidence, and that evidence might grow or change over time as you zero in on exactly what value you *provably* deliver for your customers. When you're an expert in your customers' pain *and* success, your messaging, in terms of your value props, is going to be infinitely more compelling.

That's the 80% that makes the final 20% of writing a whole lot easier. In the next chapter, we'll get hands-on again as we run through two founder-friendly checklists for that final 20% execution phase: one stepping through the think → align → execute process, and another for applying your narrative to your website, including your homepage, product page, and about page.

Notes

1. Laja, Peep. October 13, 2022. **https://twitter.com/peeplaja/status/1580530886013222912**
2. Wodtke has been a fantastic contributor to the design, tech, and business worlds for a very long time, and is probably best known for her highly regarded book on OKRs, *Radical Focus*.
3. Wodtke, Christina. "How I Stopped Worrying and Learned to Love Design Thinking," *Medium*, August 27, 2017. **https://cwodtke.medium.com/how-i-stopped-worrying-and-learned-to-love-design-thinking-f1142bab60e8**
4. Honan, Mat. "The Most Fascinating Profile You'll Ever Read About a Guy and His Boring Startup," *Wired*, August 7, 2014. **https://www.wired.com/2014/08/the-most-fascinating-profile-youll-ever-read-about-a-guy-and-his-boring-startup/**
5. Watkins, Kelly. "Taking a left at the next light," *This Is Liminality*, April 1, 2023. **https://thisisliminality.substack.com/p/taking-a-left-at-the-next-light**

14

Founder's checklist: Messaging & your website

> Across dozens, maybe hundreds of distinct pieces of user research, the most consistent piece of feedback is to literally just say what the dang product is and does. Forget positioning. Benefits. Use cases. Client pain points. At least up front. Job 1: what the hell is it?
>
> *- Bryan Casey, Director of Digital at IBM*[1]

In this chapter, we're going to stick with our simple thinking → aligning → executing (TAE) process with two founder-friendly checklists for:

1. The TAE process, in preparation for...
2. Applying your narrative to your website.

All our work on narrative and positioning only lands if your execution lands. These checklists — and the discussion that follows — are intended to help founders ensure they're covering the basics when it comes to shipping their narrative.

> **Deep in the weeds:** We're deep in the messaging weeds here, dear reader, and if that floats your boat, great, but if not, skip to chapter 15 on memes and metaphors.

Getting hands-on

Note that this *isn't* Luke's Top 10 Tips for Writing Killer Copy That Converts™. (Those lists ironically almost never have any real conversion evidence behind them, either, but that's a rant for another day.[2]) But this *is* a chance to get hands-on again as a founder or operator and look at exactly what you're currently shipping. My hope is that this will give you a tangible context for the rest of the narrative and positioning ideas we're going to discuss.

To that end, let's *very* briefly run you through these checklists that reflect how I'd expect your messaging execution — your copy — to be structured on three key pages: your homepage, product page, and about page. (For mature, later-stage folks, the same ideas apply, just on a larger scale. You might be repeating this exercise for each industry you operate in or each product you offer.)

No hard and fast rules, just suggestions.

> **Doing is thinking:** The TAE process is a cycle. The execution phase inevitably involves a lot of refinement of original ideas or messaging concepts that may have been a little fuzzy or not quite dialed in, and this is a chance to feed those refined ideas back into your original thinking. That might be in the form of catchy lines, ideas, or metaphors you come up with when your team is working on your website, for example. That is, working on how to communicate clearly to customers should always help you clarify your own thinking, too.

Thinking — clarity strategy

Let's start from the top. Here, I'm going to be annoying and check you've done your homework, because I'm a big believer in putting pen to paper and getting started with a crappy first draft.

- Do you have a sales narrative to work from, as we looked at in **chapter 9**? That's the starting point for our work. **If your idea isn't strong, nothing else really matters.**
- Do you have a clear understanding of story 1 and story 2 (we'll be exploring this in more depth from **chapter 19**)? Your homepage, for example, is where you need to start getting very specific on the story 2 side of things, with some laser-focused value props.
- Does your sales narrative include strong lines (including a catchy Hemingway story from **chapter 10**) that you can extract for broader use?
- Do you have your 5W+1H origin story and then/now (**chapter 11**) worked out, with a clear sense of where you're at today, having cleaned up any old legacy messaging?
- Do you have your key differentiators worked out? (If not, well, there's always... **gestures at the rest of this book*.)*

Alignment — surface area

This is about being aligned internally *and* aligned with how your prospects actually buy.

- Can you identify your key buyer touch points across their journey (and I mean buyers specifically, not random visitors, existing users, or even form-submitting leads)?
- Can you pull the customer-facing assets buyers see on their 'critical path'

together on a single canvas (e.g., using a whiteboard tool)?
- How coherent and aligned is your current messaging across both your assets and across your teams internally? What resources are required to do a new messaging revision? Who are the key stakeholders?
- Do you have a central source of truth for messaging? E.g., a Google or Notion messaging doc with approved messaging everyone agrees on and is relatively concise? (I.e., you've made the hard choices about what's in and what's out.)
- Do you have strong customer (or founder) interviews to draw from that highlight buyer pains, jobs-to-be-done, language, trends, or other types of buyer insight?

> **What if you *don't* have those things?** Every 'best practice' approach — including this one — tends to suggest a logical, linear approach that makes sense for folks who have done it before and know what they need, and that's why you get a neat 'gather ingredients → produce result' approach. But don't be afraid to take the chaos approach instead and just get stuck in, pulling in what you need when you need it and iterating as you go.

Execution — copy

Now, let's get into the execution side. This is, again, just a very brief checklist for founders so you can understand how to apply your narrative across your homepage, product page (or product overview), and about page. The basic formula is:

1. **Homepage:** *What* is the value you offer.
2. **Product:** *How* your product delivers that value.
3. **About:** *Why* that value matters, perhaps in a broader context, and *who* is behind the operation.

> **Balancing the abstract and the tangible:** Businesses are ultimately buying an outcome, and that's the value we're trying to make crystal clear on your homepage, especially if the buyer *is not* the user. (You can be far more product-focused if the buyer *is* the user.) So for homepages, I like to express your narrative and positioning in the form of value → tool (i.e., the emphasis is on the outcome or benefit, supported by the nuts and bolts of how it works), and then we'll flip that around for the product page, where how it works comes first. But it's important not to get carried away with value in the abstract on your

> homepage. If you're selling a tool, that's still the tangible thing folks are buying, and as our opening quote suggests, folks need to know what that is! It's the anchor for everything that follows. If folks don't get that, nothing else lands.

1. Homepage — the high-level value overview

- Does your homepage tell readers what your product is and does? (Seriously! So many SaaS sites fall at this very first hurdle.)
- Does your homepage match your GTM motion? (High-awareness PLG products may push users into the product ASAP, sales-led folks often need to do more explaining upfront.)
- Does your hero section capture the BLUF (Bottom Line Up Front) value proposition of your first narrative slide?
- Does your homepage show or visually represent the product in some way? (That's what you're selling after all — don't get so caught up in benefits or abstract imagery you neglect the product itself.)
- Does your homepage capture a single, key, compelling reason to buy? (What's the hook? Have you validated that with customers?)
- Do you connect that value prop with how the product works, perhaps as a three-step 'how it works' or some sort of flow from start to outcome?
- Can you drive the point home with from/to, before/after, or with/without language to contrast the status quo and your new way?
- Do your testimonials and social proof reinforce the core message you're trying to communicate?
- Do you land on a big call-to-action (CTA) 'closer' that tries to get folks to take a clear next step (usually a demo)?
- Do you have a secondary CTA for folks who aren't ready to commit to a demo at this stage? (Think: short explainer video, mini demo video, etc.)

> **An in-depth guide to B2B SaaS homepages:** Emily Kramer of the excellent MKT1 newsletter has a great step-by-step guide to SaaS homepages if you're looking for something to walk you through the whole process: https://newsletter.mkt1.co/p/homepage-copy.

2. Product page — features to value

- If your homepage is more value → product, does this features-first, product → value page adequately describe your key features and why they matter? (There should be a clear connection between your product's features and your overarching value proposition.)

- Does your product page walk through your ICP's jobs-to-be-done (**chapter 36**)? (Consider framing sections using jobs-to-be-done verbs — 'analyze,' 'monitor,' 'create', etc.)
- Does your product page provide adequate detail for buyers with a checklist of features in mind? (You can do this by offering details at different 'resolutions' — and not walls of text — as the box below describes.)
- Have you validated the sort of value customers are getting out of each specific feature or product area? Cute writing is nice, but the proof of the pudding is in the eating, and you've got to know how customers think yours tastes.

The anatomy of a feature block: Product pages inevitably involve a series of feature blocks, from the classic zig-zag pattern to scrolling, slide-style overviews. In terms of alignment, we discussed settling on specific, validated feature props in chapter 13, and now in terms of execution, this is what I suggest for your feature blocks:

- A small 'eyebrow' header with a one or two-word description or verb.
- A value proposition for the feature or product pillar as a skimmable header.
- A sentence or two of copy elaborating on the feature.
- If appropriate, one to three smaller, succinct bullet points highlighting particular details or benefits. (These can be turned into blocks, boxes, whatever — they don't have to be literal bullets.)
- Keep it interesting by mixing up the format with callouts, 'bento'-style grids, social proof, whatever makes it interesting to actually read.

I like this approach stylistically because it lets readers zoom in and out as they scan the page. (Screen recordings will show you that folks rarely proceed linearly through a page. Instead, they jump around, stop and inspect, and generally try and minimize cognitive energy expenditure.) But I also like it because, in a sense, this is the final, crisp articulation of all the work you've done to understand exactly what value matters to customers, and boiling that down to laser-focused copy is very satisfying. Consider the vast gulf between this sort of validation-driven marketing and the usual marketing fluff that passes for web copy. It is, in a sense, the difference between a startup that is in touch with reality and one that isn't, a difference we'll explore further in chapter 28.

3. About page — narrative without product

About pages can vary wildly between companies at different stages (if they exist at all at later stages). As far as your clarity strategy goes, however, this is your chance to lean into your big story 1 change narrative, ideally *without* your product.

- Does your about page (or 'why' page) capture your story 1 change narrative, perhaps *without* your product, to get potential new hires excited, for example?
- Does your about page explore the *why* of your narrative — why you're doing all this?
- Does your about page cover the *who* that's driving this whole thing? Are you a solo operator or a team of dozens or hundreds? (If buyers are taking a significant career bet on a long-term relationship with your company, they'll want to know!)

> **Notion's 'story 1' about page:** Notion has, as of writing, an about page (https://www.notion.so/about) that includes a 9-ish slide section that explains the big history-of-computing, story 1 rationale behind their product. The page is only linked from their footer, probably hasn't been touched in years, and includes a cute call for folks to "Check out the product! Write in! We need early adopters like you to start a movement." (It worked!) But the page also captures the future-of-tools-for-work ambition that has driven the product from the very early days. The page might not be much more than a relic from their early days, but the narrative it captures — the ideas that drive the company — is still incredibly powerful. And it's no surprise that this page also features 'Join us' and 'Our investors' sections, as they're the two audiences (employees and investors) where the big story 1 pitch resonates.

Context matters

That is, in a *very* brief nutshell, how I tend to suggest you apply your narrative to your key website pages. Messaging execution is, of course, highly contextual, so if you have good reason for doing it your way instead, go for it. Your audience's level of awareness and customer journey; your product's maturity, depth, and differentiation; and your GTM motion can all influence how you execute. As can the creativity of your team.

Nevertheless, can you see how important it is that these assets all work together to both communicate and drive home your narrative? Your narrative and positioning is ultimately one long thread that runs from first touch through the sales experience, into the product, through success, and back out into the market as word of mouth. The more coherent your message, the stronger that thread will be. That's what it means to have a strong clarity strategy. It all starts with your deck that we looked at in **chapter 9** and from there it should run through everything you do.

Goldilocks execution — how much should we invest?

How much time, money, and effort, then, should you invest on the execution side of your messaging? I tend to believe there's a Goldilocks approach to getting things just right, especially for your website, where the result is not too hot and not too cold. How do you find the 'just right' sweet spot?

The first step is to understand you're executing on a *message*, not a "design" or a "build." Everything exists to serve the message, and each function of your startup needs to work together to deliver the message. Sales and marketing run on messaging, and if you ain't got the words, it doesn't matter how much you spend, you're going to struggle. Saying nothing beautifully with blazing-fast load times doesn't move the needle.

The next step is to set your sights on meeting buyer expectations for your particular category.

'Too cold,' in the Goldilocks sense, is not meeting the most basic expectations for your site, which might include how it's structured and how easy it is to read. Buyers — brace yourself for this scorching insight — need to actually read your website. You wouldn't know it from some B2B SaaS sites, however, especially in the more sales-driven enterprise end, with long lines of tiny centered grey text that's just plain annoying to read. For the love of all that's readable, you can do better than looking like a $5 WordPress template.

On the 'too hot' side, the pendulum can swing so far beyond buyer expectations that we get these massively over-designed, over-engineered websites where hundreds of hours of work are poured into projects where the actual buyer is, ironically, something of an afterthought. What's particularly interesting about these approaches is that folks are really trying to do a good job. They're working hard and investing considerable sums in their communication, so what's the problem?

The problem is the only way they know how to execute — the only levers they know how to pull — is to max out on design and engineering, and so that's what they do. The result is a beautiful, empty, impeccably engineered box with a big hole where their clarity strategy should be. The execution might be fantastic, but there's no 'there' there. That's a great shame. (And that's why clarity strategy exists.)

Sitemaxxing

This maximalist approach to site design has come back in vogue as I write, and I admit feeling conflicted about this, dear reader. Honestly, between you and me, *I kind of love* a completely OTT website. But I love it from a *cultural* point of view (yay web tech!) as someone who once dreamed of making similarly indulgent creations, and not from a business, do-buyers-actually-care point of view.

I also feel conflicted because I believe most B2B SaaS companies significantly *underinvest* in their website, but that's more on the narrative, messaging, and measurement side — having something to say and knowing what buyers actually do — not in terms of the complexity of their tech stack or visual design.

The maximalist websites I want to touch on here are, let's face it, often driven by designers and developers wanting to have fun trying to impress themselves and/or other designers and developers. Sometimes that's fine *if that's who you're selling to,* and that is genuinely the expectation of your category.[3] Sometimes, there really is a high bar. And if you're selling to developers (Stripe, Linear) *and* you're both lucky and skillful enough to capture the imagination of your target audience with an industry-benchmark-setting site design, then great. (Your reward? Being copied to death.) In these cases, there's at least some overlap in attention between creator and buyer, that is, what you think is cool is what your buyers think is cool. That's the brand effect folks are often hoping for, but note how specific these circumstances are. (We'll see how branding really works in **chapter 45**.) The vast majority of B2B prospects outside of the tech world, however, *do not care* how cool your website is. But they do want something coherent and compelling to read!

And sometimes, even if you are aiming at tech folk, the result is still a very expensive swing and a miss,[4] or an incredible — an incredibly short-lived — site that precedes any form of product/market fit (see the examples of pretty, dead startups in **chapter 45** also).

Why does this happen? Well, making cool things is fun (be that features, websites, or whole products), even if it's strategically suboptimal. It's psychologically comfortable in a way that trying to find product/market fit in the acid test of the market most certainly isn't.

But part of it is also the difference between the parts and the whole. Focusing just on the parts — 'company X succeeded because their website had this style/used this tech/said these things' — is more of a left-brain thing, where we forget there's a much bigger context at work. That bigger context — where a startup really sinks or swims — is more of the right brain's domain of thinking about the whole. This is a theme we'll return to throughout this book (especially in **chapter 42**). Put it this way: if you bake a cake, is it the egg, or the flour, or the butter, or the sugar that makes it delicious? It's a nonsense question. The ingredients only ultimately matter in the context of the whole, the whole cake, the whole operation, the whole market context, the whole time and place. And that's the kind of thinking — about the whole, not just the parts — that I want to encourage a lot more of.

> **Attention mismatches:** Maximalist startup websites are also often another example of the classic attention mismatch that happens in marketing and sales in all its forms all the time. In chapter 45 we'll see that it happens in B2C with elaborate "brand" campaigns that are deep, creative, purpose-driven exercises for the creators where they apply loads of their active attention to 'what the brand means' but the result is absolutely baffling to the audience who is *at best* applying the tiniest flicker of passive attention to a random ad that caught their eye. Effective creative work needs to work backward from the audience's attention, not simply what's fun to make.

'Just right' messaging

There is a 'just right' happy medium here, and it's *working hard to communicate what the product is, does, and results in.* It's not rocket surgery, and the vast majority of that work happens before pixels, prose, or content platforms are ever touched. It's doing the legwork to understand your customers better than anyone else and having the right story to tell in the first place.

Execution, then, can happily err on the more minimal side where you keep things very simple and direct. If — if! — you've got the resources to spare, then sure, you can invest in a more maximalist approach of doing your best to *show* and not just *tell* folks what your product does.

But fundamentally, your approach should be in service to the reader, communicating what they want to know. The reality, however, is that many execution teams don't know what that is. I know that sounds trite, but both the lazily undesigned enterprise sites and the massively overdesigned sites fail in the same way — they fail to engage with your customer's genuine expectations, communicate in line with them, and ultimately serve the customer by helping them understand *what it is you're actually selling and why it matters.* That's why great websites are a function of great narrative and positioning, first and foremost, and everything else comes second. And that's where, if you put the time in — by studying how folks actually buy in your category and testing whether your messaging lands with them — you can outdo the competition in both customer expertise and laser-focused messaging. It's win-win.

With all that said, I'd love to hear how your site performs either way — if you have data and examples, I'm all ears! **luke@lukestevens.co**.

Next, let's look at how you make your messaging stick with memes and metaphors, and then we'll look at Stripe's messaging evolution.

Notes

1. Casey, Bryan. October 29, 2023. **https://twitter.com/bryanfcasey/status/1718638402701087148**
2. Or, what the heck, today. Somewhere along the line, we realized that *talking* about conversions — the design, copy, or messaging tricks everyone surely knows must "convert" was more important than doing the actual experiments to see what *really does* convert *in a given circumstance*. (And I include myself in that, too.) The reality is, dear reader, and take this from someone who got *super* into A/B testing back in the day, that no one tests anything, especially in B2B. And if they do, the tests are often incredibly poor, being called too quickly, focused on a superficial goal, and used for "directional" purposes, which is nonsense. This is worse than not testing — it creates false knowledge that cuts off future learning. The reality is — and the A/B testing literature tends to bear that out — that the vast majority of tests get no result because they're not testing anything meaningful. You need to be making bigger moves if you want genuine growth outcomes. There's nothing wrong, therefore, with taking a qualitative approach instead — that's what most of this book is about after all — but science-y lipstick on the anecdotal pig is a pox on the sales and marketing world and often results in the blind leading the blind. This isn't just a pedantic point: folks need to understand there isn't one true High Converting Way™ that they're one redesign away from. What matters instead is telling your story about your product in a way that makes sense to your customers. That will inevitably marry both the common features of your category (Jakob's law says most folks spend their time on other websites, not yours, so try and follow convention) and whatever is unique to you. That's ultimately what "converts," as unsexy as that may be. I'll rant more about this in **chapter 42**.
3. I suspect that the "buyer" for some maximalist sites isn't the typical B2B prospect but actually engineers the startup is trying to hire. In this case, the maximalist website signals 'this is a cool, fun, techie place to work' as sugar on top of the marketing message for the actual buyers. This could *potentially* be useful when you're solving otherwise boring-sounding challenges in HR or finance, for example, and need to attract talent, but this is purely speculation on my part.
4. One SaaS example I spotted while writing on the skillful-but-ridiculous end of the spectrum: a homepage opens with a generic, ultra broad, story 1 three-word hero headline and then segues into a wordless, parallax-scrolling hero illustration that, over the course of at least three screens-worth of scrolling (!), gracefully fades in with calming sunset hues in a retro style. As you scroll past the starry sky and canyon walls (we're in the desert for some reason) you arrive at a scene where a few tiny workers appear to be — after much squinting — working on a futuristic radar thing. Why? It's

not really clear. What does it mean? I have no idea. Are they selling communication equipment? Nope, this was for a mature B2B product analytics company. Fun to make? I'm sure. Fun for buyers to scroll through? I'm not so sure. It's a long ~~walk~~ scroll to the story 2 stuff they actually care about. Despite the tasteful execution, the lack of any kind of coherent clarity strategy was so glaring it made me *doubt the future growth prospects of the company*. When you see it — or see the lack of it, in this case — you can't unsee it.

15

Memes & metaphors

> Metaphoric thinking is fundamental to our understanding of the world, because it is the *only* way in which understanding can reach outside the system of signs to life itself. It is what links language to life.
>
> *– Dr Iain McGilchrist*[1]

The strength of your clarity strategy rests in part in your ability to create sticky, *memetic* messages. We're not telling Lord of the Rings-style epics here. We're telling short, punchy stories with a core idea we hope folks remember and, ideally, pass on. That's what word of mouth is — a shareable idea that gets passed from person to person. This means our messaging has to be Actually Good™.

We can do all the narrative and clarity work in the world, but if our stories don't catch on, what's the point? This is why I love messaging, incidentally. It's that creative challenge of boiling vast possibilities down into catchy sayings and memorable lines. And in some ways, that's what messaging comes down to — the depth of your discovery and the strength of your lines.

But what makes for *great* messaging? Well, a lot of things. But in this chapter, we're going to explore two key ingredients: memes and metaphors.

It's memes and metaphors all the way down

Messaging's job is to create shared context with our prospects. It helps folks know if they're on the same page as us, and vice versa. That means using language that resonates with them, and that's why the best messaging sometimes comes straight from your prospect's mouth. (Assuming, that is, they're customers we want more of — not much point using the language of bad-fit customers.)

Sometimes, however, it's up to us to find the language for our customers to use. We need to do the work to find those apt metaphors, memorable turns of phrase, or catchy category definitions that put a finger on a particular phenomenon — a pain, a playbook, a failure state, a success state, whatever it might be that helps convey our narrative.

And, hopefully, the message sticks. But what does it mean for a message to be "sticky"?

It means it's memetic. As in "memes." But not *those* memes, as in internet memes. "Meme" in the original sense, which was a riff on "gene," coined by Richard Dawkins. In this context, *meme* is to cultural bits of information as *gene* is to biological bits of information. Dawkins likes to use a catchy tune as an example. The *fitness* of the tune, i.e., its catchiness, determines if it's passed on, and conversely, tunes that are passed on and survive are, by definition, the catchy ones.

Therefore, the ideas we're surrounded by are, from this point of view, the ones that were catchy enough to survive.

Advertisers have, for decades, used catchy tunes — jingles — to make their own messages as memetic — as catchy — as possible, with a fair helping of Pavlovian repetition. (I can still recall the Pizza Hut delivery number from the 90s thanks to their jingle burning it into my brain through sheer repetition.) And even when there isn't an explicit tune, catchy messaging still tends to use the tricks of music, like rhythm and repetition, to roll off the tongue.

In B2B, it's less about the Pavlovian response to a jingle or recognizable sound (like a can of soft drink opening — *pssshhhh*) and more about the concept, technology, or behavior change you're advocating.

For example, when we see folks in startup land trying to 'create a category,' usually they're trying to meme, if I can turn it into a verb. They're trying to seed an idea in the broader cultural ecosystem they hope will point back to them. And that's also why a lot of 'category creation' attempts fall flat — they're not memetic because there's no real catchiness there. But I digress. (We'll explore category creation over several chapters from **chapter 32** on.)

> **The meta "meme" vs. "memetic" distinction:** The classic, ironic example of memetics is the definition of the term "meme" itself. The "internet meme" definition of meme ironically outcompeted the original, quasi-scientific definition of meme because funny, captionable images are an idea with much greater cultural 'fitness' (in an evolutionary sense) than the original idea quasi-academic of "meme" as in "gene." This both simultaneously proved the concept and muddied the waters forever more.

Information packets with attitude

What's your 'information packet with attitude'?

We evolved for many thousands of years with oral culture as the sole means of passing down information, and our brains are well-primed to send and receive catchy packets of cultural information. The philosopher Daniel Dennett, for example, described memes as "information packets with attitude," and I love that as a description of the ideas we're often trying to seed in the market.

Unfortunately, what usually makes the rounds in the business world — especially the puddle-deep social-media-driven business world — is the information packet without much attitude. Business culture is rife with memetic ideas — often pretty lame ones. But the good ideas need to compete in the same space: meaningful and genuinely helpful ideas have to stick and spread; otherwise, what's the point?

Therefore, if you believe in your ideas, it pays to refine them to the point where they're sharp and catchy, where they're memetic. That is, in part, what we were trying to do with your Hemingway story (**chapter 10**) and in naming things to flesh out your narrative (as we'll see in **chapter 17**) — we want to find that 'information packet with attitude' that contains your core insight.

Do you have that attitude-packed information packet?

Metaphors link language to life

We want our message to be memetic — sticky, catchable, spreadable. But how do we convey a lot of information in a mere few words? We need to borrow meaning from other points of reference — we need metaphors.

As one example, I'm relying heavily on the metaphors of *radars* and *laser beams* for the two kinds of attention we'll be exploring in this book.

Messaging is metaphor all the way down, and that's because metaphor is — and this is a big idea! — the foundation for understanding, as McGilchrist writes in the opening quote above ("It is what links language to life"). It also shapes our thinking as he writes below (emphasis his):

> *[W]hat we choose to compare a thing with determines which aspects of it will stand forward and which will recede. Thus comparing a football match with a trip to the betting shop brings out some aspects of the experience; comparing it with going to church brings out others. The model we choose to use to understand something determines what we find. If it is the case that our understanding is an* effect *of the metaphors we choose, it is also true that it is a* cause*: our understanding itself guides the choice of metaphor by which we understand it.*[2]

What do you call an ultra-short pitch? Think about what we did when we called your concept-in-a-few-words your 'Hemingway story.' We picked a metaphor, and now we have a name for a particular type of story that's more evocative and, therefore, memorable than its raw, functional description. Metaphor is the foundation of an evocative pitch.

Killer robot dogs

Much of this book is about positioning, and metaphor plays a big role in positioning, too. The art of positioning is often the art of metaphor — picking the right metaphor determines what aspects we bring to our attention. I love to illustrate positioning-via-metaphor with the idea of metal, plastic, and electronics arranged in a certain way — like, say, a robot dog.

I'm sure you've seen videos of Spot, the robotic dog from Boston Dynamics, over the years. A company called Unitree Robotics announced an — *ahem* — similar-looking robotic dog in 2021 at a more consumer-oriented price ($2,700 vs. Spot's $74,500).

A generic robot dog. What positioning — or what metaphor (beyond 'robot dog') — should we give it?

When reading about the robot on Hacker News, I couldn't help but notice how people reached for metaphors to position this otherwise generic piece of tech in their minds. And that's understandable, as it's a blank slate with no default use case. Everyone is free to imagine what it could be. Is it a:

- **Killer canine:** A robotic military pack dog or roaming security dog? Scary!
- **Cuddly companion:** A cuddly, friendly human companion? Cute![3]

Same platform, different metaphors, different position.

And while this is a contrived example, you can see how your choice of metaphor not only helps customers understand what you're about; it helps *you* understand what you're about. Is the next step on your product roadmap sharp, steel teeth or soft, cuddly fur, for example?

That's the power of metaphor — it both consciously and unconsciously evokes and minimizes certain qualities, and therefore they should be used thoughtfully and deliberately. You don't need to rush to pick a metaphor either — you can always try and see where early adopters take the product and double down on emergent use cases.[4]

> **Metaphors to hold messaging:** The 'message house' (metaphor!) is a popular concept for organizing your key messages — there's the overarching statement (the roof), the three or so pillars holding the roof up, and the 'foundation' of proof points and supporting evidence. Messaging really is metaphor all the way down.

B2B positioning through metaphor

In B2B SaaS, the choices are less killer canine or cuddly companion and more a matter of finding a reference point so people understand what we're about. Perhaps you are, by way of analogy, the HubSpot for X or the Netflix for Y (as we'll see in our mid-market LMS discussion in **chapter 46**).

We'll explore different forms of positioning in chapters to come, but next, let's continue looking at positioning-as-metaphor as we dig into Stripe's messaging.

Notes

1. McGilchrist, Iain. *The Master and His Emissary* (pp. 115-116). Yale University Press. Kindle Edition.
2. McGilchrist, Iain. *The Master and His Emissary* (p. 97). Yale University Press. Kindle Edition.
3. And not original — Sony was there first in 1999 with AIBO (which means "pal" or "partner" in English), a robotic pet dog that people got so attached to they ended up giving them Buddhist funerals when they could no longer be repaired. See "Japan: robot dogs get solemn Buddhist send-off at funerals," *The Guardian*, May 3, 2018. **https://www.theguardian.com/world/2018/may/03/japan-robot-dogs-get-solemn-buddhist-send-off-at-funerals**
4. Tip o' the hat to April Dunford for that wisdom. I wonder what early adopters have done with the Unitree Robotics dog, for example.

16

Stripe, mission, & homepage evolution

In this chapter, we're going to continue the theme of memes and metaphors and look at how they manifest in Stripe's mission and historical website messaging.

Let's start with mission statements.

Some folks like to noodle on mission and vision statements with big :thinking: vibes. If investors want to hear it, fine, but I think for most startups (and companies, for that matter), it's an indulgence — your mission is to build a better product and tell people about it, and your vision is your customers' success. That's it. (I mean, do you ever hear about Apple's mission statement? They let the products do the talking.)

Sometimes, however, a well-crafted mission statement can catch on and provide a sense of shared context and purpose ('Don't be evil' was a classic unofficial one. RIP.). And, in larger, multi-product organizations, sometimes there is a need for something more aspirational than We Make A Very Good Product, especially if you work in a dry area like, say, payments.

Stripe, for example, has one of the most oft-repeated mission statements I've seen, and I only know about it because, well, people keep repeating it. Here, for example, is ex-Stripe employee Brie Wolfson discussing it when describing her time with the company in the *Every* newsletter:

> *The mission of the company, they told us, was to "increase the GDP of the internet." It was a little abstract, but we believed in it enough to recite it with pride.*[1]

This metaphor, the idea of the internet having a GDP like a nation, evoked something implicit, perhaps about growth, scale, and progress, and that's why I imagine it felt "a little abstract." But it was still nonetheless meaningful, which is why it was 'recited with pride' — people will imitate memetic messages! That's a powerful leadership tool. I've seen it cited as a reason why people were interested in working at Stripe *at all*. The rationale was that the mission sounded cool.[2]

This is the power of a good metaphor — evoking implicit ideas that capture the imagination. Consider if Stripe's mission statement was "Our mission is to help businesses take payments online" instead. I mean, that's powerfully boring, right? There's nothing metaphoric or evocative there; it's too literal and therefore loses the implicit meaning of the original, like a joke explained.

Message evolution

Stripe is also a great example of messaging — and metaphors — evolving with your startup. First, let's note the clarity of Stripe's initial messaging and website copy. Take their 2013 homepage[3], for example, from a couple of years into their journey.

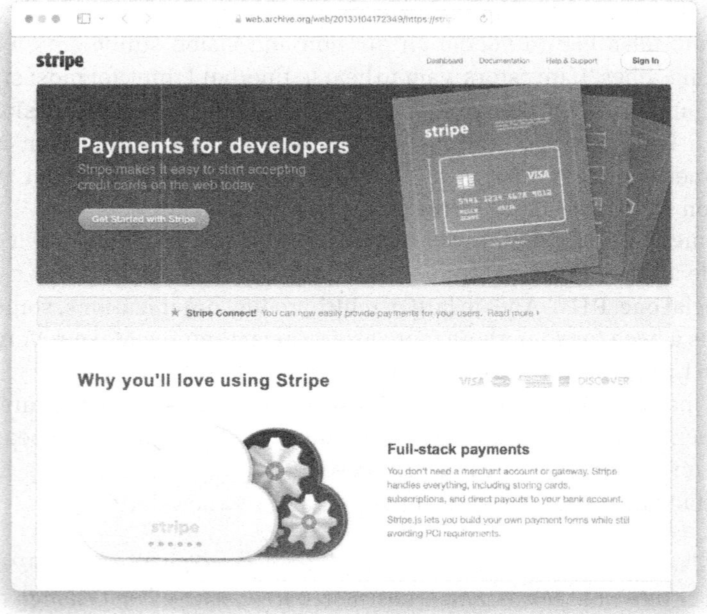

Figure 16.1: Stripe's 2013 homepage.

They kept it very direct to start with:

- "Payments for developers" was the headline, and "Stripe makes it easy to start accepting credit cards on the web today" was the subhead. (The status quo at the time was, of course, pretty awful for developers.)
- Their five-pronged value prop was (i) you don't need a gateway or merchant account; (ii) the API is nice; (iii) pricing is simple; (iv) we're developers too; (v) lots of people use us. (Or: Problem/pain, solution/delight, fair price, we're like you, let's go.)

As they grew — and boy did they grow — so did their metaphors:

- By 2016, Stripe was "Payments infrastructure for the internet," serving marketplaces, mobile, ecommerce, and subscription services. That is, their operating metaphor evolved to *infrastructure.*[4]
- By 2022, Stripe claimed to be building a "modern finance OS"[5]. The operating metaphor was now an *operating system.*

And not without good reason. Stripe has had, to put it mildly, a hell of a run, and you can see in this tiny snapshot how their messaging — and metaphors — have evolved along with their ambition.

B2B SaaS & the 'management system' metaphor

Putting Stripe and "operating systems" aside for a moment, a significant chunk of B2B software tools have traditionally been 'management systems,' which is worth thinking about as a metaphor, even if it doesn't fit all B2B startups (Stripe and payments being an exception, for example). Consider:

- Content Management Systems (CMS).
- Customer Relationship Management... Systems (originally, now just CRM).
- Learning Management Systems (LMS).
- HR Management Systems (HRMS, or HIRS).
- There are also *project* management systems, including those for developers (Jira and Linear, for example).
- I've worked with a client on an emerging Finance Management System, for example (though they prefer 'operating system').
- Even the underlying database is probably a Relational Database Management Systems (RDMS). (A management system for your management system, stretching the metaphor to breaking point.)

This kind of software is the bread and butter of the B2B world. And as diverse as these applications may be, the fundamental metaphor is much the same — there's a collection of data that has emerged that now needs a 'management system' of some description.

If this describes your startup, whether you're bringing the first 'management system' of its kind to market or whether you've got a unique take on doing it in a new way, it's worth reflecting on what the underlying metaphor means. What is the scope of the data your system manages? And what's your innovation with the management of that data? Are you innovating through experience? Underlying technology? Point of view? Organizational change and playbook (**chapter 35**)? What's your TTT sweet spot between tool, technology, or transformation, for example? (We touched on TTT when discussing Slack in **chapter 12**.) What company will you build to *deliver* that innovation? Drilling down to the first principles of language, metaphor, and meaning can be a surprising source of clarity that cuts through much of the hype, including your own.

You probably aren't Stripe

Returning to Stripe, they didn't start as an "OS". Startup metaphors are themselves memetic, and you can see how they catch on and fade as startup fashion dictates — from management systems to platforms, from suites to clouds, and now to operating systems and whatever comes next.

There's always a term *du jour* that everyone wants to imitate and milk for its implicit meaning of being *cutting edge* before that loses its edge, and everyone wants to be *bleeding edge* instead (a graphic metaphor!), until the cycle peters out and we stop talking about edges.

If your product is big enough to justify the 'operating system' label, great. But for early-stage folks, it can be a trap, because it mixes story 1 and story 2. I see early-stage startups trying to roll out investor-friendly messages like (and I'm paraphrasing) "We're building the new marketing OS," which is... not what you've got. That might be your story 1 aspiration, but your aspiration isn't much of a product. You sell aspiration to investors, but customers need something tangible — they need that story 2 wedge, that 10x improvement along some dimension that makes up for the fact you're more MVP than giant platform or 'OS' like the incumbents. That is, you need to make sure your product can cash your messaging checks.

Ultimately, you want language that resonates with the exact customers you're targeting — developers, in Stripe's case. And that's the lesson from early-stage 2013 Stripe, to have a product (and messaging) that 'just makes sense'[6]:

> *Stripe was one of those things that just made sense. As a web developer I was tired of jumping through the many hoops of authorize.net or praying that people would complete the purchase path on PayPal. When Stripe came out I [...] jumped immediately on it and haven't looked back.*

Forget the grandiose metaphors in the early days and find that initial Hemingway story (**chapter 10**) to drive your narrative in a way that 'just makes sense' to the folks you're targeting — people whose pain you know inside out. And may your Hemingway story — and narrative more broadly — be as successful and lucrative as those three magic words were for Stripe: "Payments for developers."

Next, let's round out this messaging section with a quick tip: *name things*.

Notes

1. Wolfson, Brie. "What I Miss About Working at Stripe," *Every*, July 19, 2022. **https://every.to/p/what-i-miss-about-working-at-stripe**
2. Stripe's name could also be considered metaphoric in that it evokes the magnetic stripe on the back of a credit card, but the way they landed on Stripe was more by default than deliberate choice. As former Stripe CTO (and now OpenAI co-founder and CTO) Greg Brockman describes in a Quora answer, they started as /dev/payments, decided that didn't work, looked at a bunch of other ideas, almost settled on "PayDemon" (!) but after much debate they couldn't come to a decision, so they defaulted to Stripe. See the full answer here: **https://www.quora.com/Stripe-company/How-did-Stripe-come-up-with-its-name**.
3. See Stripe's homepage from January 2013 via the Wayback Machine: **https://web.archive.org/web/20130104172349/https://stripe.com/**
4. See Stripe's evolved messaging as of April 2016: **https://web.archive.org/web/20160401025128/https://stripe.com/**
5. "Building for the next era of the internet economy." **https://stripe.com/sessions/2022/keynote-speech**
6. ivyirwin, *Hacker News*, September 29, 2021. **https://news.ycombinator.com/item?id=28699240**

17

Cheatsheet: Naming things

[Y]ou can change people's behavior by describing things

– *Visakan Veerasamy*[1]

We're going to explore more strategy in the next part of this book, but before we leave this section on messaging, I want to give you a practical tip for helping you improve your messaging and your sales narrative slides from **chapter 9**. And that is: *Name things.*

And by things I mean *concepts*. Named concepts are shareable. It's hard to discuss and communicate your insights when they're all abstract ideas; it's much easier when there's a name stuck to them. And those names help make your messaging that much more crisp, specific, and *catchable.*

Named things give folks language to imitate — we *want* people talking to their colleagues or boss using the names of things we give them. That's the only way they'll have to communicate the abstract idea we're trying to get across, be it a new way of understanding their pains or opportunities, a new playbook for the way they work, or a whole new class of tools that's now available.

Named things can also be *owned* things, at least for a while (until your name escapes into the wild), in that they have a strong association with your brand, as we see with HubSpot and Drift in **chapter 31** and Qualtrics in **chapter 35**.

When it comes to your clarity strategy, simply naming things — the pain, the problem, the change, the need, the solution, whatever it is — can make your narrative distinctly *yours.*

So what can you name?

Name it and claim it

Here are a few ideas on what to name. You don't have to have cute, made-up names for all of these things, either — an accurate, succinct description of one key idea that resonates with prospects is better than a confusing collection of concepts that prospects are never going to get their heads around.

What does it look like when this works? HubSpot's former CEO Brian Halligan, who we'll meet again in **chapter 31**, had a few catchy turns of phrase up his sleeve, not least of which was "inbound marketing" (their name for the change and new playbook in digital marketing).

But it went further than that, with his pitch often mentioning that inbound marketing was more about 'the width of your brain than the width of your wallet.' That's a catchy, evocative, aspirational line describing success. That's what I mean about naming things, and you can use a phrase, too. It doesn't have to be one or two words or, heaven forbid, a TLA[2], it can be a catchy saying, and that gives you novel bits of language for your clarity strategy *that are unique to you*.

Name one thing

Start by just trying to name or succinctly describe one thing where you have a compelling insight into your buyer's world — a pain, a solution, a playbook — something that's unique to your point of view.

Here are some suggestions to get you started:

- **Problem:** What's your name for the burning pain or impending doom you're tackling? Is there a specific trap, challenge, or hidden or missed opportunity you can name?
- **Solution:** How do you describe your solution? Is it by metaphor or analogy (X for Y)? How would you ELI5 it (Explain it Like I'm 5)? What's your name for the success? How do you describe it?
- **Secular change:** What's the name of the secular change (if any) you're riding (**chapter 30**)? Are there other trends you can tap, perhaps from adjacent disciplines?
- **Position:** What's the name of the segment you'll position around (**chapter 37**) or, instead, the attribute you want to position around (**chapter 46**)?
- **Category:** Do you have a unique category definition (**chapter 32**)? Do buyers share and understand that?
- **Your playbook:** Do you have a named playbook (**chapter 35**)?
- **Product pillars:** Give your features some structure: Group ideas, name sections, split things into three, and find a coherent flow. But use common nouns; don't try and brand them, as David Kellogg warns.[3] (Branded features dilute your product's brand and are extremely corny coming from a startup — leave that to Apple.)
- **Your key value props:** How do you describe the *value* of those core pillars in a clear, crisp way?
- **Jobs to be Done:** Can you name the top job(s)-to-be-done your customers use your product for, even if it's just by verb? (We'll look at JTBD in **chapter 36**.)
- **Behavior:** What do you call it when people do more (or less) of your

desired behavior when they buy into your approach? What's the from/to?

- **Your ICP:** Is there a name for an emerging role, buyer, or community you're targeting? (Like "Design Systems Manager" that Figma picked up on in **chapter 27**.) Or the type of buyer you *don't* want (e.g., non-early adopters)?
- **Anything I've missed:** What other shared language do you use that's unique to your organization or your market?

This is not an exhaustive list, and we will tackle many of these topics in the chapters ahead. But that's the point — much of our work amounts to *naming the critical things that matter* so we can build your clarity strategy around those concepts.

You don't have to have the answer right now, but the more you can keep an eye (or ear) out for hooks — catchy lines and turns of phrase that describe your insights — the more building blocks you'll have for your narrative.

The point is not to name everything on the list above, either. A powerful narrative isn't a product of throwing everything you possibly can into the mix, and that's always the temptation with complex products — it hurts to leave things out. But the pain is worth it — the more you can curate these ingredients and your insights, the more you'll end up with something strong and coherent.

What ultimately matters is the richness of your insights. The richer your insights into your ecosystem, the trends around you, and your buyer's world, the more accurately you can identify, name, and turn abstract ideas into catchy, relatable ideas. (Or even just one killer idea.) And the richer your ideas, the richer your marketing, sales, product, and messaging work will be.

> **Name and separate:** Naming your insights — in so far as it makes abstract ideas clearer and not muddier — helps drive your uniqueness because it puts distance between your concept and common understandings of similar concepts. To name is to separate, it's to say, 'this is a distinct thing.' That distinctness can then distance you from the competition, enhancing your differentiation. But this cuts both ways — it can distance you from customers, too. For example, if you try to make up a category name that they don't get, you've created a gap between you and the market, not just the competition. That is, in part, why using customer language where possible is always a good idea — keep an ear out in sales calls and customer interviews for the names customers use to talk about their world, and if they're good, steal them!

Notes

1. Veerasamy, Visakan. June 19, 2022. **https://twitter.com/visakanv/status/1538415210221842432**
2. A sin I've committed several times already with the TAE framework (Thinking, Aligning, Executing) and the TTT rubric (Tool, Technology, or Transformation). The shame...
3. This is timeless advice from David Kellogg, who wrote in "Branded Features: Resist the Temptation," published July 4, 2020, "[I]f you're not a multi-billion-dollar company, then maybe you shouldn't emulate the marketing strategy of one. If you take a breath, pause, and think about what it means to create branded features — to your branding, to your comprehensibility, to your industry leadership, to your international operations, to your legal strategy (and associated costs) — you'll decide to pass on branded features every single time." **https://kellblog.com/2020/07/04/branded-features-resist-the-temptation/**

PART IV

CLARITY & **PITCHING**

18

The claw game

In chapter 15 on memes and metaphors and chapter 16 on Stripe's messaging, we saw how fundamental metaphor is for understanding what we're building and what we're selling. To introduce this next section, where we're going to return to our discussion of your clarity strategy and sales pitch, I want to start by offering a handy metaphor that helps us understand the nature of clarity strategy itself.

So far, we've mostly discussed your clarity strategy as story 1 and story 2, which are my labels for a particular kind of broad change story on the one hand and narrow positioning on the other. Most pitches comprise a bit of both. The two story types are not mutually exclusive; you dial them in for your respective audience. That means knowing when to go broad (right brain, change, story) or deep (left brain, positioning, specifics). It's that fundamental understanding of the nature of our left- and right-brain attention that gives us the direction on how we use these tools.

I like to think about this dynamic as the *claw game of attention*.

You might know them as arcade crane games or "UFO catchers," as they're called here in Japan (the claw is a UFO coming down to 'catch' the prize, get it?). You move a claw across, and then down, and you try and grab a soft, fluffy, *very* precisely positioned soft toy which comes up... and... _I think I've got it!..._and... *no, don't slip... don't slip damn you!...* drops back down to the bottom of the machine. *Ugh!* Better luck next time...

X and Y axis

The claw game is a great metaphor for us because it operates on an x- and y-axis. Claw goes out, claw goes down, claw attempts to grab prize. That's the game.

McGilchrist sometimes talks about attention as mapping to a horizontal or vertical axis, too, as we'll discuss shortly in **chapter 20**. Therefore, in clarity strategy, we think about the two modes of attention as existing on the x- and y-axis. We have the x-axis for outward attention, i.e., our change story, and the y-axis as downward attention, i.e., positioning. I love this because it captures the essence of our modes of attention — the x-axis is what unfolds over time, and the y-axis is about diving down to that particular thing in front of us.

And that's what you'll see in your audience, too.

If we're going to meet your audience where *their* attention is at, and capture your audience's attention as the 'prize' — the fluffy, slippery toy — then we've got to know whether we need to go hard on the x-axis (perhaps for an executive who wants the big change story) or deep on the y-axis (perhaps for an end user who has to do the work).

In practice, that might mean:

- **More change story, less positioning:** Maybe your audience cares more about the big change story, and they don't need to hear about the nitty-gritty of your positioning (executives, conference audience, outbound sales leads wanting to know 'what's out there'). That's a lot of x-axis movement of your claw (the change story), with *maybe* a short drop down into your solution (and how it's positioned), if at all.
- **More positioning, less change story:** Maybe they're well aware of the changes going on, are in the market and ready to buy (that's why they requested a demo!), and are just applying their own laser beam of attention to the specific features and weird edge cases they have. In that case, they might want to know precisely how you're positioned relative to competing solutions: how you're different, the specific value you offer, and where you fit in their mental map of the market. These folks could care less about the big industry changes.
- **A bit of both:** Or maybe they're somewhere in between, and you 'win' their attention with both context setting with your change story (the claw moving out), then an explanation of how your particular solution is positioned (the claw moves down), and perhaps (to extend the metaphor) you bring their attention back up to the overall trends and win the prize.

But it's only with our mental model of attention and our twin tools — the radar, or change story on the x-axis, and the laser beam, or positioning on the y-axis (to mix metaphors) — that we can accurately calibrate our message to our specific audience, especially if we want to avoid the story mismatches we discussed in **chapter 6**.

In the next few chapters, we're going to explore our two modes of attention in more depth, and then we'll see how it plays out in a real-life experiment of sorts.

19

The new hemisphere hypothesis

My thesis is that for us as human beings there are two fundamentally opposed realities, two different modes of experience; that each is of ultimate importance in bringing about the recognisably human world; and that their difference is rooted in the bihemispheric structure of the brain.

– *Dr Iain McGilchrist*[1 2]

So far in this book we've covered:

- **The intro:** The tl;dr introduction to Dr. Iain McGilchrist's new hemisphere hypothesis and the hierarchy of attention (**chapter 4**) that we're expanding upon here.
- **The hands-on part:** A basic deck where you may have applied the ideas of story 1 and story 2 to your own narrative sales deck, and you've hopefully given your Hemingway story and origin story (**chapter 9** and **chapter 10**, respectively) a bit of thought too.
- **The operating metaphor:** The claw game as our metaphor for 'winning' attention — go out, go down, grab attention.

It's now time to peel another layer from the clarity strategy onion. In this chapter, and indeed the rest of Part IV, we're going to further unpack what right- and left-brain attention means for your change story and your positioning. We'll also be drawing further on McGilchrist's work, dragging ol' mate kicking and screaming into the icky world of business.

> **Choose your own adventure:** We're spending the next few chapters in theory town. If that doesn't float your boat and you want something more practical, skip ahead to chapter 22 where we'll see how this theory played out in one B2B SaaS company's sales operations, or check out chapter 23 where we return to pitching different kinds of B2B SaaS products and look at three alternate deck formats.

McGilchrist saw me coming

First, let me briefly reintroduce McGilchrist who, as mentioned in **chapter 4**, almost single-handedly reintroduced the new hemisphere hypothesis into the public consciousness, particularly through his 2009 book *The Master* *and His Emissary*, which he also recaps in his 2021 multi-volume magnum opus, *The Matter with Things*. (I'm drawing largely from the 2019 edition of the former in this book, but I'll quote from TMWT, too.)

Those are some big, deep books. Fortunately, he also has extensive videos, podcasts, interviews, lectures, and even a documentary you can sink your teeth into if you're so inclined.

Let me stress again that I'm not *remotely* going to do justice to the science, or McGilchrist's decades or work, in this business-oriented tome. We're splashing in puddles while McGilchrist takes a submarine down to depths few humans have gone before. McGilchrist is a serious public intellectual, and we're very fortunate that his work is just *so* irresistibly perfect for our clarity strategy work here.

It turns out past management and marketing writers couldn't resist the lure of a juicy story of brain differences either, as McGilchrist tut-tuts in *The Master and His Emissary* (my bold):

> *Despite the recognition that the idea [of brain differences] has been* ***hijacked by everyone from management trainers to advertising copywriters****, a number of the most knowledgeable people in the field have been unable to escape the conclusion that there is something profound here that requires explanation.*

As someone sitting in the middle of the Venn diagram overlap of 'management trainers' and 'advertising copywriters' (I'm a copywriter turned consultant), I have to say that is pretty funny. Sorry, doc[3]

But I digress. Back to hijacking McGilchrist's ideas to tell better stories...

The bunk version

As we've previous discussed, the original left brain/right brain ideas you may be familiar with emerged in the '60s and '70s and turned out to be, as McGilchrist tells it in one short video, bunk.[4]

Everybody in the science world was kind of embarrassed by the theory and stopped talking about it, despite it living on as a zombie pop-culture meme, which I've found included corny-sounding business books about *left* or *right* or *whole brain* business. (And that includes this book, too, I guess. Seriously, Iain, apologies!)

Here, I again stress that my simple explanations are intended to be more illustrative of observations than descriptive of scientific reality, with the goal of extracting some value from this profound mental model of attention. A model that we can understand through the example of... a bird?

Eating without being eaten

> *Every animal, in order to survive, has to solve a conundrum: how to eat without being eaten. It has to pay precisely focussed, narrow-beam attention that is already committed to whatever is of interest to it, so as to exploit the world for food and shelter. Put at its simplest, a bird must be able to distinguish a seed from the background of gravel on which it lies, and pick it up swiftly and accurately; similarly, with a twig to build a nest. Yet, if the bird is to survive, it must also, at one and the same time, pay another kind of attention to the world, which is the precise opposite of the first: broad, open, sustained, vigilant attention, on the lookout for predators or for conspecifics, for friend or foe, but also, crucially, open to the appearance of the utterly unfamiliar – whatever may exist in the world of which it had no previous knowledge.*
>
> – Dr Iain McGilchrist[5]

Attention evolved in nature like everything else, and nature is a hostile place. Our brains developed to survive in that hostile environment, including the ability to *eat without being eaten*. This has driven — so the theory goes — the development of two very distinct hemispheres in our brains, hemispheres that have only become more asymmetric over the course of human evolution.

So if there are two different hemispheres, they must be doing at least somewhat different things, right?

By studying everything from how animals feed to how humans respond after a hemisphere-specific stroke, researchers have been able to infer not just *what* each hemisphere does, but *in what way*. McGilchrist, as per the quote above, likes to use his illustration of a bird to convey this idea. Imagine you have a small bird picking at seeds on the ground. The bird has to do two things simultaneously:

1. **Narrow attention:** Maintain a narrow focus on picking out seeds from the rocks and pebbles on the ground, i.e., *eating*.
2. **Broad attention:** Maintain a wide, open awareness of their environment, especially for the job of *not being eaten by predators*. (And perhaps finding a mate while they're at it.)

They are the two very different modes of attention at work: one highly specific, focused, and manual, the other much broader — connected to (and vigilant about) what's going on in the environment the bird finds itself in. While the human version is vastly more complex, it turns out that we manage the same *kinds* of attention, and that is one of the key differences between our hemispheres:

- **Radar:** Our right hemisphere gives us a broad, global, vigilant awareness of the environment.
- **Laser beam:** Our left hemisphere gives us a sharply focused, local, narrow attention to detail.

That's the global and local *hierarchy of attention* that we're simplifying here, tongue somewhat in cheek, to right-brain radars and left-brain laser beams. (Seriously, Iain, I'm so sorry.)

From insight to mental model

> *Whatever else, the hemisphere hypothesis helps us become aware of who we are, what it means to be a human being at all.*
>
> – Dr Iain McGilchrist[6]

"What it means to be a human being at all." That's how profound of an insight the new hemisphere hypothesis is — it speaks to the very nature of the human experience. But what's the connection to our more mundane work around clarity strategy? What makes it not just a general insight but a mental model we can use to drive our narrative and positioning?

The more I heard McGilchrist talk about hemispheric differences — the mechanical nature of the left hemisphere and its orderly, sequential task focus, its focus on tools, reductive abstractions, and precision, versus the connected nature of the right hemisphere, where the synthesis and, well, *story* (my description) happens — the more I couldn't escape the thought that this described the two story schools of thought — positioning and change stories — I had been wrestling with in my own work.

Two schools, two brains

> *[T]he most fundamental difference between the hemispheres lies in the type of attention they give to the world.*
>
> – Dr Iain McGilchrist[7]

If, with our stories and narratives, we're all targeting humans with similar brains, but those brains are split, and that results in two different forms of attention, then it stands to reason that two modes of telling those stories would arise, wouldn't it?

And if we're trying to tell stories that resonate, shouldn't we frame our story to match the way our audience is listening, per the kind of right- or left-brain attention they're using?

That, to me, is the art of developing a clarity strategy, as I outlined in the very first chapters of this book, the art of which is, in some ways, just connecting the dots between McGilchrist's work and the two broad, pre-existing schools of thought in the B2B SaaS GTM world, which we'll explore next.

What's amusing is that each of these schools of thought can give the other some serious side-eye. (Sometimes deservedly so!) But I believe this mental model of attention both:

1. Explains why the two schools exist.
2. Helps us *direct* our attention internally in how we work and *engage* our customer's attention externally, meeting them where they're at.

That is, our mental model of attention sheds light on a whole class of problems bigger than any one approach to story and startups. Nevertheless, we need to understand specific approaches to understand the whole. Therefore, the next two chapters briefly introduce each school, and we'll continue fleshing them out as we go. First: the world of change stories.

Notes

1. McGilchrist, Iain. *The Master and His Emissary* (p. 3). Yale University Press. Kindle Edition.
2. McGilchrist goes on to say, "It follows that the hemispheres need to co-operate, but I believe they are in fact involved in a sort of power struggle, and that this explains many aspects of contemporary Western culture." I'm deliberately avoiding McGilchrist's philosophical and cultural arguments in this book, but they're nevertheless worth your time if you're curious!
3. This is all very tongue-in-cheek. McGilchrist is a very good sport about these things and understands that his theory, if he is right, should affect everything. He has appreciated other folks riffing on his ideas, like Orlando Wood in the context of advertising, for example, as this 2023 interview on the Uncensored CMO demonstrates: **https://uncensoredcmo.com/108**
4. I'm paraphrasing the short 2011 video "The Divided Brain" by RSA ANIMATE here, which you can check out at: **https://www.youtube.com/watch?v=dFs9WO2B8uI**
5. McGilchrist, Iain. *The Matter With Things: Our Brains, Our Delusions and the Unmaking of the World* (pp. 36-37). Perspectiva Press. Kindle Edition.
6. McGilchrist, Iain. *The Matter With Things: Our Brains, Our Delusions and the Unmaking of the World* (p. 52). Perspectiva Press. Kindle Edition.
7. McGilchrist, Iain. *The Master and His Emissary* (p. 4). Yale University Press. Kindle Edition.

20

Story 1 — change stories & 'strategic' narratives

> The right hemisphere understands narrative. The left hemisphere, if offered a story whose episodes are taken out of order, tends to regroup them so as to classify similar episodes together, rather than reconstruct them in the order that has human meaning.
>
> *- Dr Iain McGilchrist*[1]

"The right hemisphere understands narrative" is a profound handful of words that helps explain why narrative — i.e., a story of events unfolding over time — is so powerful for engaging attention in a way that bombarding people with facts and figures is not.

"Narrative," in B2B startup storytelling, manifests in a variety of ways, the simplest of which is "we built a thing." Not a very engaging or profound narrative, granted (!), but it represents an event in time, in that your thing now exists.

The next step is "we built a thing *for your specific problem,"* and that is, for those that have that problem, inherently more interesting. If you can *position* your product around that need, you can unlock considerable demand. That's what we will explore in the next chapter, and that's where most startups should focus.

But for you extra-ambitious types, you probably want to go bigger still. You may have heard folks say there are ways to tell a story that's bigger than a simple problem/solution. A way that might make your startup more relevant, more necessary, and more desirable for a larger or more lucrative segment of the market. That's what we're going to explore in this chapter.

> **Description vs prescription:** In Part II and Part III of this book, where we looked at your narrative, origin story, messaging, and names for things, I was being very prescriptive. There was even a checklist! But I want to be clear that here in Part IV I'm attempting to be much more *descriptive,* and set up some deeper ideas about story 1/story 2 that we'll use as a platform to explore how you might stretch, extend, or refine your narrative and positioning in a more comprehensive way.

Narratives about narratives

As your startup matures and you get better at selling, you'll realize you need a bigger story to close more deals and win bigger accounts. You need a bigger 'so what?' for those juicy upmarket deals — a story big enough to capture the attention of execs with the authority to roll out tools across their entire function or company.

Or perhaps you want the market to come to you. You want a story that transcends your product, a story that's so visionary it forces people to rethink the way they've always done things, and may well suggest not just a new tool but a whole new *category* of tools... just like yours.

Either way, this requires a bigger story; a grand narrative about *change* happening in the world, change that demands attention, new ways of working, and new solutions to survive and thrive.

Or, at least, that's how folks who tell these kinds of stories sell it.

Flavors of change

These kinds of change stories come in a variety of flavors:

- **The world:** The change that's happened *in the world* that creates the conditions for your offer, which we will explore in this chapter. (Perhaps AI is taking over the world, or new legislation is creating an urgent need for a new compliant solution, for example.)
- **The category:** The change *you* are bringing about thanks to the new category you're creating, with a new playbook to boot, off the back of your innovation. (We'll look at category creation in **chapter 32**.)
- **The transformation:** When you sell enough software, you can generate your own transformative wave of FOMO, perhaps from bottoms-up adoption from business phenomenons like, say, Slack. If everyone else is using Slack (or the current tool *du jour*), why aren't you? (We looked at Slack's famous strategy memo back in **chapter 12**).

That's not an exhaustive list. Just note that *external* change is the common denominator. That change is instrumental in narrative, and narrative is a fundamental way of capturing right-brain attention. This is about telling stories where change is coming for *you*, dear prospect. Get on board or get left behind.

The strategic narrative

These kinds of change stories are sometimes referred to as *strategic narratives*. The term "strategic narrative" appears to have been borrowed from politics and international relations, but it has been popularized in business contexts too. For example, in his 2016 *HBR* piece[2], Mark Bonchek wrote about strategic narratives in business, arguing:

> *It's not enough any more to say "we make widgets." [...] You want a story that inspires employees, excites partners, attracts customers, and engages influencers. A story that is concise but comprehensive. Specific but with room to grow. One that defines the company's vision, communicates the strategy, and embodies the culture. [...]*
>
> *A strategic narrative is a special kind of story. It says who you are as a company. Where you've been, where you are, and where you are going. How you believe value is created and what you value in relationships. It explains why you exist and what makes you unique.*
>
> *This doesn't come out of the usual competitive landscape, customer interviews, and whiteboard sessions. It takes a different approach and a shift in thinking led by the leadership team.*

As examples, Boncheck cites IBM's "Building a Smarter Planet" campaign and its associated CEO thought leadership; Nike's mission "to bring inspiration and innovation to every athlete in the world"; and Starbuck's on-and-off relationship with being a "third place."

I can see the point of these examples — a strong, shared sense of what a company is about can be a powerful tool. That said, if an enterprise technology company, a shoe manufacturer, and a coffee chain walked into a bar, I doubt they'd see much common overlap in their respective 'strategic narratives.'

Nevertheless, there are some high-level insights tease out here:

- **Narrative as identity**, where a master narrative[3] brings clarity to what the organization does, or as Boncheck puts it, "Why you exist and what makes you unique."
- **Narrative as change over time,** or as Boncheck writes, "Where you've been, where you are, and where you are going."

There's a dual narrative of narrative at work here. When we speak about sales narratives that capture your story, as we looked at in **chapter 9**, we're often trying to create something of a master narrative that speaks to your fundamental identity: something that captures *both* story 1 and story 2 aspects in a tight pitch that makes sense to buyers (so it's not internal navel-gazing).

Some folks, however, lean far more heavily on the story 1 aspect of that master narrative when it comes to B2B technology, where external change is the fundamental driver of the master (or 'strategic') narrative.

Selling bigger

In our little B2B SaaS neck of the woods, Andy Raskin is perhaps the biggest proponent of this approach, having done considerable work to popularize the concept of the strategic narrative as a sales and leadership tool for technology companies and beyond. Raskin's approach has the ambition of Boncheck's ("a story that inspires employees, excites partners, attracts customers, and engages influencers," per Boncheck) but draws heavily from sales decks, especially Zuora's, as we'll see. This 'strategic narrative' approach is perfect for highlighting the pros and cons of 'story 1' change stories in your clarity strategy.

The origin story

Raskin has written extensively about his particular formula for a strategic narrative, including this anecdote about its origin and its effectiveness[4]:

> *[T]im called to say he was seeing promising changes in how prospects at large enterprises were reacting to his new deck, which we drafted together based on Zuora's template. For one thing, prospects opened up much faster about the challenges they were facing. Also, he said the new pitch was more effective at engaging CFOs and other senior gatekeepers.*

Note two critical details:

1. These 'promising changes' the seller was seeing was in the context of selling to *large enterprises.*
2. The pitch was effective at *engaging CFOs* and other senior folks.

That is, the context wasn't, say, high-velocity inbound SMB sales. That's a crucial distinction for reasons that will become clearer as we go.

Instead, the folks Tim, in Raskin's story, was selling to were folks who were open to what I call *whole stories,* the whole three acts — setup, conflict, resolution — and that's what the strategic narrative proposes, a whole new narrative. That can make them powerful story 1 tools. But like all powerful tools, using it the wrong way can be dangerous.

> **Skip to the deck:** If you want an example of how B2B sales pros tell these kinds of change stories, skip to chapter 23 ("3 alternate decks"), where we'll look at one pitch outline I call 'The Ray of Sunshine,' as well as completely different approaches that may be more appropriate in high-velocity situations.

Strategic narrative in B2B SaaS

Raskin, as he describes it, modeled his 'strategic narrative' approach on Zuora's mid-2010s sales narrative. Zuora coined the idea of the 'subscription economy,' and their pitch really was a whole story about a big change in the world. It was the full multi-act play: a new subscription era was here with big stakes, strong urgency, winners and losers, a promised land, ways to get to that promised land (i.e., Zuora's platform), and proof the company (Zuora) could help you succeed.

As Raskin tells it, Tim, Raskin's friend selling to enterprises in the anecdote above, described Zora's deck as the "greatest sales deck" he'd ever seen. Raskin ran with that line ("The Greatest Sales Deck I've Ever Seen") as the title for his Medium piece[5], the article went viral, and the strategic narrative for B2B SaaS was born.

The strategic narrative formula

There's a delicious bit of internal logic in how Raskin describes how he sees a strategic narrative working and how he narrates finding it. A strategic narrative should, in Raskin's view[6] (and I'm paraphrasing):

- **Change:** Start with identifying an 'undeniable change in the world' that's independent of your product or company and creates big stakes and strong urgency.
- **Promised land:** Describe a 'promised land' awaiting prospects who will thrive if they're on the right side of this change (with loserdom and despair awaiting those that don't).
- **Magic gifts:** Offer the prospect 'magic gifts' by way of your product and its features that help the prospect on their journey to overcome their obstacles and reach the promised land.

And in Raskin's own narrative about this kind of sales narrative, he describes:

- **Change:** He starts with an external event — cutting-edge companies (at the time) like Zuora were selling with a particular formula, according to his salesperson friend. (I.e., this isn't just Raskin's opinion.)
- **Promised land:** Drops strong hints of a promised land — enterprise deals are on offer, and the deck landed 'the biggest deals' of the seller's career.
- **Magic gifts:** Offers an explicit 5-step formula to follow — the 'magic gifts' to help you get there.

That's all very clever. Raskin is a *fantastic* communicator, and it's worth noting how straightforward and relatable he makes this approach[7] through metaphor (promised lands, magic gifts) and near-universal references (popular Hollywood movies, as we'll see). We touched on the importance of metaphors (**chapter 15**) and naming things (**chapter 17**) earlier, and you can see it in full effect here. One company, one deck, one point in time, and it's spun out into its own cinematic universe. And why not: narrative is an effective way to engage right hemisphere attention, after all.

There's much to learn from this approach, and I use parts of this approach when crafting sales narratives all the time. 'What's changed in the world' can be a very powerful framing device and I have enormous respect for Raskin for identifying, developing, and popularizing this idea of the strategic narrative in SaaS in particular. It's a fantastic example of the 'story 1' approach that targets that right-brain radar, and he deserves huge credit for almost single-handedly blazing that trail in the tech world.

But there are inherent limitations with this approach that we need to be aware of, too.

> **Pitch by stage:** Your sales pitch will vary greatly depending on your stage as a startup, too. Copying mature, we-close-enterprise-deals-on-the-reg approaches wholesale can, as an early-stage startup, be a bad idea simply because you haven't done the learning — or product development — to get to that point. We'll discuss this more in chapter 38 when we look at message testing and early-stage sales.

Epics and fairy tales

The most obvious limitation to this approach is the context I highlighted above — this is a narrative technique derived from enterprise sales to C-suite folks, and if they're not who you're selling to, then you need to tread carefully. That's not to say you can't use it in other contexts — we'll see how it lands with outbound rank and file leads in **chapter 22** for example — just that it needs to match prospect attention.

The other limitations are found in the metaphors Raskin uses. Raskin, writing in the same Medium piece, says that:

> *If it's not clear by now, successful sales decks follow the same narrative structure as epic films and fairy tales. Your prospect is Luke, and you're Obi Wan, furnishing a lightsaber to help him defeat the Empire. Your prospect is Frodo, and you're Gandalf, wielding wizardry to help him destroy the ring. Your prospect is Cinderella, and you're the fairy godmother, casting spells to get her to the ball.*

Here, dear reader, I might gently suggest that we're selling corporate software to businesses, not blockbuster movie tickets to pimply teens, and we shouldn't get too carried away with the metaphors.

Again, I have enormous respect for Raskin and his insights into this work, insights he very generously shares publicly through his writing and his podcast[8], but let me also very gently suggest that we take generalizations about 'successful sales decks' with a grain of salt. (Do *all* successful sales decks follow the epic-film-and-fairy-tale formula? Really? Across all of B2B? And do epic films and fairy tales all have the same narrative structure anyway?[9])

Sales decks are highly contextual, as we'll see, and as much as I love a good narrative, we need to have some immunity against narratives about narratives, too, especially as Raskin's is a very good one.

> **"The dumbest and most effective sales pitch I've ever heard":** In the next chapter, we'll look at the bizarro story 2 twin of the 'greatest sales deck' — the 'dumbest sales pitch' or, perhaps most *obvious* sales pitch, from a ~$50B cloud juggernaut. This pitch, from a sales powerhouse, demonstrates that successful sales decks certainly don't all follow the epic-films-and-fairy-tales formula.

Story 1 fundamentals

Now, you might think it's a bit rich for your humble author, with a book boiling the brain down to 'radar' and 'laser beam' attention, to be picking nits about the metaphors of others. I take the point. But what we're interested in here is the underlying fundamentals that make a strong, context-setting right-brain story 1 narrative tick.

And those fundamentals are fairly simple:

- **Change over time:** Movie analogies aside, there's a lot to be said for simply putting your solution in a broader context with a narrative of events unfolding over time. That helps explain 'why now' in a way that simply announcing your concept and dropping it on a prospect doesn't.
- **Scale of impact:** Your tool solves a specific problem, sure, but what are the broader, second-order implications in light of this broader context? What's the more existential, change-is-coming impact for a given role, function, or company?
- **Infectious ambition:** Less tangibly, a good change story can carry an infectious ambitiousness that gives folks access to a category of thinking they'd never otherwise be exposed to, freeing them to think thoughts they simply wouldn't have otherwise. There's something to be said for the "high-return activity of raising others' aspirations."[10]

This is the kind of first-principles, nuts-and-bolts approach to change stories that make them work in specific B2B contexts. That tends to mean enterprise accounts and C-suite buyers, but it can be any context where folks have their attention attuned to those sorts of story 1 narratives. We want to match our pitch to the mode of attention we're trying to engage. That's our fundamental starting point.

> **Stretching your change story:** We're discussing the basics of change stories here, and this is just the tip of a much larger iceberg we're going to explore. For example, in chapter 30 we'll see how *secular change* is the driving force of change (and category) stories, and in chapter 31 we'll see how secular change combined with a *new playbook* drove the narrative success of Salesforce, HubSpot, and Drift in CRM.

Have you got that story to tell?

There are two very simple questions you need to answer if you want to run with this kind of story 1 change story:

1. Is that where your audience's attention is at?
2. Do you actually have that story to tell?

If it is, and you genuinely do, great, more power to you. If you're the subject matter expert, and your audience is looking for what's 'out there,' then buyers may well find the new mega-trends you've identified insightful.[11] Crank up your change story, give them the three-act narrative, and see what happens.

But if you don't, or your audience's attention isn't there (and we'll see an example of this in the narrative experiment in **chapter 22**), simply confecting some made-up macro change to hitch your wagon to isn't likely to move the needle. Plenty of deals are done without generic stories about how the economy has changed, or technology has changed, or the way we do business has changed, or whatever. In fact, if you try and sell with a narrative that's overly ambitious for a given role in an org you're targeting, the narrative can be an *impediment* to getting deals done.

Let's talk about being real for a minute, too.

If you're going to talk about *the future of work*, or *the future of commerce*, or *the future of selling*, or whatever, then you better actually have the necessary expertise from what you or your customers have witnessed firsthand. If you're telling everyone to get on board with the new game, the new change, the new trend — *we're creating a new category here, baby!* — then there needs to be an actual, real trend there.

I know that sounds like I'm pointing out the bleeding obvious, but founders often get trained to pitch VCs on big change stories that are, shall we say, *highly speculative*, and then they let that bleed into their sales pitch, often resulting in a story mismatch as we discussed in **chapter 6**. Specifics sell, and you need to calibrate your story to the actual insights you have at your disposal, which may actually be a 'story 2' positioning pitch, as we're about to see.

There's a time and place for big change stories. Change stories resonate in the right context for a reason — they can open the customer's eyes to what's possible, give them more to aspire to, and create a greater goal than "buy another tool."

But you've got to get the story right.

> **Product story or management story?** The story 1 approach comes with risks — it's not a magic win-deal deck formula. One risk, especially in complex sales, is pitching a change story that's more about management than product. I imagine there's a big graveyard of category ideas about X intelligence or Y collaboration that were really management problems, not tool problems. Sometimes putting the management change before the product is putting the cart before the horse. Instead, consider doing it the other way: have folks come for the tool, but stay for the (management) school.

Exploding causality

So far we've established that change stories, in all their forms (i.e., not just the strategic narrative formula), are very powerful but require nuance and a dose of realism if you're going to use them effectively.

But the fact they can be so powerful[12] means that nuance often goes out the door and folks run to the biggest story possible, sometimes to the benefit of their ego but to the detriment of their chances in the market.

Therefore, I have another analogy that suits these kinds of stories, inspired by the Very Expensive Agency I alluded to back in **chapter 7** when we discussed investing in your clarity strategy. In my view, big, the-future-of-X change stories are a bit like explosives — they're very powerful, should only be used sparingly, and if you don't know what you're doing, you're likely to blow yourself up.

And the most common way people blow themselves up with change stories is by getting causation back to front. It's true that successful companies often ride a change in the world to fame and fortune, and they often get credit for naming the trend or coining a category name. We'll see how this works when we explore secular change and category creation in **chapter 29**.

The trouble is that finding the right trend to ride is very hard — there's a massive degree of luck and timing involved, for one, and it can take a lot of resources to sell the new trend *and* your solution. Often it's better to start small with a wedge and work your way into the market from there.

But so often, we get monkey-see, monkey-do imitation instead, where people end up imitating the *artifacts* of successful companies rather than trying to understand the *conditions* that drove their success.

Nowhere was this more evident that in the farcical put-it-on-the-blockchain cryptocurrency era, where folks seemed to want to LARP what a big shift in the world would look like without stopping to think about whether there actually was a 'there' there. (There wasn't.) This was an industry project that was almost as narrative driven as it was devoid of meaningful use cases (excluding the classic trifecta of crime, fraud, and speculation). All story 1, no story 2. And it's telling that those that remain influential in the crypto world, such as it is, are those that are best at narrative ('the glorious future still awaits!'), not building.

The same has happened in startups, albeit with far fewer deleterious social effects, with a litany of failed 'category creation' attempts, too, but obviously few folks hear about those because they never catch on. Founders and investors went for the big ego-boosting story 1 change story when they should have been getting into the weeds to understand story 2 reality.

This can be the danger when narrative escapes the lab — and the humble sales pitch — and becomes an all-encompassing *raison d'etre* untethered from reality.

That's the danger, but don't forget the nuance.

Change stories are everywhere, and they're a crucial pattern to understand because you'll start to see them in all kinds of places (including in later chapters of this book!). They can be incredibly useful for setting context and helping buyers understand why your solution matters. I borrow from the "strategic narrative" formula in decks all the time. Sometimes new categories *do* emerge and I recommend folks think in those terms. And some level of external change has to be afoot for us to drive interest in our software, especially if we're targeting right-brain folks.

But if you start breathing your own exhaust, you risk blowing up both your company, a lot of capital, and almost a whole industry.

So what do we do? Do we abandon narrative and change stories? Not at all — they'll keep coming because technology keeps changing (hello generative AI!) and the world (and businesses in particular) will inevitably want to both adopt and adapt. But when it comes to *your* change story, you've got to relentlessly verify the reality of it, not just imitate formula. And that means focusing locally.

> **Vision without verification is vanity:** That's the bluntest way I can put it, and this plagued the ZIRP era. The irony was — and is — that the lack of willingness to truly test and refine a vision is what stops founders from achieving greater success. If you want the big result, swallow your ego and build an identity on being curious, not 'correct' when it comes to your original vision .

Verify locally, pitch globally

> *The importance of the global (right hemisphere)/local (left hemisphere) distinction cannot be overstated. [...] The right hemisphere is on the lookout for, better at detecting and dealing with, whatever is new.*
>
> – Dr Iain McGilchrist[13]

'Whatever is new' is fundamental to what we're discussing here. Sometimes breaking prospects out of their original expectations and giving them the full, right-brain, three-act play gives them something new — not just something new that makes you sound smart, but something that changes the way they think about *their own situation* and how they should therefore act (and buy).[14]

That can be very powerful, and it that doesn't have to be that extreme, either. I'm all for sprinkling a bit of external, why-now change in your clarity strategy and sales narrative where appropriate. The nature of the attention claw game is to go at least a little bit out on the x-axis before going down the y-axis after all, and why-now change was an early slide in our introductory narrative deck in **chapter 9** for that reason.

But context is a two-way street.

Rather than thinking about your change story in purely top-down, external-in terms, think about what it looks like bottoms-up. What experiment have you run? What have you personally lived or experienced? What have you verified with your customers or through your own experience that extrapolates back out to a greater change, or to a common need or pattern for folks in similar circumstances? Think about Stuart Butterfield's grand narrative for Slack, as we saw in **chapter 12**. This wasn't pure vision where he woke up one day and sketched out a plan for world domination on the back of a napkin — his team stumbled across the usefulness of the internal tool they built, and therefore they had local verification. The vision came next.

Or take Loom, as we saw in **chapter 10** ("The 'Hemingway story' micro-pitch"). They spotted an interesting unintended use of their tool and built a vision around that. Vision is built on verification, not empty speculation. But how do you know what to verify? You've got to deploy your own left-brain attention. You've got to understand your *positioning*.

Objection!

This brings us back to the world of the left hemisphere.

When I encountered the world of story 1 change narratives — be they strategic narratives, category creation stories (as we'll explore over four chapters from **chapter 32**), or just general transformation stories — I knew there was another school of thought that would strongly object to the characterization that 'successful sales decks follow the same narrative structure as epic films and fairy tales,' i.e., that successful sales decks *must* be story 1 change narratives. This school of thought would see things very differently. Almost as though they were seeing things from a completely different perspective...

This is where things get a little spicy — well, spicy for a business book.[15] Two different, seemingly opposed schools of thought, two different modes of attention, two different hemispheres in our heads... what's going on? And what can we do if we, you know, just want to know what works?

Change stories on the x-axis

> *[As humans] we need to maintain the broadest experience of the world as it comes to us. We need to be going out into the experiential world along the horizontal axis, if you like. By contrast, in order to control or manipulate we need to [...] map the world from the vertical axis.*
>
> –Dr Iain McGilchrist[16]

In this chapter, we've been exploring a high-level introduction to the 'horizontal axis' of your clarity strategy, to borrow from McGilchrist.[17] And while external change *is* a vital ingredient for the horizontal x-axis of your clarity strategy, clarity strategy is a claw game; it's not *all* x-axis. There's a y-axis too.

But how does the y-axis of attention work? Going deep on a specific problem is how plenty of startups start and succeed after all. Next, let's explore that other school of thought: the positioning school.

Notes

1. McGilchrist, Iain. *The Matter With Things: Our Brains, Our Delusions and the Unmaking of the World* (p. 49). Perspectiva Press. Kindle Edition. (Note: I've expanded McGilchrist's RH/LH abbreviations in this quote.)
2. Bonchek, Mark. "How to Build a Strategic Narrative," *Harvard Business Review*, March 25, 2016. **https://hbr.org/2016/03/how-to-build-a-strategic-narrative**
3. The idea of a 'master narrative' has been discussed in the context of the individual in psychology. See, for example: McLean, K. C., & Syed, M. (2016). Personal, master, and alternative narratives: An integrative framework for understanding identity development in context. *Human Development, 58*(6), 318–349. **https://doi.org/10.1159/000445817**
4. Raskin, Andy. "The Greatest Sales Deck I've Ever Seen," *Mission.org*, September 16, 2016. **https://medium.com/the-mission/the-greatest-sales-deck-ive-ever-seen-4f4ef3391ba0**
5. ibid.
6. ibid.
7. Something your humble author could learn from!
8. You can find Raskin's podcast, *The Bigger Narrative*, here: **https://andyraskin.com/podcast/**.
9. As an aside, dear reader, there's a whole genre of storytelling books out there by dudes that *love* to claim they've found the universal key to *all* stories humankind has ever told — it's three acts, or five acts, or it absolutely must hit these points, or the hero must go on this kind of journey. Story *tropes* are definitely a thing, but where this obsession around finding the universal rules of storytelling comes from, I don't know. The "universal" rules all turn out to be different and are usually so abstract as to be meaningless, effectively saying a story must have a start, middle, and end, which is not exactly a revelation. I mention this weird little genre just to say that you shouldn't be afraid of telling your story your way. If you can find a format that you can bring real energy and conviction to, and prospects respond, awesome. Take the "rules" — including anything I write here — with a grain of salt.
10. Cowen, Tyler. "The high-return activity of raising others' aspirations," *Marginal Revolution*, October 21, 2018. **https://marginalrevolution.com/marginalrevolution/2018/10/high-return-activity-raising-others-aspirations.html** (via Patrick McKenzie **https://twitter.com/patio11/status/1254069302589710336**).
11. It's worth noting sales classics *The Challenger Sale* and *The Challenger Customer* both discuss the importance of 'commercial insight' in the sales process, and the hierarchy of attention gives us a framework for what *kind* of insight works in what context.
12. We'll see in later chapters how powerful narrative is even in a meta sense — ideas like category creation and 'evidence-based' marketing are founded on powerful change narratives.
13. McGilchrist, Iain. *The Matter With Things: Our Brains, Our Delusions and the Unmaking of the World* (p. 48). Perspectiva Press. Kindle Edition. (Note: I've expanded McGilchrist's abbreviations in this quote.)

14. Tip o' the hat to *The Challenger Customer* for that insight.
15. Which is, let's say, about 0.01 on the Scoville scale.
16. McGilchrist, Iain. *The Master and His Emissary* (p. 22). Yale University Press. Kindle Edition.
17. Yes, this book is essentially me stealing McGilchrist's ideas and giving them silly names like claw games, lasers, and radars.

21

Story 2 — positioning

> The pitch we heard from Snowflake was both the dumbest and most effective sales pitch I've ever heard. We were told that it was the same as Redshift—and really, the same as Postgres—but big, fast, and stable. [...] There were no forced mentions of digital transformations, distributed systems, or, God forbid, the blockchain. They never even told us what we could use Snowflake for. We were told to just keep doing what we'd been doing, except if we did it with Snowflake, they'd host the whole thing, back it with bottomless storage, and run our queries faster than ever before.
>
> — *Benn Stancil*[1]

"The dumbest and most effective sales pitch I've ever heard" is the perfect way to introduce the other school of SaaS sales storytelling — the positioning school.

In the previous chapter, we looked at Andy Raskin's account of the 'greatest sales deck I've ever seen' from Zuora, the structure of which was apparently successful when pitching enterprise and C-suite buyers, CFOs in particular. Raskin uses this as the basis of his "strategic narrative" — a story 1 change pitch.

However, I knew that wasn't the only kind of sales deck that could win deals. I knew there was a positioning school of selling, too, where the point of your sales pitch is to position your product in the mind of the buyer such that it is, to borrow a phrase, the *obviously awesome* choice. (IYKYK. We'll get there.)

Therefore, in this chapter, we're going to go from 'the greatest sales deck I've ever seen' to 'the dumbest and most effective sales pitch I've ever heard' and lay the platform for a very deep exploration of story 2 positioning in B2B SaaS.

Remember: two brain hemispheres, two modes of attention, and two approaches to story (story 1/story 2) for engaging that attention.

Let's explore the story 2 approach by going back to our opening quote from engineer, CTO, and co-founder Benn Stancil, who was looking for a database vendor and found Snowflake.

The Snowflake pitch

As Stancil tells it[2], he ran into trouble with Amazon's Redshift data warehouse, so he and his team decided to go shopping for an alternative. They take a call from Snowflake "out of obligation" and hear the pitch, "the dumbest and most effective sales pitch" he had "ever heard."

The pitch works — they go with Snowflake, drop it in as a 1:1 replacement, and then they start finding new things to do with it, just as, they imagine, Snowflake "undoubtedly anticipated."

Why did it work? Well, Snowflake had the tech for one, but why did Stancil and his team choose them and not a competitor? Especially when Stancil says that Snowflake's major competitor, Databricks, may well have been a better fit? Why didn't they see that? Stancil says they "never connected these dots during our evaluation because the story Databricks told was buzzwords. Snowflake's was boring, and all we wanted—at least at first—was boring."

Catch that? It came down to the story the two companies told. Databricks' was "buzzwords"; Snowflake's was "boring", but boring was what they wanted. But what does boring mean here?

- **Boring meant it *wasn't* a major change.** There was no behavior shift required — it was positioned as a 1:1 replacement but "big, fast, and stable."
- **Boring meant there wasn't a new 'whole' story.** This wasn't a whole new narrative, a whole new three-act play with new winners and losers; it was just a new ending: use us, not them, for what you're already doing.
- **Boring meant it was easy to buy:** There's a lot to be said (as we'll see in later chapters) about simply being easy to buy. To Stancil, Databricks looked like it "was built for people smarter than me. Instead of a demo or product screenshots, I found technical papers explaining something I didn't understand. The case studies talked about real-time enterprise workloads, streaming Spark applications, and distributed, highly-scalable data science model development. I needed a database."

That is, "boring" meant that Snowflake had positioned themselves, through story alone, as the *obvious* (it was a drop in replacement) and *awesome* ("big, fast, and stable") choice *for buyers just like Stancil.*

That is incredible story 2 positioning at work. It's miles away from the story 1 change narratives we discussed last chapter.

But let's keep a few things in perspective:

- **The buyer:** In the previous chapter, the pitch format apparently succeeded with enterprises and CFOs; here, we've got a hands-on buyer who is using the product.
- **The pitch:** This was Snowflake's pitch *to this particular buyer.* I have no idea if they have different style pitches for different types of buyers (though I'd love to know!).

- **The competition:** It's not as though Databricks is doing poorly — they're a very successful business, too,[3] with a $43B valuation as of September 2023[4].

That is, we need to be careful not to over-interpret this story. I'm setting up the story 1/story 2 binary not to pit them against each other, but to show that both can work in certain circumstances. The simple point here — and the art of clarity strategy more broadly — is that our pitch needs to be derived from (a) the mode of customer attention we're targeting and (b) what's true about your startup.

Do your buyers just want boring done better, or do they want buzzwords and big, fresh concepts *if* you have that story to tell? It's a crucial distinction that will drive how you go to market.

Left hook

> *In humans the left hemisphere is designed for grasping, controls the right hand with which we grasp [...] and helps us manipulate, rather than understand, the world.*
>
> — Dr Iain McGilchrist[5]

If our right-brain change stories are about helping folks *understand* the world in a new way — through narrative — then our left-brain positioning stories are often about showing folks how to *manipulate* the world with a new tool in a new way.

This is great little test for your clarity strategy. Are you trying to give folks:

1. **A new way to *understand* the world**, with a right-brain, story 1 change pitch? That was the idea behind Zuroa's pitch — the world had changed, the subscription economy was here, and they had the product to help you play this new game.
2. **A new way to *manipulate* the world**, with a left-brain, tool-focused, story 2 positioning pitch? That was the idea behind Snowflake's pitch — just want a better database? That's us. Nothing has changed; this is just better technology for managing your data in the exact same way as you were doing it before.

Some situations call for the big change story — people need to understand something new about the world, and that's our right-brain, story 1 pitch. But some situations are about selling the boring-but-better, drop-in replacement — the tightly positioned story 2 pitch. The art of your clarity strategy is knowing when to use which type of story, knowing how to use each story type to its fullest (the rest of this book is about stretching out on each axis and what that means for your GTM program), and, if you have to choose, knowing which to make your strength.

> **What do we do with a breakthrough insight?** Indulge me for a moment of immodesty, dear reader, but it's worth noting that this is a breakthrough insight. No one (as far as I'm aware) has framed the humble sales pitch in this way before. Consider how many billions of dollars turn over each year on the basis of sales pitches that might be missing one or the other mode of attention. But now we know there are two kinds of attention and two types of stories for those modes of attention. Ultimately, that should have profound effects on not just pitching but the way GTM is run and the way startups on the whole are organized. I mean, I *think* it should, but it's really up to you. This is a new insight and fertile ground. There's loads of room for experimentation and ambition in how far you take it. I'd love to know what you find: luke@lukestevens.co.

Picking the right story

There are real stakes here. Imagine if Stancil and his team got the big, macro, three-act, change-in-the-world, in-a-galaxy-far-far-away, new-category, story 1 pitch. The seller would have lost the deal. This wasn't a buyer who was looking *out* at trends in the environment; they were looking *down* for the kind of solution that did exactly what they wanted, that met their specific needs. That's very story 2.

The verse can be true, too. Imagine pitching a CFO who is happy enough with their current solution but open to new possibilities. Giving them the "boring" 1:1 drop-in replacement story 2 pitch isn't going to be a huge motivator for them to switch and to go through the pain of change — pain that only gets bigger as their organization gets bigger. They may need a bigger story 1 change story, and without it, that might be a deal lost, too.

Therefore, it's not that Databricks (or Snowflake) were 'wrong,' perhaps they were just playing to their strengths. Snowflake was, at the time, run by the legendary, no-nonsense, sales-first Frank Slootman; Databricks was, as Stancil puts it, founded by "brilliant computer scientists turned reluctant entrepreneurs [who] were reinventing big data, which itself was reinventing the entire world." Perhaps they're both just selling the way that makes the most sense to them. But when there are tens of billions of dollars of market cap on the line, can you really afford to ignore one entire mode of attention, or half of folks' brains?

> **April Dunford and B2B tech positioning:** None of what I write on positioning would be possible without the work of the B2B positioning G.O.A.T., April Dunford, whose book *Obviously Awesome* I alluded to above. We will explore Dunford's how-it's-really-done-in-B2B approach to positioning later in chapter 37. From there, we'll launch into a deep exploration of positioning and GTM strategy, peeling off more layers of the clarity strategy onion as we go. In a choose-your-own-adventure sense, don't be afraid to jump straight to those chapters after this one if you want to stay focused on positioning.

Positioning-as-story vs. positioning-as-GTM

Before we get too far into the positioning weeds, indulge me for a brief digression, dear reader. Let's draw a distinction between positioning-as-story and positioning-as-GTM (Go To Market). Positioning is a huge lever for any business, one that can cut across the entire organization in a variety of interesting ways. While we're focused on positioning-as-story in this chapter, in the story 2 sense, we're ultimately building towards a fuller view of positioning as fundamental to your GTM strategy, given ultimately your positioning must:

- Match your best-fit buyers' jobs-to-be-done (**chapter 36**).
- Match your distribution strategy (**chapter 42**).
- Match the early/late adopters you're encountering (**chapter 40**).
- Match the market you're targeting (i.e., SMB vs. enterprise, as we'll see in **chapter 43**).
- Match your fundamental competitive strategy (**chapter 49**).

That is positioning as a GTM and core management challenge. It's something the founder or CEO needs to lead, and it's not something you can just delegate down the org to Product Marketing as they work up a new deck or messaging.

That said, positioning as story — as story 2 in your clarity strategy — may well be more of a product marketing challenge. In the early days, that still might be the domain of the founder, but as you grow and expand into new industries and use cases, you might delegate positioning-as-story *for specific verticals* down the org chart. And that's fine, just don't forget there's a vast strategic side to positioning, too, one that deserves serious consideration if you're going to get the whole GTM orchestra — and indeed the entire company — playing in harmony.

Arrows and positioning by refocusing your product: If positioning-as-story is ultimately about your sales narrative and messaging, positioning-as-GTM might be more about shaping the product or company to better fit the market. For early-stage founder Daniel Zarick, CEO of customer onboarding SaaS Arrows, tightening Arrows' positioning meant *lopping off 1/3rd of their product* and going all-in on HubSpot as the platform they supported.[6] That's a big management and GTM bet that illustrates why positioning, especially on the GTM side, is one of the founder or CEO's most important jobs. After all, they're usually the only ones with the authority to make those choices.

Story 2 positioning basics

Digression over; let's return to positioning-as-story, or story 2. Threading the story 2 needle first requires a basic articulation of your current position, and that's why we got hands-on as early as possible in **chapter 9** with your narrative deck. That's the starting point, and that lets you do things like test if your message resonates (as we'll explore in **chapter 38**).

Next, it requires understanding where you're catching prospects:

- Are they early or late in their journey?
- Do they understand your value?
- Do they match your ICP (Ideal Customer Profile)?

We want to derive the best positioning story that matches exactly where our prospect's attention is at. For some folks, by the time the prospect reaches sales, it's too late:

> *Our sales calls these days are often super specific questions. "I watched a bunch of your YouTube videos... can I do X and Y via the integration?" We still do a structured sales demo, but it's almost like solutions engineering half the time.*
>
> — Daniel Zarick, discussing sales on Twitter[7]

That's Zarick, co-founder of Arrows, again. (We touched on their big positioning bet above.) As far as positioning-as-story goes, if you're catching folks this late in their journey, your window for giving them a fresh insight *in the sales call* into what might work is very limited. This is an extreme example, but in the SaaS world, it's obviously not uncommon for folks to do their homework well before talking to sales. While not every prospect will enter your sales process as late in their journey as the Arrows example quoted above, they might not enter it much earlier, either.

This a broader trend that those who study B2B sales have noticed, too. The B2B seller's seller, Brent Adamson, co-author of *The Challenger Sale* and *The Challenger Customer*, wrote in HBR that (my bold):

> *For years, however, B2B buying has dramatically evolved to a far more digitally dominant buying behavior, rendering much of [the traditional, marketing-to-sales] commercial model not only out of date, but nearly obsolete.*
>
> *Consider the following data from Gartner research: In a pre-pandemic survey of 750 B2B customer stakeholders involved in complex "solutions" purchase within their organization, customers reported spending only 17% of their total buying time interacting directly with supplier sales teams. [...]*
>
> *As small as it is, however, that 17% of purchase activity allocated to supplier interaction (both virtual and in-person), represents all suppliers, not each supplier. So, if three suppliers are competing for the same opportunity, one can assume customers divide that time roughly equally across all three,* ***leaving any given sales team with a vanishingly small window of opportunity to interact directly with that customer — perhaps 5% or 6% of total buying time if they are lucky.***

This, Adamson says, prevents sellers from "materially impact[ing] purchase deliberations and bend[ing] customer preference toward their company's unique offering."

I don't know whether that's true for you (collecting qualitative pipeline insights about what stage folks are at in their buying journey might be a good idea), but can you see how *where* you catch folks in their journey dictates what kind of story might resonate? (And remember that even if you are catching folks late, you can still sell! You can ask what the prospect is trying to achieve and then try and reframe the conversation around your positioning and perspective. You don't have to simply order-take.)

More broadly, though, what does this mean for positioning?

Job 1 for story 2

The reality is that positioning is inevitably more than slides in a sales deck. It has to be, even from a communication point of view, if you're going to influence the buyer's journey, which may well almost be over by the time they talk to sales.

Therefore, while positioning-as-story is great, it's positioning-as-GTM where you'll really move the needle. That's why I've kept teasing the latter chapters in this book — that's why they exist, to help you understand and tackle this broader positioning GTM challenge.

This is, ultimately, about using positioning to find product/market fit (PMF). Even if positioning only addresses the M in PMF it still deserves just as much scrutiny as your product if you hope to find fit. In that sense, positioning, as part of your clarity strategy, is a tool to help you not just drive deals but drive the entire GTM approach of your company, and it must be reflected in everything you do — from product and R&D to marketing, sales, and success.

Positioning positioning

I'm trying to reposition "positioning" here so you can see how it's something far more powerful than a fill-in-the-blanks marketing exercise.

To go deep for a moment, let's think about what "positioning" really means. In this chapter, we've already seen how "positioning," as a term, gets used in various ways. Sometimes, it's the big bet-the-company strategy; sometimes, it's just refining your point of view and language in your messaging. (And sometimes, it's a practical set of steps, as we'll see later in **chapter 37**.) But this range of meaning stems from the fact that "positioning" is, at a very high level, a metaphor. It's a metaphor for finding a spot, or a 'position,' *relative to something else in some abstract space.*

We can then think about how we might collapse — or create — distance between those two points. We might try and move closer to our customers' needs while trying to create distance from our competitors through differentiation, for example, collapsing space and creating it. Both are acts of positioning.

This is an incredibly broad way of looking at positioning, but that's the point. There are lots of ways to move in this abstract space. When you think about a company's activities, many of them could be considered acts of positioning — getting closer to customers and further away from competitors. That's the art of GTM strategy. You might, therefore, 'position' your product or company along a variety of different dimensions, including competitive strategy (**chapter 49**), category (**chapter 32**), brand (**chapter 45**), attribute (**chapter 46**), and of course, market segment (**chapter 37**). There are as many ways to "position" a company as there are meaningful dimensions to compete on, and we're going to do our best to touch on most of them!

> **Positioning by metaphor:** Don't forget there's also positioning by metaphor that we looked at in chapter 15 when we discussed robot dogs as either killer canines or cuddly companions. Same tech,

> different positions, and therefore wildly different paths to market, product directions, and sales pitches.

The Seed Finder 10000®

Ok, let's take a breath, that was the deep stuff. We got a little abstract there. Let's come back to our more straightforward world of positioning-as-story, and return to our theme of right-brain, story 1 change pitches vs. left-brain, story 2 positioning pitches. This time, with a theoretical example that articulates those exact differences.

Imagine we're selling products to the pigeon we used as an illustration of left- and right-brain attention in **chapter 19**. (This will make sense, I promise.) Let's call our pigeon, err, Peta. (Stay with me!)

Peta has two main jobs: to eat and to not get eaten. Not getting eaten requires right-brain awareness of the environment. In a sales conversation around a security tool, for example, a typical pitch might begin with outreach (because this is news to the prospect) and focus on the increasing number of threats pigeons like Peta may be seeing in their environment. (Perhaps there's a new predator in town?) The narrative, in this case, is grounded in a trend that's unfolding over time and that's happening 'out there' — very right-brain. This story may be entirely new to Peta and other prospects, and therefore you have a lot of story to tell. That's the change story we looked at in the previous chapter — a 'whole story' about what's happening and why they should care.

But what if you're trying to help Peta with their other job — eating?

Let's say you've developed the Seed Finder 10000®, and it's the perfect tool for Peta's very specific job of finding the best seeds. Here, the story isn't focused on a grand, unfolding narrative (though there might be a sprinkle of new tech advances or decreasing seed availability). Instead, it's all about drilling down into *how* Peta finds seeds and how the Seed Finder 10000® compares to other alternatives — its superiority to doing it manually and its differentiated features relative to similar seed finders. That is, this is about *manipulating* the world, not *understanding* it in some new way.

Your pitch, in this case, is about integrating with existing behavior (like the Snowflake pitch), not establishing new behavior (like dealing with a new security risk). Peta has to find seeds either way; it's just a question of if and how your tool will help. Therefore, your clarity strategy is proposing a new ending to their existing story — that using the Seed Finder 10000® will solve the pain of manual daily seed foraging and perhaps increase the rate of seed discovery — especially the really tasty ones. In supply and demand terms, positioning here is about getting as close to, or as 'tight' with, some existing demand. The closer you get, the more you sell.

On the one hand, this may mean there's less total story to tell. But that doesn't mean there's less work to do. If you want to sell something truly differentiated, then you have to work extra hard to align your product and your story with the narrative Peta is already living. You need to deeply understand the challenges this specific type of potential customer faces with this very specific job. You need to come up with those ultra-specific, verified value props we discussed earlier (**chapter 13**). And that's going to require a lot of your own left-brain, narrow-focus, laser-beam attention. You can't just coast on The Big Idea and dreams of money raining down. You've got to put the work in, and if you're a founder, that means building a team — and broader product and GTM organization — that can apply this kind of forensic attention and not just hand-wave instead.

> **Positioning & JTBD:** This is why Jobs to be Done (chapter 36) and positioning tend to go hand-in-hand — the better you know your buyer's JTBD, the closer you can position your solution in functionality and communication, closing the supply/demand gap and improving sales efficiency. Think of this more like "making a thing people want" rather than trying to "make people want a thing." In the latter case, you're trying to move the market to you, which requires far more resources.

Creating a position

We've established that this kind of left-brain positioning can just be positioning-as-a-story, where you align what you've got with a customer's existing needs in a sales and marketing story. That's what Snowflake did very effectively. Or it can be positioning-as-GTM, where you make big strategic moves across your entire organization.

But what we haven't considered is what positioning looks like from the customer's point of view. You can have a deeply considered internal sense of what your "positioning" is internally, but does it mean anything if that's not how prospects see you? I don't want to get too if-a-tree-falls-in-a-forest here, but I'm a big believer in the work only being done once you get a result, and for positioning, that means finding a slot in your buyer's mind.

The battle for your mind

> *The left hemisphere has a much more extensive vocabulary than the right, and more subtle and complex syntax. It extends vastly our*

power to map the world and to explore the complexities of the causal relationships between things.

— Dr Iain McGilchrist[8]

Positioning in your clarity strategy is about hitting left-brain attention. That means drilling down to your customer's needs and creating a narrative that intersects with their left-brain attention on whatever it is they're doing, be it finding seeds, juggling data, onboarding customers, or whatever it is you sell.

The company-centric perspective here is that we map our product to our customers' needs. That's how we 'do' positioning. But we can look at this from a customer-centric perspective, too, and consider how they might map companies in their mind, which in turn helps us unlock the fundamental nature of this specific kind of customer attention.

Mapping the world is, it turns out, also the domain of the left hemisphere, as McGilchrist describes in the quote above. Left-brain attention is how we map the world, and it's how buyers map the market. That's a powerful insight. Companies have been fighting for a spot on that map — on your map — for decades. This is where positioning kicked off in the 1980s as a marketing discipline, where positioning was, per the title of the famous book by Ries and Trout, a 'Battle for Your Mind'.[9]

If Coke was #1 in the consumer's mind for soft drinks, and Pepsi was #2, where were you going to fit? The top spots are taken. Where would you find your place — your position — instead? To find success, you must find a different position, a different mental slot you can occupy in a consumer's brain. And that's going to require some creativity and lateral thinking!

> **The challenge of getting buyers to think of you at all:** There's a prominent school of thought in the B2C brand marketing world that says this Ries and Trout approach to positioning — finding a unique segment in the market you can occupy — is a mistake, and growth comes from going for reach, not niche. That is, you need to *maximize* the number of people who can buy you, not limit it through segmentation, and you should worry more about getting buyers to think about you at all. This is, in the sales-driven B2B world, highly contentious! We'll unpack this spicy debate in chapter 43 and look at the niche-to-reach framework in chapter 44.

Maps not statements

In the B2B world, we tend to fall back on the company-centric approach to positioning and focus on our competitors — they have X feature, but we have Y feature. They serve A, but we serve A+B. But it's worth asking if customers ever map the market in the terms you've chosen. Those feature-focused positioning ideas were a product of *your* attention — attention that's often focused on your competitors — after all.

In reality, how your organization considers its 'position' internally may have little to do with how customers mentally map the market, or map solutions to their job-to-be-done, which may involve entirely different classes of solutions. But that's the map we want! That's the map where X marks the spot of the treasure we're chasing. That's the map we need to understand and navigate.

This is why guessing your way to a solid position doesn't work. You've got to break out of your own bubble, break through your own 'brain putty' (**chapter 28**), get down to the reality layer, and understand the buyer's map you're trying to work your way onto. That can involve a lot of pattern-matching and customer analysis, as we'll see in **chapter 37**. Either way, this is why positioning statements like 'we do X and unlike competitor Y we offer Z' are often meaningless exercises — there's little connection to whether this actually matters to customers and their left-brain mental map of the market.

Market maturity

When you consider the positioning map of the market you're operating in, it's worth considering how much room there is to move. What describes your market? Are you in a mature market or an early market? Are you in a competitive B2B enterprise space where customers have their left-brain, narrow-focus attention fully engaged and have put you in a feature-for-feature bakeoff with the competition? Or are you trying to put your Ultra Cloud OS™ on your buyer's radar for the very first time?

For mature markets, competitor-focused positioning work is somewhat inevitable — if buyers are coming to you trying to pick the difference between you and the competition with narrow-focus attention, then chances are internal folks are spending a lot of time fretting about and analyzing peer competitors. The result is a lot of sales battle cards, secret shopping, and attempts to match folks feature for feature. (While this is understandable in the hand-to-hand combat of enterprise sales, ironically, it makes it even harder for customers to tell you apart because you're all competing to be the same, which we'll explore in **chapter 49**.)

But for early startups, it's a different game. It's about finding any good reference point *at all* so folks start to understand the new thing you're doing. That's the whole reason why you exist, after all — to do something new. There's so much room for creativity in how you make that connection.

There's no need to try and blend in with the incumbents copying each other to death — you don't need to be the kid in an oversized suit pretending to be an adult. Own where you're at — own the fact you're for early adopters (**chapter 40**), own the fact you've got an edge that delivers that 10x improvement, and own the concept that differentiates you from the way everyone else does it.

Blaze your own trail with positioning that's a customer-centric creative act, not one of pure competitor-focused comparative analysis. You don't beat the market by being just like the market. Care more. Communicate differently. Cut through the cruft that's built up. Be a little weird, and find that position that's uniquely yours.

Choose your fighter

We've covered our introduction of the two types of story in your clarity strategy — right-brain change stories and left-brain positioning. But which one should you start with? When? In what context?

This was, dear reader, an itch I couldn't scratch for *years*. And I saw what happens when it goes wrong, too — if you mismatch change stories and positioning with the type of attention you're targeting, you'll have a bad time. But if you get it right, you'll open up new opportunities, from C-suite and outbound folks to niches filled with (hopefully!) riches.

How do I know this? Well, thanks to a strange confluence of events, I got a chance to observe a natural experiment where a classic 'story 1' change narrative was unleashed on a mix of B2B SaaS prospects.

The results were fascinating. It was proof that these two schools of thought — and two modes of attention — weren't just elaborate theories. They really were showing up in the sales trenches. We'll explore exactly what happened next.

Notes

1. Stancil, Benn. "The end of Big Data," *benn.substack*, April 9, 2022. **https://benn.substack.com/p/the-end-of-big-data**
2. ibid.
3. I should note, however, that the premise of Stancil's piece is that, in his view, "Databricks is a $38 billion dollar mistake" and that it should adopt Snowflake (and others') "boring" approach. I'm not qualified to offer an opinion on whether that sounds like a good move or not, and I imagine Databricks would argue they have 38 billion reasons (or, indeed, now 43 billion reasons) not to change. I do wonder, however, if there's a degree of attention preferences at play in Stancil's argument, in the form of 'Stop with the right-brain narrative stuff and just give me the left-brain tool,' but that's

just speculation.

4. Per TechCrunch's report, "Databricks raises $500M more, boosting valuation to $43B despite late-stage gloom." **https://techcrunch.com/2023/09/14/databricks-raises-500m-more-boosting-valuation-to-43b-despite-late-stage-gloom/**
5. McGilchrist, Iain. *The Matter With Things: Our Brains, Our Delusions and the Unmaking of the World* (p. 43). Perspectiva Press. Kindle Edition.
6. Zarick discussed the change on *Relay*, June 2022. **https://gorelay.co/t/how-we-deleted-a-1-3rd-of-our-product-and-made-it-10x-better-with-arrows-co-founder-daniel-zarick/764**
7. Zarick, Daniel. April 3, 2023. **https://twitter.com/DanielZarick/status/1637484374349484034**
8. McGilchrist, Iain. *The Master and His Emissary* (p. 70). Yale University Press. Kindle Edition.
9. Ries, Al, & Trout, Jack. *Positioning: The Battle for Your Mind.* Warner Books. 1981

22

The narrative experiment

Let's say you have a sales-led B2B SaaS product for the mid-market and enterprise. You need to pitch everyone from rank-and-file workers, mid-level managers, and C-suite executives. Some are coming to you through inbound lead gen, and some you're finding through outbound lead gen.

So far, so typical.

Here's the question. Do you pitch:

1. The precise positioning of your product?
2. A broad change story about their industry?

I've seen this tested in the wild, almost by accident, and it turned out to be a very revealing natural experiment. In this chapter, we will unpack this experiment and see how left- and right-brain attention played out in a mature B2B SaaS sales operation.

Hang out for the twist in the story, too.

The change story

Let's start with the right-brain change story first.

The SaaS company in question — let's call them Company X — hired a leading Silicon Valley consultant to develop a change-driven narrative with the founder. This was intended for the founder to use when pitching the C-suite. Those C-suite folks, according to the founder, apparently liked it.

Here's where it gets interesting.

This same pitch was also tested with mid-level and low-level folks in the org chart — think managers and workers in tiny teams. For these prospects, the feedback from sales and marketing was that the narrative just *did not* resonate. It hit themes that were well above their pay grade. The narrative proposed a new way of working for their function, but they simply weren't in a position to effect that level of organizational change. They had a specific job to do, often with very limited resources. I imagine the idea that their little team of 1-5 people could reinvent not just their department, but how their whole company operated, came across as a little absurd.

For these folks, the narrative (which, to be clear, wasn't originally intended for them) was of a promised land with no bridge to get there. This is the classic story 1/story 2 disconnect we touched on back in **chapter 6**, and it can happen with category creation stories, too. It's all well and good to come up with grand visions that sound great in theory; it's quite another thing to create the product, resources, and support systems to help folks actually get there. In this case, the rank-and-file folks wanted a new way to *manipulate* the world (a left-brain pitch about the tool) but were being pitched a broad, right-brain narrative on how to *understand* the world instead — a world they had no control over. That's the disconnect.

So far, so obvious, right? Obviously the C-suite and the rank and file have different priorities and responsibilities in that regard. Execs need to understand the world; workers need to, well, do the work.

But!

For some rank-and-file employees, the narrative *did* resonate, much to my surprise. These weren't folks close to the executives, either, sometimes quite the opposite.

What was going on?

Where's your head at?

> *The right hemisphere yields a broad, vigilant attention, the purpose of which appears to be awareness of signals from the surroundings.*
>
> – Dr Iain McGilchrist[1]

When I played back these calls between sellers and rank-and-file prospects in Gong, I was *sure* these folks, resource-strapped and with little authority, would balk at the big-picture narrative that I *knew* their fellow practitioners didn't like.

But they didn't. They stuck with it, agreed with the general premise, and set up a call to get a demo of the product. That was a huge surprise to me — it was the complete opposite of what I expected.

What made these people different from their peers?

The difference was this: they were taking sales calls after being contacted by outbound prospecting.

Radar outbound

These folks were using sales reps as their eyes and ears for what was happening in their industry, to keep tabs on 'what's out there' as one prospect put it, or to get right-brain 'signals from their surroundings' as McGilchrist puts it above. It was their way of staying up to date with their industry.

That is, their attention was directed outwards, to their environment — it was all right-brain radar.

And the same is generally true, I imagine, for executives. They're not the ones doing the in-the-trenches work; they're the ones setting the plan that has to work in a given competitive environment. Their concern is the broader fate of the enterprise and sustaining its competitive advantage, not so much whether you can shave some % of time off a given task for some role buried in the depths of the org chart.

Laser beam inbound

Let's think back to the new hemisphere hypothesis we discussed in **chapter 19** and our bird illustration in particular — the bird that has to keep one eye out for predators so as not to get eaten, and one eye focused down on eating. It turns out *both* rank-and-file outbound leads *and* execs are the ones with an eye out for change in the environment. They're watching for proverbial predators. I don't mean 'predators' as in 'competitors' necessarily, but changing trends in tech, culture, or customer preferences — keep up or get left behind.

And in B2B, we're often selling change on an organizational level, not just a user level. Recall that when we discussed change stories and strategic narratives in **chapter 20**, the original context was enterprise deals. 'Category creation,' too, is often discussed in the context of enterprise sales because that leans more on external change. Enterprise deals require a lot of consensus, and that consensus needs to be driven by a narrative strong enough to bring everyone together.

> **Story mismatches and second meeting syndrome:** A great story 2 product story might appeal to end users, but if you lack a compelling broader story 1 business case — one that's bigger than just raw ROI for a given role — you risk suffering from 'second meeting syndrome.' Here's what happens: the first demo is great, you've made the shortlist, the next demo is scheduled with upper management or C-suite folks, you demo, and then... crickets. What captured the imagination of your would-be users fails to capture the imagination of upper management, and it's game over for the deal. What was missing? A right-brain story for the folks with right-brain attention — a narrative about gains that are commensurate to the pains they're experiencing *as an organization.*

Playing the claw game

A narrative that helps drive consensus around the *why* for an enterprise buying group — like responding to a big shift in the world that's bearing down on you — can certainly help get deals done.

But what about the rank-and-file folks and middle managers who didn't like the narrative? The ones who thought it was above their pay grade? The ones head down doing the work?

Broader changes in the environment didn't concern them; it wasn't their duty to worry about, and they didn't have the power to make any org-level changes anyway. Instead, their attention was focused on the task at hand — finding the seed between the rocks, so to speak, to use our earlier bird analogy. They were using their laser-beam attention for their actual job. And for them, they very much *did* care about the tool, where it fitted in the market, and why they should choose this tool over that tool. They wanted the specifics, the details, the feature lists, the mechanics of it all. They cared about the static issue of their needs at that point in time, not unfolding narratives around them.

That is, they cared about the product's position relative to other possibilities. And fair enough, too. But let's not throw the baby out with the bath water here. Remember our claw game metaphor from **chapter 18**? Capturing attention isn't a matter of right-brain narrative *or* left-brain positioning. They're not 100% mutually exclusive. It's more a question of the degree of how far we go out on the x-axis of change and down on the y-axis of product positioning. Maybe that's 80% change/20% product for right-brain folks and 20% change/80% product for left-brain folks. I'm just making those numbers up; the balance in your approaches — plural — will be unique to you.

The point is that big change stories still need to lead back to your product, otherwise you're just doing trend consulting for free. Likewise, narrow product stories need some context-setting around *why* your solution matters, and that may include a change story of some description, some reason why your approach deserves to be on the prospect's radar *today*, at this exact moment in time. And that's especially true if folks are running straight to extremely narrow questions about integrations, for example, as we discussed in **chapter 21**. You've got to sell your story, too. Don't be afraid to catch folks before they jump all the way down the y-axis and discuss the x-axis of change, and not just the change you see, but what's on their radar and what they're responding to, as well.

Always be diagnosing

Let's wrap up this chapter by discussing the difference between the *form* of your sales story and the *content* of your sales story.

The form of your sales story is dictated by the shape of your audience's attention. In light of that, there's a simple, obvious application for your clarity strategy here: execs care about the big picture; users care about the tool... duh.

But it's not that simple. Sometimes, users care about the big picture too

— like our outbound contacts in the story above. And not just in outbound calls — one way SaaS companies get famous in the market is when they have a story to tell that's bigger than their product.

To that end, your clarity strategy is about getting the form of your story right, otherwise nothing else really matters — you might do a lot of talking, but you'll never really be *heard.*

To solve issues with the form of your clarity strategy, let's briefly return to our diagnostic rubric from **chapter 6**, where we identified the following common issues:

1. Weak story 1 or story 2
2. Mismatched story 1 and story 2
3. Disconnected story 1 and story 2

I won't rehash that chapter here — jump back and check it out if you like — but it's worth considering these common issues in light of the natural experiment above. Looking back, here's what I saw:

- **Story 1/2 strength:** Company X at least *had* a sales story that encompassed the two different kinds of stories — a change/market story and a positioning/product story. While their narrative wasn't revolutionary, it was unique and gave them a meaningful position to occupy in the market.
- **Story 1/2 mismatches:** This was the most interesting aspect of their clarity strategy, as we've discussed — inbound rank-and-file was a mismatch (right-brain story 1 for left-brain attention), outbound rank-and-file *did* match, and it apparently worked when selling to C-suite folks, who then moved the conversation down from the market-level story to a product-level story with folks further down the org.
- **Story 1/2 disconnect:** Whether your story 1 and story 2 connect — whether your product really does help you achieve your big change — depends, ultimately, on what kind of success customers achieve. As far as I could tell, some of Company X's customers *did* have success that mirrored the story 1 narrative, but that narrative never really caught on or opened up new markets; it just greased the wheels of a phenomenal amount of ad spend.

This brings us to the *content* of your clarity strategy.

Theories of market

Your clarity strategy is also as much about the *content* of your sales story as it is about the *form* of that story.

Every tech company has a clarity strategy by default, and that clarity strategy reflects their underlying theory of their market. The question for me is just how rigorous this theory — and the resulting clarity strategy — actually are.

In this natural experiment, Company X was doing their best to get their sales story right with a big story 1 change narrative. Props to them for that. They invested in it, put in the work, and rolled it out, and while that experiment had mixed success, it revealed a lot about what does and doesn't work. We learn by doing, and they *did*.

That's great. But what was their narrative actually about? What theory of the market did it contain? It might have been story 1 in form, but what was the actual story? I'm not at liberty to reveal its exact contents, but I will say that this was Company X's second bite at the category creation apple.

Company X wanted to *play bigger* (ahem) by creating a new category in their segment. Part of the reason why we'll dissect category creation at considerable length from **chapter 32** is that narrative, category creation, and positioning, for that matter, are often tools companies reach for when they want to try and win a market. These are tools and concepts founders effectively bet their company — and a decade or more of their lives — on. Investors often bet millions — or hundreds of millions — of dollars on these theories of market, too.

But are these tools very good? Are they fit for purpose? Are they rigorous? Do they stand up to scrutiny? The rather shocking answer to that question that I've found, dear reader, is often: not really! That's why this book is so meaty — if people are betting *so much* on these theories of how B2B tech markets work, shouldn't there be a sound theoretical basis behind them?

Again, I don't blame Company X, or any other company for that matter, for reaching for these tools — they were the best that were available. But I believe clarity strategy provides a far more rigorous framework for this kind of work, and that includes getting the founder's vision into shape.

Mystical visions

A big reason why story — especially story 1 change stories — *isn't* strategy is that these stories often neglect the rigor of the narrow-focused, story 2 attention on what success really looks like for customers in a given market.

In this natural experiment with Company X, what was particularly striking to me, as an outsider looking in, was just how, shall we say, light on detail the vision from the founder was.

Don't get me wrong — good on them for having the vision in the first place, doggedly pursuing it, and growing their company around it. That's awesome. But at some point, if you're expecting hundreds of people to pour years of their lives — and investors to pour many, many millions of dollars of funding — into your vision as a founder, you have to put some meat on the bones of your early skeletal vision.

Without that — and with a vision that's stuck in story 1 mode, where the glorious future is always just over the horizon — the entire bet hinges on one very small small packet of information that is the founder's vision. Is that a healthy way to build world-class software companies?

Given so much rides on that core story, that core narrative, you'd think we, as an industry, would be good at treating that vision quite rigorously, but it's instead often given a degree of mystical reverence. Sometimes that's justified — ideas are fragile, and right-brain vision is inherently hard to articulate — but sometimes it's not, especially when it's ultimately about corporate software on the road well traveled.

Today, in the third decade of SaaS, we have a lot of examples to draw from and learn from. While running a company is still *exceptionally* difficult (and I doff my cap to those who do it!), the experiments have been run, and the mechanics of success, while still oxymoronic (success being emergent and non-determinant), aren't a *total* mystery.

And now we understand the hierarchy of attention, we have a framework for knowing how to apply rigor to a founder's vision in a way that strengthens it on either axis — either out as a change story and/or down as a positioning story.

That is, we can develop a clarity strategy — including the founder's vision — in a way not dissimilar to developing the product itself. It doesn't have to be the domain of mystical vision and magic one-size-fits-all deck formulas or iffy ideas from pop business books. But it does require putting in the work, and the back half of this book is about doing just that — examining examples and demystifying some common approaches (and mistakes), including category creation, to stretch your narrative out on the x-axis and down on the y-axis.

Culture and opportunity cost

And it pays to do that sooner rather than later. I can't help but wonder what situation Company X would be in if they'd spent their early years rigorously refining their vision while they built the company. There's a tremendous opportunity cost to not doing so as you go, as it means:

- You never build the vision-refinement muscle in the org, i.e, the ability to go from point solution to infrastructure to an industry 'O.S.', to use Stripe's evolution from **chapter 16** as an example.
- You *do*, however, bake in the fuzzy vision, building your startup empire on an alarmingly sand-like foundation, and you end up in a Red Queen situation where you have to run harder and harder just to stay in place.

- Everyone in the company kind of cottons on to the fact that the appearance of success matters more in your company than honest and painful scrutiny, and ends up doing make-work exercises to project an image of success, compounding the problem.

This is fundamentally a cultural issue: either you're committed to rigorously digging down to reality and uncovering what success looks like, from the founder down, with your vision as a bootstrapping starting point, and you build a company to match, or you're not, and you prioritize the appearance of success and muddle along hoping you still get the big exit along the way. (And maybe you do! But to what end?)

In my view, dear reader, life is too short to fake your way through it. Reality is in the here and now — either you're working to get closer to reality every day, or you're not. Either you're making it happen, or you're not. And if you are, more power to you. It can take an *inordinate* amount of thinking and effort to come up with a genuinely unique perspective — ask me how I know! — but that's what makes the world go 'round.

And that is my hope for your clarity strategy, too — that it's not just window dressing in form around a fuzzy vision to provide the appearance of success, but that it's a means of refining the content of your vision as you *drill down to what customer success actually looks like* so you can replicate it again and again with the folks that buy you — often after months of deliberation — in turn making the businesses that run on your tech more productive, profitable, capable, and less painful. (We'll explore the reality layer vs. the speculation layer further in **chapter 28**.)

Split brains, whole stories

Let's close this chapter with a simple recap of clarity strategy. Consider this approach to your story 1/story 2 narrative as:

- **Market/change story to drive interest** — outward and interested in the change that's happening in the broader environment (like the bird watching for predators). Think transformation, trends, secular change, and playbooks. This is story 1.
- **Product/positioning story to drive decisions** — downward and position-based, looking narrowly at *how you do the thing* (gotta find those tasty seeds!). Think user experience, needs, integrations, jobs-to-be-done. This is story 2.

Which of those two stories do you lean into now? Are folks showing up inbound to hear the product story? Are there big accounts knocking on your door (or, more likely, you're knocking on theirs) that need a strong narrative to drive consensus?

This is the art of developing your clarity strategy. It's about playing the claw game, trying to 'win' our audience's attention by going out on the x-axis (maybe a lot, maybe a little) and down the y-axis (ditto). Different contexts call for different approaches — going wide enough when appropriate (outbound, keynotes, customer conferences) and into the specifics (discovery and detailed demos) when required.

This ideally gives you full coverage for big-picture coverage from your media engine (owned or otherwise) that will drive your marketing, and it gives you the cards to play in the cut-and-thrust of sales discussions around needs and objections.

I am, dear reader, incredibly keen to hear stories of how these two approaches do (or don't!) show up for you in your work. Let me know: **luke@lukestevens.co**.

Next, with this deeper understanding of story 1 and story 2, we're going to keep things practical and look at pitch deck outlines for different ends of the story 1/story 2 spectrum.

Notes

1. McGilchrist, Iain. *The Master and His Emissary* (p. 27). Yale University Press. Kindle Edition.

23

Pitching I — three alternate decks

Back in chapter 9 we peeled off the first layer of your clarity strategy onion as we put together your initial narrative sales deck. We've covered a lot of ground since, peeling off more layers as we've gone, including our look in the previous chapter at how the theory of right- and left-brain attention plays out in real sales conversations.

At this point, it's probably clear that your earlier deck — or your current sales pitch — may need to be tailored to more story 1/right-brain or story 2/left-brain attention. The next three-chapter series covers how to do that, including:

1. In this chapter, we'll look at more real-world(ish) pitch deck outlines, starting with the extremes of right- and left-brain sales narratives before meeting somewhere in the middle.
2. In **chapter 24**, we'll continue looking at our three deck outlines, and zoom out to see what your narrative needs to do across the entire buyer journey, including the five stages of awareness and the concept of sensemaking.
3. In **chapter 25**, we'll look at the dynamic nature of attention, and how expert sellers move prospects out of their narrow-focused attention and put new insights on their radar.

(If this is all too in-the-sales-pitch weeds for you, jump ahead to **chapter 26** where we look at deploying your attention internally using the right/left/right jump for insight generation.)

> **Your current sales narrative:** Which way does your current sales narrative lean — radar or laser beam? If you're a sales-led organization, dig out your current sales deck (or the deck you made in chapter 9) and take a look. Where is it strong or weak in terms of change stories and narrow-focused positioning and insight? What would it look like if you pushed it in either direction? Consider experimenting with alternate approaches to see what gets folks opening up in discovery, perhaps with their eyes lighting up (chapter 38) or just agreeing more, relating more, and essentially selling themselves on your solution.

1. The right-brain 'ray of sunshine' pitch

Let's start at the extreme story 1 right-brain radar end, where I'm going to borrow from one of the authorities on modern B2B selling.

It might be a surprise that people have been studying B2B sales for decades. It's easy to miss in the startup world, where folks often have to learn that sales is a thing in the first place (as we saw with Segment in **chapter 8**), let alone that there are different types of B2B selling. But folks do study sales, and there's one particular loose group of researchers that have been making waves in B2B sales since the late '00s, sharing their insights in books, podcasts, and HBR articles. They are the original authors of *The Challenger Sale,* and we touched on some of their work when looking at positioning and late sales engagement in **chapter 21**.

Matt Dixon is one of those authors. Dixon is not just an expert in B2B selling, he's an expert in *selling sales insights to B2B sellers,* a rare combination of both theory and practice. In a 2018 interview for HubSpot[1], Dixon shared his framework for selling to CEOs and revenue department leaders — folks who lean pretty heavily on their right-brain radar. That's a great resource, and that's what we're going to use here as an archetypal right-brain story 1 pitch example.

Dixon calls this approach "The Lift," but I'm going to dub it the Ray of Sunshine pitch. You'll see why in a moment. Here's my paraphrase of Dixon's 8-slide framework as the Ray of Sunshine:

1. **The secular change(s) happening in the world:** Dixon starts with these 'mega-trends' and might rattle off several. Stacking trends is a powerful way to set the context for your pitch. (We'll explore secular change further in **chapter 30**.)
2. **The impact for people like you:** I.e., these trends are a wave, and it's coming for, to quote Dixon, "business leaders like you," with some specific implications.
3. **The pain it'll cause:** This is where the painful second-order effects happen, and here we dial up the pain.
4. **Stuck in the maze:** Today, "most companies" are trying to find a way out.
5. **The dead ends:** But, as Dixon says, these folks are getting mediocre results, i.e., hitting dead ends.
6. **Glimmer of hope:** However, some folks are doing things differently.
7. **Ray of sunshine:** And getting far better results than the stuck-in-the-maze folks.
8. **Blinding light:** Here's how it works, segueing into your solution.

This is a neat three-act play. In act one, darkness descends. In act two, everyone is lost and stuck. In act three, a ray of sunshine appears and turns to a blinding light as hope returns.

Now, to be clear, I made up those titles to describe Dixon's slides,[2] but it helps highlight the flow of the story for a right-brain story 1 pitch, and that deck outline is an excellent example of what a strong, right-brain, radar-attention pitch looks like. Big trends, big implications, big stakes.

Three act pitch

Let's flesh out this three-act story structure a tiny bit more. We've got setup, confrontation, and resolution.

1. **Setup:** The setup comes via the mega-trends bearing down on the protagonist (the executive and their company).
2. **Confrontation:** The protagonist now faces a variety of obstacles (as illustrated by how others have struggled to deal with the same trends).
3. **Resolution:** The story segues from crisis and despair to a flicker of hope — there is evidence *outside of the vendor's pitch* that others have overcome these obstacles, and you can, too.

This is a universal approach to story that's useful for your narrative sales deck (and case studies!) precisely because it makes it more of a story and less of a dry monologue of facts and features.

You can also see echoes of the 'strategic narrative' story 1 approach we discussed back in **chapter 20**, both in form and, as we're about to see, the type of attention it's aimed at (C-suite folks looking 'out there'). Again, it's the right hemisphere that understands narrative, and that hemisphere has been around for a while, so we'd expect multiple people to reach similar conclusions on what a solid right-brain pitch looks like, which is what we have here.

Give me tension

There's one key ingredient that separates boring right-brain pitches from engaging ones. That ingredient in this approach — and indeed any story — is *tension generated from conflict.*

Take a moment to go back through that outline and get a feel for the tension it creates, from multiple external environmental changes through to pain, uncertainty, and the risk of failure. That's what gets you invested in the story and sets up the emotionally satisfying pay-off in the final act. The ray of sunshine at the end is only interesting if it's tension-relieving. If it's already bright outside, who cares?

Often, in B2B sales, we're so quick to try and jam our company and product solution down the prospect's throat (awards! funding! customers! features!) we forget to establish why the solution even matters. But if folks don't agree with us on the problem they face — if they don't *feel* that tension

— there's little reason for them to agree on the solution, either.

> **Outside-in radar insights:** Also note the indirect nature of Dixon's pitch — it's *not* "I think X / therefore you should Y." It's not trying to force your opinion on the prospect and getting them to acquiesce. It's far more about evidence and observation — the trends we see, the struggle we see, the success we see. And by offering what you see, you're giving folks a chance to respond with what *they* see. You're laying out your evidence about the environment and giving the prospect space to form a judgment and respond in turn. The more credible your evidence and the more common ground you can find, the more likely folks will accept your suggested way forward. But it's a matter of guiding *via their radar*, not strong-arming via your opinion.

Fear vs. indecision

The pitch formula here is not rocket science. As we discussed in the last chapter, formulas and templates for pitch *form* are great, but getting the form of a pitch down is fairly straightforward. What's less straightforward, however, is the *content* — the insight, evidence, and messaging that drives the pitch. That's the hard stuff at the heart of your narrative. But it still needs to be packaged in such a way that it actually sticks, and there are nuances to that.

In Dixon's HubSpot interview, he describes typical B2B pitches that can range from old-school, insight-free, "show up and throw up" pitches to open-ended, long-winded solution selling ("What are your biggest pains/priorities right now?"). Dixon suggests an insight-driven approach instead, and in 2022, he went on to elaborate what that looks like based on a COVID-era study of millions of sales calls in *The JOLT Effect*, a book he co-authored with Ted McKenna.

The main finding of their research was that *customer indecision* was a bigger enemy than the status quo when it came to deals falling through, and *in those cases,* dialing up the fear of missing out was counterproductive. That is, these buyers are more scared of messing up than they are of missing out (to use *JOLT's* highly memetic line), but most sales efforts are focused on agitating FOMO, not resolving FOMU, if I can put it that way. As Dixon and McKenna write:

> *And yet, our research also shows—in very stark terms—that none of this [FOMO agitation] works. In fact, these time-honored sales tactics that have been passed on from leader to manager to seller for decades aren't just unproductive; they're actually counterproductive*

to the goal of getting the customer off the fence.[3]

If overcoming the status quo is all about describing how the prospect will succeed with your solution, in the *JOLT* world, overcoming indecision is all about reassuring the prospect they won't screw up by choosing you and helping them narrow their choices so they can actually make a decision.

It's worth pausing for a moment to reflect on that insight. How many deals do you lose not because you lack feature X or Y, but because the prospect simply couldn't make a decision? We develop B2B products to solve pain points in folks' work lives, but what would it look like to develop a sales process that solves the pain of indecision for prospects? That's what *JOLT* attempts to outline, with a scorecard for assessing a customer's level of indecision, for example.

This kind of methodical approach to selling is, let's be real, several galaxies away from where most startups are in their sales process, and that's understandable — you've got to learn to crawl before you can walk. But the simple takeaway here is that when we talk about tension in a narrative, it needs to be *constructive* tension that gives folks the confidence to move forward, not tension overload that cranks up the fear, uncertainty, and doubt to the point where they can't make a decision.

In terms of resolving that constructive tension, *when facing prospect indecision* (i.e., not every buyer!), high-performing sellers, *JOLT* suggests, offer more prescriptive recommendations in a 'Here's what I would do if I were you' sense. That is, they recommend clear ways to move forward.

In my view, sellers need to be equipped with meaningful insight they have confidence in so they can make that kind of recommendation. But where do those insights come from?

Insight, a 'whole product,' and MAYA: We're just touching on insight as a solution to prospect indecision in the sales process here, but in a world of conservative B2B buying, sometimes the solution for startups, especially early on, is simply not to sell to indecisive, conservative buyers. When sales resources — like the founder's time — are scarce, you need to be extremely picky about who you engage with, and it can be much easier to sell to edge-seeking nerds than risk-averse normies, as we'll see in chapter 40. Over time, when you have a consistent pipeline of deals, you'll ultimately need to find the sweet spot between two things — being 'most advanced,' (i.e., having that unique product, perspective, or tech edge), while being 'yet acceptable,' (i.e., having a 'whole product' that B2B buyers want — support, integrations, ecosystem, etc). This is the MAYA dilemma we'll explore in chapter 49.

Deploying your attention

This is where your clarity strategy again goes well beyond a simple pitch. The confidence sellers have in your sales narrative — and the confidence they can instill in buyers — is a function of their confidence in the insights your organization is built on.

Generating those insights is something every B2B outfit should treat very rigorously, especially as it goes to the heart of your credibility in sales, but few do. The good news is that if you've been around for a while, you've got at least one good source to mine: what you've learned in-market, especially from customers who are at the forefront or very edge of their industry or practice. That's often where these insights come from. You're the one observing an industry across *N* customers. You and your organization are the ones who should be the subject matter experts in your customers' problems, having credible evidence of where they struggle and where they succeed.

Now, you might think, 'Wait a minute, we hear from our customers all the time! They're always telling us what features they want and what we're missing!' And that's true — it's true to a fault. Many SaaS companies tend to get stuck on the parts and pieces of the next feature or the next deal and aren't thinking deeply about industry-level insights on the one hand or true best-fit customers on the other.

That's not to say it never happens, but directing right-brain attention to the implicit whole or really stretching your left-brain narrow focus down to who succeeds with you and *why* happens far less often than just focusing on what's immediately in front of you, and it shows when things like generative AI, for example, gets bolted on to a product rather than truly built-in in a coherent, thoughtful way.

Insight generation is something you need to operationalize. This requires you and your organization to be actively deploying both modes of attention, both your collective right-brain radar and left-brain laser beam, to your customers' world. But what are you looking for?

1. Secular change and emerging playbooks on the one hand (or hemisphere).
2. Specific patterns around best-fit customers on the other hand (or hemisphere).

They're the big topics the back half of this book tackles — skip straight there if those ideas tickle your fancy! This is the big flip where we go from trying to *engage* with the attention of prospects to the question of how you *deploy* your attention collectively as an organization, which is a theme we'll return to in coming chapters.

The point here, however, to bring us back to our pitch outlines, is simply that this right-brain pitch approach is wholly dependent on the quality of your insights in terms of what you see 'out there,' both in trends, failures, and success. Become the expert in both your customers' successes *and* failures so you can help them succeed far beyond their current status quo *and* make them confident enough to know they won't screw it up.[4] This way, you'll have much better odds of coming up with an insight-rich pitch that generates real, relatable tension *and* an emotionally satisfying — and safe and reassuring — resolution in act three.

> **Learning through selling pre and post-PMF:** Insights come from learning, and pre-product/market fit, a startup's product might be pretty slim, but they will be very good at learning. And for good reason: they have to be if they want to achieve PMF, and every interaction with folks in the market is a chance to accelerate up the learning curve. "Sales" at this stage often looks more like getting feedback, as we'll see in chapter 38 when we look at early-stage pitch testing in contrast to established, mature, late-stage sales. In those later stages, however, the organization's ability to learn in an insight-generating way gets weaker. You lose that beginner's mind. "Discovery" gets formalized as a function of product management, perhaps, but sales can still be a great place to keep learning about the market, especially if you're collaborating with buyers on a new way of understanding their world and what their work looks like. But if selling just becomes preamble, demo, click, click, click, deal please — which may be appropriate in left-brain situations, as we'll see! — then you need to find new ways to keep *exploring* new segments and opportunities while *exploiting* the market you know. This is the classic explore/exploit tradeoff,[5] and if you neglect it, don't be surprised if growth gets harder as competitors catch up — they're out there learning, too.

CxO = radar

Continuing with the theme of where right-brain insights come from, let's return to Dixon's deck outline and note the context: selling to CIOs. The seller, in this case, as Dixon describes it:

- Talks/sells to 100 CIOs a year.
- Has unique insight because of that position, as they can see, as Dixon says, "what other CIOs are doing to solve their problems, but where they're falling short and how they're moving forward."
- Needs to tell stories to sell that insight, and the deck outline above is a means to do that.

That's a *very* specific sales scenario: C-suite selling, C-suite insights, and, presumably, C-suite deals. And that's why it's so trend/radar-driven. The seller is the *aggregator of radar-level insights,* and that's what makes their perspective uniquely valuable to other CIOs who are otherwise flying solo. That's why they pick up the phone or accept the calendar invite. This is all about giving folks that new, right-brain *understanding* of the world, not the specifics of manipulating the world with your tool, and that's a critical insight into how this sort of selling works.

In this case, the seller is uniquely positioned to win the attention claw game on the x-axis — being able to go a long way *out* with their narrative regarding the secular change/mega-trends well before they head *down* into the left-brain world of tools and solutions. But what does the opposite look like?

SMB = laser beam

If the one-to-one, executive-focused, insight-driven pitch is at one end of the spectrum, then the other extreme is the SMB one-to-many product-focused pitch. (Here, we see the limits of the term "B2B", by the way. Lumping SMB and enterprise/C-suite selling into the same "B2B" category is kind of silly, and we'll explore those differences more in **chapter 43**.)

In SaaS SMB land, it has become popular to use on-demand demos to deliver pitches in a way that wouldn't be economically viable otherwise. The math of having sales reps give every account the white-glove treatment at scale simply doesn't work out, so you need to find ways to be ruthlessly efficient instead, hence on-demand demos. Fortunately, that also makes them easy for us to study.

SMB-focused SaaS Keap (formally Infusionsoft), for example, is a marketing automation and CRM platform that has experimented with this tactic. Their homepage CTAs are, as of writing, "Try free" and "See demo," the latter leading to a gated on-demand demo. That demo is a five-minute, story 2, what-the-tool-can-do overview with a *hint* of more right-brain opportunity tacked on at the end, specifically that there are experts who can help you *after* you've gotten to grips with the tool. That sequencing — try the tool first, get the insights afterward, or manipulate the world first, understand it in a new way afterward — is very obviously the other way around to our earlier CIO pitch.

Keap has also done live webinar demos in the past, which are similar in a one-to-many sense but give us a bit more to unpack in terms of pitch, which we're about to do. The context for this pitch is: SMB folks are looking for CRM and/or automation tools, they're curious if Keap is a fit, and they (I imagine) want to know exactly how it works and if they can see themselves using it. We're talking narrow, left-brain, laser-beam, story 2 vibes here. Let's break it down.

2. Pitching on (SMB) positioning

The product-focused Keap pitch includes just a couple of slides to set up the demo, and then the rest of the time is spent in the tool itself before ending on an offer to try and close folks at the end of the call. Here's the outline:

- **Who we're for:** The rep explicitly positions Keap as a 'small business growth tool' to set the context. (Note the lack of jargon — it's just a 'growth tool,' not marketing automation or an OS/cloud/buzzword of the day.)
- **The problems we solve:** The rep outlines eight key areas Keap addresses, from collecting new leads to making appointments, processing payments, and sending emails.
- **The tool:** Then it's on to the tool, including walking through the nav, looking at what's included in given plans, 'what we do best,' and some very clear walkthroughs around the advantages of automation and so on, literally taking prospects step by step through what they can do.
- **The offer:** A discount for attendees to get them to sign up in the moment.
- **Q&A:** In the chat after the session ended.

That's a very straightforward, very tool-focused pitch. Consider the distance between this approach and the C-suite radar approach above — it's a world away, with Keap focusing on features (literally: 'check out this feature, this is really cool!') and closing very warm leads *right then and there* with a discount offer.

It was all very left-brain laser beam — how to use the tool, manipulate the tool, and get results out of the tool. That's what left-hemisphere attention tends toward — manipulating the world, pecking the seeds between the rocks, understanding the particular parts. And therefore we get selling that looks a lot more like product explaining vs. insight delivery. (It's worth asking how much of your org's selling is really *selling* in an insight-driven, value-creating way vs. just on-demand product explanations.)

Now, let's be clear that Keap doesn't do this because they *want* to; they do this because they *have* to. And yet many SaaS companies pitch like this even when they have time with the prospect to do more discovery and introduce insights that give the all-important *why* to the product's *how*. What does *that* kind of pitch look like?

> **No three-act story:** Notice there isn't a grand, three-act story here

> because prospects are already in their third act. The story, such that there is one, is what's happened in the prospect's life prior — they've already had some 'inciting incident' (to use storytelling jargon) that has propelled them on this journey and through their challenges, and now they're here for the resolution. This pitch, with its brief positioning setup, is just intended to demonstrate that this is, indeed, the resolution you're looking for.

3. Pitching on position: the headshot

Our third example pitch is what I call the positioning headshot. It sits between the two extremes of C-suite story 1 insight selling and SMB story 2 show-me-the-solution-now selling with a particular focus on tight positioning, i.e., the narrow-focus story 2 approach.

This format will, I imagine, suit most B2B SaaS folks serving the mid-market, and it's similar to the initial sales narrative format we explored back in **chapter 9**, but this time we're giving it a bit of a real-world spin.

This outline is based on a client I worked with several years ago, with a slight twist. Imagine you're the seller here — a B2B SaaS with a search marketing solution for retailers with large catalogs. Prospects — retailers — know they need more organic traffic, they know SEO tools exist, but they don't know about yours. Therefore, we've got to both get on their radar (story 1) but then pivot to why this very specific approach is a perfect fit for them (story 2). What would a pitch here look like?

1. **Name the problem:** To get on their right-brain radar, we'll lead with an insight into a problem and opportunity *for this specific segment* they haven't paid much attention to. We might give it a name like 'paid spend dependency' as a framing device and then discuss our diagnosis of missing search traffic — search traffic that *should* find them but isn't. (Note that this *isn't* a broad observation about change in the world in general; it's a relatable insight into their world right now, and the more specific, the better. We'll discuss what this pitch *isn't* below.)
2. **The alternatives:** Retailers could solve for this missing organic traffic in a variety of ways, but they're all bad options for a variety of reasons, and we might discuss why this hasn't been solved and why the pain persists. (Specifically, the pitch might be: Retailers can have massive catalogs; most SEO tools are targeted at manually created written content; retailers can't create relevant pages manually at the scale; it's therefore an unsolved problem.)
3. **Agitate the pain:** The traffic you should be getting is missing and significant money is being left on the table. (Ideally, the prospect can

relate to the difficulties they've faced if they've tried to address this in the past.)

4. **The tension:** (This is the part most folks skip!) We might sprinkle these initial insights with slightly awkward or uncomfortable questions about the customer's story to establish consensus on the problem, perhaps digging into whether they *really* need this solved and where they've struggled in the past. (Too many folks just accept buyer skepticism and plow on with the pitch without switching gears to take a genuine interest in the customer's actual situation.)
5. **The target:** (Note how similar the first four slides are to the Ray of Sunshine pitch. Now we pivot to story 2.) We want to switch gears and zero in on this very specific problem — we're focused specifically on solving for retailers with large catalogs missing out on search traffic, so what does this look like specifically for this prospect? We need the prospect to give us the target to aim for here, establishing consensus on the exact problem we've diagnosed, paving the way for...
6. **The headshot:** Now we go for the headshot at the target we've established, introducing our fits-like-a-glove solution for the prospect's very specific challenge, resolving all that awkward tension from before.
7. **Proof and pillars:** You then might work through product pillars to demonstrate the 'how' of your solution and the outcomes you've achieved for others, knocking down objections as you go.
8. **Next steps:** The pilot, trial, live demo, whatever that might be — just keep it as low stakes as you can to minimize buyer fear, and take them step by step on the journey towards the deal.

The right-brain CIO pitch was more about selling an approach or transformation that might necessitate a tool; the left-brain SMB approach was pure product explanation with an offer to close. In this case, we're using our insight into the very specific challenges of a very specific group of prospects to get on folks' right brain radar, and then we lock onto their problem with a very tight positioning-driven story 2 approach to the product.

In claw game terms, this is more 50/50. We're hitting the prospect's radar with our novel insight on the x-axis, but then we're moving quickly to go further down the y-axis of positioning, reaching down to a very specific issue or pain point to zero right in on the prospect's specific challenges. It's similar to the narrative sales deck we explored back in **chapter 9** and captures how both right- and left-brain attention can come together in a single pitch.

This third outline also highlights the fact that broad change-in-the-world narratives aren't the only way to engage right-brain environmental awareness. Very focused insights can work, too. If you rock up to a prospect's inbox or feed and say, "We're addressing your very specific challenge," then that's obviously going to land on a prospect's radar and get their attention.

It'd be hard for them *not* to be curious. Once that context is established on the right-brain x-axis you can then dive deep on the y-axis into the gory details of the prospect's issues and how your product might solve them.

> **Knock objections down:** Another nuance to pitching: it's important to rebut (or pre-but!) objections as you go. In this example of an SEO SaaS solution, prospects would obviously have concerns about what Google would think about the pages the tool generates. That, like any objection, needs to be hit directly and perhaps preemptively. Especially when addressing objections is vital for win rates. Dixon writes in *JOLT* that objections in sales calls were "one of the biggest drags on conversion, and yet, with just one rebuttal win rates go up to 31 percent. When no rebuttal is offered, conversion rates drop by nearly half, to 17 percent."[6] That's quite a drop!

No more 'trends & stats' pitches

Let me stress what the positioning headshot pitch *isn't*. This is especially for folks who think they're already doing this kind of right-to-left-brain attention pitch but aren't getting results.

This pitch is *not* a naive combination of broad mega-trends or industry statistics on the one hand and then an awkward segue to your product. These trends-and-stats pitches tend to plague the industry. With our understanding of the hierarchy of attention, however, we can see what folks are *trying* to do. They're trying to go from right to left-brain awareness with big-picture story 1 context setting followed by a story 2 product pitch. But why does it fall flat? Isn't that what we're supposed to do?

The key to this positioning headshot pitch is in, well, your positioning. It *has* to be dialed in. For most folks, it isn't. Most startups and SaaS companies are afraid of missing out on any sale to anyone, so they go super broad. In my client's case, their technology could be used for retailers with large catalogs, but it could also be used for media companies, marketplaces, and more — anyone with a lot of niche content online, basically.

If you want to sell to all those people, you end up with a typical trends-and-stats pitch that's equally broad, with intro slides trotting out some generic stats like "Online businesses are seeing traffic growth slow 22%" or the classic survey stat "Over 75% of online businesses report that it's getting harder to grow traffic year over year" or a market trends stat "Online spending is predicted to grow 39% in the next year" and so on. We seem to instinctively know we need to engage folks with some kind of story 1 change narrative, but this sort of context setting is so broad as to be meaningless.

It's generic and insight-free. This is where both tight positioning, sequencing, and knowing your GTM strategy matter enormously:

1. **Positioning:** Knowing who your best-fit customers are, which we'll discuss in **chapter 37**.
2. **Sequencing:** Knowing which vertical or ICP you're going to tackle first as you work your way across the adoption curve that we'll touch on in the next chapter and explore further in **chapter 40**.
3. **GTM strategy:** Knowing whether you're fundamentally taking a marketing-driven 'cast and catch' approach or a sales-led 'segment and sell' GTM approach, which are important concepts (and exclusive to this book!) that we'll also discuss in **chapter 40**.

With our SEO tool example, this might play out by:

1. Doing analysis on ACV (average contract value), conversion rates, and market size for retailers vs. publishers vs. marketplaces.
2. Looking at our existing logos and customer industries to plan a vertical-by-vertical approach that leverages existing customer references and results. (Maybe you've had success with not just retailers but fashion outlets specifically, and can target more of those businesses.)
3. Understanding where your organizational strengths lie (or need to be built up) in terms of broad marketing and advertising across search and social vs. more targeted sales outreach. Can you reach retailers in a broad sense, or do you have connections that can get you a meeting with CEO or CMO of target apparel accounts, for example?

This is where your GTM machinery and your pitch need to work together to drive growth. Too many folks get scared by the thought of picking a niche to start, thinking they need to find exponential growth in the broadest possible market from day one. In B2B, that's rarely how it works. That's hope, not a strategy. What's more, we've known this for decades! That's why we're going to spend a considerable amount of time exploring the GTM side of positioning and sequencing in the back half of this book — we know how this stuff works, and I'm going to keep beating that drum until we meet in those chapters. If you're serious about growth, you've got to put in the work to make the hard decisions about how you're going to get there. It's not easy, but it's what separates the pretenders from the pros.

Good pitches vs. lame pitches

If you do put in the work, your reward is hard-won and hard-to-copy insights and GTM decisions that give you the ingredients for a high-quality, insight-rich, confidence-building pitch.

From there, a good pitch is like a good joke — all about the setup, punchline, and knowing what's funny (and relatable) for your specific audience. There can be a lot of tension between setup and punchline, too — the relief and surprise of the punchline is part of what makes a joke funny.

Lame pitches have a weak setup, a self-serving punchline, and no tension. They stumble through some stats, rush to the end, get attention wrong, and fail to agitate the problem in a way that gets folks invested in a solution. Instead, they just mindlessly rattle off facts, figures, and features like a talking brochure with no pause button. Sometimes a very gloomy one at that, with the Morbo-esque rep implying if you don't buy their solution, you're DOOMED!

There's no need to do this. No one cares about the boring facts about your company or generic, pseudo-scary, and often plainly fake stats about the market. People know they're nonsense. Likewise made-up tosh about 'the way things are today' or 'what today's buyer wants' want or zzzzzzzzzzz.

Kill me now.

No energy, no accuracy, no expertise, no genuine insight, no focus, minimal relevance, and the buyer is left to do all the work to try and understand how this boring sales presentation translates to their pains, their jobs-to-be-done, their success, and alleviates their risks.

The good news is that the poor quality of most pitching makes it very easy to stand out when you do actually care about the buyer's success in both a FOMO and FOMU sense. When you can prove you've done the learning and you're armed with insights that change *how the buyer sees themselves*,[7] you can stand out as the credible alternative in a sea of very boring Blah2Blah pitches.

Just like there are a lot of ways to land a joke, there are a lot of ways to land your pitch, and just like in comedy, the effortless joke is usually the one that took an inordinate amount of work and iteration to get just right. And while it might take a lot of work to get it just right, when it lands, *it lands*.

In the next chapter, we're going to stick with our three-deck approach, looking at the five classic levels of awareness and how they apply to pitching, and then we'll zoom out to look at how your clarity strategy applies across the spectrum of buyer attention as we examine another hot trend in sales research: sensemaking.

Notes

1. Signorelli, Brian. "Insight Selling: The 8-Slide Framework for a Better Pitch," *HubSpot*, March, 2018 (updated September 8, 2020). **https://blog.hubspot.com/**

sales/insight-selling-better-pitch

2. Check out the original outline in Dixon's full interview, linked above.
3. Dixon, Matthew; McKenna, Ted. The JOLT Effect: How High Performers Overcome Customer Indecision (p. xv). Penguin Publishing Group. Kindle Edition.
4. Again, thanks to *The JOLT Effect* for that insight.
5. Per Wikipedia: "The exploration-exploitation dilemma, also known as the explore-exploit tradeoff, is a fundamental concept in decision-making that arises in many domains. It is depicted as the balancing act between two opposing strategies. Exploitation involves choosing the best-known option based on past experiences, while exploration involves trying out new options that may lead to better outcomes in the future. Finding the optimal balance between these two strategies is a crucial challenge in many decision-making situations, where the goal is to maximize long-term benefits." **https://en.wikipedia.org/wiki/Exploration-exploitation_dilemma**
6. Dixon, Matthew; McKenna, Ted. The JOLT Effect (pp. 95-96). Penguin Publishing Group. Kindle Edition.
7. That's another deep cut from the *Challenger Sale* folks, this time from 2015's *The Challenger Customer*.

24

Pitching II — awareness, content marketing, & sensemaking

In its natural development, every market's awareness passes through several stages. The more aware your market, the easier the selling job, the less you need to say.

– *Eugene Schwartz, Breakthrough Advertising*[1]

In the previous chapter, we kicked off this little three-chapter series by looking at the following deck outlines:

1. **Right-brain:** The 'ray of sunshine' pitch aimed at CIOs.
2. **Left-brain:** The SMB pitch for folks who are just about ready to buy.
3. **Right-to-left:** The 'positioning headshot' that sets up a specific problem and zeros in on it.

Just knowing that this spectrum exists can open up a lot of doors for your sales process — perhaps you can pitch C-suite folks more effectively on the one hand, or perhaps you can engage targeted prospects with a tighter 'positioning headshot' on the other.

But if we step back a little bit, we can see that there's a general approach from broad to narrow awareness, from the introduction that some problem or need exists at all to the very specific details of your offer to solve it.

This journey from broad to narrow awareness is obviously a common phenomenon. That shouldn't be a surprise, given we all share the same brains with the same universal nature of attention — it's something we all experience. Nor should it be a surprise that other folks have discovered (and keep rediscovering) a similar transition, as we'll see in a moment.

But despite it being part of the universal human experience, we often fail to address the spectrum fully in our clarity strategy across sales and marketing, going too broad (general trends) or too narrow (minute feature details), often failing to join the dots in between. And that's a bummer.

In this chapter, we're going to flesh out this progression from broad awareness to narrow awareness by exploring a classic framework — Eugene Schwartz's five levels of awareness — and what it means for your pitch and broader clarity strategy. We'll also look at what happens when prospects face too much awareness in the form of information overload and how a sensemaking narrative can help them make a confident purchase decision instead.

No One True Pitch

First, let's note that our three common deck outlines demonstrate that there's never really One True Pitch™ for all narrative situations, be that in sales or beyond, like pitching investors or potential employees. Instead, there's a claw game to meet your prospect where their attention is at, which requires tailoring your decks towards insight and trends, product and features, or something in between.

In practice, however, we need to start somewhere, so you still might start with a pitch described in a narrative deck (and in my work, I often do) because the first job of your narrative deck is to capture the fundamentals of your offer in a compelling way. But sellers — and founders — need the freedom to lean more into one side or the other, so when we talk about archetypal pitches here, they're intended as examples of what it looks like to push on a certain axis, not a rigid straightjacket that should be applied in all sales situations.

There's no need to be constrained by a deck, either. If you're product-led, maybe your sales narrative exists as a Slack-style 'We don't sell saddles here' memo (**chapter 12**), and perhaps the only sales pitch most customers see is a five-minute on-demand video (like Keap in the previous chapter) if they see one at all. Or perhaps your pitch — and broader clarity strategy — can be found in a strategy doc in Notion, or diagrammed in a FigJam/Miro canvas, or as the result of a positioning workshop (**chapter 37**), or even as a mission statement and web messaging (see Stripe's early messaging in **chapter 16**, for example).

Either way, the point is that this isn't about having the perfect deck, it's about thinking through your broader clarity strategy on a bigger scale and at greater depth than you may have otherwise. That includes zooming out from the sales pitch, especially when, as we discussed in **chapter 21**, B2B buyers are engaging sales later and later in their buying journey, leaving just 5% or 6% of their buying time for sales discussions with you as a vendor.

What does it mean, then, to look at your clarity strategy across the other 95% of a buyer's journey?

Levels of awareness

In the 1960s, ad man Eugene Schwartz described his five levels of awareness that some marketing folks have been using ever since. (Direct response copywriters, in particular, who have to sell through writing alone.) These levels of awareness are also immensely helpful for fleshing out your clarity strategy beyond the pitch itself. They are typically summarized as[2]:

1. **Unaware:** The prospect isn't aware they have a problem.
2. **Problem Aware:** The prospect knows they have a problem, but doesn't know if a solution exists.
3. **Solution Aware:** The prospect knows solutions exist, but may not know about yours.
4. **Product Aware:** The prospect knows about your product, but they're unsure they want it. They need encouragement that your solution is the best one.
5. **Most Aware:** The prospect knows your product and knows they want it, they just need a price and offer.

While Schwartz was an ad man through and through — these levels of awareness were mostly about finding ad headlines that worked — he was fundamentally a *seller*. He was trying to find the message that would get a sale — nothing more, nothing less. While a headline and some body copy were mostly all Schwartz had to work with — and he made an art of wringing every last bit of sales power out of them — for us, there's so much more to work with.

Clarity strategy is bigger than sales, and it's bigger than marketing — it's about selling with all your GTM ducks in a row. Nevertheless, the same rules of awareness that Schwartz identified still apply, and for Schwartz, the rule was that the more 'aware' you had to make prospects, the more you had to say, and therefore the harder it was to sell.

If you struggled to nail your Hemingway story back in **chapter 10**, for example, this may have been why. The more aware your audience, the simpler, shorter, and more straightforward your message, and the easier it is to sell. But if you've got a less-aware market and more to say, finding that pithy story in a handful of words can be quite tricky. The same is true for your pitch — we've touched on the idea of selling with a three-act play (story 1) vs. just the final act (story 2), and the more you have to say, the higher the risk, but the higher the potential reward, as well, simply because you're operating in a less contested space.

Pitching by level

While your clarity strategy might be bigger than just your pitch, it's instructive to see how these levels of awareness might inform your sales narrative, too. For example, if we apply these levels of awareness to our three pitch outlines from the previous chapter, we see:

- **The CIO pitch:** This targeted folks at 1 and 2 on the awareness scale. Prospects weren't necessarily in market, and therefore the right-brain radar pitch was intended to take them from 1 (unaware) through 2 (awareness of a problem) to perhaps 3 (knowing solutions exist).

- **The SMB pitch:** This pitch, on the left-brain side, was for folks at stages 4 and 5 in awareness — they just needed some convincing this was the right product for them and an offer to create urgency to buy right there.
- **The positioning headshot pitch:** This pitch was more in the middle — picking up folks around level 2 or 3 who might know they have a problem or that broad solutions exist, but they don't know about yours or what solution is best.

> **What's a "mouse"?** The maturity of your market plays a big role in determining how much you have to say, too. For example, in 1984, Apple had to introduce the concept of a *mouse* with its original Mac ads. ("If you can point, you can use a Macintosh.") The market itself was at such an early level of awareness that advertising had to teach folks how you use this new computer.[3] Fast forward to the introduction of Apple Vision Pro in 2023, almost forty years later, and Apple has a computer that obviates not just the mouse but any touch surface altogether. And yet, Apple can obviously focus much more on selling the experience because of the decades of awareness that already exists in the market around VR/AR headsets, apps, and modern computing. That is, there's far more they can take for granted. How mature is the market you're selling into? How much do you take for granted? How much *should* you take for granted?

Different level, different game

B2B SaaS covers a vast spectrum of selling, tech, and business, and it's important to know what game you're playing, as your clarity strategy work can look vastly different at each end of these levels of awareness.

As SaaS legend Jason Lemkin writes in his article "Why SMB and Enterprise Sales Have Nothing In Common"[4]:

> *SMB SaaS companies are overloaded with "leads." They often have so many that they cannot follow up with all of them. [...] Most SMB SaaS companies I'm working with, after $2m-$3m in ARR, leads aren't the issue. SQLs may be, but really it becomes an efficiency, capacity, and closing ratio game. The key is closing as many leads as possible, as quickly as possible.*

Think about leads on our levels of awareness scale from 1 to 5. In SMB (like our Keap example in the previous chapter), it's an efficiency game: turning 4s into 5s, and 5s into customers with an offer. That's a very different game to C-suite selling (like our CIO pitch), where you're prodding folks at stage 1 or 2 with radar-targeted insights, or even prospects at stage 2 or 3 with positioning insights about a very specific problem they face. We'll pick up this topic of the difference in SMB, mid-market, and enterprise SaaS again later from **chapter 44** on.

Pick your levels

What's fascinating to me in the least-aware-to-most-aware process is, as always, this transition from right-brain to left-brain attention — from targeting folks' right-brain radar and then getting more specific — zeroing in on category, product, and price.

While Schwartz had his levels of awareness, other highly influential folks have had their own riffs on the concept, too. Let's take Everett Rogers, for example. Rogers popularized the *Diffusion of Innovations*, which is usually represented in a chart like this that should be familiar to anyone with a passing interest in tech market strategy:

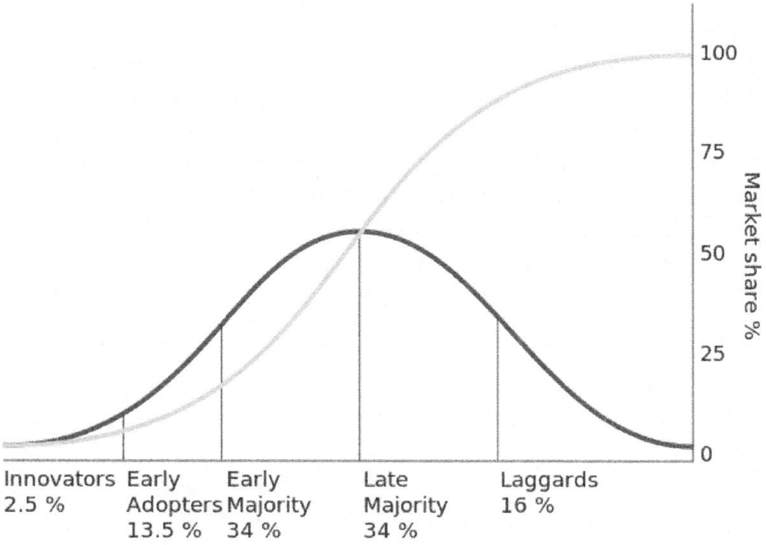

Figure 24.1: "The diffusion of innovations according to Rogers" from Wikipedia.

(The technology adoption curve represented by this chart, courtesy of Wikipedia,[5] is the foundation for tech market strategy classic *Crossing the Chasm*, which we'll explore in-depth in **chapter 40**.)

Rogers was of a roughly similar era to Schwartz, and his stages of awareness through to adoption were "awareness, interest, evaluation, trial, and adoption."[6] To my ears, those labels neatly describe the enterprise software buying journey, 60-odd years later.

Ad folks, on the other hand, have mini frameworks like AIDA (Attention, Interest, Desire, and Action), and these days in sales, we might consider the stages of awareness as buyer jobs-to-be-done, which, as we'll revisit in our JTBD primer in **chapter 35**. Those buyer JTBD include:

1. Problem identification
2. Solution exploration
3. Requirements building
4. Supplier selection

Whatever the levels, jobs, or stages, the question is: can you take folks through them? Do you have the story to pick up folks at one level and generate enough interest to reliably take them through to being 'most aware'?

Most SaaS outfits don't, and they end up stuck on #4 in the jobs list above (supplier selection). These prospects are the easiest to sell to and require the least story — if someone has already made a decision to buy and is just trying to select a vendor, then that's very different to someone who doesn't know they have a problem in the first place. But this is where competition is fiercest and opportunities for breakout growth are few and far between.

A major question for your clarity strategy, then, is where you want to operate on the scale of awareness. Are you trying to improve sales win rates for most aware folks, or are you trying to open up new sources of demand from unaware or less aware folks, perhaps by 'creating a category'? (We'll unpack category creation from **chapter 32** to **chapter 35**.) That's a strategic decision with profound consequences in terms of how you go to market and the resources you may need to have an impact.

Pitching MagicAIDB

Let's look at a theoretical example where we've mapped our clarity strategy against these levels of awareness for, say, an AI-powered database called MagicAIDB.[7]

If you were to segment *all* potential customers in your market by awareness, from the most tuned-out to the most tuned-in, you might describe each segment like this:

1. **Unaware:** They have no idea that your product or technology exists *and* no idea that the ensuing wave or trend will affect them (e.g., they have no real idea about today's hot-hot-hot flavor of AI).
2. **Concept aware:** They know the core trend or technology exists (e.g., they've read a brief piece about it in the mainstream media, so they get the concept), and they have a vague sense of the implications, but they

don't know what to do about it.

3. **Category aware:** They're aware there's a category of emerging tools that operate within this new concept, but they have no idea which one is best or why they should pick you.
4. **Product aware:** They understand the concept (new AI wave), category (a new breed of AI tools for, say, databases), and that you're a player (MagicAIDB), but they're not convinced you're for them.
5. **Offer aware:** They want to buy something, they know what you offer, and now they want a deal on MagicAIDB's pricing and to finalize the details.

This is just a simple example, but consider where your company targets folks along those five levels. How much story do you need to tell to get folks knocking on your door at #5?

This is where your ambition, resources, and narrative have to be carefully calibrated to the quantity of information you're not just trying to impart but persuade people with. The higher you go, the more you need to say, and if you have ambitions to compete at the concept or category level, your product either needs to speak volumes in and of itself and/or you need the resources to have your content (paid and organic; on and off your site) do that work for you. (One trick is to pick a wave to ride so the wave does the awareness work for you. We'll discuss waves and trends in **chapter 30**.)

If you can't do that, that's fine — just target 'more aware' folks further down the scale and calibrate your ambitions, GTM strategy, and perhaps even your product accordingly.

Content marketing and laddering up

> *Gonna get the banhammer from SEO Twitter if I keep posting like this - but SEO content is becoming a commodity. It all looks/feels/ reads the same. To be meaningful in B2B, you have to execute in a way so it reinforces your strategic narrative, otherwise it's just a carbon copy.*
>
> *The underlying issue here: until you have your strategic narrative figured out, and you can turn a random booth visitor at a conference into an interested account that wants to demo, you're not going to be able to do that with content.*
>
> *All of your problems are the same problem.*
>
> – JH Scherck[8]

We've established that B2B buyers often engage with face-to-face sales very late in their sales journey, which means a lot of 'selling' has to be done digitally beforehand. This is where marketing folks put content marketing to work, either for search or social.

On the search side, this can be done well, and JH Scherck of Growth Plays (with the very apt motto of 'pipeline not pageviews') is one of the most insightful voices on what that looks like in B2B SaaS. To make it work, as he says above, you have to have a meaningful narrative in place if you want to stand out in a sea of highly commoditized content.

What does that mean in practice? Note how Scherck describes taking folks through the stages of awareness, from a "random booth visitor at a conference" to an "interested account that wants to demo" on the back of your narrative and pitch. That kind of narrative is gold, especially if it's been tested and proven in the field. If you have that, you have the core ingredients for a playbook that generates pipeline through content. At that point, it's mostly an execution problem.

But folks often get this backward and start with the execution, forgetting the narrative. Indeed, one of the first marketing activities most B2B startups do is to start writing or commissioning content — articles, blog posts, whitepapers, and so on, all with the hope of generating sales one way or another, but often without the narrative to actual move folks through the necessary levels of awareness.[9]

Missing rungs

This kind of narrative-free content marketing is often done very naively, with the hope that if you cast your net wide enough, you can catch less aware folks who will land on your website and magically become "aware" of you, perhaps in a "brand awareness" way.

While this approach is often well-intentioned, it's ultimately from the underpants gnomes school of strategy, where the idea is:

1. Broad awareness pieces ('What AI will do for your {role/industry/future}').
2. ???
3. Profit!

Maybe this worked in 2010, but in the 2020s, the idea that you can even rank, let alone convert customers with such a simplistic approach when you're not an established or fast-growing brand is, well, far-fetched, to put it mildly. Hit-and-run traffic doesn't create awareness, and unless there's some action there — like a newsletter signup — you're unlikely to have another shot at taking them a step further up the awareness ladder.

To do that, though — to ladder folks up through the stages of awareness — you need to have enough rungs on the ladder in the first place. One generic rung at the bottom and one at the top with nothing in between won't cut it. This bottom-rung content might generate visits (or views on social media), but it rarely generates pipeline. How could it, if there are no more rungs for visitors to climb? This is why "content upgrades," free tools, product trials, and the ever-important newsletter help. They give you more chances to engage and provide more rungs towards your full product. And if nothing else, they can help build mental availability (**chapter 43**) so you come to mind *at all* when a buyer is ready to buy.

> **Product-led content:** Some SaaS companies, of course, do content marketing very well. SEO tool maker Ahrefs (disclosure: a former client), for example, does an exemplary job of weaving their product into their content marketing, so there's a natural ladder from interest in a topic to solving it, potentially with their product, in a single article. (Makes sense that an SEO tool company would be good at genuine SEO, right?)

The missing middle

The typical inbound marketing and sales SaaS playbook is built around the idea that the more you can show up in a buyer's journey, the more you'll be able to direct folks to your solution. Therefore, if you want people to buy your product, the more information you need to provide, and the higher the quality of that information, the better. Leads will follow the path you've laid out for them and, hopefully, arrive at your product, taking you up on your offer.

The missing middle rungs in most content marketing campaigns, however, have typically been due to a lack of subject matter expertise. Most marketers aren't subject matter experts outside of their respective disciplines, so the next move has been to rely on (and interview) folks who are.

Hey presto: quality content.

That's a great step towards fleshing out the middle rungs in the content marketing ladder, and true B2B content marketing leaders develop quite sophisticated approaches for both showing up when they need to in search and tailoring their narrative to folks at different stages of the buying journey.

If that kind of content marketing is something you've been able to pull off, awesome. More power to you. But remember that competitors get to have a say, too. And they might have quite a lot to say. What happens, then, when every player in a given category follows this same awareness playbook? What happens when there's an overwhelming amount of "expert" opinion floating around from every vendor out there?

The awareness arms race

While educating folks is still an inevitable part of the inbound marketing playbook, it has led to something of an awareness arms race. Buyers, especially in competitive markets, are now often faced with a high *quantity* of high-*quality* content.

This, if you follow the B2B sales literature, has become something of a problem for sales.

We've gone from the missing middle of quality content — those missing rungs on the awareness ladder — to having so many ladders with so many rungs that buyers have no idea which ladder they should even be on.

So what can we do?

We can turn to the sales literature because some smart folks have already thought about this. The next stage in the awareness arms race is helping folks make sense of all the information out there. Reps — and marketers, too, I'd suggest — need to bring *sensemaking* to their sales (and marketing) game.

This is the view of Brent Adamson, another co-author of *The Challenger Sale* (along with Matt Dixon, whom we met in the previous chapter) and one of the sharpest minds in the B2B sales world. (We first met Adamson in **chapter 21** on late sales engagement, and we'll meet him again when we touch on sales JTBD in **chapter 36**.)

Selling with sensemaking

In 2011, Adamson, Dixon, and their co-authors encouraged sellers to 'challenge' buyers with unique insight (as we'll see next chapter), hence the name of their book *The Challenger Sale,* and everyone dutifully did. Fast forward a decade or so, and now we've got Dixon making the case for helping prospects overcome their fear of messing up and giving them the confidence to make a decision *at all,* as we saw in the last chapter, and we've got Adamson making the case for sensemaking.

"Sensemaking" is a weird word, but the definition is just what it sounds like. Adamson writes in HBR[10] that "sensemaking reps" are the ones that "help customers make sense of the information they've encountered to make confident purchase decisions," decisions that result in a "high-quality, low-regret deal."

These B2B reps, Adamson says, citing Gartner research, close far more of these high-quality deals (80%) relative to "telling" reps (50%) and "giving" reps (30%).[11] Those are huge discrepancies in outcomes if the Gartner research is to be believed.

I find these insights into specific reps very revealing, and for us, the real value is extrapolating these insights back out to your broader clarity strategy. Selling is, of course, more than what a rep says; it's the whole point of view of the organization. It's the product and narrative. It's the marketers and success folk. It's your whole commercial GTM operation, and having that all geared to help folks make sense of their work and their world is what generates confident and successful customers.

Awareness overload

For folks with truly novel innovations, basic awareness of your new weird thing may still be the name of the game, but for those of you in hotly competitive B2B markets where buyers will make heavily researched, considered purchases, it's not so much *awareness* that's the problem as it is *awareness overload*.

This is a new reality that the typical levels of awareness frameworks don't capture, because being 'most aware' (stage 5 in Schwartz's 5 levels of awareness) is just about when the purchase happens and the buyer rides off into the sunset. But it turns out there's a sixth level of awareness — one of overload and indecision!

The SaaS world is nothing if not competitive, and when buyers are overwhelmed with choice, their ability to neatly zero in on one product becomes much harder due to the sheer amount of information and conflicting opinions out there. It's the paradox of choice writ large. This all conspires to undermine the prospect's confidence in making a sound buying decision. Sensemaking is about restoring that confidence, and your clarity strategy is fundamental to doing so. Or, as Adamson puts it:

> *Sensemaking sellers are particularly adept at clarifying complex problems, explaining technical information, and turning abstract concepts into understandable, shareable, and compelling insights. They give customers just enough information and then help them meaningfully interpret and simplify that material, creating a consistent, coherent, convincing narrative that bolsters customer confidence.*

That's a great description of what we're trying to develop here with your clarity strategy. "Creating a consistent, coherent, convincing narrative that bolsters customer confidence" nails it in delightfully alliterative fashion.

However, rather than just developing sensemaking *sellers*, I want you to think about developing a sensemaking *organization* where everything you do is an expression of how you help customers make sense of their choices, their problems, their risks, and what success can look like for them.

This is, in a sense, the ultimate form of awareness we want buyers to develop — not just awareness of a problem and our solution, but awareness that translates to confidence that they've found the best one for their particular needs, one that delivers the kind of success that was promised, success they're more than happy to tell others about, delivered by a team they believe in and trust.

> **Sensemaking and the right-brain whole:** In a McGilchristian sense, this is the more right-brain whole we're after in terms of how you see your company, product, and impact in the business world. It's not about *just* the product or *just* the pitch or even *just* your company — it's about all of it coming together between you, your customers, and the folks they serve, too. It's about those relationships and connections and that implicit thread that ties it all together as one interconnected whole.

Creating confident customers

Ultimately, you're in the business of generating confident customers, from low awareness to high awareness, from indecision to a confident decision, from being overwhelmed to things making sense, from random conference booth visitor to someone who wants a demo.

Remember, this doesn't mean just *giving* folks information that says your solution is the best or *telling* them this is the best solution (remember, information givers and tellers don't do so well, as Adamson suggests), but guiding folks so they feel confident *they* have made sense of *their* options and how *they* want to move forward.

This can be done from a right or left-brain perspective. We'll look at a bunch of examples of SaaS companies that helped buyers make sense of the world in a new way when we look at concept-driven clarity strategy in **chapter 31**, for example, and in **chapter 37** we'll look at how you can pattern-match and drill down to serve those that are truly best-fit customers for what you offer, i.e., the customers who most make sense *for you*.

Clarity strategy, then, isn't just a tool to help buyers go from unaware to aware. Ultimately, in a broader sense, it's a tool for helping buyers make sense of their situation, and not just as a tactic for sellers, but as an expression of your insight as a company, one that needs to run through everything you put out there that covers the 95% of the buyer's journey that happens before they talk to sales and the 5% that happens when they do talk to sales. And you can only do that if you've figured out how you make sense of your own market, too.

In the first part of this mini-series on pitching, we covered our three different pitches for different modes of attention, and in this chapter, we've zoomed out to the broader buyer's journey. In the next chapter, we're going to zoom back in and continue building our sales-fu as we explore the dynamic nature of attention, unpack more insights from the *Challenger* folks, and further explore the implications of the hierarchy of attention.

Notes

1. Schwartz, Eugene. *Breakthrough Advertising: How to Write Ads that Shatter Traditions and Sales Records* (1966).
2. These are derived from page 16 and on of Schwartz's book *Breakthrough Advertising: How to Write Ads that Shatter Traditions and Sales Records* (1966).
3. See the vintage ads here: **http://www.macmothership.com/gallery/gallery3.html**
4. Lemkin, Jason. "Why SMB and Enterprise Sales Have Nothing In Common," *SaaStr*, February 7, 2021. **https://www.saastr.com/why-smb-sales-and-enterprise-sales-have-nothing-in-common/**
5. From **https://en.wikipedia.org/wiki/Diffusion_of_innovations**.
6. Later, the labels became "knowledge, persuasion, decision, implementation, and confirmation," though the contents stayed the same, as Wikipedia details: **https://en.wikipedia.org/wiki/Diffusion_of_innovations#Process**.
7. Note that your broader GTM (go-to-market) strategy might segment the market using the Diffusion of Innovations buckets of innovators, early adopters, early majority, and so on, but here we're just interested in what we have to say to folks to take them from unaware to aware.
8. Scherck, JH. April 21, 2022. **https://twitter.com/jhtscherck/status/1516885680747540481**.
9. Scherck goes on to say "The other thing that makes it tough [in B2B]: vendors will tell you "it's never too early to start producing content and ranking in search" and that's mostly bullshit designed to get budget from companies that are way too early to invest in the channel but don't know any better." **https://twitter.com/jhtscherck/status/1516886713796218880**. Focusing on a narrative that works rather than just spraying and praying with content tends to be a much more productive exercise in the long run.
10. **https://hbr.org/2022/01/sensemaking-for-sales**
11. Adamson's rep definitions: '"Telling" reps are reps who speak from experience and their own expertise; "giving" reps are reps who fetch information for buyers, opting for comprehensive information.'

25

Pitching III — dynamic attention & Challenger selling

> The 'functions' are not arbitrarily housed together in this or that hemisphere: they form, in the case of either hemisphere, aspects of two whole ways of being in the world. [...] This is where the narrow focussed beam of left-hemisphere attention is concentrated: what is clearly seen.
>
> *- Dr Iain McGilchrist*[1]

In the past two chapters, we looked at tailoring our pitch to prospect attention and moving folks through the levels of awareness from broad awareness to narrow-focused attention on our product or offer.

These are both helpful ways of considering how you can engage prospects — assuming you can catch them at the right time in their journey. That is, we're assuming the buyer is on a fairly linear journey where some inciting incident starts their journey, they assess options, and select a solution. You catch them either early on (like our CIO or positioning headshot pitch) or towards the end of their journey (Keap and the SMB product pitch), pitch appropriately, and that's job done.

Sometimes it's that easy, but often it's not. Often, as we've established, buyers engage sales *very* late in their journey when they're already in laser-beam attention mode. That can make it hard to do anything but order-take. The prospect thinks they know what they want, they've got you locked into a feature bake-off as just another commodity vendor, and they only ask for information on the finer points of your particular features or integrations. (At the extreme end, this can make demos look more like solution engineering, as we saw with Arrows in **chapter 21**.)

But you don't want to be an order taker or a talking brochure; you want to sell. You've got a narrative and insights to deliver. You want to get the prospect thinking, have them open up, and engage with them as a thought partner on their real problems (which is unlikely to be the finer points of integration X).

The good news is that buying journeys are rarely as linear as our simple descriptions might suggest. Prospects bounce around levels of awareness and their buying jobs to be done, moving from vendor selection back to problem identification all the time. You can work with that. But to do so, you've got to break them out of that left-brain laser beam.

In this chapter, we'll wrap up our three-part series on tailored pitching by looking at the dynamic nature of attention and how you can use insights to move folks out of their narrow-focus laser-beam attention and back to the right-brain whole.

> **Choose your own adventure:** This isn't the final word on pitching in this book — in chapter 38 and chapter 39 we'll look at a couple of examples of early-stage startups who used founder-led sales to work their way into their market, testing their pitch and positioning and discovering their market as they went. And remember, the coming chapters are all about making your right- or left-brain pitch stronger. Make a beeline for chapter 29 if that's more relevant to you.

Right *and* left

So far, we've treated attention as static, something you either meet or miss. There's a lot of truth to that when it comes to engaging prospects using primarily one mode or the other. But attention is, broadly speaking, dynamic, with both modes working to help us 'attend' to the world around us all the time. And this is where we need to sprinkle some nuance on our simple (and, perhaps, metaphoric) right/left-brain binary.

For example, I've been describing positioning as a left-brain activity, as it's often a product of that left-brain laser beam: precise, logical, methodical, mechanical — *engineered* even. And that's true. However, a lot of the tools we use in *communicating* a position are right-brain ones in a McGilchristian sense: pattern matching (right brain), metaphor (right brain), and newness (right brain). The 'positioning headshot' pitch outline we looked at in **chapter 23**, for example, attempted to use both forms of attention to convey our pitch — some new insights to start before zeroing in on the prospect's specific situation and needs. What if we could move flexibly between those two modes of attention?

Moving between modes

As humans, we don't go through life using just one hemisphere, as though we're all collective stroke victims. The two functions — the two modes of attention — act as "two whole ways being in the world," as McGilchrist says in the opening quote. But we are — if McGilchrist is correct — prone to getting stuck in left-brain narrow focus. Sophisticated sellers — and startups — therefore, need to find ways to move folks between both modes of attention.

We know that, on an individual level, prospects may show up in sales calls in left-brain, narrow-focus mode, taking a logical, methodical approach to their purchase. We might oblige with detailed specifics about how our tool helps them, in a left-brain sense, 'grasp' or manipulate something in a new way. But we're not obliged to *only* respond narrowly with specifics and leave it at that.

And that's true on both an individual rep <> buyer level *and* a company <> market level. What unlocks the ability to move folks out of that laser beam? Well, awareness of the two modes of attention on the one hand and insight on the other.

Frame-breaking insight

"Insight" is such a broad term. It's not far off saying "good idea." Imagine if I said, 'It's ğood ideas~ that help unlock the ability to move between modes of attention.' I mean, wow, what a revelation!

But contrary to the idea that "insights" are just vague, nice-sounding, wishy-washy ideas, some folks have developed a fairly sophisticated take on what *kind* of insights make a difference in B2B sales specifically. It's not thought leadership (folks will think you sound smart, but they won't change their buying behavior), and it's not generic trends (they're too broad to be relevant). It's something much more specific, and that brings us back to B2B sales classic *The Challenger Sale* from the authors we met in the previous chapters (Matt Dixon and Brent Adamson).

The Challenger Sale gives us plenty of clues about what kind of insight can help us break out of the left-brain sales jail, introducing the idea of *reframing* sales discussions with a new insight that gets prospects thinking about their world in a new way[2] (their emphasis):

> *The act of delivering a teaching pitch is a skill, to be sure, but the* content *of a teaching pitch—the business issues you teach customers to value, the idea around which you reframe how the customer thinks about their business—must be scalable and repeatable, and as such, must be created by the organization (in most organizations, this is the job of marketing).*

(As an aside, it's interesting they point to marketing as the function responsible for developing this insight. This obviously has to be done in tandem with sales, but I'd love to see marketing be tasked with taking on rigorous insight development more often.)

This approach to insight gets further developed in *The Challenger Customer* as "Commercial Insight," which involves breaking down a buyer's existing frame of reference before introducing a new idea and 'upending the status quo'[3] (their emphasis):

> *What we find is that [commercial] insight is something else entirely. It's designed to upend the status quo. [...] It doesn't just convey an idea of what the customer could be doing (like thought leadership), but also conveys a story around what the customer is* currently *doing, explicitly laying out why that current behavior is costing the customer time or money in ways they never realized. That's the key. The contrast.* It's the cost of current behavior juxtaposed to the potential of alternate action. *Implicit in any good insight is the simple message: "Hey ... you're doing it wrong!" And done well, it causes your customer to say, "I have to change what I'm doing!" Thus the term "frame breaking."*

In terms of breaking down folks' existing frame of reference, you might recall how, in **chapter 8**, Segment's early rep would quiz customers to break down their assumptions with a directness that bordered on uncomfortable, asking why they couldn't just build this themselves (and leaving the prospect a little discombobulated) but, having had their frame broken, the prospect would then essentially sell themselves on Segment's solution.

In the same way, the idea is that this use of insight, done *carefully and respectfully* (!), appears to break folks out of their narrow, laser-beam attention and shift them back to a more open-minded, right-brain radar mode of attention, where they're open to what's 'out there.' That might be your solution, but more likely, it's *the reason why your new solution matters at all.*

Insight and behavior change

On the one hand, simply being aware of the kind of attention you're interacting with and the fact that you can nudge folks into a more right-brain mode of attention is a huge advantage relative to being unaware that you could do anything at all.

On the other hand, this is where it's critical you understand what kind of behavior change you are — or aren't — asking from your prospects and perhaps the market itself.

In *The Challenger Customer* quote above, their focus is very much on meaningful behavior change. You break through the status quo with a message that essentially says, to use their words, "Hey ... you're doing it wrong!" Why? Again, from the quote above, you want "your customer to say, "I have to change what I'm doing!" and that's why they use the term "frame breaking."

Now, *The Challenger Customer* is from 2015, and it's interesting that the focus on bringing fresh right-brain insights to your prospects has been so universally adopted in the B2B world in the years since that both Dixon and Adamson these days now temper that with recommendations on when to activate more left-brain attention, too, e.g., focusing folks' attention so they can get off the fence and make a decision (Dixon) and trying to make sense of all the competing right-brain ideas by zeroing in on what makes sense to the prospect in their particular situation (Adamson). It's a dance, and the dance evolves as we swing between insight and focus, perhaps bouncing back between modes of attention in terms of how B2B sales is done at all.

The Challenger Customer is also full of highly relevant insights on working with a champion on the buyer's side[4] (see the footnote for more), but for our purposes, let's keep our focus on insight and behavior change.

A little change or a lot

Back when we looked at the story 1 approach in **chapter 20**, we discussed 'the greatest deck ever' which was all about the big right-brain narrative and resulting behavior change — the world has changed, winners now act differently, and we can provide the tools to help you get to the new promised land.

Then we looked at story 2 in **chapter 21** and discussed the 'dumbest and most effective sales pitch I've ever heard' from Snowflake, which was explicitly about *avoiding* behavior change — just keep doing what you're doing, but use our tech, and it'll be better and faster.

Prior to both of those chapters, we looked at your Hemingway story in **chapter 10**, where I encouraged you to distill your proposed behavior change down to a pithy few words, a simple 'from/to' (like 'use chat not forms') that captured the idea at the heart of your narrative.

Whichever approach you take, if you're going to pitch on insight, you have to be *exceptionally* clear in your own mind about the kind of behavior change you're asking from prospects, how you describe it, and what they can expect in return.

A lot of startups want to have a bet each way and say 'It's a revolutionary new way to do X!' while also saying 'But don't worry, you can keep just doing what you're doing!' Well, which is it? If you're serious about a new behavior — if that's the foundation of your right-brain insight — think through exactly what you're asking and what the pain of change looks like for prospects. How can you support every step of their transition to this new approach? How can you reassure them about their FOMU — fear of messing up? What do your recent customers say was hard about making the change? What was most rewarding? And how successful were they, ultimately, in adopting this new behavior?

Likewise, if your pitch is that you really can just keep doing what you're doing, but with our new tool or tech, you still need to make the customer's switching costs — the pain of change — worth their while and know in exactly what way the prospect's life will be better because they switched to you. You'll probably still have to put your unique positioning attribute on their radar (**chapter 46**) and support their migration, but the outcome should be more immediate and quantifiable.

Whatever approach you take — the big or minimal behavior change, the big change story or the obvious solution, consider working on your nudge, or your 'challenge,' if you like, to get folks who show up very late in the sales process out of their left-brain attention — out of that narrow focus on very specific parts and pieces — and to get them to step back and think about the whole of what you're offering and what they're trying to achieve.

The nudge

Buyers in left-brain mode are not immune to being switched back into right-brain attention to think about the reasons driving their purchase. That is, in my view, the whole point of the *Challenger* concept. Mastering your ability to do that attention-switching can potentially improve win rates and lead to much more productive sales discussions.

However, if you think about this just on the rep level, you're only going to be operating within the relatively narrow window of opportunity a rep has when a prospect wants a demo. Ultimately, if you're a venture-backed startup, you don't just need incremental gains, you need to start challenging folks who aren't necessarily in buying mode but may well be open to *either* a bigger story 1 pitch *or* a laser-targeted story 2 pitch.

That is, we want to think about your clarity strategy as taking what works in individual deals and extrapolating that back out to your approach to the market as a whole.

In individual deals, I'm sure the best sellers move folks very naturally from left-brain attention back to right-brain attention, perhaps almost socratically (to borrow from Adamson), with probing questions and skill and thoughtfulness. When done well, I suspect the left- to right-brain attention switch is a bit of a (very gentle!) judo flip, with pro sellers able to engage with prospective buyers where their initial attention is at and take them *into and out of* the two respective modes of attention to get them seeing a given problem or potential solution in a new way, as a new whole, and then demonstrating why that matters with highly relevant specifics.

But can you do that on a bigger scale?

From nudge to shift

Those win/loss rates you might be trying to improve on a rep level don't include all the deals you're not seeing, and that's where the bigger game is played: at the market level. There, you need something more than a gentle nudge or flip. You need a serious shift.

What kind of frame-breaking can you do to get folks interested in your offer in the first place? This is where narrative *development* — the back half of this book — is, again, so fundamental to your clarity strategy. If you're still feeling like "insight" is a fuzzy concept here, don't worry, I've got you covered. We're going to get very specific on right-brain concepts like secular change + playbook + tool starting from **chapter 29** and left-brain insight from detailed customer analysis in **chapter 37** as we look at different ways you can stretch your narrative *out* and *down*.

Finding your insight

> *It might then be that the division of the human brain is also the result of the need to bring to bear two incompatible types of attention on the world at the same time, one narrow, focussed, and directed by our needs, and the other broad, open, and directed towards whatever else is going on in the world apart from ourselves. In humans, just as in animals and birds, it turns out that each hemisphere attends to the world in a different way – and the ways are consistent.*
>
> – Dr Iain McGilchrist[5]

If the modes of attention are universal — if, as McGilchrist says, humans attend to the world in two different, incompatible, but consistent ways — then it shouldn't be a surprise that what B2B sales researchers find (like the *Challenger* folks) has a left-brain/right-brain aspect to it. We're all dealing with the same brains, after all.

And if you and your sellers can experiment with moving folks in and out of the two modes of attention — sometimes you want prospects to hear your new idea; sometimes they need to stop fretting about all the competing ideas and zero in on a decision — then I believe you'll have a powerful skill at your disposal.

The real power move, however, is baking this attention to, well, attention into your organization as a whole. That's what we're going to briefly look at next, as we examine the right/left/right-hemisphere jump in the next chapter, before we look at how Figma applies its attention to building product and the lurking dangers for founders of left-hemisphere brain putty.

Notes

1. McGilchrist, Iain. *The Master and His Emissary* (pp. 83-84). Yale University Press. Kindle Edition. (Note that McGilchrist is talking quite literally about the high-resolution centre of our gaze here, but this quote highlights where I got the left-hemisphere 'laser beam' metaphor from — yet again, I nicked it from McGilchrist.)
2. Dixon, Matthew; Brent Adamson. *The Challenger Sale: How To Take Control of the Customer Conversation* (pp. 33-34). Penguin Books Ltd. Kindle Edition.
3. Dixon, Matthew; Adamson, Brent; Spenner, Pat; Toman, Nick. *The Challenger Customer* (p. 67). Penguin Books Ltd. Kindle Edition.
4. We're just focusing on attention here, but it'd be remiss of me to not briefly mention how the use of insight gets fleshed out in *The Challenger Sale* (TCS) and *The Challenger Customer* (TCC). The thrust of TCS is that B2B sales research revealed that 'Challenger' reps tend to be the successful ones, far more so than 'relationship builders' (to most folks' surprise). The authors present research and ideas about why this is, suggesting Challenger reps 'teach, tailor, and take control' in the deal, including with the use of insight, as discussed. This gets further fleshed out in TCC, where the authors suggest you need a prospective *customer* on the buying side to fulfill that role of teaching, tailoring, and taking control to drive consensus in the buying group (they call these folks 'mobilizers'), hence the name of the book: The Challenger *Customer.* This playbook was adopted so broadly, directly or indirectly, that it led Adamson, for example, to land on sensemaking, because customers were getting overwhelmed by sellers with a high quantity of high-quality insight, as we saw in the previous chapter, and then the challenge was to make sense of it all. How this plays out for your organization and your market will be unique to you, but the joy of SaaS and startups is that we're not trying to find ways to squeeze ever-diminishing increases in sales efficiency out of a static commodity B2B market; we're free to invent whole new products, new narratives, and new insights to help drive sales and improve how businesses work.
5. McGilchrist, Iain. *The Master and His Emissary* (p. 27). Yale University Press. Kindle Edition.

PART V

CLARITY &
VISION, ALIGNMENT & INSIGHT

26

Alignment & the right/left/right jump

[T]he right hemisphere tends to ground experience; the left hemisphere then works on it to clarify, 'unpack' and generally render the implicit explicit; and the right hemisphere finally reintegrates what the left hemisphere has produced with its own understanding, the explicit once more receding, to produce a new, now enriched, whole.

– *Dr Iain McGilchrist*[1]

Welcome to the insights section! We've been going deep on pitching, i.e., how we meet prospect attention on the x-axis of outward radar attention and/or on the y-axis of downward laser-beam attention. But the hierarchy of attention is useful not just for understanding how we *meet* attention in others, it also explains how we can *direct* our attention, too. That's the flip side we're going to look at in this section, which includes:

- **The right/left/right jump & how insight generation works:** In this chapter, we'll look at the hierarchy of attention in terms of how our brains operate in processing information and how we can basically derive — and debug — the startup journey from those principles.
- **Figma & how product management works:** In **chapter 27**, we'll briefly look at how Figma does product as an example of the right/left/right processes.
- **Brain putty & how our thinking works:** In **chapter 28**, we'll look at how our brains can undermine our quest to understand customer reality due to the insidious nature of left-brain brain putty. Founders, this is particularly relevant for you.

The insights for your clarity strategy — the information that helps you either work with, or break out from, left- or right-brain attention — have to come from somewhere, but where? Well, from how *you* 'attend' to the world around you, both as a leader individually and as a company collectively.

In the coming chapters, we're going to look at different examples of these kinds of insights. In this chapter, however, we're going to look at more first principles of attention and how the brain works and, as mentioned, essentially derive the startup journey from how we process information. I think this is kind of mind-blowing, but I'd love to know what you think, too, so let me know: **luke@lukestevens.co**.

The jump

It turns out McGilchrist has a neat model for how we deploy our attention: the right hemisphere → left hemisphere → right hemisphere jump, which works as he describes in the opening quote.

Recall that each hemisphere, while involved in everything, has its own tendency, and our right hemisphere is, to glibly simplify McGilchrist, concerned with the whole, while our left is concerned with the parts and pieces as it "unpacks" our experience. That means we tend to use the right half of our brain to find a thread in the broader environment, we take that observation to our left hemisphere where it's chopped up and analyzed, and then we put the parts back together in our right hemisphere to reintegrate into the whole.

That's the process we're going to unpack here. It sounds abstract, but it describes how a comedian might work a joke into their set or how a musician might learn a piece of music. We encounter something, we go through a process of learning, processing, and refining it, and then we reintegrate it so we can do it effortlessly and intuitively.

We can consider developing your clarity strategy in the same way:

1. **Right brain:** We start with intuitive and perhaps implicit ideas from broad observations.
2. **Left brain:** We nail down the narrow, specific parts of our pitch, for example.
3. **Right brain:** We then weave our pitch or insights back into the bigger strategic whole and our broader understanding of our product, company, and market.

To look at it another way, imagine you're a comedian writing a joke:

1. **Right brain:** You have an intuitive idea for something that could be funny — a thought, an observation, an experience.
2. **Left brain:** You write it out, iterate on it, run it by people, test reactions, and iterate some more. That is, you figure out the mechanics of the joke.
3. **Right brain:** You weave it back into some intuitive way of telling the joke in your set, getting the timing just right.

Ultimately, we want your narrative to land with the same reliable impact as a joke effortlessly told (but endlessly iterated on). But there's a process to these things, and that process reflects the way startups fundamentally come into being.

No to neurobabble: Let me stress again that this is but a humble business book and not a neuroscience paper and that I'm acutely aware of the blight of neurobabble in popular books, where folks hand-wave evidence for just about any claim with lines like 'and scientists have shown that's just how the brain works!' Therefore, we're discussing 'how the brain works' with our eyes open here, and I'm obviously greatly simplifying what I'm sure is an infinitely more complex process. But that simplification gives us a mental framework to work with, and that framework unlocks deeper insights into clarity strategy, startups, and, indeed, how we see the world, so I believe it's worth it.

Startups and the R/L/R jump

Consider the right/left/right-hemisphere (R/L/R) jump and the process most startups follow on their journey to the market, whether they realize it or not:

- **Broad awareness:** Startups begin with some founding insight — some intuitive initial thought, observation, or experience.
- **Narrow focus:** That gets turned into very explicit, left-brain lines of code, docs, pixels, features, etc.
- **Integrated whole:** Those parts get rolled back up as some implicit whole — the product, the brand, the experience.

In very broad strokes, then, startups themselves seem to follow this R/L/R pattern that we individually use to assimilate our experience of the world.

If we have steps that are necessary and near-universal, we can then debug what might go wrong in terms of our application of attention at each step. That is, some founders might get stuck on one step or another, e.g., thinking vision is everything or that they should skip straight to the second step and run headlong into the weeds. (This is the fat vs. lean startup distinction we touched on back in **chapter 4**.) The reality, however, is that you've probably got to go through the whole process to have something coherent. That is, so our brains make sense of reality in a coherent way, so your startup needs to make sense of the slice of reality you're tackling in a coherent way, too.

Selling and the R/L/R jump

Where else might we see this R/L/R jump? I suspect there are shades of this in the journey great sellers (including founders and CEOs) take their audiences on, too:

- **Broad awareness:** They introduce a new idea or insight, perhaps on the back of broad trends or a specific finding relevant to a narrow group of people.

- **Narrow unpacking:** They get down into the specifics of how it works.
- **Reintegrated whole:** They then come back up to reintegrate it into the whole, a new way of working, living, or Doing The Thing, whatever that might be.

There's something so universal about this experience and process that I suspect if you looked for it, you'd see it everywhere. What does it mean for B2B SaaS and tech companies?

> **Narrative, positioning, brand & the customer's R/L/R jump:** Back in chapter 5, we used a 2x2 chart to map narrative, positioning, and brand to three viable combinations of *mode* and *quantity* of attention — active right-brain attention (narrative), active left-brain attention (positioning), and passive right-brain attention (brand). Notice a pattern there? R/L/R. To be clear, in that chapter, I wasn't suggesting you have to move linearly *through* those contexts of attention. However, when we're talking about introducing new products and technology to a B2B market more broadly, then perhaps ultimately we *do* want to mirror this R/L/R loop, that is, mirror the way our customer's brains fundamentally attend to the world and *integrate* their experience. That is, from the customer's point of view, they need to: (a) hear about something genuinely new and compelling and have it land on their radar as such (narrative); (b) interrogate it in detail to ensure it maps to their specific needs (positioning); and (c) integrate it into their memory and view of the world as a whole as the brand associated with a specific need (brand). That is, the product must register as new, do what it promises, and be integrated into the customer's world such that they can't imagine life without it. Easier said than done, of course, but it's fascinating to see the R/L/R loop show up both in the micro of a sale, the macro of customer attention, and, as we're looking at in this chapter, the way startups can operate.

The vision → details → integration loop

> *The left hemisphere [...] is never an endpoint, always a staging post. It is a useful department to send things to for processing, but the things only have meaning once again when they are returned to the right hemisphere.*
>
> – Dr Iain McGilchrist[2]

In a world that easily strays into left-hemisphere dominance, as McGilchrist believes, it's worth noting that the left's place, as it were, is really as a "staging post" and "never an endpoint," as McGilchrist says.[3]

I think this is a profound insight into how companies are run, especially in terms of insight generation (or the lack thereof). Consider not just the three stages — right, left, right — but the two jumps in between:

- **Step 1 — Start with (but don't get stuck on) vision:** There must be some original right-brain vision or bet about what could be 'out there' in the future, based on the success of your product or technology, as a starting point. It generates the hypothesis that gets unpacked and tested by the market.
- **Jump 1 — turn the implicit explicit:** The first jump from vision to specifics is the hardest. It's a big part of what your clarity is about — trying to articulate the vision in the first place, perhaps as a Hemingway story (**chapter 10**), so you know exactly what it is you're trying to prove in the market. (We discussed the implicit vs. explicit when discussing Slack back in **chapter 12**.)
- **Step 2 — Left-brain deep dives:** This is where you and your team obsess over customer behavior and outcomes. Note that this requires narrow-focus attention, a skill vision-focused leaders aren't always blessed with in abundance (and vice versa). This is where insight is found, but the fact it *can* be found is because the vision — and the activity that ensued — generated it.
- **Jump 2 — turn the explicit implicit:** If the first jump is about going down from vision to specifics, the second jump is about going back *up* with specifics to inform the vision — do we stick with it or pivot? Do we go all in on the original bet, refine our positioning, or look at smaller, subsequent positions we can take?
- **Step 3 — Reintegration:** Your discovered insights need to be reintegrated not just into 'the whole' as far as the founder or CEO's thinking goes but ideally into the fabric of the company itself. It has to loop back into the implicit right-brain context that everyone is operating from. That is very hard to do! That's why simplicity in narrative goes a long way in keeping everyone on the same page.

This process demonstrates what makes startups so hard — you're both trying to sprint ahead from the starting point you've chosen while simultaneously trying to look for insights that might suggest a far better starting point — or even just a better lane to be in — instead. That would put a lot of demand on anyone's, or any team's, collective attention.

Nevertheless, we can see that McGilchrist's work and the hierarchy of attention give us — if we run with it — not just a way of meeting buyer attention, but a fundamental mental model for the operating loop for companies on the whole, especially when it comes to insight generation and integration.

> **Principles vs. practice:** If you need insight, the R/L/R jump is, in the abstract, how it's found. This is the kind of powerful first-principles theory that can help you diagnose your own situation, and we'll see how it appears to play out for Figma in the next chapter, too. In terms of insight driving incredible success more broadly, we've covered some examples in earlier chapters (Loom in chapter 10, Slack in chapter 12), and we'll look at plenty more in the chapters ahead, including our trifecta of CRMs in chapter 31 and category-creation darling Qualtrics in chapter 35.

Alignment states

The R/L/R jump also sheds light on the #1 management challenge most startups face as they scale: alignment.

> *The teams aren't aligned. How do we align? We need to be more aligned. Are we aligned?*
>
> – Every founder trying to scale

Alignment is one of the most overused words in modern office speak, subbing in for agreement, awareness, understanding, acceptance, and more. Nevertheless, 'alignment' has become a catch-all because it captures the core challenge of any growing organization: coordination.[4]

Clarity strategy work isn't just about dialing in a story 1 or story 2 pitch for sales to find alignment with customers; it's often more about the necessity of the team, especially the leadership team, being aligned *at all*. Having a narrative *at all* to 'align around' (i.e., coordinate around) is where much of the value comes from.

That is, the value of your clarity strategy is often in taking leaders through a collective right/left/right brain loop: starting with threads from the environment, chopping them up and analyzing them collectively, and reintegrating them as a whole. But not just an abstract, on-paper whole; a *shared whole* in the form of a narrative everyone accepts and can get behind.

That gives you real narrative alignment — shared context, shared language, shared strategy.

Debugging alignment

Now we've got a sense of the right/left/right process and the need to arrive at an aligned, integrated whole, we can debug each stage.

As we do so, let's remember that your clarity strategy isn't just about having everyone agree to a given narrative we've come up with; we still want that narrative to be driven by insight that accords with customer reality. It needs to be found as much as it is created. That means there can be a lot of aligning — of threading the needle — required, including:

1. **Language/vision alignment:** Finding the language for that initial right-brain vision can be extremely difficult. Sometimes, founders are years into their journey before *finally* getting the language that clicks with their vision. (And I often can't help but wonder how much more efficient their years prior would have been if they were underpinned by a clearly articulated narrative.) You need a name and language for what it is you're aligning around, both internally and with customers, which is part of why we discussed the importance of naming things back in **chapter 17**.
2. **Vision → details alignment:** This is the first jump — a classic startup killer. Having a big vision is one thing; being able to accurately assess if the details on the ground reflect that vision or not is another. It requires enough confidence to back the vision in the first place and enough humility to refine it — at speed — in the face of contradictory information from contact with reality. That's a rare combo. (If you want to see what that kind of founder humility looks like in practice, watch the former CEO at Segment, Peter Reinhardt's interview with Logan Bartlett.[5])
3. **Discovery/customer alignment:** The narrow-focus attention has to align with customer reality. There's a lot of 'discovery theatre' that happens out there. These are the assumptions you build your company on — better be sure they're not garbage.[6]
4. **Details → vision alignment:** The flip side of the first jump — it's easy to get stuck in the weeds. Think about left-brain consultants churning out massive, detail-riddled decks that end up as dust-collecting shelfware, for example. People are just not going to integrate that. *(Not that your humble author would know...)* Don't be data-rich and insight-poor — develop insights you can socialize.
5. **Narrative/team alignment:** Reintegration needs to happen on a company level, too. You might have the perfect clarity strategy, for example, but it's pretty useless if you can't socialize it internally. Selling a narrative internally *and* externally is how you ship it, and the more memetic it is (**chapter 15**), the more likely it will stick.

This is how you get alignment between vision, customer reality, and team direction. It's vastly (!) easier said than done, of course, but it's so vital for the health of any company that it's worth taking seriously.

Working backward

If insight discovery, alignment, and integration is the goal, and if we know what, at a very high level, the process looks like, then we can work backward by asking: what *can* be reintegrated? And how do we make sure alignment sticks?

This is where simple wins. If you have to take folks through a 30-slide deck *(or an absurdly long book...)* explaining the ins and outs of your core insight, they're going to struggle to get it.

Thinking is a social activity (that's why we like movie or music reviews: to see what others think), and we need to calibrate our efforts and expectations accordingly. We need to calibrate our narrative to what we can all 'think' socially and collectively, if I can put it that way. Whatever you come up with needs to be succinct enough and compelling enough to stick in your mind a month from now, and doubly so for those one step removed from this process.

The one thing

This is why, for all the ideas we're working through here, you need to keep coming back to the one thing — that one hook to take to your team, your customers, and your market. That one thing might be expressed in your narrative snapshot deck (**chapter 9**), your Hemingway story (**chapter 10**), what you name (**chapter 17**), the specific ICP your target and position you take (**chapter 37**), or your category concept (like Drift in **chapter 31**, Yammer in **chapter 32**, or Qualtrics in **chapter 35**).

There are plenty of ways to arrive at that *one thing* because business success is largely emergent — that's what folks mean about luck and timing. Therefore, the point isn't that you have to mirror one of the above approaches exactly, it's just that you have your own awareness calibrated to see meaningful insight when it appears, luck and timing included. Then, when you have that one thing, you know what you're naming, what you're validating, what you're testing in the market, and what, ultimately, you're aligning around.

That's the discipline working backward encourages — zeroing in on the one thing that helps you thread the needle between your vision, team, and customers. That's the founding insight that matters, and it's that insight that will drive your clarity strategy.

Next, we'll see what the R/L/R jump looks like in management practice as we observe how Figma builds product.

Notes

1. McGilchrist, Iain. *The Master and His Emissary*. Yale University Press. Kindle Edition.
2. McGilchrist, Iain. *The Master and His Emissary* (p. 195). Yale University Press. Kindle Edition.
3. That is actually the whole thrust of McGilchrist's book, *The Master and His Emissary*, that is the philosophical basis for this book. McGilchrist's view is that the left hemisphere should be the 'emissary' to the right hemisphere's 'master,' or the staging-post to the right's integrated whole. Instead, however, we live in a world where the left hemisphere has usurped its position and become the master, leading to a variety of social and civilizational ills that plague us today, ills that are well beyond the scope of this humble tome!
4. Specifically, overcoming coordination headwinds. See this piece from Venkatesh Rao, for example, **https://studio.ribbonfarm.com/p/coordination-headwinds** on this fantastic deck on coordination headwinds by Stripe's Head of Strategy, Alex Komoroske: "How Organizations are like Slime Molds" **https://komoroske.com/slime-mold/**.
5. This is the interview I quoted from when we discussed Segment back in chapter 12, and I love his down-to-earth retelling of their journey despite their success: **https://www.youtube.com/watch?v=gPGVuL1KLhg**
6. It doesn't take much to quantify it, however you choose to do that. You might write up formal job statements (we'll look at Jobs to be Done in **chapter 36**) or just rate issues you discover by frequency and severity. Perhaps you can add a confidence rating about the insight, too. A little bit of rigor goes a long way when doing discovery, and there are plenty of more formal product-management-oriented approaches out there, too.

27

How Figma builds product

Great execs have exquisite control over when to dive into nuance and when to keep things high level. The best can switch instantly between the two, and understand the value of both. The worst see the world as binary (you're dumb if you see details / dumb if you're too high-level)

- *@staysaasy*[1]

The tweet quoted above perfectly captures what this chapter is about. In recent chapters, we've been deep in clarity strategy theory, thinking about dynamic attention, shifting buyers (and markets) from left-brain to right-brain attention with your insight, and finding and aligning around that insight through the right/left/right jump.

That's fun to nerd out about — the hierarchy of attention really does have profound implications for how we deploy and direct attention — but it can be quite theoretical. In this chapter, we're going to switch gears and think about what these ideas look like in practice. To do that, we'll briefly examine Figma as an example of a successful, highly competent, well-managed product team and see if we can spot the right/left/right pattern in how they work. (You might be able to spot it in your own team's rituals and processes, too.)

Figma's management radar and laser beam

After Figma was almost acquired by Adobe for an astronomical $20B, Sho Kuwamoto, Figma's Director of Product for their Editor, outlined on Twitter how Figma works as a company, at least on the product side.[2]

Let's unpack Kuwamoto's observations in his insightful Twitter thread[3] and look for the hierarchy of attention in action. First, on the executive level, here's how Figma operates:

> *At the company level, prioritization is about strategy and resource allocation. As a company, we decide on the most important areas of focus at the exec level, and staff teams accordingly.*

That is, of course, the domain of the right-brain radar. Executives look at strategy, make their bets, and allocate resources accordingly. They dedicate a large portion of their attention *out* — out at the competitive landscape and out at the progress they're making and the reception they're getting in the market. They watch their radar. When they see something that looks interesting, they push it down for left-brain attention — the left-brain attention of their team.

Here's Kuwamoto again (emphasis his):

> *Each team has a rough sense of purpose, and is not* overly *metrics driven. Instead, each team focuses on user happiness. For example, the design systems team makes sure that design system managers really like working with Figma.*

At the team level, the radar is a bit fuzzy ("a rough sense of purpose"), but that's fine; they're there to do the left-brain laser-beam work. But what do they focus their proverbial laser beams on? Well, not just KPIs, but *actual* customer success — "user happiness"!

That is, the team has to take the right-brain thread of narrative insight (the 'rough sense of purpose'), chop it up into whatever it is they'll build or solve, and know their work is done *when it's integrated as right-brain happiness.*

> **The fractal R/L/R jump:** This is pure speculation on my part, dear reader, but I like to think of the right/left/right jump in fractal terms. At an executive level, folks are mostly using radar attention, and then they task a team to execute on a certain bet or initiative. It would be tempting then to grossly simplify and say the team pushing pixels or writing code then just focused on the narrow, left-brain parts, builds the thing, and calls it a day. But of course, that's not what happens. It's cycles within cycles, and the Figma R/L/R cycle on the team level is particularly instructive because it speaks to something implicit and right-brain — user happiness.

Contexts of use

I love that focus on user happiness. Making users happy seems painfully obvious, but sadly many product teams will substitute some abstract, left-brain metric for happiness instead and confuse the map for the territory. Not because they don't want to make users happy, but because *they don't really know what that is*. Happiness isn't a number; it's implicit, and it's hard to put a KPI on the implicit.

Many software teams lack a deep understanding of the implicit idea of user happiness for their specific customers. Why? Because the team lacks clear contexts of use. As Andy Matuschak writes in his short, sharp post "Tool-makers usually lack connection to a serious context of use"[4]:

> *People making Tools for thought often say they're trying to help people do math, or make art, or whatever. But in reality, the people making these tools are rarely connected very deeply with the actual creative practices they're trying to amplify [...] Such tools might look mathematical (or whatever) on their surface, but they're not seriously trying to answer hard problems in those domains—often because the creators don't actually know what those problems are or understand how to solve them. Effective system design requires insights drawn from serious contexts of use.*

To really fire up the laser beam of narrow-focus attention, it has to be aimed *at* something, i.e., at a serious context of use. Without that, how will you ever develop an intuitive sense of user happiness?

Users first

Back to Figma. Kuwamoto continues[5]:

> *Within this broad umbrella of making users happy, teams have to prioritize between bug fixes, performance, new features, etc.*

At the team level, that laser beam of attention gets applied very carefully and deliberately, as every product team knows. But what's fascinating about Figma's approach is that, according to Kuwamoto, they won't, by default, prioritize what would make "a good story," the flashiest launch, or even what would attract the most new users — 'self-centered' goals that would make the team look good but might not make users happier.

> *In practice, [this] means we end up working on a bunch of stuff that isn't flashy.*

Instead, they drill into those contexts of use asking "What do our users want us to make better?" and putting effort into non-sexy work, which can mean spending "months working on copy and paste because we heard it was frustrating."

And they really do go deep. Kuwamoto calls out *not* going deep as a trap:

> *Trap #2 [is] not going deep enough on problem solving. For big, important problems, we try to give ourselves time and space to think deeply. There was a point in time when people were complaining that the assets panel was too hard to navigate. This was especially true of our biggest customers, who had thousands of components.*

The point of the left-brain laser beam is to go deep. And the more time and space you give a team to apply their collective laser beam, the more chance they'll have of getting to the root issue. In this case, it was a prominent issue that customers were complaining about, one that got worse as the customer got bigger. Ouch. How did they apply the laser beam there?

Solving the real JTBD

They drilled right down to the real job-to-be-done:

> *An obvious "solution" to this would be to add a way to organize the assets panel. Or maybe improve search so people could find their component among the hundreds of components. But that would be solving the wrong problem. [...] When we looked at real world component libraries, it was obvious why the assets panel felt overwhelming.*

They got down to the real issue and discovered the root problem, which was that "there might be twenty different versions of a button." Versions... variants... too many variants that should be grouped together... aha! They needed a variants feature, which is what they built, and that solved the fundamental problem. (A solution, Kuwamoto notes, they continue to refine rather than always chasing the shiny new thing.)

That is, they were able to solve the *class* of problems (handling variants), not just a superficial *instance* of the problem (organizing the asset panel), but drilling down to the core job-to-be-done.

Figma does that over and over; designers and others who might spend eight+ hours a day in the app come to love it, and away they go. Rinse and repeat. That's the beauty of having the resources and inclination to really solve the problem, and not just in a juice-the-metrics way, but in a solve-the-real-job-to-be-done way.

Ok, so they solve nitty-gritty problems well... so what?

The mega-trend

On its own, that dedication to product development is admirable, but what does it mean on a strategic level? How does it connect back up to Figma's radar-level bets? How do they do the reintegration jump in the R/L/R loop?

Here's Karri Saarinen, co-founder of buzzy Jira competitor Linear, talking about his observations of the Figma team when he was working at Airbnb as a designer[6]:

> *Figma [...] listened to what Silicon Valley design and product teams*

> *actually needed. They were discovering the product for designing products [...] For example, they came to the Airbnb design team to ask about what we would need for the design system and then they basically built that. [...] It was way better than what [competitor and forerunner] Sketch [offered] and closely matched where the industry was at. They addressed the whole process, not just designing the UI. They built simple prototyping tools [and] web sharing and basic handoff tools [that made other previously necessary tools] obsolete.*

Did you spot the mega-trend there?

Large teams — like Airbnb — were building digital products with design systems at scale. That was a big secular change: a collaborative approach to digital products (i.e., websites and apps) at scale. These kinds of industry-level trends create new needs, new methods, and the desire for new tools.

Macro to micro

Recall Kuwamoto's comment that "Each team focuses on user happiness. For example, *the design systems team makes sure that design system managers really like working with Figma.*" (My italics.)

There's a big clue in that statement, and it goes beyond just drilling down to the job-to-be-done.

The emergence of design systems, especially in large companies, and the "design system managers" who oversee them was an emergent trend along with digital product design.

The emergence of a new role — like design system manager — is a big hint that a significant change is underway. (The emergence of the customer success manager, for example, is the trend Gainsight rode to fame and fortune, as we'll discuss in **chapter 30**.) That's where category-level opportunity comes from, and that's something we'll explore in much more depth in the coming chapters.

But back to Figma. As Kuwamoto mentioned above, the Figma editor team also solved for ultra specific jobs-to-be-done, spending months getting copy and paste right, for example. That requires very highly attuned laser-beam attention focused on the *micro* of customer success, too.

Right/left/right

See the pattern here?

- **Right-brain radar:** Figma's management radar was plugged into the *macro* of their customer's success — new ways of working, new systems, new roles (design system manager).

- **Left-brain laser beam:** Figma's team also had permission and encouragement to go deep on *micro* details like getting copy & paste right to remove frustration and keep customers happy.
- **Right-brain reintegration:** The virtuous loop of working hand-in-glove with their ICP customers meant they could integrate the macro and micro insights into a compelling whole — a product that superficially competed with existing products but really was a new, distinct entity that rendered a whole ecosystem obsolete.

This is the radar and laser beam working beautifully together — global and local attention, x- and y-axis, story 1 and story 2 — whatever term for attention we use, you can see it clearly in action here.

The *way in which*

I want to make two observations here. First, I understand there's an obviousness to this, to a degree. *A company works on major initiates for their customers and small quality-of-life improvements, too. Big deal!*

But the point here is that it's *the way in which they actually did it* that matters. They engaged on the radar level to build something "way better" than the competition that "closely matched" where the best in the industry were at, to quote Saarinen. And they engaged on the laser-beam level in such a way they could spend *months* getting the details right *without* intense KPI pressure or even the pressure to tell the best story.

That combination is, sadly, exceedingly rare. One wonders what kind of software utopia we might live in if it wasn't.

The second observation is more subtle. I don't think it's too much of a stretch to say that as an organization, Figma really was able to do the right/left/right jump, from right-brain awareness down to left-brain laser beam focus, and back up to *integrated* right-brain awareness.

That enabled Figma to solve the *whole* job-to-be-done of designing digital products. That involved fantastic execution and problem-solving, too, let's not forget. But to enable that work, management had to conduct the right-brain/left-brain orchestra in such a way that lets folks engage both broadly and deeply on stuff that matters.

The end result is, as Saarinen says in reference to how Figma has inspired his work with Linear, a product where "the individual features can often look very simple, but the magic happens when all of it comes together."

That's all well and good for users and fans of the product, but what does that look like in competitive terms? Absolute, crushing market dominance[7]:

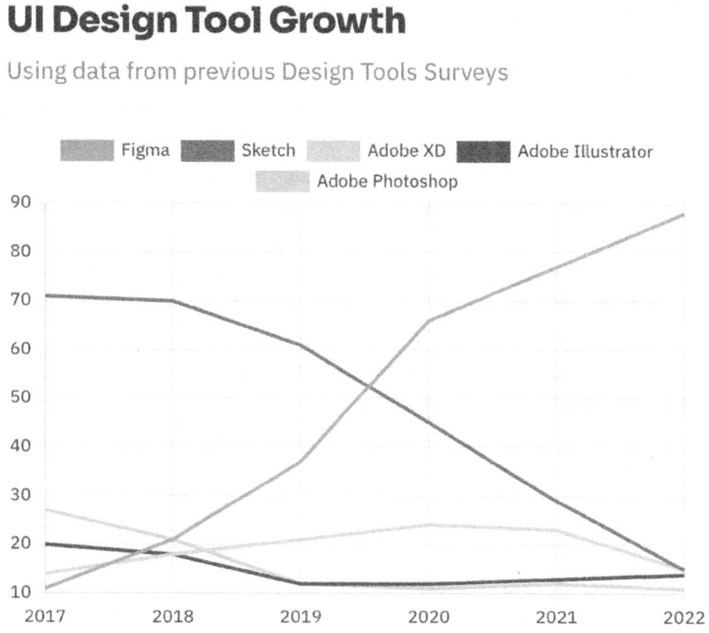

Figure 27.1: Figma's unstoppable growth as captured in UX Tools' annual Design Tools surveys.

What's on your radar?

Way back in **chapter 5**, we discussed clarity strategy as the custom framework for our NPB tools — narrative, positioning, and brand. The positioning part of that equation can be a function of your competitive business strategy, and Figma, as Saarinen (as I'll draw from here) and many others have noted, is quite unique in that regard. Consider the unique aspects of Figma's strategy that have contributed to them being such a competitive success:

- They had a cloud-first, WebGL-based architecture they spent four years (!) building before shipping, and no one has really been able to match it since.
- Their GTM and pricing model was (as Saarinen notes[8]) built for whole-org collaboration and user growth in a classic PLG play.
- They rode the secular change of digital product design *itself* — they didn't outcompete Adobe, for example (as Saarinen also notes) by fighting yesterday's battle and building "Photoshop in the cloud."[9])

- They didn't subject their teams to total metric/OKR hell[10], instead letting them solve for customer success both in the macro by studying their ICP and in the micro by focusing on user happiness.

The result was something that was greater than the sum of its parts... $20B greater, at one point, after Adobe agreed to acquire them in September 2022. (And then un-agreed in December 2023 due to antitrust concerns, necessitating a couple of hasty edits to this chapter!)

The macro lesson here is that it's this multiplicative approach that pays off, as we'll see when we move into discussing the *strategy* in clarity strategy. Spotting a trend and positioning yourself around it is nice as far as your *story* goes, but executing that as a *strategy* across product, leadership, and growth wins markets.

Doing that, however, to revisit our opening quote, requires execs who "have exquisite control over when to dive into nuance and when to keep things high level" and can go from speculation and theory to execution and reality, dodging left-hemisphere brain putty along the way, as we'll see in the next chapter.

Notes

1. Anon, @staysaasy. March 29, 2023. **https://twitter.com/staysaasy/status/1640808749160976394**
2. There have been other insider takes on how Figma operates; this is just intended as a neat example of directed attention in action, and it is just one slice of one person's experience at Figma. For other perspectives, see, for example, this 2022 interview with Figma's CPO Yuhki Yamashita by Lenny Rachitsky: **https://www.lennysnewsletter.com/p/how-figma-builds-product**
3. Kuwamoto, Sho. September 24, 2022. **https://twitter.com/skuwamoto/status/1573418806323843072**
4. Matuschak, Andy. "Tool-makers usually lack connection to a serious context of use," February 2020. **https://notes.andymatuschak.org/zs5uUEv9iJH7JuAmsCChxBEMP2zW6CRbiAWE**
5. Kuwamoto, Sho. September 24, 2022. **https://twitter.com/skuwamoto/status/1573418806323843072**
6. Saarinen, Karri. September 21, 2022. **https://twitter.com/karrisaarinen/status/1572353519205888001**. Lightly edited for clarity.
7. "2022 Design Tools Survey," *UX Tools*, 2022. **https://uxtools.co/survey/2022/ui-design**
8. Saarinen, Karri. September 21, 2022. **https://twitter.com/karrisaarinen/status/1572353541251158016**

9. That was done by Ivan Kuckir, an at-the-time 20-something solo developer who put 7+ years (!) into his web-based Photopea project, as discussed in this interview: **https://www.lunadio.com/blog/the-story-of-a-unicorn-solo-founder-making-dollar500000-arr/**

10. Though they did for a while, and seem to have found a decent balance since. In his 2022 interview with Lenny Rachitsky, Figma's CPO Yuhki Yamashita said, "When I first arrived at Figma, I actually deprecated OKRs. We had dreadful companywide OKR meetings where we went through a spreadsheet tracker of hundreds of OKRs that were really more like tasks assigned to individual contributors. [...] I asked teams to instead define headlines—essentially, claims that they'd like to make by the end of some time period. [...] This helped me understand what a team was philosophically going for. [...] It recognized the reality that some things, like the core experience of Figma, are hard to measure and can't be reduced to a single metric [...] Once we hired a data-science leader, we brought OKRs back and have iterated on this format a bit. In some areas, we've become more sophisticated with metrics. In other areas, they still can be a bit more qualitative." **https://www.lennysnewsletter.com/p/how-figma-builds-product**

28

Brain putty & the reality layer

> [T]he left hemisphere needs certainty and needs to be right. The right hemisphere makes it possible to hold several ambiguous possibilities in suspension together without premature closure on one outcome.
>
> *- Dr Iain McGilchrist*[1]

Let's wrap up the first half of this book with one slightly unpleasant fact: even if you follow the best narrative examples and use the best narrative formats, it's not going to work if the insight at the heart of your narrative turns out to be brain putty — stories from your left-hemisphere your brain tells you are true but aren't.

In this chapter, we're going to inject a little humility before we hit the big ideas to come, where there will be plenty of temptation to not just *speculate* about narratives that *might* be true, but *confabulate* about narratives that go beyond what we know and... well, there be dragons.

This humility is necessary to temper the tendency of the left hemisphere, which, on its own, "needs certainty and needs to be right," as McGilchrist says in the quote above. This can manifest in some truly bizarre beliefs in extreme cases, as we'll see in a moment. But the more insidious form of faux-certainty is what I describe as "brain putty," and the antidote is intellectual honesty and a healthy culture of investigation, doing everything you can to avoid a cult of confirmation.

> **Choose your own adventure:** We're about to embark on part two of this book, sailing the seas of narrative development, both on the story 1 side of concepts and categories (chapter 29 on) and the story 2 side of positioning in all its varied forms (chapter 37 on). There are *loads* of practical ideas and examples ahead, but in this chapter, we're going to take one last big swig of McGilchrist's ideas and apply them to *how we see reality.* It's pretty deep. If you want to stay focused on the practical story stuff, skip ahead to chapter 29. But if you want to nerd out on some more profound left-brain/right-brain concepts, read on!

Right- or left-brain founders?

Healthy cultures are built by founders and operators who have a unique skill: balanced, flexible attention. They have "conviction without delusion," as one investor put it.[2] They can maintain that broad view and dip down — right down — into the specifics with ease. They can hold a vision in their minds while also staying firmly grounded in reality.

We saw what this looked like with Figma in the previous chapter. But what might a less healthy attention balance look like?

- **Too right-brained:** Narrative is right-brained, and the right deals with the implicit. Founders can get stuck in their implicit vision and fail to articulate it clearly, leaving everyone guessing about what they really mean.
- **Not right-brained enough:** Per McGilchrist's quote above, the right hemisphere "makes it possible to hold several ambiguous possibilities in suspension together," but if you're not right-brained enough, you may succumb to the left's faux-certainty, which has "a tendency to premature over-interpretation."[3]

Or, on the flip side, you could be:

- **Too left-brained:** You get overly fixated on the parts, not the whole, and fail to integrate your discovery or technology into some broader cohesive whole.
- **Not left-brained enough:** Your left brain helps you drill down to customer reality and focus narrowly on what's in front of you so you can manipulate it (in a tool-wielding sense) and find the seeds, so to speak. Fail to do that, and you end up in product no man's land.

Instead, we want a healthy balance that avoids the extremes while making the most of each mode of attention:

- **Maintaining right + left balance:** You need to manage ambiguity (right brain) while narrowly focusing on what's in front of you (left brain) to get a reasonable grasp of reality.
- **Resisting right/left extremes:** You need to avoid ignoring the details or failing to communicate your ideas explicitly (too much right) and avoid failing to sufficiently understand the world around you as it's unfolding over time (too much left).

That is, it's not just the *type*; it's also the *quality* of attention you bring to the world that matters as a leader.

And no one did it better than you-know-who[4]:

> *"[He would be] really pushing you... relentless on the details...*

> *challenging you, for the right reasons. It wasn't bullying, it wasn't demeaning [...] it was really that attention to detail. "*

That's Tony Fadell on the *Lex Fridman Podcast* responding to Fridman's question about what Steve Jobs was like to work with. Fadell continues:

> *"When he would make a [opinion-based] decision... and you're trying to work with a team to implement those decisions, you have to really tell the* why *of those decisions... so you can feel part of that; so you can understand the tradeoffs... so that you felt really good."*

Relentless attention to detail *and* a deep understanding of the 'why' *and* the ability to explain that to the team in a way that meant they all felt good and were invested in the direction. All this from a guy with a famous 'reality distortion field.' Big narratives *and* extreme attention to detail might be products of different modes of attention, but the ability to wield *both* expertly at very high intensity is an exceptionally rare gift that few develop in the way Jobs did.

It's amazing to me, dear reader, that the new hemisphere hypothesis sheds light on everything from how we pitch to the product and leadership genius of Steve Jobs[5], but that's what makes it such a powerful mental model. We do have better and worse ways of attending to the world, and when trying to do something very hard — like building, running, and scaling a high-growth tech company — it pays to try and err on those better, balanced modes of attention and to be aware of where things can go wrong, especially when we *aren't* Steve Jobs.

And where can things go wrong? When our left brain *makes stuff up.*[6]

Left-brain dominance

> *Finally, it would appear to be a literal truth that, as a society, we are becoming more like individuals with right hemisphere deficits.*
>
> – Dr Iain McGilchrist[7]

To understand why brain putty is a problem, we need to understand why left-hemisphere dominance is a problem. And McGilchrist, for what it's worth, very much does see left-hemisphere dominance as a problem. It is, in fact, the basis of his entire scientific and philosophical project. He thinks we should return to a healthier balance of attention, with the right hemisphere back running the show.[8]

While McGilchrist's philosophical project is well beyond the scope of this book, it would be unfair not to mention it. Likewise, I should mention that we in the tech world are probably guilty, in McGilchrist's view, of this left-hemisphere view of the world (my bold)[9]:

> *[T]he left-hemisphere view offers simple answers. Its mode of thinking prizes consistency above all, and claims to offer the same mechanistic models to explain everything that exists. This thinking is common to those who espouse naïve reductionist science ('scientism'),* ***enthusiasts for technological solutions to what are complex human problems****, and designers and implementers of bureaucratic systems.*

Nervously tugs on collar. I think it's fair to say that in tech, a little more right-brain attention would go a long way[10]:

> *The right hemisphere is also more realistic about how it stands in relation to the world at large, less grandiose, more self-aware, than the left hemisphere. The left hemisphere is ever optimistic, but unrealistic about its short-comings.*

What can we do about this? Our brains don't come with volume dials where we can manually dial the right hemisphere up and the left hemisphere down. Instead, we need other ways of approaching the world. In my view (and in the spirit of trying to name things), to be more realistic and more self-aware, we need to be more in touch with what we'll call the reality layer.

The reality layer

A refrain of this book is that your clarity strategy is only as good as your discovery; that is, your clarity strategy is only as good as your engagement with this reality layer.

The reality layer doesn't really care about your ideas, your vision, your dreams, or whatever you hope *might* be. The reality layer is simply what *is*. It's where the market in product/market fit exists. And while we can engage with it and shape it — it's not static, and we're trying to do something new, after all — it deserves appropriate deference.

It deserves deference because it's the reality layer — and the market that exists within — that generates the information we ultimately want to capitalize on. The reality later is where real, observed customer success lives. It's where strong 'whys' live, whys that drive behavior, dive product/market fit, and may ultimately drive your narrative. The startup learning curve is, if nothing else, a process of learning what's real — what *really* drives buyer behavior, what features customers *really want* (and use), and what outcomes your product *really* achieves.

That we have to work so hard to get down to what's *really, really real* (apologies to K. Lamar) speaks to our limited grasp of reality, which is, dear reader, a little unnerving. We are, on the one hand, in a very privileged position where we can indulge our curiosity about what *is* real, but on the other, we're still fallible humans who are subject to narrative bias and brains that will make stuff up, as we'll see.

> **Your customers' reality layer:** This is true for customers, too. Customers may have their own left-brain dominance where they can't handle ambiguity and want to be certain and right despite not having an adequate grasp on what their situation really requires. In some ways, this is why bumping folks back out to right-brain attention, as we discussed in chapter 25, can help prospects escape their own narrow attention. And then, together, you may be able to work on creating a new reality with the customer — a new vision of what's possible, supplying the tools and resources to help them get there.

The speculation layer

Startups ideally start with a fair dose of insight from the reality layer, but there's often a fair helping of speculative vision, too. This gives us two planes of existence, in a sense:

- **The speculation layer.** This is what we hope *might* happen. This is where we guess. We might have a vision, or a hypothesis, or a hunch, or a wish... but it's not demonstrable reality right now. While this vision might sound cool, it usually results in a weak *why* as far as buyers are concerned, simply because it remains to be proven.
- **The reality layer.** This is what *is* happening. Not what *might be* or *could be* or even *should be* happening. What *is*. You know it, I know it, everyone can see it. It's what's real *now*. Narratives that start here are built on a strong foundation, with a strong, proven *why* that can drive change.

As I write, dear reader, the tech industry hype cycle has shifted from crypto to AI. Crypto was a classic case of narrative that existed almost entirely in the speculation layer — a lot of wishful thinking about what might happen, with very little proven out in reality (apart from, you know, all the fraud and crime).

Generative AI and LLMs, on the other hand, took over because there was a 'there' there, if I can put it that way. The tools and utility were real *now*. Not in some speculative future; *today*. (Of course, there was an incredible degree of frankly insane speculation-layer thinking that came with it,[11] but that's par for the course.)

Shared reality

Startups are, by definition, attempts to go from speculative vision to some discovered, profitable reality. That requires iterating through the right/left/right loop refining your vision as you go, radar and laser beam in action as you hopefully build a clarity strategy that takes you from speculative futures to observable customer realities.

It's much easier to sell customers if you operate from a shared understanding of reality, and that requires a balance of deeply understanding what *their* reality layer looks like while having the vision to know where they can go. And that vision matters. We need to paint a picture of new futures, so we know where we're going. Leadership is, by definition, leading people to this new future.

That is, we can't live in the speculation layer forever. But sometimes, our brains — and our narrative — can work against us, especially when it comes to that oft-dominant left hemisphere.

The temptation of narrative bias

It is, dear reader, very easy to say, 'Start with a vision and build toward something people want' — that's about as banal as startup advice gets. But why is the 'build something people want' part so hard *to do*? Why do our puny human minds struggle so much to drill down to the reality layer and understand what customer success looks like on our customers' terms, not our imagined ones?

Well, it turns out that our brains don't always make it easy to do the ol' right/left/right jump. Sometimes, our left hemisphere will jump in with a story for us instead. To give you a taste of the *extremely weird* kinds of stories that our left hemispheres will make up when perception has been altered by damage to the right hemisphere, here are a few anecdotes McGilchrist cites in *The Master and His Emissary*[12]:

- **The half-wooden body:** Following a right hemisphere stroke, one patient perceived the left side of his body (the one controlled by the damaged right hemisphere), specifically his torso, as a wooden plank, where food gets sucked through scaffolding and falls out the bottom. (He perceived the right half of his torso, controlled by his undamaged left hemisphere, as entirely intact.)
- **Doubles:** Right-hemisphere deficits can result in patients believing in doubles of people and things because they can no longer sense a unique whole. One patient, McGilchrist recounts, believed there were eight fake

duplicate cities of his own, each with duplicates of his wife, children, and indeed himself. He believed he'd spent eight years "wandering between them, without finding the real one."

- **Not my arm:** A faulty right hemisphere has led to patients believing their left hand or arm simply doesn't belong to them, and perhaps belongs to someone else entirely, even their mother, or they believe it's inanimate or dead, or it's being controlled by aliens.

With these stories, McGilchrist is trying to tease out the nature of the left hemisphere and what happens when it's left to run the show due to (very sad) injuries to a patient's right hemisphere. These stories, while extreme, offer some fascinating clues to those differences, specifically the false narratives our often all-too-dominant left hemisphere will conjure up when called upon.

While these are extreme cases, they demonstrate that our left hemisphere can make up some really weird stuff when it has to fill in gaps. That is, rather than just saying *IDK shrug emoji* in the face of missing information, it will confabulate instead, as McGilchrist suggests, simply confecting a story that seems plausible and consistent (even if it's consistent with a damaged perception of the world as we see above). What's more, the left hemisphere then appears "completely convinced by it."[13] It is, in that very specific sense, the original ChatGPT that exists between our ears.[14]

The balancing act

In these highly unusual cases McGilchrist cites, the fake-ness of the story is, from the outside at least, self-evident.

But it also speaks to the fact our (left) brains are reality-patching machines, and this brings us to (in a much less extreme sense) the balancing act founders have to do — to try and let their brain patch reality *just enough* to create conviction about what is possible, but not so much as to slip into delusion and get cut off from the reality layer.

That balancing act is omnipresent in clarity strategy. We want to draw as much as we can from the reality layer while still balancing that with the vision of what could be for our customers. But the temptation is to double down on delusion and think that the more certain you appear, the more grandiose your claims, the more optimistic you are about outcomes, the more likely customers will buy into your story.

It's a temptation because it can work, for a while. Most folks assume good faith, after all, and there certainly are plenty of hucksters out there who prey on this assumption of good faith with extreme, unearned, and undeserved confidence (hello Theranos *et al.*). But I think in B2B, most prospects tend to see through that. They're often in laser-beam mode for a reason.

The real risk, and more subtle problem, then, is when folks *telling* stories *don't* see through their own shtick and end up believing it all the more. They become trapped in a web of their own making, unable to get back out to the reality layer, and that is a genuine risk that we'll see in the chapters ahead, especially when folks tie themselves into knots with ill-conceived category creation narratives.

On the flip side, if you're aware of this risk, then developing your clarity strategy can be an excellent chance to question the assumptions in what you 'know'; to go in with an open mind, holding on to the 'known knowns' but being flexible about the rest.

> **Golden information vs. proving hypotheses:** Proving your speculative ideas true or false in a traditional hypothesis/experiment/result process is obviously useful in startups, but we're not here to just prove something true or false in a strict binary sense. Action generates information, and it's that *generated* information that can be gold, precisely because it came from outside our reality-patching brains. To use Loom as an example again (discussed in chapter 10), they experimented with a user-testing marketplace and tool and went nowhere, but by putting a tool in users' hands, they could generate information about ways it *could* be used. That information — that folks could send short video updates instead — turned out to be vastly more valuable than their earlier hypothesis. It's precisely because no one would have guessed that this particular kind of tool would be valuable that creates the opportunity. It's not the *only* way to succeed by any means — sometimes initial bets pay off (like Figma). But it is the nature of learning by doing, and it's that doing that generates the information *from the reality layer* that we can learn from and capitalize on, finding that unique edge.

Reality patching

There's a subtle point here: it's not that we, as humans, knowingly and mischievously bullshit about reality (though that obviously happens a lot, especially when money is involved). Nor is it that we often have to hold assumptions that we need to test against reality to see if they're true.

They're both conscious actions, malevolent and benign.

What's difficult is the *unconscious reality patching* that our left hemispheres may be doing for us. That is, a perfectly honest and earnest person still has a left hemisphere that's prone to confabulation. That is, in part, why our grasp of the 'reality layer' can be much more tenuous than we might think, which is why intellectual humility and curiousness are so vital for founders and startup culture as a whole.

> **Confabulation, please:** Startup culture, for better or worse, can sometimes seem like it *demands* confabulation. We all know the cliche of the 'visionary founder' who can weave grand narratives about entirely speculative futures. Mix that stereotype with a strong left-hemisphere need for certainty and to be right, and you have a situation ripe for the left hemisphere to go a little crazy with confabulation. Again, I'm not talking about the obvious frauds and liars here, but when ideas or beliefs that started as speculation slowly morph into certainties with no validation in between. It might happen on an individual level, but it can happen on an organizational level, too. This is why humility is a virtue, and the solution is to always and forever be working hard to get back down to on-the-ground customer reality as much as possible, embracing the inherent ambiguity of startup life as you go.

Brain putty and product/market fit

In startup culture, the visionary leader is supposed to have all the answers. But what happens when we have gaps in our knowledge, and everyone is demanding correct answers?

We patch the gaps with confabulated information just like putty, which is how I landed on the term. It's the filler our left brain likes to use to patch those reality-layer holes. If nature abhors a vacuum, evidently, our (left) brains do too.

And this can be a particular issue when folks hold on to a story for too long, not pivoting fast enough or looking hard enough for new sources of product/market fit. All of this means we must be very careful with narrative and story, too. Our brains can love stories to a fault.[15]

> **Speculation ain't all bad:** Let me be clear that this is, again, a subtle point: my argument is not that speculation in and of itself is wrong or bad. It's speculation that there *is* value to be created and a pot of gold that will exist as a reward that drives startups in the first place. That's great. Our brains have to do some degree of gap-filling, presumably so we can function and take risks at all. The point is to be clear about what is speculation and what isn't, and to not blur speculation with reality, which is much harder than we intuitively appreciate. What started as theory can easily get accepted as fact, especially in larger

> organizations where folks may not have a connection to the original source of insight.

Busting brain putty

Brain putty is the reason for so much of all the Twitter, LinkedIn, and business book (*ahem*) advice out there. It's not that folks are trying to wilfully lie to you, it's just misguided left-brain certainty and the need to be right. It's just brain putty.

In tech, brain putty is the reason why empiricism matters and 'talk to your customers' is the #1 bit of startup advice. If there was no brain putty, the gaps in our thinking would be obvious, and we wouldn't need to be *told* to go talk to customers. We'd do it intuitively, without fear of not being right and certain.

But because our left brains will simply make up all kinds of stories about our customers to paper over our ignorance — what they want, what they need, what they'll use — we have to be reminded to go back to the reality layer to check ourselves with some right-brain appreciation for ambiguity and multiple possibilities.

This should keep us up a little at night. It's an endemic problem. But it's not impossible to overcome. In the early days, especially, it's easier to bust through the brain putty — you *know* you don't really know that much, it's easier for everyone to see customer behavior and share insights, and you can collectively do the right/left/right jump to find a thread, unpack and analyze it, and reintegrate what you learned. Brain putty be gone.

But in bigger companies with more distance between a team and its customers, more diversity in users and user needs, and more incentive to just be Very Confident and do random acts of whatever (product and marketing especially), it gets much tougher to escape collective brain putty and get back to that reality layer.

But getting back to that reality layer is the only solution, and that might mean more sharing of context, more chance to interact with customers, more end-to-end thinking about the whole of what you're trying to achieve, and more fundamental interest in the behavior your startup or company is trying to facilitate for your customers. (And that's what Figma was great at, incidentally, as we saw in the last chapter.)

> **Busting brain putty by watching users:** People have been busting through brain putty with user testing for decades (see Jakob Nielsen in 1995, for example[16]), but not many folks make it a regular team ritual. It may be worth taking a page out of PLG coach Andrew Capland's

> playbook, who, when at Wistia, instituted a program they called FullStory Fridays[17]. "Every Friday we would order lunch, hop in a conference room, and watch 20 real people use the product for the first time." They'd watch user session recordings in FullStory (hence the name) augmented with user tests and customer interviews. "That ritual," Capland says, "was a game-changer for our team." No more guessing and speculating with cobbled-together data (i.e., brain putty). Instead, they could literally see what was and wasn't working. Testing your narrative is important for the same reason, a topic we'll pick up again in chapter 38.

Balanced attention & big ideas

As we hoist the sails, cast off, and, err, do other sailing-related things in preparation for the next section on clarity strategy development, I hope that we can bring balanced attention, a keen appreciation for the reality layer (and our often tenuous relationship to it), and an awareness of the risks of left-brain brain putty with us as we go.

And while we're going to look at a bunch of interesting and inspiring stories as fodder for your clarity strategy, ultimately, what matters is finding *your* specific insights into the reality layer of your customers that only you can bring about. Your company, in that sense, is a lab you're building to generate information about what works in that reality layer and not, as we mentioned earlier, a cult to confirm brain putty.

The power of narrative cuts both ways — it can make brain putty seem truer than it is, or it can give the value you've discovered the clarity it deserves. With great power comes... well, you know the rest. Next, let's explore the ways we can further develop your clarity strategy, starting with the three-act story, category concepts and secular change.

Notes

1. McGilchrist, Iain. *The Master and His Emissary* (p. 82). Yale University Press. Kindle Edition.
2. Tobias, Martin. February 20, 2023. **https://twitter.com/MartinGTobias/status/1627488271411425283**
3. McGilchrist, Iain. *The Master and His Emissary* (p. 82). Yale University Press. Kindle Edition.
4. Fadell, Tony. "Tony Fadell: iPhone, iPod, Nest, Steve Jobs, Design, and Engineering |

Lex Fridman Podcast #294," *Lex Fridman Podcast*, June 16, 2022. **https://youtu.be/4oDZyOf6CW4?si=AV4D82qdVlwCSEem&t=4827**

5. The same could be said of Elon Musk in his prime, where he demonstrated an incredible intensity in his application of both modes of attention, too — an intense laser beam of attention applied to technical detail on the one hand and broad, expansive, captivating narratives about the future on the other. As for where Musk's attention is post-Twitter acquisition (and post years of all-narrative-no-detail announcements), the less said, the better. Let's just be glad Jobs never became addicted to posting.

6. I want to stress the brain is, of course, a mysterious and complex organ, and there's a certain irony in saying 'be careful about being overly certain' in a business book, a genre where being overly certain is the genre's staple, its *raison d'être*, and where I'm boiling attention down to simple binaries. But I hope awareness of overly certain brain putty at least provides some defense against it.

7. McGilchrist, Iain. *The Master and His Emissary*. Yale University Press. Kindle Edition.

8. McGilchrist writes in *The Master and His Emissary*, for example, that "These two asymmetries indicate where the interhemispheric balance of power ought to lie, and indeed needs to lie: with the right hemisphere. But it does not." (p. 227)

9. Ibid.

10. Ibid., p. 84.

11. We seemed to oscillate between predictions of either fully automated luxury communism on the one hand or the entire extinction of humankind on the other. Unfortunately, this kind of high-IQ hysteria seemed to distract from the real, tangible harms of a tsunami of web spam, deep fakes, and copyright abuse. That doesn't mean generative AI is inherently 'bad,' it just highlights how out-of-control left-brain attention was in AI discourse, full of absolute certainty about AI being either a savior or a destroyer, with little nuance in between.

12. From pages 55, 54, and 68, respectively.

13. McGilchrist, Iain. *The Master and His Emissary* (p. 81). Yale University Press. Kindle Edition.

14. With all the hype around artificial general intelligence (AGI) supposedly being just around the corner, it's interesting to ponder which hemisphere's intelligence any such AGI will mirror. I suspect, sadly, it's not the right's.

15. So much so that we'll believe stories about stories that aren't necessarily true, proving and disproving 'people love stories' at the same time.

16. Web Design Museum, "Usability testing case study of SunWeb by Jakob Nielsen in 1995! UX design is older than you might think :-)", April 6, 2023. **https://twitter.com/webdesignmuseum/status/1643988178997743621**

17. Capland, Andrew. "The growth team ritual that changed everything," *Delivering Value*, March 8, 2022. **https://deliveringvalue.co/growth-essays/growth-team-ritual-that-changed-everything**

PART VI

CLARITY & CONCEPT VS. CATEGORY

29

Concept: three-act vs. final-act stories

Welcome to the second half of *Clarity*, where we'll explore ways to develop and stretch your clarity strategy. High five for making it this far! You absolutely deserve that next cup of tea and/or a brief lie down.

In the second half of this book, we're going to return to our left-brain/right-brain two-stories approach, with chapters grouped loosely under each story type:

- **Story 1 — concept:** The big change, category, or concept story that lets you hit folks' right-brain radars.
- **Story 2 — positioning:** Your narrow-focus positioning that needs to thread the needle between your prospects' jobs-to-be-done (JTBD) and your go-to-market (GTM) strategy.

While we're still going to have these two story types in mind throughout the coming chapters, we're going to continue expanding on what they look like in reality, peeling off some deeper layers of the clarity strategy onion.

So far, we've covered:

- **A story:** Having a narrative *at all* with your narrative snapshot (**chapter 9**).
- **Two stories:** Expanding our understanding of narrative to story 1 (change stories, **chapter 20**) and story 2 (positioning, **chapter 21**).
- **Degrees of the two stories:** Looking at our pitch as a claw game (**chapter 18**) that results in different types of decks with different *degrees* of story 1/story 2 (the three alternate decks, **chapter 23**).

And now we're about to cover:

- **Two stories in depth:** What it looks like to stretch your narrative, both for right-brain change stories, as we expand story 1 to whole *concepts* (of which the change story is the foundation, but just the start of the full three-act story) and what deep, left-brain, story 2 positioning looks like, and how that works with your GTM strategy and sales motion.

This is how we'll stretch and develop your clarity strategy — going *bigger* with full three-act stories and/or going *deeper* in terms of zeroing in on the right customer segment in the right way.

If this all sounds a bit abstract, don't worry, we'll look at plenty of examples as we go.

What's relevant to you?

Like the rest of this book, the second half has choose-your-own-adventure paths through it. You can:

- **Go big:** Keep going with the full, three-act, big-concept change story that we'll start exploring in the next chapter (**chapter 30**).
- **Go specific:** Jump to our primer on Jobs to be Done (**chapter 36**).
- **Go deep:** Skip ahead to the positioning and GTM chapters (**chapter 37**).

You can also jump to specific chapters if there's an area or question that interests you, for example, on the story 1 side:

- **Secular change:** How secular change drives story 1 narratives in **chapter 30**, especially in CRM, as we'll see in **chapter 31**.
- **Category creation:** How category creation works in our mini-series on unbundling category creation from **chapter 32** on, looking at the difference between literal categories, new *concepts*, and category-of-one Aggregators. (Don't miss **chapter 33** as we pick apart *Play Bigger,* too.)
- **Playbook as brand:** How Qualtrics embedded a playbook in their brand as we'll see in **chapter 35**, and how despite being the poster child for 'playing bigger,' they started by playing narrower.

And on the story 2 and GTM side:

- **JTBD:** How Jobs to be Done sets up our look at positioning in **chapter 36**.
- **Positioning & pattern matching:** How finding your best-fit customers can drive your positioning with April Dunford's approach in **chapter 37**.
- **Early-stage sales:** How you should test your pitch and value proposition early on in **chapter 38** and how extremely focused selling gets the initial flywheel going in **chapter 39**.
- **Nerds, normies, and distribution mismatches:** How targeting the right stage of the market can be the difference between your company sinking or swimming in **chapter 40**, and why distribution mismatches can result in advertising money bonfires in **chapter 42**.
- **The B2B spectrum:** How positioning, marketing, and selling differ between SMB, the mid-market, and enterprise. This mini-series explores SMB SaaS and positioning for reach in **chapter 43**, positioning by attribute in the mid-market in **chapter 46**, and positioning for, and selling to, enterprises (perhaps with 10x pricing) in **chapter 47**.
- **Mental availability:** Don't miss our discussion of mental availability, the challenge of getting people to think about you at all, and whether niche

positioning is actually a mistake (!) when we look at *How Brands Grow* in **chapter 43**, too.

- **The MAYA sweet spot:** How a seventy-year-old B2C industrial design concept sheds light on the fundamentals of B2B positioning as we look for the sweet spot between the new but not *too* new, or the Most Advanced Yet Acceptable option, in **chapter 48**.
- **Fundamentals of competition:** Finally, we'll look at whether you're positioning your company to *compete to be unique* (what everyone thinks they're doing) or *compete to be the same* (what everyone is actually doing) in **chapter 49**.

Phew. When I said back in **chapter 2** ('How this book works') that this book was two books in one, I wasn't kidding. There's loads of meaty content and insights waiting for you to explore, *and I really don't expect you to read every chapter.* Pick whatever is of interest and jump around.

But how do you know where to start — concept or positioning? Should you go for the big change story or the ducks-in-a-row positioning approach?

How much story do you have to tell?

To answer that question, I like this simple, clarifying question that we touched on previously (in **chapter 9** and when looking at our three deck types in **chapter 23**), which is: how much story do you actually have to tell?

Let's think about a basic three-act story structure:

1. **Set up** — there's an instigating incident at a certain time and place.
2. **Conflict** — stuff happens; there are obstacles, challenges, action, adversity, etc.
3. **Resolution** — there's a conclusion, either tragedy or victory, or some bittersweet combination of the two.

That's a useful model for your pitch when you're trying to engage prospects in a story 1, right-brain way, but when we talk about clarity strategies with three acts, the three acts are more like:

1. **Change in the world** — there's an external trend playing out 'out there' in the environment that will impact you.
2. **New playbook** — the trend means you need to 'manipulate' (in a McGilchristian, left-brain sense) the world in a new way.
3. **New tool** — our platform/tool/solution is how you efficiently execute that playbook.

Some companies really do have a whole story to tell in that regard — change, playbook, *and* tool — while others have just built a better tool that meets existing needs and have done very well that way. I like to group these as:

- **New whole concept:** Where you're proposing a new three-act story, from change to playbook to tool.
- **New final act:** Where you're hoping to catch buyers who are in the middle of their own story and just provide a better ending.

Which are you?

There's a big difference between trying to finish someone else's sentence, so to speak, compared to getting them to buy into an entirely new story. Proposing a completely new concept is high risk, high reward. It's about new ways of working, new playbooks, new categories, new concepts. That's a lot more story to tell.

> **Why not something halfway?** You can, of course, land with a little bit of right-brain change and a little bit of left-brain narrow focus, and many companies do. But this is often the mushy middle — it's not where decisive competitive advantage lies. This half of the book is all about trying to push you to one extreme or the other — either really going for the full three-act change story or really diving down to serve best-fit customers better than anyone — to get you out of that same-y middle ground that offers no edge for your clarity strategy.

Dialing in your story

Everyone loves the *idea* of these big, right-brain, all-new stories, and often founders think if they try and paint themselves as bigger and more revolutionary than they are, then more people will throw money at them. Investors, maybe; customers, not so much.

That's why pulling this off is exceptionally hard, not just in a skill and resources sense — or brain putty sense, as we saw in the previous chapter — but in a timing sense. You need the right conditions. Otherwise, you're attempting to set sail when there's no wind, which will just leave you paddling frantically as you bob around in the dock.

Plenty of savvy folks, however, position themselves around known problems and known concepts. Sometimes, that's by *hunting deals* by being vastly more specific than your competitors and building a laser-focused GTM machine that can hunt for just the right deals. Other times, it's by *harvesting demand* by building a superior product as the answer to a broad set of customers' needs, as we'll see with Squarespace. Both of these approaches are a 'new final act' rather than a three-act story, capturing demand through hunting or harvesting.

Remember, startup culture, as we touched on in the previous chapter, tends to celebrate the visionary with the big ideas that (supposedly) bring the masses to them. That's great when it works, but don't forget a *lot* of business is done on the back of left-brain hunting or harvesting, too. Neither approach is right or wrong; it all comes down to how much story you actually have to tell.

With that said — I really don't want you to miss out on the later chapters! — we're going to start with the big change stories, and we'll look at some incredible examples as we go. I'm stoked we can start stretching your narrative on one or both axes, from the *particularly big* big-picture change stories down to the *very specific* specifics of positioning, looking where such approaches do — and sometimes don't — work.

It's powerful stuff, so let's start by seeing what makes big change stories tick.

30

Secular change: Gainsight vs. Hopin

> Things happen fairly slowly, you know. They do. These waves of technology, you can see them way before they happen, and you just have to choose wisely which ones you're going to surf. If you choose unwisely, then you can waste a lot of energy, but if you choose wisely it actually unfolds fairly slowly. It takes years.
>
> – *Steve Jobs*[1]

Hitching your wagon to a *secular change* is one of the most powerful things you can do in developing your clarity strategy. In this chapter, we'll look at why secular change is the secret to strategic narratives, category creation, and other high-level, mega-trend-driven narratives.

Riding a secular change is also one of the hardest things to do in B2B. I see a lot of founders and CMOs trying to cargo-cult their way to a made-up secular change or new category. But as we're about to see, 'category creation' is something of a misnomer — categories aren't *created* so much as they are *co-opted.* While the former sounds sexier, it's the latter that reflects reality.

But what generates this movement to co-opt in the first place?

Hitting the right-brain radar

In previous chapters, we've seen that there are specific times when a right-brain, story 1 change story is particularly effective, simply because your audience is in right-brain radar mode. That includes times when:

- You're approaching C-suite folks or *outbound* rank-and-file folks in complex sales (as we saw in our natural narrative experiment in **chapter 22**).
- You want to pitch a new general playbook, not your specific product, e.g., in a keynote speech, at a meet-up, or in a webinar, etc.
- There really is a wave sweeping your industry (that's not one you dreamed up, and it's not reality-patching brain putty), and just about everyone has their eye on it (generative AI being a classic example).

In these cases, it's secular change that drives the narrative — one focused on what's 'out there.' It works because it matches folks' attention and lands on their right-brain radar.

What do these right-brain stories of secular change look like in practice? We're about to look at another natural experiment the COVID pandemic dished up for us that neatly illustrates the difference between *secular* and *cyclical* changes (i.e., permanent vs. transitory) and, in the next chapter, we'll see how secular change drove the narrative for big winners in CRM, specifically Salesforce, HubSpot, and Drift, respectively. But first, let's think about capitalizing on change in technology more broadly.

Energy comes from the bottom up

The energy that drives a secular change comes from the bottom up and, as Steve Jobs says in the quote above, can unfold over years.

Jobs obviously surfed waves of technology like none other. When the surf was up (personal computing, personal music players, smartphones), he led the development of some of the greatest products of all time (Mac, iPod, iPhone), but when it wasn't (NeXT, which made computers targeted at businesses and universities), even his genius wasn't enough to manufacture the energy needed to create a market.

When it comes to market-defining winners, it's not brilliance *or* environment; it's brilliance *and* environment, and even someone as brilliant as Jobs — selling the same product, personal computers, no less — couldn't manufacture success when the market conditions weren't there.[2]

Harnessing secular change — technological or societal — provides that market-making energy, energy that's far greater than you could ever generate on your own. But given these waves can take years to arrive, you have to, as Jobs says, choose your timing wisely. For example, a wave might:

- Spring up relatively quickly (the incredible power of LLMs and generative AI).
- Be many years in the making, off and on, and never quite arrive (VR).
- Or it may be a hype-driven ZIRP wave (whatever web3 was supposed to be).

If you can get your timing right, ride that momentum, and help popularize the trend, you can do very well. But if you don't get your timing right, or you try and confect a secular change that's just not there, then you end up looking like some slightly deranged surfer trying to create a wave by splashing about in a big, empty, placid lake on a quiet summer's day. No amount of splashing is going to make a wave happen. Not even if you're Steve Jobs.

This may seem like a strange thing to harp on about, but many startups are coached or inclined to tell what are fundamentally fake change stories, and it's a mistake. (Humane's launch of their Ai Pin in 2023 being a classic case in point of assuming your narrative.[3]) We're going to explore what real change stories look like instead.

> **The bootstrapper and the API change:** While we're looking at big, venture-scale narratives here, if you look carefully, you'll see this same phenomenon play out on much smaller scales, too. For example, one self-taught SaaS bootstrapper built a SaaS around Instagram that fairly quickly ramped from almost nothing to $50k MRR, which is awesome. However, prior to that, their SaaS spent years going nowhere. What changed? Instagram opened up a new API that enabled new functionality. The bootstrapper was — to their credit — in the right place at the right time, and their MRR took off. That's fantastic, but the lesson was: See? "Anyone can build a SaaS." That's technically true — anyone can *build* a SaaS and spend years going nowhere — but what's harder to replicate is the environmental conditions for success — the change in the world that creates the opportunity, like Instagram creating a little wave to ride with their API change. That's easy to miss if we only look down at the activities and not out at the environment.

Secular vs. cyclical change

Let's clarify what secular change actually means. We're using it in the financial sense. Here's a definition from Investopedia[4]:

> *In finance, secular is a descriptive word used to refer to market activities that occur over the long term. Secular can also point to specific stocks or stock sectors unaffected by short-term trends. Secular trends are not seasonal or cyclical. Instead, they remain consistent over time.*

Some trends are flashes in the pan. Some trends stick. It's those the-world-has-changed trends, including technology itself or consumer or B2B buyer behavior that we're interested in here.

When you look at the biggest technology companies, they tend to be the ones that successfully rode the biggest technology-driven secular changes. Think:

- Microsoft and personal computing.
- Apple and the smartphone.
- Google and the web.

And so on. Big trends, big companies.[5] Secular and cyclical trends can make or break B2B SaaS companies, too.

Gainsight vs. Hopin

COVID created a neat natural experiment that demonstrated just how fundamental — and brutal — these waves can be in B2B SaaS. To understand what happened, let's compare these two companies:

- Gainsight, founded in 2009, which is famous for helping create the Customer Success category.
- Hopin, founded in 2019, which started as a virtual events company.

On the face of it, there's not much similarity between the two companies, except for one thing: they shared the same very talented CMO, all-round nice guy Anthony Kennada, who worked for both. What's more, Kennada happens to be an expert on wave-riding category creation, having written a book on the topic based on his experience at Gainsight.[6] That makes these two companies an ideal study in the power of trends and change.

> **Talent wasn't a variable:** I have loads of respect for Kennada and the amazing things he achieved at Gainsight and beyond, far more than anything your humble author will ever achieve. He's a humble dude who has done incredibly impressive things, and that's what makes this natural experiment so fascinating. It's *because* he's a talented guy that we can rule that out as a variable, like Steve Jobs and NeXT, and that allows us to think about the broader environmental conditions instead. Again, it's talent *and* environment that matters, and while I'm sure there's much more to the stories we're about to look at than we can cover here (different CEOs, products, and markets, for example), I think the stark differences in outcomes highlights the sheer power of secular change as a force for startup success.

Gainsight

Gainsight is considered a canonical example of category creation, and it's easy to take a superficial view that they just willed the Customer Success category into existence. But reality looked quite different, as Kennada documents in *Category Creation*, writing[7]:

> *We recognized, however, that one of the early users of the [...] platform was a persona with the job title of Customer Success Managers (CSMs)—a role created (and named) by Marc Benioff and the team at Salesforce almost 15 years ago. The role had been slowly picking up momentum in the eight years before Nick and I joined.*

That's a wave that had been slowly building, to riff on the opening Jobs quote, for *eight years* before Gainsight came along. Kennada is clear that categories aren't created out of thin air so much as they're nurtured into being by harnessing an existing trend. He writes, for example, about Gainsight *finding* — not *creating* — their tribe:

> *We knew we had found our tribe, and by championing their success and making them feel like heroes, perhaps we could build a great company along the way. ... [W]e decided that we would devote our marketing energy to spreading broad awareness that churn was the massive problem for subscription businesses that no one was talking about, that Customer Success was the business discipline with the charter to own it, and Gainsight was the technology platform to operationalize it.*

That's also a neat summary of what a big, right-brain change story looks like:

- **Problem:** There's a massive problem "no one" is talking about.
- **Role:** There's a specific function, role, or discipline whose job should be to address it.
- **Solution:** We have the product to help you solve it.

In practice, for Gainsight, this meant[8]:

> *We made an early bet to focus less on positioning our solution, but to put forth more effort to advance the interests of the Customer Success profession by facilitating the exchange of best practices, building community, and ultimately creating industry around the profession that Salesforce is credited with establishing.*

That is, they marketed the change, not just their tool. And that can require significant resources. Gainsight raised $156.3M in total, according to Crunchbase, with rounds of $9M, $20M, $25M, $25M, $50M, and $52M after their seed funding.[9]

But when I say Gainsight marketed the change — i.e., they focused their narrative on the big change story — what does that actually mean? What's the formula? I like to think of it as:

> *Change → playbook → tool*

We'll take the tool piece of the puzzle for granted here, as we're focused on the right-brain side of the story, but those three pieces add up to a 'whole concept,' a whole three-act change story, with the tool/solution as the final act.

Before the tool, however, is the change and playbook, and those elements are the ones we're going to keep coming back to over the next few chapters because they're fundamental to these big narratives. They're the critical pieces that take folks from 'unaware' to 'problem aware' and perhaps 'solution aware' in our levels of awareness from **chapter 24**.

In Gainsight's case, while they undoubtedly did much to educate the market (they raised $150M+ for a reason), both the change and even the playbook already existed in some form, for example:

- **The broad change:** 'Customer Success' came to exist in part due to the realization that churn was symptomatic of a bigger problem for subscription companies — that traditional customer support alone wasn't enough. You couldn't just reactively *support* your customers; you had to proactively help them *succeed*.
- **The specific playbook:** The Customer Success playbook, generally speaking, emerged involving teams working with clients to get them set up, engaged, active, and growing on the platform with things like Quarterly Business Reviews. This helps drive adoption, reduce the likelihood of churn, sell upgrades, and increase NRR (net revenue retention). Gainsight would go on to codify this playbook in a book of their own, 2016's *Customer Success: How Innovative Companies Are Reducing Churn and Growing Recurring Revenue.*

When it comes to the broad change and specific playbook, you can see the switch from right-brain radar attention to the left-brain laser beam. The change is 'out there' in the environment; the playbook is about how you manipulate the world in light of the change. And, broadly speaking, you need both:

- If your narrative consists of a playbook with no right-brain discovery driving it, it may lack context or a rationale.
- If your narrative consists of a broad change with no playbook, then it isn't actionable in a novel way, as it doesn't suggest a new way of 'manipulating' the world (in a left-brain, tool-wielding sense).

It's easy to confect speculative changes + playbooks, and maybe you get lucky and your speculation turns out to be true. But the risk of brain putty is every present, which is why it's so important to try and get beyond the speculation layer and down to the reality layer to see what genuine trends and emerging playbooks *already exist*. That's the irony of category *creation* — it's very rare that you actually want to create a category from scratch. Instead, you're trying to tap some emerging category, as that's where the market power comes from, and the power of that wave divided by your competition is probably directly proportional to the size of your opportunity.

Therefore, a change that *hasn't already* generated a nascent playbook in your target industry probably isn't meaningful in a market-shaping sense, and a playbook that isn't driven by a change probably isn't useful. These are handy things to keep in mind when formulating your own narrative.

Educating the market

> *In category creation, the majority of effort and resources are spent developing early stage content, educating the market on a problem they don't know they have, why it's important, and how to solve it.*[10]

Even when you have all the necessary change + playbook ingredients, it's still a long road to educate the market, as Kennada says above. Educating an *entire market* on a problem *they don't know they have* is a very hard task, as you're starting at the bottom of the awareness ladder. There *must* be external momentum to help you (with funding to match), and that's what Gainsight tapped into — years of development of Customer Success, and an understanding of churn in particular. This meant:

- Hey presto, you've got a specific metric to attack (churn).
- You've got a new function or role to sell to (the Customer Success Manager or head of CS).
- And there's a very compelling business case on why it matters. (Revenue!)

For Gainsight, then, not only was there an underlying trend and emerging playbook, there was a very specific metric to target that could be tied directly to revenue. That makes for an incredibly compelling three-act B2B change narrative, because it's driven by an equally compelling revenue opportunity — solve this, make $. That's the kind of opportunity that unicorns are built on. And that's what happened, with Gainsight selling to Vista Equity Partners for $1.1B in 2020. Happy days.

Hopin

Kennada's book *Category Creation* was published in October 2019, Gainsight sold in 2020, and by February 2021, Kennada was CMO at virtual events company Hopin, a year into the COVID-19 pandemic.

Hopin couldn't have been better positioned to ride the COVID wave to a new future — both in concept (virtual events), bubble-era funding (raising over a billion dollars *in a little over a year,*[11] making Gainsight's ~$150M over four years look decidedly modest in comparison), and talent, now with the guy who literally wrote the book on category creation in B2B SaaS on board.

The change + playbook Hopin had to fuel its success was:

- **The broad change:** COVID = no in-person events.
- **The specific playbook:** Run them online instead.

Pretty simple!

That led to explosive growth. Hopin, "a pandemic sensation and Europe's fastest growing start-up ever" according to the *Financial Times*, was, at its peak, valued at $7.8 billion.[12] At one stage, per the *FT*, the company was boasting of more than 15,000 monthly events listed on the platform and that virtual events were "here to stay." At the height of the pandemic, it must have seemed like this was the new normal. The world was going remote, and why *wouldn't* COVID accelerate the trend?

But COVID waned, and so did the idea of virtual events, with *FT* writing in April 2022 that there were less than 500 virtual events listed at their time of publication — a drop of almost 97%. Ouch. Hopin became the poster child for the COVID-era SaaS bubble, and Kennada headed off to work on his new startup, AudiencePlus.

Trends giveth, trends taketh away

The contrast here couldn't be more stark. Same CMO, all the same lessons learned on category creation, and no shortage of funding (especially in Hopin's case), but, critically, different trends:

- Customer Success was a *secular change* that Gainsight rode to great success, and CS, as a function, is very much part of the furniture in any established high-ticket SaaS operation these days.
- COVID (and virtual events) was a *cyclical change*, and while it may have looked like a secular change for a while, it wasn't, and Hopin bombed as a result when the tide went out,[13] with their events platform selling for $15M to $50M.[14]

Secular and cyclical — the trends both made and broke the respective companies. The bottoms-up energy sustained one and deserted the other. Live by the trend, die by the trend.

And again, let me stress this is not to take anything away from Gainsight's fantastic execution or Kennada's talent or skill.[15] It's just that brilliance alone isn't enough. It's brilliance *and* environment that matters when attempting to pull off these big right-brain change stories.

And it was ever thus for successful clarity strategies, as we'll see next when we look at that staple of B2B SaaS, the CRM market.

Notes

1. Jobs, Steve. "Steve Jobs speaks out," Betsy Morris, *Fortune*, March 7, 2008. **https://money.cnn.com/galleries/2008/fortune/0803/gallery.jobsqna.fortune/index.html** via **https://allaboutstevejobs.com/verbatim/quotes**
2. And we're all the better for it, of course, with NeXT essentially reverse-acquiring Apple, and that, as the history books will show, worked out, shall we say, quite well.
3. Humane, founded by two ex-Apple employees and staffed by dozens more, raised $241M ostensibly on the narrative that we spend too much time staring at our screens, and a more 'humane' future lies in devices that operate without traditional screens. Their first product was a voice-and-gesture controlled broach that was intended to be the first step into this post-screen world. After considerable hype, the device reveal was decidedly underwhelming, and I suspect we'll be staring at our screens for some time to come. And yet, the thing that struck me throughout all of this was the risk they were taking based on their narrative. On the one hand, it helped them raise hundreds of millions of dollars and recruit two hundred employees, many from Apple. On the other, if you're going to bet years of your life and your reputation on a very specific narrative, you'd want to make sure it was, you know, actually real.
4. From **https://www.investopedia.com/terms/s/secular.asp**
5. Of course, these companies don't have a perfect record of wave-catching, either. The Mac didn't end up in every home like Windows, and all of Microsoft's many bites at the mobile computing cherry (Windows CE, Windows Mobile, Windows Phone and the Nokia acquisition, etc.) amounted to very little. Spotting a wave and successfully riding it are two very different things, but if you do catch a wave big enough, it can evidently drive decades of market incumbency.
6. Kennada, Anthony. *Category Creation: How to Build a Brand That Customers, Employees, and Investors Will Love*, 2019.
7. Ibid., p. 15.
8. Ibid., p. 16.
9. See: **https://www.crunchbase.com/organization/gainsight/company_financials**
10. Kennada , Anthony. *Category Creation* (p. 78).
11. Hopin's founder also (in)famously pocketed $195M in secondary stock sales, a true testament to the excess of the times.
12. "Hopin: virtual events start-up struggles as real gatherings return," *Financial Times*, April 25, 2022. **https://www.ft.com/content/312acbb3-eb72-4d2f-b81f-649dcb3583ca**
13. Note that Hopin lives on as a company through its acquisitions of StreamYard and other video-oriented products as a "community engagement" company. A billion bucks in capital buys you a few goes on the startup merry-go-round, but it's hard to imagine that any of this resembles what dozens of big-name investors FOMO'd into the original company for.
14. "RingCentral To Pay $50M For Startup Hopin's Events Management Assets," *CRN*, August 9, 2023. **https://www.crn.com/news/networking/ringcentral-to-pay-50m-**

for-startup-hopin-s-events-management-assets.

15. It's also worth noting that CMOs aren't all-powerful, either. Far from it, as is often the case. But CMOs are often the ones excited by, or tasked with, 'category creation,' and these examples serve to highlight the reality about what is and isn't possible.

31

Secular change and CRMs: Salesforce, HubSpot, & Drift

[O]ur SaaS concept, by which a customer handed over his or her proprietary data to be housed on our server—and then "rented" access to our software—sounded like lunacy to most people.

- *Marc Benioff*[1]

CRMs are the bread and butter of the B2B SaaS world, which makes them an interesting study in clarity strategy. It's not like "use software to manage customer relationships" is a novel concept, so you better have a strong narrative if you're going to make a dent in the market. Fortunately, that's where secular change comes in, providing the energy to take you to the top of your segment or, at least, to a very attractive foothold in the market.

In this chapter, we're going to continue looking at secular change and *concept* creation (not category creation *per se*) as the lifeblood of a big right-brain change story. (CRM is a well-established category, after all.) We're specifically going to look at the CRM babushka dolls, financially speaking, of Salesforce, HubSpot, and Drift and how they each harnessed a secular change — and, crucially, a new playbook — to ride to success.

Salesforce — the SaaS wave

According to a neat 2001 time capsule in the form of an agency case study[2] Salesforce was, in the beginning, an "unknown entering a crowded market" who believed "enterprise software [was] a thing of the past, while Internet services will shape the future." Salesforce, or salesforce.com as it was known back then (it was the dot-com era, after all), was "going head-to-head with Oracle and Siebel, both of whom were making billions on enterprise software that provided that same type of solution as salesforce.com's inexpensive web-based solution."

That "inexpensive web-based solution" is what we might describe today as 'product-led SaaS' or just 'cloud software.' And therein lies the tight change/playbook combo for the grande old dame of the SaaS world, Salesforce:

- **The broad change:** Salesforce was to ride *the concept of SaaS itself* —

'rented' web software — as its secular change. And that was a hell of a wave to ride.

- **The specific playbook:** Bottoms-up adoption from end users, paying out of their own budgets, not top-down-decreed software installed on every machine. In a left-brain sense, the playbook was a new way to use and manipulate software itself, not just the domain-specific data (i.e., customer records).

This secular change was the foundation of Salesforce's famous 2000 "The End of Software" campaign, i.e., this trend was so severe it was the end of (enterprise) software as we knew it. That was one powerful narrative, driven home by marketing stunts like their fake protests at a competitor's user conference, which went down in the, uh, hallowed halls of B2B software marketing folklore, and the rest is history.

And I can't resist this tidbit from the earlier 2001 agency case study:

> *The military-themed Salesforce.com official launch party carried the theme further to celebrate "The End of Software." [...] The game area included "Throw the Disk in the Toilet" and other similarly themed games.*

Wild times.

As I write, post-pandemic highs, Salesforce's market cap is just shy of $200 billion, and they employ 70,000+ people. And, "In yet another indication that the cloud is the greatest growth market the world has ever known," as Bob Evans writes in *Acceleration Economy,* more than two decades after Salesforce's 2000 campaign heralding the end of (on-prem) enterprise software, "cloud-native Salesforce has displaced iconic SAP as the world's largest provider of enterprise applications."[3]

That's one helluva multi-decade wave to ride, with Salesforce developing one very successful playbook to ride it, one that's been reincarnated several times in the decades since as bottoms-up, self-serve, or product-led SaaS.

But most folks who look back fondly on the early, disruptive days of Salesforce often miss this foundational reality of the power of the trend — SaaS itself — that made all the rest possible.

Time marches on, however, and new waves create new opportunities. Consider HubSpot and the digital marketing wave, for example.

HubSpot — the digital marketing wave

> *"An easy-to-use CRM. Who knew that'd be revolutionary?"*
>
> – HubSpot's homepage, early 2022

Before HubSpot was *HubSpot,* it was a general-purpose platform for legal firms called — wait for it — LegalSpot, built by co-founder Dharmesh Shah and two Egyptian contractors from Elance (now Upwork).

After a lukewarm reception from legal firms, but repeated requests about some aspects of the platform folks liked (the lead gen aspect, the website grader), the early HubSpot team decided to essentially give up on the original idea, pivot from a vertical play to a horizontal platform, build on the aspects people liked, and sell to all their startup friends instead.

That is, at least, how HubSpot co-founder and former CEO Brian Halligan explained the company's history in an insightful interview on the *My First Million* podcast.[4]

HubSpot is particularly interesting in that they rode several waves of secular change to become a major CRM player and is, as of writing, worth around $20 billion. For example:

- **The first wave** was, in the view of your humble author, organic digital marketing itself, which at the time was SEO, blogs, email, and late-2000s social media (it was a different landscape back then!).
- **The second wave** was marketing automation — Halligan recounts in his interview that they saw their customers buy HubSpot and then buy a big automation platform like Pardot or Marketo, so they built that too.
- **The third wave** was CRM. Halligan, for what it's worth, actually sees this as their second major pivot (the first being the pivot from LegalSpot), with HubSpot going from a "marketing apps company" to "a CRM platform company".

If they were the waves, i.e., the change, what was the playbook?

- **The specific playbook:** HubSpot famously defined the early content marketing playbook — or "inbound marketing" as they called it, summed up in the title of their 2008 book *Inbound Marketing: Get Found Using Google, Social Media, and Blogs*. (A playbook that your company may still be following, wittingly or not, to this day.[5])

HubSpot rode those waves on the strength of their original insight, which they could capture in a single word: inbound. Halligan, in his podcast interview, said:

> *[We were] right about inbound, [so we thought] we've got to raise a lot of money and go fast, or people will start to copy us, and people did copy us, like crazy.*

"Inbound marketing" was, in a sense, both the broad secular change they put their finger on *and* the playbook. It captured where the leading edge of the market was, and the playbook that more or less every business would have to follow if they wanted to do digital marketing. That was the story 1 right-brain radar narrative. And it also articulated HubSpot's specific playbook that they named and claimed, spelling out, in a left-brain, McGilchristian sense, how you now manipulate the world in light of this new change.

That is, they saw the wave coming. Plenty of folks were already doing these modern forms of marketing, but lacked a unified platform. HubSpot just had to make the most of their timing, so they raised capital and went hard. Wait too long, and the wave will pass you by.

"Inbound" captured the HubSpot folks' imagination and, eventually, the market's imagination too. In that regard, HubSpot is sometimes referred to as a "category creator." But that only really scratches at the surface of what they offered, which was a whole concept, from change to playbook to platform, *built to address the trend they saw in the market.* That was their three-act story with a catchy name — a single word, really — which became an integral part of their brand. (Finding those catchy words was what we were attempting to do with your Hemingway story back in **chapter 10**.)

And here's the kicker — *they couldn't repeat the playbook.* Here's Halligan in 2024, writing about his journey from co-founder to CEO of HubSpot (his emphasis):

> ***Creating a* category *is harder than it looks***: *HubSpot created the "inbound marketing" category. Pulling that off involved writing about zillion blog articles, giving a jillion speeches, writing a book, running a conference, etc. We invested way more energy in creating the inbound marketing category in the early years than we did in marketing the HubSpot product. ...So, when we wanted to go into the sales category, we thought we could just re-run the same playbook for "inbound sales." Failed. When we went into CRM, we thought we'd create a new category called CMR, "customer managed relationship" software. Failed. When we released our CMS, we thought we'd create a new category called COS, "content optimization system." Failed. In retrospect, we caught lightning in a bottle with "inbound marketing."*

They "caught lightning in a bottle" with inbound marketing but then failed doing the same in sales, CRM, and with their CMS. What was the difference between the first success and the other failed three attempts? I'd suggest it was the wave — "inbound" captured the story 1 change over time that *really was* happening in the world. "Inbound sales," and especially "customer managed relationship" and "content optimization system," on the other hand, are far more left-brain hair-splitting of existing categories. There was no real story 1 to drive those would-be categories — and no need for a new playbook as a result — and those category labels, therefore, never took off.

The other interesting observation here — and I say this with the utmost respect for Halligan who has achieved infinitely more than your humble author ever will — is that people who succeed aren't always perfect narraters of their own success. Halligan adopts the "category creation" lens because, as we're going to see, that's the language the startup world adopted to describe this kind of success. But language dictates how we see things and, in this case, it dictated how HubSpot saw their own success in a subtly wrong way, which is why they couldn't repeat the dose. Not that it stopped them growing into CRM anyway — when they were big enough to make that pivot, *they were the wave.*

> **Concept transcends category:** Interestingly, the actual *generic* software category that emerged around HubSpot and its ilk was 'marketing automation.' That was the *software* category that emerged, and while it's related to HubSpot's concept, it's not the same as HubSpot's concept. "Inbound" — their actual concept — transcended simple software categories. We'll explore the difference between software category and concept in the next chapter, but for now, keep in mind that a fundamental lesson for this kind of right-brain narrative development is: *concept transcends category.*

Execution is more than a playbook

Nevertheless, when HubSpot was in startup mode they had all these big pieces to work with, but they still had to execute, and execute they did, with:

- **Partner-friendly playbook:** They wrote the playbook, used the playbook, evangelized the playbook, and critically shared the wealth by equipping loads of channel partners (marketing agencies) to sell the playbook — and their software — as well. The agencies were affiliates, acting as both sales reps and success teams, while HubSpot built (and sold) the software, which was a powerful GTM accelerant.
- **Tight pitch:** Having the big vision is one thing; nailing it as a tight pitch is another. HubSpot could scale their story from keynote down to a very tight, right-brain friendly 'old way is broken / new way is here' pitch, contrasting the old, "interruptive," outbound way with the new, market-how-people-want-to-buy, "inbound" way. It was concise, high contrast (old/new, from/to), and relatable. (Halligan gives the pitch very briefly in his aforementioned podcast appearance, and you get the sense he's delivered it approximately 9392892737459 times in his career — a good

pitch sticks!)

- **Strong messaging:** They had cute, catchy turns of phrase too — inbound was, per Halligan, "much more about the width of your brain than the width of your wallet," i.e., anyone with some smarts could do this, budget or no budget. Unique messaging helps reinforce a unique playbook.

I know it sounds corny, but this really is something of a masterclass in clarity strategy. From the market insight to the macro change story, from the catchy, sticky messaging to the channel partnerships, it all came together fantastically well.

HubSpot wasn't built in a day, of course, and I'm sure there were many ups and downs along the way. But their core insight eventually gave them the basis to own their customer's entire customer relationship within their platform. HubSpot now proudly declares they're a "CRM platform," and the emphasis really is on "platform." As the secular change continues to play out — with just about all businesses doing some sales and marketing online — HubSpot continues to build out their platform to match, now as a broad multi-product stack, including CRM, CMS, email, automation, and more.

HubSpot's real playbook

HubSpot saw the secular change happening, named it, owned it, wrote the playbook, built the solution for it, expanded on it, reinvented themselves in the process, and rode to a dominant position in a mature category — CRMs, the market that was "crowded" when Salesforce entered back in 2000.

How's that for a strategy?

So I guess that's it for CRM. I mean, Salesforce exists on the sales side, and HubSpot exists on the marketing side. How do you compete with those all-encompassing platforms?

> **Playbooks — brain putty vs. behavior change:** Having some half-assed playbook document has become *de rigueur* for many B2B SaaS companies. Maybe it's an ebook, white paper, or guide, but whatever it's called, my question for you, dear reader, is whether this came from your careful observation of behavior change in your target market (or indeed in your own prior experience), or is it a product of a content writer who has, let's be honest here, a bit of knack for speculative business fiction and can make up whatever "playbook" you need? Doing the latter is easy — imitation is always easier than invention, but monkey-see, monkey-do content strategies don't drive competitive advantage.

Drift — the chat wave

How do you compete in a crowded market, a red ocean like CRM? You find a new wedge.

David Cancel arrived at HubSpot as Chief Product Officer in 2011 after HubSpot acquired Performable, a startup he co-founded. In 2014, Cancel left HubSpot to start Drift, with angel investment from the HubSpot co-founders, among others. By late 2021, Drift was valued at over a billion dollars, with a majority stake owned by PE company Vista Equity Partners.

Drift is particularly interesting from a secular-change point of view because it was, like HubSpot, explicitly built on awareness of the secular trends that were happening at the time.

By the time he co-founded Drift, Cancel was a seasoned exec with a background in software engineering. As he told VC Matt Turck in Turck's podcast[6], his first ventures were "started from a very engineering standpoint — [they were] all about building the thing." Drift, on the other hand, would be different (quote lightly edited for clarity):

> *I went to the other end when I started Drift and said "What are the undeniable trends that are happening in the world? How are those changing user behavior?" [...] We believed that we were undergoing this paradigm shift — moving from a world where the company controlled the buying process (in the B2B space) to a world where the buyer had all the control.*

That's some great right-brain, radar-level insight there. But what does that mean for the product and company?

> *What we've been trying to do from the beginning, which is invisible to everyone, is build a new type of CRM. We're doing that [with] a different type of data model, which is largely driven by conversational data. But no one sees that. You have to start with a wedge in the market and then you have to expand from there. Our wedge was, and is, chat.*

It means chat. Specifically, chat *as opposed to* forms and email for interacting with customers.

That's a very simple, very catchy idea, where the right-brain change and the left-brain playbook are two sides of the same coin. And in this case, it had almost universal appeal for online businesses. That's S-tier clarity strategy.

Drift was riding emergent waves in both trends and solutions. Drift didn't invent the now-ubiquitous chat bubble that floats on the bottom right of millions of websites. That had been pioneered by the smart cookies at

Intercom, but as a support tool. In 2017, Intercom was one of the fastest-growing SaaS companies at the time, second only to Slack[7], so there was a clear trend there.

That shift to messaging was the secular change that had already happened in the consumer world. Cancel had noticed that messaging had become an integral part of our lives, replacing in-person meetings and phone calls.

The question was whether he could bring that to business — and sales and marketing, specifically — too.

Run for the red ocean

SaaS was crowded. Chat solutions existed. What does this mean for Cancel's strategy? I love his response in the same podcast interview, where he spells out exactly what it meant (my bold):

> *When I was at HubSpot, we would religiously read* Blue Ocean Strategy *and talk about blue oceans versus red oceans. I think everyone got obsessed with this idea of finding these perfect blue oceans where there's zero competition. My argument, even back then, was most of the time, a market that has no competition is not a market. [...] So my view [when starting Drift] was, look, I want to go into the most commoditized market possible.* ***Really, we're trying to go into the CRM market. I want to go into that red ocean.***

Straight into the big, red ocean. But how do you avoid the bloodbath and win?

> *But [in a] red ocean, we need a fundamentally different paradigm and technical shift, so that we can re-segment and re-categorize this market, so that we have an opportunity to win. That's a very different approach than saying, "It's a red ocean. I'm just going to build a me-too product, and somehow I'm going to win because it's nicer."*

There it is — the need for a secular change, a "fundamentally different paradigm and technical shift." (Also note the ambition, genuinely playing to win in an insanely competitive market.) They had to name the change:

> *[Going into] this commoditized market with a wedge, [we had to figure out how we can] expand that wedge over time. Can we create a category? What is that category? We created a category called "conversational marketing," a name I hated because it was so long that I didn't think anyone could spell it.*
>
> *But then we woke up one day, we did that in 2016-'17, and then we*

> *woke up one day and then Gartner recognized it, Forrester, SiriusDecisions, all those people. Wiley asked us to write a book. We wrote a book on it in 2019. And now, everyone from Salesforce to whoever calls themselves conversational marketing in some way.*

Selling a lot of software tends to do that.

There's a lot to unpack here, but we'll keep it brief. "Conversational marketing," as Drift coined it, was, like "inbound marketing" before it — a new playbook. That's why they could write a whole book on it, and it highlights again why "category" is usually more synonymous with "concept," or, as I like to say, change + playbook + tool.

In Drift's case, that meant:

- **The broad change:** The shift in power from seller to buyer in B2B and the rise of messaging as a primary communication medium.
- **The specific playbook:** "Conversational marketing," engaging prospects in real-time with bots or chat, and moving away from the old forms-and-email approach.
- **The tool:** The chat bubble that lives in the corner of your website on the one hand, and the backend to support that, on the other.

The HubSpot founders put their finger on a secular change that was happening — digital marketing, or "inbound marketing," wrote the playbook, and built the tool. Drift, several years later, essentially does the same thing with the shift to messaging, or "conversational marketing."

Like HubSpot, Drift transcended their category (chat tools) with a concept they owned, and that fast became an integral part of their brand.

Like HubSpot, Drift's playbook was copied to death, and they moved on to new playbooks to further expand on the secular change they saw in the market.

And like HubSpot, who started with marketing tools and built a platform and eventually a CRM around that, Drift started with chat and built a platform and eventually a CRM, too.

> **Execution:** Cancel is surely one of the best, most customer-obsessed execs in the business. And Drift's execution, it's worth emphasizing, was phenomenal. We have to remember that Cancel talking about how he'd win in the CRM market is akin to a great football coach talking about how they're going to win the Super Bowl that year. They're going to skip over the part about how they're already an incredible coach with years of experience under the belt and top talent at their disposal, and just give you the tip-of-the-iceberg strategy that sits on top. But

> we can't forget the execution underpinning it all.

The strategy explanation test

Walter Kielchel, in his excellent book on the rise of corporate strategy and consulting firms, *Lords of Strategy*, offers this test for the kinds of case studies we're looking at here:[8]

> *[The] question that should haunt readers of business literature whenever they encounter an author using a company example to illustrate a point about strategy [is:] "Is that what the company thought it was doing? Did it know that it was performing this alleged act of strategy?"*

Cancel is an excellent example of someone who passes the 'strategy explanation' test — their explanation *is* what they believed they were doing at the time. For Cancel, that meant:

- **Market:** Attacking a red ocean with an otherwise commoditized product (CRM).
- **Wedge:** Doing so by riding a secular change with a technical angle (chat).
- **Playbook:** Naming the change and the playbook, owning that name, and using it to drive fame in their market ("conversational marketing").

That's exceptional execution of an exceptional clarity strategy.

This is not the only path to success by any means. There are plenty of cases where impeccable product design and engineering meet bottoms-up adoption with network effects — think Slack, Figma, and Notion, for example. And there are, of course, plenty of successes that were, or at least began as, niche tools with highly targeted positioning, as we'll explore in coming chapters.

But we're interested in the right-brain narrative game here — the three-act variety in particular — and Drift *is* a fantastic example of the role a clear internal narrative plays in success. The change and playbook they identified and created amounted to a story they could tell in both short form ('use chat not forms'), or long form (a whole book on conversational marketing). And if you can get that story out there — Cancel had a very influential social media and podcast presence — then you can, if Drift is any indication, use your narrative as part of your strategy to ride those trends to a very lucrative position in a fiercely contested market. A market where there already was a Salesforce and HubSpot happily competing, too.

Story 1 tools

In these last couple of chapters, we've seen the power of secular changes, those market-shaping mega-trends, or paradigm shifts, that savvy founders can ride to fame and fortune.

These are powerful — if rare — tools of right-brain, story 1 narrative. They give your story incredible momentum because you have a big, obvious change 'out there' to point to, and that sets up your more left-brain playbook and the need for new tools. Buyers believe it because they can see it for themselves — they're not taking your word for it — and if you execute well, you can scale your story up and down from tweets to books to podcasts and back again.[9]

But the trend alone isn't enough — you need the playbook and, obviously, product to make it count and complete out change + playbook + tool trifecta. When you can bring all of these aspects together as one coherent whole, one unique concept, then you're well on your way to creating an incredible company.

The power of secular change

Let's wrap up this chapter with a few additional thoughts on secular change as we set the stage to untangle typical rah-rah category creation advice in the next few chapters.

- **Right brain, big change:** This is all about the right-brain side of your story — the big x-axis market-level story. (But it's not the only way to succeed, and I'd be bummed if you stopped reading here thinking it was!)
- **Talent + environment:** Gainsight succeeded because Customer Success was a strong, enduring, secular trend. Hopin struggled because virtual events were not — the tide went out, and they had to pivot.
- **Concept = broad change + specific playbook + tool:** People conflate *category* with *concept*, but concept is the real driver, the real 'whole,' as it captures attention across the whole spectrum of awareness.
- **The underlying trend is productivity:** Incidentally, the *real* underlying trend in all these cases is about productivity. SaaS was more productive than on-prem. "Inbound" digital marketing is more productive than blind ad spend. Chat is more productive than forms and email. And soon, AI-everything is going to be more productive than everything that went before.
- **Well-funded whole > cheap insight:** Having an insight into a secular change or emerging playbook is one thing, but you've still got to execute the whole concept, from market education to product development and beyond. If your narrative is only embodied in a blog post or webpage,

consider how quickly folks will copy that, because they will. What's hard to copy is expensive to build, both in product, awareness, and community.
- **Leading through narrative:** A genuine right-brain narrative — built on a secular change + specific playbook — can be a fundamental pillar of market leadership. The founders we've looked at thought long and hard about the change they were seeing and how they would attack it, and their narrative was, I imagine, a powerful coordination point for all the work that followed — from product to go-to-market and beyond.

We've touched on category creation here, looking at it from a concept point of view. But what happens when you really do create a new generic category of software, particularly in enterprise sales? That's what Yammer did, and that's what we'll look at next.

Notes

1. Benioff, Marc. Adler, Carlye. *Behind the Cloud: The Untold Story of How Salesforce.com Went from Idea to Billion-Dollar Company-and Revolutionized an Industry,* p. 105.
2. Holmes, Paul. "The Launch of Salesforce.com and the End of Software," PRovoke Media [sic], f.k.a. OutCast Communications, July 8, 2001. **https://www.provokemedia.com/latest/article/the-launch-of-salesforce-com-and-the-end-of-software**
3. Evans, Bob. "Salesforce Leapfrogs SAP as World's Largest Enterprise-Apps Vendor," *Acceleration Economy,* August 29, 2022. **https://accelerationeconomy.com/cloud-wars/salesforce-leapfrogs-sap-as-worlds-largest-enterprise-apps-vendor/**
4. Halligan, Brian (Guest). Parr, Sam (Host). Puri, Shaan (Host). "Near Death Accidents, Hitting $1B ARR, and Selling Like the Grateful Dead with Brian Halligan, Founder & Executive Chairman of HubSpot," *My First Million,* October 14, 2021. **https://www.mfmpod.com/near-death-accidents-hitting-1b-arr-and-selling-like-the-grateful-dead-with-brian-halligan-founder-executive-chairman-of-hubspot/**
5. For a taste of the narrative they'd developed by 2014, see this neat time capsule of a deck: **https://www.slideshare.net/HubSpot/what-is-inbound-marketing-29665969**
6. Cancel, David (Guest). Turck, Matt (Host). "In Conversation with David Cancel, CEO, Drift," September 24, 2020. **https://mattturck.com/drift/**
7. Bort, Julie. "How drinking Guinness with Biz Stone launched one of the fastest-growing startups in the Valley today," *Business Insider,* February 23, 2017. **https://finance.yahoo.com/news/cold-call-biz-stone-launched-210855431.html**
8. Kiechel, Walter. *Lords of Strategy* (p. 131).
9. It's interesting to note these companies with strong concepts *really could* write whole books on them — Gainsight, HubSpot, and Drift, for example, all published books

about their concept. That is perhaps a clue as to whether this approach is for you, i.e., whether your insight is fundamental enough that you could spin a whole how-to book out of it.

32

Unbundling category creation I — Yammer & enterprise category creation

> No objective is more important for SaaS founders than early category leadership. Over 75% of the market cap in a software category typically goes to the category leader. VCs know this, so funding disproportionately flows to startups that can lay claim to early category leadership. Perception often becomes reality, as the startup with the most resources to spend on sales and marketing builds the largest subscriber base, which gives it more resources. This virtuous cycle causes most SaaS categories to accrete to a big winner over time.
>
> *- David Sacks*[1]

Is category creation essential for VC-backed startups? Is it something you should aspire to? Does it even exist? We've been tip-toeing towards 'category creation' as a strategy, and now we're going to tackle the idea in all its messy glory in this four-chapter mini-series on the topic.

It won't surprise you, dear reader, to hear that I think a higher-leverage strategy is to have a well-calibrated right-brain/left-brain story 1/story 2 combo that's bigger than 'category creation' alone, simply because attention is more fundamental than category, and clarity strategy helps you go big *and* deep. It's not that category is unimportant — we're exploring it here because it may well be part of your clarity strategy — it's just that it's a second-order issue, and one that's in desperate need of unbundling.

"Category creation" folks get tripped up trying to either talk in grand story 1 terms ("We're creating a whole new category that's going to change the world, maaan!") or in very laser-beamy story 2 terms ("We've created this finely sliced niche category, just for you!"), without knowing which is relevant, if it's relevant at all. That both get muddled together is a source of endless confusion. This chapter, and particularly the next, are intended to resolve some of that confusion.

First, though, we're going to look at *literally* creating a new software category.

The investor view

Why do we talk about 'category' at all? There's a view that category *leadership*, if not outright creation, is essential, as David Sacks of Craft Ventures demonstrates in the quote above from his Medium piece "The One Who Defines the Category Wins the Category." The headline captures the argument: if you want to be the big winner, you better define the category, because the winner takes all. (Sacks has first-hand experience of this, as we'll see shortly.) Category leadership, in this view, becomes self-perpetuating, with those who tell the strongest category leadership story:

- Getting the most funding.
- Acquiring the most customers with that funding.
- Attracting even more funding, more customers, and accelerating their flywheel.

There are a lot of assumptions packed into this view about what category creation actually is, and we're going to tease them out as we go, but for now, I want to highlight the self-perpetuating nature of *investor* belief.

If investors all believe there's something about a type of company that makes it more likely to win, they'll pile into that company, making it (theoretically) more likely to 'win.'[2]

Founders know this too, so in the competition for capital, they're incentivized to start spinning very particular kinds of narratives — that they're the category leader or, better yet, the category creator. But what does that even mean? What *is* the 'category' in 'category creation' or 'category leadership,' anyway?

> **Perverse incentives:** In my view, dear reader, investor memes around category creation create perverse incentives where investor assumptions drive founder behavior (and narratives) around a concept that, as we'll see, ***may not be particularly true in the first place***. Nevertheless, if you're a founder, and this describes the environment in which you want to attract funding, it pays to know how it works. Just don't internalize it, which is easier said than done. As we'll see, while the concept may have reflected some market dynamics in the past, it doesn't necessarily universally reflect them today, and your investors may not realize that.

Category vs. concept

If you have — or are being encouraged to — install the idea of category creation into the foundation of your mental model for startups, then it's a good idea to understand how it works. In my view, "category creation" in startup discourse blurs three different ideas, specifically:

1. **Traditional software categories,** i.e., literal *groupings of similar things*.
2. **New category *concepts*** that can transcend a literal category (as we touched on in the previous chapter, where CRM players transcended the generic category definition with their own unique concept).
3. **Truely unique companies and products** that become cultural touchstones in a category-of-one sense.

We'll unpack this further as we go, but for now, let's note that most discussions of category creation tend to conflate a mix of these things, which makes it very easy to grab the wrong end of the stick for the uninitiated.

But if it's confusing, why is it used? Let's start with category 101.

Where do software categories come from?

> *Companies don't name categories, analysts do.*
>
> – David Kellogg[3]

Tech companies make a lot of software.

Buyers find it hard to know what to buy.

Analysts (like Gartner, Forrester, and so on) put things in buckets and try and make sense of the market.

Those buckets are analyst-created 'categories.'

Categories then have leaders and challengers, visionaries and risky bets, all mapped out in Magic Quadrants and quarterly Waves.

Buyers — particularly enterprise buyers — go to the analysts wondering what they should buy. Analysts give them their take on the market, err, I mean category. Buyers use that to build their requirements checklists. It's intended as a clarifying exercise for the buyer. Understand that, so long as there are an unholy number of very expensive products for buyers to choose from, there will be a market for advice about what fits where so that a buyer knows which aisle they should be pushing their shopping cart down in the enterprise SaaS supermarket.

With this in mind, before you fret about creating your own category, you should first make sure customers know which category you're in (or want to be in) because vegetables don't tend to sell in the confectionary aisle, for example. It also pays to understand the category language your buyers speak.

> **'Where to shop' before 'what to buy':** Buyers want to know what aisle

> you're in as well. I've seen plenty of demos for clients where buyers show up confused or have guessed incorrectly about the category a SaaS product fits in. A big part of buying is trying to mentally map the market, and that's hard to do. Trustworthy reps (and founders) can help prospects do this[4] because they're more in touch with the *reality layer* of the market than the buyer is. Do you help prospects map your market?

Categories as left-brain maps

Speaking of maps, back in our introduction to left-brain, story 2 positioning in **chapter 21**, we discussed how, to quote McGilchrist, "the left hemisphere [...] extends vastly our power to map the world"[5]. We have also learned that the left hemisphere, "if offered a story whose episodes are taken out of order, tends to regroup them so as to classify similar episodes together, rather than reconstruct them in the order that has human meaning."[6]

That is, categorizing things in groups helps us map the world *in the abstract*, and that is a very left-brain thing to do.

It's also a very time-consuming thing to do, and that's why a market for analyst-created market maps (i.e., categories) exists, especially for highly considered enterprise purchases. Buyers just want to buy the right solution; they don't want to become market experts, and that's why a market for analyst insights exists.

Choose your bucket

But back to the analysts.

When analysts draw up their categories, tech companies suddenly find themselves swimming in a very specific bucket — one not of their choosing — with a grab bag of other competitors. That might be nice if you're a leader; it's not so nice if you're just a tiny challenger or you're genuinely trying to do something unique. If you're a new majestic tropical fish, why get lumped in with all boring old salmon, say, just because you're all fish? Why not compete separately in the majestic tropical fish category, where your colorful majestic-ness can shine through?

Or, if you're a B2B tech company, why let the analysts set the terms of competition in the first place? Why not **drum roll** create your own category instead? A category you can *position yourself* in.

Creating an actual category

Sometimes this can work. Yammer is perhaps the canonical example.

Yammer, co-founded by David Sacks (quoted above) in 2008, started out as the internal communication system of Geni, an earlier startup Sacks co-founded. *Inc.* reports that Sacks "liked the tool so much, he spun it off into a new company."[7] (You can see shades of Stewart Butterfield and Slack there, which again speaks to the emergent nature of startup opportunity.) Four years later, in 2012, Yammer was acquired by Microsoft for $1.2 billion.

Sacks, one of the original "PayPal Mafia," would go on to co-found venture capital firm Craft Ventures in 2017 and therefore provides an interesting perspective as both a founder and investor. In his Medium piece on category creation that we quoted from earlier, he walks through how it played out for Yammer in the enterprise software world of the late 00s. (And this *is* very enterprisey: analysts, requirements, magic quadrants, Fortune 500 deals, etc.)

This makes for an excellent case study of category creation that actually happened, so let's unpack it. First, to summarize:

- **The existing category:** Yammer attempted to enter the 'social workplace software' category, competing against late-stage Jive. (Jive was founded in 2001, went public in 2011, and was acquired in a 'fire sale' in 2017 for $462 million.[8])
- **The incumbent:** Jive was Gartner's Magic Quadrant winner for that category — recall that analysts define categories — and had spent years building features to meet Gartner's definition of the category so they could use Gartner's endorsement to win Fortune 500 enterprise deals.
- **The setback:** Gartner *excluded* Yammer from the category because they didn't meet the list of requirements — features like wikis and blogs, and so on. Yammer was bummed.
- **The opportunity:** Yammer realized they would have just been a small fish in a big, outdated pond. They were about social networking, which was a new paradigm and hence a new opportunity.
- **The new category:** They instead came up with their own category definition, which was Enterprise Social Networking (ESN).

What did this new category definition mean for Yammer? According to Sacks:[9]

> *This was more intuitive to almost everyone, and it allowed us to ride the media coattails of what was happening in the consumer space. Freed to define our own product roadmap, we could determine what ESN meant. Our north star was user engagement and bottom-up freemium adoption by employees. On-premise competitors deployed by IT struggled with actual usage and could not compete on these dimensions.*

Category and clarity

To tease out some of the key aspects of this strategy:

- **Clarity:** Sacks said their new category definition was "more intuitive to almost everyone." The new category definition was clarity *creating*. Many faux-category creation attempts are clarity *reducing*. It wasn't about creating a web of buzzwords; it was about creating clarifying language.
- **Frame breaking:** There was an existing frame Yammer wanted to break, a clear old way/new way they wanted to establish. Their category worked as *both* a right-brain change story *and* as a meaningful new spot on the buyer's map. For the former, it was: 'Hey, we're the cool new thing, not that old school, blogs-and-wikis stuff.' They broke the old frame with positioning by association — creating a mental hook using a known (and, at the time, very hot) reference point. (We discussed frame-breaking insight, Challenger selling, and the right-brain radar in **chapter 25**.)
- **Secular change:** Ah, our old friend, secular change. Yammer was, as Sacks says, riding the secular change of consumer social networking (back when Facebook was truly on fire and not in a this-is-fine sense), and therefore, they could "ride the media coattails of what was happening in the consumer space." (Twitter, for example, was founded two years earlier and had 100 million users by the time Yammer sold in 2012.)
- **GTM strategy:** Yammer didn't just coin a category name; they had a unique (relative to their competitors) GTM motion behind it too, which in Yammer's case was bottoms-up adoption from employees with all the juicy network effects that entails.

You can see the strategic benefits of riding a scorching-hot trend in one area (consumer tech), using it as a positioning and narrative hook to fuel hype around your push into the other end of the market (enterprise), and making your product offering much clearer in the process. The "Enterprise Social Networking" category name made all of that instantly apparent. I mean, "social networking" was literally in the name!

> **Lightning in a bottle:** It's worth noting the conditions that allowed this category opportunity to emerge. Ubiquitous social networking was in the air: Facebook opened up to everyone in 2006; the iPhone launched in 2007; Yammer launched in 2008. Sacks was already interested in social networking, founding Geni in 2006, where Sacks said they were "trying to create a family tree of the whole world."[10] The team was able to build and dogfood the internal communications tool that would become Yammer, spin it out, fight their way into the enterprise, figure

> out sales,[11] and get acquired for over a billion dollars, all in four years after Yammer's founding. Calling that "category creation" *is* true in the fullest sense, but consider how emergent many of those conditions were. While the category has to be created by someone, it's still that *energy of secular change* that drives the opportunity. Full credit to those able to capture that opportunity, but if you're trying to replicate that kind of success, remember it's spotting the opportunity in the first place, not trying to cargo-cult the eventual outcome (i.e., a category name), that matters.

Right *and* left

While Yammer rode a powerful secular trend just like our CRM examples, in this case, Yammer really did 'create a category,' coining a meaningful new generic software category. Enterprise Social Networking is now an established, analyst-blessed category. According to Gartner, for example, Google, Meta, Microsoft (with their Yammer acquisition), Salesforce, IBM, and others all have products in the category to this day.[12]

This combination of a powerful change story *and* redrawing of the market map puts this act of category creation right in the middle of left- and right-brain attention. You can see it in the way Sacks uses "category" interchangeably with "positioning," suggesting founders (my italics): "Shape the category requirements to *position yourself* and *de-position* competitors." (That is, if you can position as the new hot thing, then the flip side is that your competitors now own an outdated, obsolete position in the buyer's mind.)

But the fact it was a *new* category, and not just a case of mapping in an existing one (or a sub-category thereof), meant it was more than left-brain positioning. New categories means narrative, i.e., something unfolding over time, and that hits the right-brain radar. That's why reporters will report on it. If it's not new, and it's just a static reference to an existing category, then that pushes it back to left-brain laser-beam mode as something that will get categorized and filed away. That's probably not going to have the incoming-change-in-the-environment urgency you're after.

Put it this way: if you were launching an enterprise social networking platform today in the '20s, that's not a new category story and, therefore, not much of a narrative. People still want to know where you fit, so positioning still matters, but the category term alone obviously doesn't generate market-level, right-brain interest. This is why the trend, rather than the category name, is often a more meaningful unit of analysis.

Categories and attention mismatch

There is an intuitive aspect to this — we're all using the same two modes of attention, and we all, as humans, naturally want to know where stuff fits (the left-brain mental map) and what's new (the right-brain radar of environmental awareness). Therefore, startups are often encouraged to try and come up with a new place (right-brain interest) that describes where stuff fits (left-brain mapping), hence a new category.

But where founders and teams often get stuck is in knowing which kind of attention they're addressing, and that results in the kind of story 1/story 2 attention mismatch we touched on way back in **chapter 6**. For example, do you:

- **Focus on the name:** If you try and come up with the perfect name, you're focusing on the left-brain mapping side of the equation.
- **Focus on the change:** If you're trying to sell how new you are, you're focusing on the right-brain side.

That might be appropriate for your startup, or it might be a problem, for example:

- **If a buyer wants to know where you fit** and you respond with why you're new and exciting, they'll be confused because they're in left-brain market-mapping mode.
- **If a buyer wants to know what's new** and you respond with a precise category definition, they'll be confused because they're in right-brain environment-awareness mode.

You can see how trying to do ostensibly the one thing — figure out your category — can be looked at from either mode of attention and, therefore, can have two valid but separate purposes and result in a lot of crossed wires and time spent spinning your wheels if you mismatch modes of attention.

> **Creating the 2097th category:** G2, as I write, currently lists 2096 categories of tech products[13] (not just software, to be fair), so consider that the utility of creating the 2097th category, in the literal-category-of-things sense, may be less decisive than it once was, especially if you don't have a hot trend to ride. There's always room for new 'change → playbook → tool' *concepts*, of course, and there's always opportunity in being laser-focused *within* an existing category, but the opportunity for creating a new actual category isn't going to be what it was in the earlier stages of the Cambrian explosion of cloud software.

When the trend fits

We've been looking at how secular change drives narratives, and a modern equivalent might be the rise of 'creators,' for example. To pluck an idea from the deepest pit of hell, imagine (Father, forgive me) TikTok for the enterprise, where the 'creators' in your company (I have sinned) can share videos. In that case, maybe you *should* coin a new category name (left as an exercise for the reader; I will sin no more) to differentiate your hot new product from the stodgy old enterprise social networking tools in the same way Yammer did to 'social workplace software' in the '00s.

But the point is the change, not the name. All Yammer did, in terms of naming, was stick "Enterprise" in front of "Social Networking." That's... not that hard. The really hard part was to be astute enough — and fortunate enough — to be able to ride the social networking trend into the enterprise in the first place. And you can see why it worked in enterprise sales: 'Social networking for the enterprise' is self-explanatory in a way that '*Well, *deep breath* we're social workplace software like Jive too, but even though we don't have features X, Y, and Z, we have new features A, B, and C*' is not.

Buyer attention & enterprise sales

This focus on what 'category' actually means can seem a little academic. Does it really matter? Well, it does if it matters to your buyers. Think about where your buyers' attention is at. Are they asking where you fit, or are they asking what's new? And do you have a compelling answer, particularly along one axis of attention or the other? Are you really new and novel, or are you a really tight fit for their exact needs?

One speaks to your insight on the radar level, and the other speaks to your position on the laser-beam level. And if you're running with a new category definition based on a secular change, that may open the door, but you've still got to strike. Think of story 1/story 2 as the headline and the details of your narrative in this case:

- **Headline:** A meaningful new category, or category definition, is like a headline — it captures attention. That's great for hitting folks who are in right-brain radar mode of attention.
- **Details:** Then you've got to switch gears into what this thing is and why it matters — you've got to get narrow and specific in a plausible way, and if you can't, your headline was just proverbial clickbait, which is not going to win you the market.

In Yammer's case, their broad narrative got them *into* the game, but it's worth noting that they still had to fight it out in the market with Jive. They just now had a position to fight *from*. Sacks continues in his Medium piece:

> *Yammer's competitor Jive was run by top enterprise sales executives and was excellent at [explaining what functionality was critical]. We always knew we were in a bake-off with Jive when the prospect asked us about analytics.*

Sacks describes how Yammer argued in deals for the importance of measurement, and Yammer responded with their own good-enough analytics package. "And so on it went, both sides looking for their next trump card."

In these cases, deals are still hand-to-hand combat. Having the narrative gets you *into* more deals, because you're on more folks' radar, but it doesn't mean you win them by default. Competitors will respond, and you need to respond in turn. (This can result in a degree of 'competing to be the same,' which we'll look at in **chapter 49**, *but at least you are competing.*)

This is also why it can be risky to be too enamored with the big we're-going-to-the-moon story 1 change story, as category definition or otherwise. It's great if you've got it, just don't neglect your story 2 positioning. Writing headlines can be fun and gratifying; getting the details right can be an unending slog, but you need both. You still need to articulate why — to use April Dunford's definition of positioning — only you can deliver some specific value that a particular set of buyers cares a lot about. That's what clarity strategy often is: looking out *and* down.

Narrative vs. strategy

When we talk about clarity strategy and your narrative, it's important to note that there's a relationship *between* your narrative and your clarity strategy; they're not the same thing. Story alone *is not* strategy.

In Yammer's case, for example, while they had a very strong category and trend narrative (social networking for the enterprise!), that wasn't the only strategic lever they had at their disposal. In their competition with Jive, as Sacks tells it, the decisive factor was that Yammer was, in fact, *SaaS,* and product-led SaaS at that (the "architectural" difference, which was combined with enterprise sales):

> *In the end, the deciding factor in our competition was architectural. Yammer was born in the cloud whereas Jive was on-premise. Eventually our viral freemium model swamped them, in both leads and popularity. If you can position yourself around an architectural strength and de-position your competitor around an architectural weakness, it will be very hard for them to respond. Architecture is destiny.*

Sacks uses 'positioning' again, this time in reference to their GTM model, but the point is that Yammer made fundamental decisions that on-prem Jive couldn't respond to.

This is why discussing category creation as though it's just a matter of coining a new term misses the point — to truly create a new category definition, you may well need a unique business model to support it. Everything else can be copied — the features, the messaging, the name. Competitors can and will copy it.

When Sacks says "Architecture is destiny," I take that to mean it's the depth of *compounding strategic choices* that can't be copied that helps you beat incumbents. In Yammer's case, relative to Jive, they had a:

- Unique product (social networking focus vs. blogs and wikis).
- Unique architecture (SaaS vs. on-prem).
- Unique positioning/category ('enterprise social networking,' i.e., not going head-to-head in Gartner's old category).
- Unique GTM strategy (the "viral freemium model" *and* enterprise sales).

That's a very strong model, especially given the network effects involved. I like to describe this kind of business model innovation as "& + & + &" — *and, and, and,* i.e., multiplicative strategic advantages. And for Yammer, it paid off, with a $1.2 billion exit.

> **It's the model that wins:** We can get too fixated on the one left-brain piece that we can manipulate to find success — the novel category name, the unique feature, the next hire — while forgetting that it's the right-brain whole of multiplicative, fundamental ('architectural') choices that makes for truly competitive business models, and the category label is often just the cherry on top.

Playing bigger with category creation

This, dear reader, has been the sane introduction to category creation as an idea.

In the previous chapter, we saw how secular-change-driven concepts *transcend* category, and in this chapter, we've seen how secular trends (like social networking) can, occasionally, create genuine new categories of business software.

Yammer's category definition clarified what they were about, tapped a powerful, zeitgeist-y secular change, made them newsworthy, and was all supported by their, at the time, unique business model. Yammer was a rare success insofar as all mega exits are rare successes, so it shouldn't be a surprise that they got there with a rare combination of multiplicative ingredients.

And yet, for some folks, category creation can be even more. For these folks, every innovation in human history was, it seems, an act of category creation. I know that sounds silly — because it is silly — but some investors take this approach to category creation very seriously, and not entirely without reason, either. We'll explore this investor story of category creation in the next chapter.

Buckle up, because it's going to get weird.

Notes

1. Sacks, David. "The One Who Defines the Category Wins the Category," *Craft Ventures*, December 17, 2019. **https://medium.com/craft-ventures/the-one-who-defines-the-category-wins-the-category-245fee85bfbb**
2. Even 'win' is an ambiguous term here — 'win' can simply mean 'gets big and exits' and it doesn't necessarily mean 'becomes a valuable company that keeps growing post IPO.' That is, it may 'win' for early investors and no one else.
3. Kellogg, David. "Congratulations, You've Created a Category. Now What?," *Kellblog*, June 7, 2020. **https://kellblog.com/2020/06/07/congratulations-youve-created-a-category-now-what/**
4. Shout out to April Dunford for this insight.
5. McGilchrist, Iain. *The Master and His Emissary* (p. 70). Yale University Press. Kindle Edition.
6. McGilchrist, Iain. *The Matter With Things: Our Brains, Our Delusions and the Unmaking of the World* (p. 49). Perspectiva Press. Kindle Edition. (Note abbreviations have been expanded.)
7. Welch, Liz. "The Way I Work: David Sacks, Yammer," *Inc.*, November, 2011. **https://www.inc.com/magazine/201111/the-way-i-work-david-sacks-yammer.html**
8. Miller, Ron. "Collaboration software company Jive to be acquired by Aurea for $462 million," *TechCrunch*, May 1, 2017. **https://techcrunch.com/2017/05/01/collaboration-software-company-jive-to-be-acquired-by-aurea-for-462-million/**
9. Sacks, David. "The One Who Defines the Category Wins the Category." *Craft Ventures*, December 17, 2019. **https://medium.com/craft-ventures/the-one-who-defines-the-category-wins-the-category-245fee85bfbb**
10. Sarno, David. "Fertilizing the family tree on Geni.com," *Los Angeles Times*, August 5, 2007. **https://www.latimes.com/entertainment/la-ca-webscout5aug05-story.html**
11. Sacks, in the *Inc.* piece above, said at the time on enterprise sales, "When we first started, I had this naive view that sales would just take care of itself—people would just pull out their company credit card and buy. But companies, especially large

enterprises, like to talk to somebody. They've got concerns about security, privacy, and compliance. So you really need salespeople to engage with them. We didn't have any salespeople when we started. Now we have 30."

12. See the list here: **https://www.gartner.com/reviews/market/enterprise-social-networking-applications**
13. See the full list in all its glory here: **https://www.g2.com/categories?view_hierarchy=true**. Hat tip to David Kellogg for surfacing it: **https://twitter.com/Kellblog/status/1603434395255427073**

33

Unbundling category creation II — the Play Bigger narrative

[T]he left hemisphere's world view is easier to articulate. The left hemisphere is the speaking hemisphere: the right hemisphere has literally no voice. The attempt to make the implicit explicit radically alters its nature; as a result, finding the language to put across the way of being of the right hemisphere is simply harder than doing so for the naturally explicit left hemisphere.

- Dr Iain McGilchrist[1]

Just how big should your story 1 change story be? We're in the middle of our right-brain narrative development section here, and if we're thinking about stretching your narrative *out* on the x-axis, just how far can we stretch it? Where is the breaking point? What's the upper limit? How big can we go?

There's one approach that has become popular in VC and startup leadership circles in recent years that takes us to — and beyond — that breaking point. It's popular because investors, founders, and marketers want to make a splash in the market, to be worthy of attention, and to land on folks' right-brain radar in a big way. They want the biggest story 1 narrative possible. They want to *create a category.*

Playing bigger

We've already seen how both literal categories of software (last chapter) and new concepts based on secular change (**chapter 31**) get lumped together under this banner of "category creation." But some folks take this even further. For these folks, it would seem category creation describes anyone who did anything big *ever*. Because they did something big, and their idea, product, or concept caught on, they *ipso facto* must have 'created a category.' Therefore, if you want to tell the biggest story 1 change story possible, you better be talking about how you're creating a category in the most expansive way possible.

Or so the theory goes.

This view of category creation goes well beyond our simple ideas of generic software categories and novel concepts that we've explored over the last couple of chapters. It's a much more implicit, right-brain, hard-to-define, do-what-big-players-do kind of approach.

It's also more than a little bananas.

This is the world of investor and CMO favorite *Play Bigger*, published in 2016, a book we're about to dissect. Over the next two chapters, we're going to explore "category creation" as *Play Bigger* tells it, to see just what maximalist story 1 narratives look like, and then we'll continue unbundling category creation as a concept to understand exactly how it does (and doesn't) work.

The category meme

This is a pretty meta chapter, dear reader, so skip ahead if this doesn't float your boat. (There are still loads of good positioning insights to come.) In this chapter, we're going to look specifically at:

1. **Category confusion:** Where the most expansive, maximalist view of 'category' comes from, why it confuses so many folks, and how the hierarchy of attention helps us resolve that confusion.
2. **Clarity strategy in action:** How the narrative *around* this particular version of category creation works, because it happens to be a masterclass in narrative itself (even if the substance is more vibe than reality).

The narrative behind category creation — even this somewhat cooked version — is fascinating because it *worked*. It achieved what many folks (your humble author included) dream of in terms of narrative success: having a whole industry buy into your story, your framing, and your language, all pointing back to you as the authority with the answers.

This is especially relevant from a story 1 marketing point of view. We've been looking at story 1 largely from a sales point of view so far, and while it's great to have that change story trump card up your sleeve for prospects, what if you could seed your story in the market at large instead? What if you could deploy your story in such a way that it became the default approach that everyone 'just knew'?

That is, what if you could meme your market?

Memeing your market

We've discussed the concept of seeding your idea in the market before. Back in **chapter 15** we discussed *memes* as 'information packets with attitude,' to use Daniel Dennett's term. These little bits of *cultural* information catch on and are passed on in the social environment in the same way genes are bits of *biological* information that are passed on in the natural environment. There is, so the theory goes, a survival of the fittest aspect at work — the catchier the meme, i.e., the more adapted it is to its environment and the more it can reproduce, the more likely it spreads.

When we talk about "category creation," therefore, we're often actually talking about catchy ideas that we want to spread — to reproduce and be passed on in a given social environment. And if they're really catchy, they might take over a whole industry... like the VC industry.

That's because the concept of "category" or "category creation" is, itself, memetic. It has, for reasons we'll discuss, survived and persisted because it was better adapted to the environment than alternatives. That doesn't mean it's an inherently good idea — it's often highly confused, as we'll see. But it has, to date at least, satisfied a particular job-to-be-done better than alternatives. Therefore people use the language, the language sticks (particularly in niches where it's best adapted, i.e., with CMOs), and it continues to evolve.

Radar bait

Memeing the market means getting a large segment of people to use your language independently and on their own terms. You're not just broadcasting your idea into the market; you're watching it spread as folks happily tell each other about your idea, using your terms, labels, and phrases. This can sound a little sinister — and sometimes it is — but if your ideas are genuinely good, then that's what we want to happen. We *want* good ideas to escape the lab of your marketing and run riot in your target market.

We've seen examples of this already, for both literal new categories on the one hand and new concepts on the other:

- **Literal category labels** in the analysts-and-requirements world of enterprise software, like we saw with Yammer and the rise of "Enterprise Social Networking" in the last chapter.
- **Explicit concepts** for software in B2B SaaS driven by secular change, such as customer success (Gainsight, **chapter 30**), inbound and conversational marketing (HubSpot and Drift, respectively, **chapter 31**), and as we'll see in **chapter 35**, experience management (Qualtrics).

These two forms of language boil down to:

- **Category:** A whole new type of software exists, so pay attention and tell your colleagues.
- **Concept:** A whole new way of doing *this particular activity* exists, so pay attention and tell your colleagues.

That's the memetic angle to this kind of so-called category creation — seeding language and ideas so folks pay attention and tell their colleagues. That is, the audience uses your terms, your metaphors, and thus your way of seeing the world themselves.

> **Radar-worthy memes:** Working backward from this "pay attention" behavior, it becomes obvious that our story has to be genuinely radar-worthy if we really do expect people to catch our language, adopt our ideas, and ultimately change their behavior. That's the litmus test, and it's a high bar to pass. This is also why secular change is so important — it provides a reason *why* people should pay attention that's bigger than you. That's also why having a playbook matters, as it guides folks in *how* to act. That is, it's much more than a cute idea or a new TLA.

Turn it up to 11

The *Play Bigger* folks succeeded in turning the concept of category creation itself into something of a meme, as it became everyone 'just knew' you had to be doing as a startup, for a while at least.

But the idea of the importance of category *leadership* goes back at least to 1991's *Crossing the Chasm,* so what did *Play Bigger* do differently to give it its own spin? They dialed up their narrative to eleven, creating a framework that (in theory) covers everything from IPO windows to CMO strategies.

Unfortunately, in doing so, this dialed-up-to-eleven-with-an-acid-tab-or-two world of *Play Bigger* made category creation *highly* confusing. We're not just talking about enterprise software categories anymore, oh no. We're talking about something much, well... *bigger.*

> **Choose your own adventure:** We're going to do a running triple backflip into the category creation weeds here, dear reader, and if that's a level of GTM nerdiness you can't abide, I totally understand. Skip ahead to our Jobs to be Done discussion in chapter 36 or positioning in chapter 37 if you'd rather keep moving.

Explicit/implicit

Everyone with an entrepreneurial bone in their body wants their company or product to take over the world — or at least their segment — and if someone claims to have the secret sauce that lets you do just that, then that's worth considering, right?

That's what *Play Bigger* purports to offer, and that's what VCs have, in some firms, taken very seriously. But when a variety of quite different approaches are lumped together as "category creation," we get a mess that needs to be teased apart.

For me, dear reader, the key to untangling that category mess was found, yet again, in Iain McGilchrist's work, specifically on the explicit to the implicit.

Here's the key insight. When we discuss 'category,' especially in the maximalist, *Play Bigger* sense, we're often actually talking about a *spectrum* from the explicit to the implicit:

- **Explicit:** The literal generic category of software (CRM, for example) in the map-the-market, collection-of-things sense.
- **Explicit/implicit:** The more concept-driven approach ("customer success," "inbound marketing," "experience management") with implied behavior change, which is more closely tied to a single company and tends to *transcend* a literal, generic category (like customer support, CRM or marketing automation, and surveys, respectively).
- **Implicit:** The unique company or product that's big enough and unique enough to create its own center of gravity, its own unique 'whole,' and defies generic categorization altogether. (IKEA, Apple, Notion, ChatGPT, etc.)

Note how different all three approaches are:

1. **Explicit categories** are generic terms and don't suggest a particular product.
2. **Concepts** are change + playbook, usually an extension of one company's brand, and imply specific behavior change.
3. **Iconic cultural touchstones** may be unique businesses built over decades or era-defining products that defy reductive explanations.

There's a strong left-brain/right-brain connection here between the left-brain explicit and the right-brain implicit. When those two approaches are mashed together as one concept, we get confusion. Part of that confusion comes from explicit, left-hemisphere-friendly language being easy for us to use, but implicit, more right-brain ideas are much harder to describe.

McGilchrist, in the opening quote, writes that "finding the language to put across the way of being of the right hemisphere is simply harder than doing so for the naturally explicit left hemisphere," and that's what we see here. We see this *ease* of explaining the explicit and the *difficulty* in explaining the implicit across these so-called 'category' approaches. For example:

- **Category:** Coining a generic software term to fit in our map of the market is very explicit and requires a degree of left-brain precision, especially in B2B, as we fret over exact labels and category names.
- **Concept:** Coming up with a concept name that identifies a secular trend and a playbook to match is harder, but by no means impossible. It still

gets a name, even though the exact meaning might be broader and more implicit. It might feel looser.

- **Company/cultural touchstone:** Creating an era-defining product or company that exists in its own orbit, independent of the generic category label (in that the iPhone transcends "smartphone") is so rare and unique — and such a right-brain implicit 'whole' — it defies simple explanations, despite making for fascinating study.

In the *Play Bigger* world of maximalist story 1 narratives and market strategies, all these approaches get lumped in together. Sometimes, folks pick out what works for them — maybe they do need an explicit category term when selling to enterprises — but often, teams just get twisted into knots trying to square the circle between an explicit category term and dreams of being the next era-defining tech company.

McGilchrist says that any "attempt to make the implicit explicit radically alters its nature," and that is the source of endless confusion in category creation. Trying to come up with an explicit term for, say, Apple ("computer maker") or IKEA ("furniture retailer") is like explaining a joke — making it explicit destroys the richness of meaning those things have accrued. But in *Play Bigger's* world, category creation is *both* those things — both the iconic, category-of-one company or product *and* the explicit category label (like Data Security Cloud in the enterprise).

It's super confusing to mash such different things together under the single banner of "category creation." These are vastly different approaches with different modes of attention and need to be treated as such.

> **Implicit vs. explicit and the art of software:** We discussed the implicit/explicit distinction back in chapter 12 in the context of Slack, where we also took a detour through Apple and Microsoft's relative implicit/explicit, right brain/left brain approaches to business. We touched on how consumer software (and products) tend to succeed in a right-brain way (because the user can get hands-on and *feel* matters), while enterprise software, in particular, tends to encourage left-brain thinking because purchases have to be analyzed by committees, who buy using explicit lists, not implicit feel. The same lessons there are true for what we're discussing in terms of category creation here, too.

Memeing ain't easy

Play Bigger wraps a very successful narrative around these confusing concepts, but I want to highlight just how deep the confusion and muddy thinking runs.

We touched on explicit labels and implicit companies and products being lumped together, but the *Play Bigger* view of category creation also lumps together *consumer product use* and *enterprise software sales* together as well.

Consider how different these are:

- **Ad-supported & PLG-driven products in B2C/B2B:** Users get hands-on with the product, either as a free service (Google, Facebook) or in a freemium, PLG sense (Figma, Notion). Here you don't have to have explicit category labels because the implicit product experience does the talking for you.
- **Sales-led in B2B/enterprise:** When folks can't get hands-on, or at least can't use the product in a fully org-embedded way, you *do* have to be far more explicit and communicate your concept up front in the sales process before folks ever touch the software.

Some companies land in the middle (like Drift), too. But either way, you could squeeze several solar systems between:

1. The implicit experience of using a free service like Google.
2. The much more explicit, we-not-me, analysts-and-requirements world of buying enterprise software.

Again, mashing these approaches together is super confusing. And this is just tech — *Play Bigger* merrily skips across basically all of human achievement, too (it gets a bit silly), as we'll see next chapter.

Create what category, exactly?

Now imagine investors with the category creation bug insist that you, as a portfolio company, must "create a category," because that's what winners do.

What do they mean?

Usually, they mean something down the implicit, market-defining product end of the spectrum. But founders and CMOs can hear that message and often pick up the opposite meaning, going straight to #1 explicit categories and market maps.

One group is speaking about implicit meaning; one starts working on explicit execution. One group has their attention attuned to what's 'out there' (right brain); one has it focused on what's in front of them and what they need to manipulate (left brain).

I say all this to highlight that there's a lot to unpack, and unpack we will. This is the confusion these chapters exist to resolve. But where does this confusion come from? Why do we even use "category" to describe these things?

Well, despite all the confusion, *Play Bigger* also contained an extraordinarily successful narrative, as mentioned, that really did meme a market. Let's take a step back and look at what *Play Bigger* is about, in its own words, including its narrative that has been so successful.

> **Psst... CMOs... in here:** I know a bunch of you love *Play Bigger*, and I know your job is often to champion the 'category' you're attempting to create in the marketplace. Don't confuse my disdain for the sillier parts of *Play Bigger* for what you're doing. I hope that all the left-brain/right-brain clarity strategy tools we're discussing helps *accelerate* what you're doing with sharper messaging, a stronger playbook, a clearer idea of your concept, and an appreciation for the two fundamental types of attention that you're trying to engage with a story 1/story 2 approach. If you want to call that 'category creation,' be my guest. If you've found what works for you, that's fantastic. But some folks get *very* confused by all the category talk, and we have to save them from that confusion!

Crossing chasms to creating categories

> Play Bigger *hit the Sand Hill Road scene in a big way after its publication in 2016. Like Geoffrey Moore's Crossing the Chasm some 25 years earlier, VCs fell in love with the book, and then pushed it down to the CEOs and CMOs of their portfolio companies. [...] Why did VCs like the book? Because it's ultimately about value creation which is, after all, exactly what VCs do.*
>
> – David Kellogg [2]

Earlier, we asked: Why do some VCs use the term 'category creation' at all? The answer is: because someone gave them that language and that story. Someone gave them that meme, that information packet with attitude. That someone was actually someones — consultants who authored *Play Bigger* and put category creation on the map.

Play Bigger might mash a lot of disparate ideas together, but on another level, it's also an extremely successful demonstration of the *techniques* of clarity strategy, and that's the other lesson here: that people adopt language and concepts — even if they're not always very rigorous ones — because other people created them in such a way to *make* them stick.

Language vacuum

This isn't, of course, some magic Jedi mind trick. For that 'supply' of language to take hold, there has to be demand. There has to be a vacuum needing to be filled. And that vacuum was, in the world of tech startups, the need for VCs, founders, and CMOs to have a common language that describes what they want a given startup to be — the market-shaping, segment-creating, category-defining mega-hit. That's the business VCs are in, after all. That is, in the opinion of your humble author, the real job-to-be-done of 'category creation' from an investment point of view — to align investor and founder on *ambition*.

On the one hand, that makes total sense. If you take funding, it's go big or go home. There's no point playing just to play. Either go for the big win or get off the proverbial pot. I can see why a concept that attempts to capture that level of ambition, that need to *play bigger*, as it were, would be popular in those specific circumstances.

Starting narrow

On the other hand, in reality, many great successes — including cultural juggernauts like Facebook and category-creation darlings like Qualtrics — don't start by playing bigger. They start by playing *narrower*. They explicitly *did not* start with world-beating, category-defining ambition. They started with a niche position that created leverage and grew from there. That kind of story 2 positioning is glaringly absent in *Play Bigger* (though we'll pick it up in **chapter 37**), and that is a huge black hole in the theory that sucks startups in. Big successes often start from narrow, focused positions, not hype-fueled dreams, and that's why we're going to keep exploring in the positioning chapters ahead.

Nevertheless, to *Play Bigger's* narrative credit, it not only gave investors the *language* of category creation, it also gave them a whole, pre-packaged mental model for value creation, as David Kellogg notes above, playbook included (there's literally a "Category King Playbook" chapter). I don't think these are always *good* mental models, dear reader, but in a vacuum, flawed beats non-existent.

Category as everything

Let's unpack the narrative of category creation. Here's where *Play Bigger's* own story 1 narrative lands[3]:

> *Today the powerful transforming force is the category—creating a new market for a new product, often (but not always) from a new company. A great message, a great product, a great innovation—these things are no longer enough on their own. Now it's critical to develop a great new market category in concert with*

building a great company and product.

You thought product, innovation, and/or a message was enough? Nuh-uh. You have to create a "great new market category" along with everything else.

Some investors really believe that this maximalist, category-is-everything approach is where the rivers of gold are, and you, dear founder, better start swimming in that direction and learn its secrets. And to be fair, secular change will, from time to time, drive new *generic* categories, and some folks inevitably ride that trend up the S-curve all the way to becoming a big, profitable, category-defining company. Think Microsoft and desktop computing, Apple and mobile computing, Google and web search, Facebook and social networking. Big trends = big categories = big companies, as we saw in **chapter 30**.

Play Bigger admirably tries to pump people up to do similarly big things — it's in the name, after all, and that's great. I'm all for the ambition. And while I have plenty of bones to pick with the particulars of this approach, it's worth acknowledging that having a *market* strategy (not just a market*ing* strategy), which is what *Play Bigger* is fundamentally about (creating that "great new market category"), is a vital and often overlooked part of a tech company's broader strategy. I can see why that aspect resonates with CMOs in particular.

Framing the framework

The venture capital industry is, however, like Soylent Green, made of people, and people use memetic ideas, narratives, and frameworks to try and explain how they see the world.

VCs exist to invest the money of their LPs (pension funds, endowments, institutions, individuals, etc.), so the LPs get exposure to high-risk, high-return outcomes they couldn't get in other asset classes. These folks need a mental model with appropriate language to use, and *Play Bigger* delivers in spades with:

- **Owned language:** "Category" operates both as metaphor and meme (we looked at both in **chapter 15**). It's essentially synonymous with "market," but "create a new market" is not an own-able term, while "create a new category" ostensibly was, and it also functioned as a metaphor for the *implicit* ideas we discussed earlier, becoming, for better and worse, the operating metaphor for 'building a big unique company' as well.
- **Owned concept:** By going all in on "category" as a term, the *Play Bigger* folks could spin their own extended universe of category everything (category creation, category thinking, category design, category kings, etc.), providing the language to flesh out this new world.

- **Owned playbook:** *Play Bigger* has its own story 1/story 2 construct in the form of the broad change + specific playbook approach we've been looking at in recent chapters, which we'll unpack below. That is, there's a right-brain 'strategic narrative' of sorts, along with a left-brain, here's-how-you-do-it playbook.[4]
- **Owned position:** They insert themselves into the history of classic market strategy books, placing themselves explicitly as the successor to books like *Positioning, Crossing the Chasm,* and *The Innovator's Dilemma* as we'll see below.

That is, there's very little left to chance. All the language and concepts are provided for you. It's a mental model with batteries included. That is a very deliberate — and successful — clarity strategy at work. And given we're deep in right-brain, story 1 change narrative in *this* book, it's particularly fascinating to see how *Play Bigger* uses mega-trends and secular change to hit the right-brain radar, too.

The strategic narrative of category creation

Here's how the *Play Bigger* story 1 change story breaks down.

First, as mentioned, they establish the narrative as a timeline, placing themselves in the canon of modern positioning, starting with 1980's *Positioning: The Battle for Your Mind.*[5]

> *Thirty-five years ago, at the height of the TV age, two marketing experts published* Positioning *[...]*
> *In 1991, as the microprocessor brought computing to the masses, another book,* Crossing the Chasm, *focused on the marketing challenges of innovation [...]*
> *In 1997 [...]* The Innovator's Dilemma *introduced the concept of disruptive innovation [...]*
> *Today the powerful transforming force is the category [...]*

Next, they layer on the mega-trends, noting that there are vastly more products, more information (Google, the web), more connectedness (Facebook, social media), more computing power (AWS, the cloud), and more devices (iPhone, smartphones). That's a lot of change, and they lay it on thick[6]:

> *In this century, the landscape has changed dramatically. Google, founded in 1998, got momentum around 2000. Facebook was founded in 2004, giving the entire world a common online social network. In 2006, Amazon.com essentially invented public cloud*

> *computing when it unveiled Amazon Web Services (AWS). Apple introduced the iPhone in 2007. Together those products and services laid the foundation for a new era driven by mobile devices, social networks, cloud computing, and data about everything. In the 2000s, Agile Software Development caught on, accelerating the ability to get new code out the door. New forms of financing such as crowd funding and angel investor networks coupled with the rise of start-up accelerators made it easier than ever to raise money for new companies, while at the same time the cost of everything involved in starting a tech company (aside from insanely expensive office space in Silicon Valley) fell through the floor.*

Everything happens so much. And it doesn't stop there[7]:

> *Meanwhile, a new generation of customers—Millennials, born since the 1980s—entered the market. They have lived their entire lives in the digital era and demand inexpensive, mobile, connected products and services. By 2020 this generation will make up half the workforce, driving business purchase decisions.*

Finally, us ~~snake people~~ millennials have arrived, and we apparently demand "inexpensive, mobile, connected products and services."[8] Now that "for every well-known problem, hundreds of solutions can instantly appear out of nowhere," what are *you*, dear business person, going to do about it?

The authors have the solution[9]:

> *A company that best frames a problem is the company that often comes to define and take the category. Put another way: winning companies today market the problem, not just the solution.*

There it is, a neat solution for this vastly different world we (apparently) find ourselves in. If, in the past, positioning was all about finding that small niche you could find in the buyer's mind, in the *Play Bigger* world, it's all about, well, going bigger: creating whole new categories you can occupy instead.

To get there, though, note just how many layers of mega-trends and secular change were stacked on top of each other to create the need for this new approach.

And look, let's be real. This is, of course, one part consultancy spin ("define and take the category") and one part vanilla B2B business advice ("market the problem, not just the solution") replete with a corny 'winning companies today...' line.

Corny as it may be, it nevertheless *does* get folks thinking more strategically about their market — the M in PMF matters — and, again, that's great. Marketing, in tech, is too often reduced to being a lead machine when it could be a far more strategic function.

And the *Play Bigger folks* really do go bigger. Lots of people talk about 'category creation' in terms of hair-splitting generic labels. The *Play Bigger* folks, on the other hand, talk about whole new disciplines, including new roles in the C-suite. They really aren't — for better *and* worse — afraid to dial it up to eleven.

But is it real?

That's all great as narrative goes, but there's one fundamental question that should be nagging away at us, and that's: is it real?

Do 'winning companies today' really market the problem, not just the solution'? Well, Gainsight did (**chapter 30**), and a similar story 1 approach to narrative at least created a foot in the door for outbound sales back in **chapter 22**.

But here's where the confusing category hydra raises its many heads again. For all the vanilla B2B enterprise tech stuff, *Play Bigger* also anoints plenty of so-called 'category kings' like Google, Facebook, Netflix, and LinkedIn.

Do these folks 'market the problem' or even create a category at all? (Did Google create the category of 'web search'? Did Facebook invent 'social networking'?)

This is where it again drifts into the absurd. We'll unpack what those winners really are in the next chapter, but it's clear there are limits to just how big a story 1 narrative can be before it loses coherence.

Remember, we're ultimately trying to create "a consistent, coherent, convincing narrative that bolsters customer confidence," to borrow Brent Adamson's line from our sensemaking discussion back in **chapter 24**, and if you go too big, you risk your narrative becoming neither consistent, coherent, nor convincing.

Three-act story

Nevertheless, let's stay focused on their narrative for a moment. If we consider *Play Bigger's* use of mega-trends as a pitch, it might flow like this:

- There was conventional wisdom (which you probably believe or believed).
- It's not that it's wrong (i.e., we're not saying you're an idiot); it's just that everything has changed *so much* (and now we'll drown you in mega-trends).
- This creates a crisis (too much choice, no one can remember anything

because "hundreds of solutions can instantly appear out of nowhere"[10]).

- There's a simple solution — "winning companies today" (we have special knowledge) do *this particular thing* instead (market the problem, not just the solution, allegedly).

You can see similarities with our 'ray of sunshine' pitch outline we looked at back in **chapter 23** (darkness descends, and then a ray of sunshine appears...) in the way it echoes a typical three-act story:

1. **Setup:** At a certain time and place, things were like this (the '80s, the '90s, the '00s).
2. **Conflict:** Change after change has hit since, creating all these challenges and a vastly different world.
3. **Resolution:** Fortunately, we've got the cure for what ails you.

That's the broad change story *Play Bigger* introduces. That's the pitch. That's how *Play Bigger* sells category creation — a memetic idea that took over an industry.

Unbundling category creation

If that's the category creation package, what's inside? Are the actual ideas any good, or is it all wrapping and no gift?

This is where things get messy... well, messier. Remember our three levels above that ran from explicit to implicit, from generic categories to concepts to unique companies? *Play Bigger* never draws those distinctions. Instead, it throws them all into a blender and calls the result market strategy, err, I mean category creation. It's never clear if we're dealing with generic categories or inspirational category-of-one companies. (I'm not sure it's clear to the authors, either.[11])

In the next chapter, we'll do what we can to unscramble the category creation egg, but for now, take a moment to reflect on just how powerful clarity strategy can be, as evidenced by the meta-story of *Play Bigger*. They crafted a narrative, memed their market, and, for a time, defined the language of an industry. Next, we'll dig beneath the narrative and look at what's at the heart of their pitch.

Notes

1. McGilchrist, Iain. *The Master and His Emissary.* Yale University Press. Kindle Edition.

2. Kellogg, David. "Playing Bigger vs. Playing To Win: How Shall We Play the Marketing Strategy Game?," *Kellblog*, August 27, 2022. **https://kellblog.com/2022/08/27/playing-bigger-vs-playing-to-win-how-shall-we-play-the-marketing-strategy-game/**
3. Ramadan, Al; Peterson, Dave; Lochhead, Christopher; Maney, Kevin. *Play Bigger: How Rebels and Innovators Create New Categories and Dominate Markets* (p. 36). Little, Brown Book Group. Kindle Edition.
4. They published a much more pragmatic, CMO-oriented guide, or, as they call it, their *"Lightning Strike Mobilization Kit"* in 2019, too: **https://www.playbigger.com/book**
5. Perhaps taking that book's somewhat dubious advice to heart: "Truth is irrelevant. What matters are the perceptions that exist in the mind. The essence of positioning thinking is to accept the perceptions as reality and then restructure those perceptions to create the position you desire." *Positioning: The Battle for Your Mind, 20th Anniversary Edition* (p. 10). McGraw-Hill Education. Kindle Edition.
6. Ramadan, Al; Peterson, Dave; Lochhead, Christopher; Maney, Kevin. *Play Bigger: How Rebels and Innovators Create New Categories and Dominate Markets* (p. 28). Little, Brown Book Group. Kindle Edition.
7. Ibid.
8. Ibid., p. 37.
9. Ibid., p. 38.
10. They sure don't spare the hyperbole.
11. The book had three authors contributing ideas and one writing it up, so I'm sure there are a variety of competing points of view in there, especially given it's true of this book, and it has a mere 1/4 of the authors.

34

Unbundling category creation III — categories, concepts, & Aggregators

If you thought we were getting deep in the weeds last chapter, dear reader, we're getting dirt under our fingernails as we dig right down to the roots in this one. If you want to keep things practical, choose your own adventure and skip to the Qualtrics case study in **chapter 35** or our primer on Jobs to be Done in **chapter 36**. If you want deep insights into the world of category creation, however, read on.

In this chapter, we're going to continue our attempts to unscramble the category creation egg. Why? Because *Play Bigger* attempts to, well, play bigger by creating a universal theory of everything. And I mean *everything*. B2B and B2C, modern tech and century-old ideas, whole companies and specific individuals, music, sport, food, religion, protest movements... *it's all category creation, baby!*

It's wild.

Seriously. *Play Bigger* defines category winners as literally anyone or anything who has done anything remarkable ever, from Apple to Elvis, from Birds Eye to Peter Drucker[1], and from Communism to Christianity. I mean, monkey flippin' *Gandhi* gets turned into a 'category creator' who — and this is a direct quote — "conditioned the market (Indians) to believe he could solve their most dire problem (getting rid of British rule) in a new way."

That is, as they say, a yikes.

It is, at times, very silly. Some of you may have no problems overlooking this and still getting value out of their approach, and that's great. Many others have, I'm sure, tried to read it only to get overwhelmed by the ridiculousness of it all. I feel your pain.

In my view, it's a worry that this is the thinking of the same folks who memed an industry — our industry — as we saw in the previous chapter. More seriously, we'll see that there's effectively junk science at the heart of their claims. And yet billions of dollars change hands in part based on these folks' ideas of category creation.[2] So what's going on?

> **The *Play Bigger* Gorilla:** One more anecdote from the book. I'd be remiss if I didn't mention that *Play Bigger* also contains troll-ish

> sections like "Great Category Design in World History," where there's a timeline of so-called category creation that runs from — wait for it — the Apostle Paul to Karl Marx to Marc Benioff, the CEO of Salesforce. (The guy considered the father of modern surfing gets a nod, too.) To me, dear reader, this is like that famous Invisible Gorilla test of attention[3] where folks get so hyperfocused on what their attention is dialed into they don't see the person in a gorilla suit walking right through. It's right there!

Category (un)defined

Silliness aside, there are two ways you can look at this maximalist, everything-is-category approach. Either you:

- Dig the vibe and just enjoy the craziness of it all, and perhaps come away inspired by the stories of success (as randomly selected as they are). That's fine.
- Think critically about what they're saying and tear your hair out. That's where your humble author landed. (Though I respect that many folks don't read business books that way!)

If we take the latter path and think critically about *Play Bigger's* core argument, we quickly hit a fundamental flaw, which is the definition of 'category creators' (or 'category kings'), or the lack thereof.

As far as I can tell, their definition is entirely tautological. If you define 'category creators' as simply big winners (from Facebook to Pauline Christianity to surfing), then big winners will always be, *ipso facto,* category creators. As the *Play Bigger* authors put it, to them, it's not just about business, but simply folks who "define and develop something new in our way of life."[4] That is painting with a brush so broad it could block out the sun.

If you run with this logic, if big winners are category creators, to be a big winner, you must therefore create a category. Obvious, right? But that's confusing cause for effect. It gets causation backward, and that leads to cargo-culting.

Fatal flaws

This is a critical point that gets to the heart of what folks get wrong about category creation. If everything becomes about your story 1, then everything story 2 — the precise, focused, left-brain market map — is forgotten.

But story 2 — that tight, narrowly focused positioning pitch — is often how startups *start.* It's what kicks off and sustains the growth flywheel, as we'll see in **chapter 39**, and again next chapter with *Play Bigger* client

Qualtrics in **chapter 35**.

If you forget about that initial flywheel and simply try to jump to the top of some made-up 'category,' you're going to miss what we've known about how markets develop for decades (which we'll explore in **chapter 40**). That's a potentially fatal mistake for startups.

To be fair, *Play Bigger* also rehashes plenty of general business advice about being unique and different, not just "better." Strategy O.G. Michael Porter has been telling folks to do that — to 'compete to be unique' — for decades (as we'll see in **chapter 49**). I'm all for that. But bundling everything up under the 'category' banner and then running around like a trigger-happy child with a sticker gun labeling everything in sight as 'category creation' actually *obscures* how these feats of value creation were achieved. But that's what we want to know!

Nevertheless, some folks take *Play Bigger* very seriously, which is why we're dissecting it here, frog in a science class style. And, if we narrow our scope, there are valuable clarity strategy insights we can glean from the book, too.

Origin stories

There are, thankfully, some tech-related examples of clarity strategy work in *Play Bigger* that we can inspect and learn from, including this origin story of sorts (we touched on the importance of origin stories back in **chapter 11**)[5]:

> *In the early 2000s, all three of us were working with the leadership team at Macromedia, which became most famous for making the Flash software that brought video to the Internet masses. Macromedia's CEO was Rob Burgess, and he was looking over the company's hodgepodge of products, and said to us, "I've got a bag of doorknobs. I keep adding to the doorknobs, and the revenue doesn't move. We need to drive growth, so help me figure something out." The three of us were asked to help Macromedia explain its unique place in the tech universe and to reposition the company.*

Explaining a company's unique place in the tech universe through story and positioning is exactly what we're doing with your clarity strategy as we address the two modes of attention. We're explaining both where the product or company fits over time (right brain) and what that means for how you manipulate the world (left brain). That's our story 1/story 2 combo in a nutshell.

In the previous chapter, we saw some of the narrative tools the *Play Bigger* authors like to use, and they're here in effect in this Macromedia case study, too. Keep your eyes peeled for the catchy names (**chapter 17**) and memetic taglines (**chapter 15**) to better articulate a company's unifying idea (**chapter 16**). These are all familiar to us here in clarity strategy land. The case study continues[6]:

> *At Macromedia, experience design brought together many seemingly disparate functions, including visual design, user experience (known as UX), product management and development, branding, and marketing. Macromedia concluded that it needed to take the lead in experience design and become known for it. And from that, a company tagline emerged: "Experience Matters." Macromedia used experience design to invent a new category that it called Rich Internet Applications, which delivered multimedia to users, and the company packaged its disparate "doorknobs" into Macromedia Studio MX. Those products fundamentally transformed the company, its position, and its revenue. In 2005, Adobe bought Macromedia for $3.4 billion. A big part of that valuation came from Macromedia's "Experience Matters" position atop the Rich Internet Applications category it had invented.*

While I can report that "experience design" never caught on as a unifying concept in the industry, it apparently worked on another level, giving Macromedia enough clarity to *unify and bundle* their offering. That's a crucial point — Macromedia was a multi-product company, and they evidently lacked a unifying concept to bring all their products together. That was the vacuum that was filled, in terms of product strategy, by Studio MX, a bundle of Macromedia's suite of products (including Dreamweaver, Flash, Fireworks, and FreeHand[7]).

This bundling and unified messaging was apparently transformative. The new software package was well received, with one relevant publication at the time describing it as an "incredible bargain."[8]

Let's take stock of what we've got:

- A mature, multi-product company in need of a unifying concept.
- An *implicit* new idea of "experience design."
- An *explicit* generic category of software — "Rich Internet Applications."

The actual category piece, "Rich Internet Applications," was just a small part in the mix of a broader market strategy involving a unifying concept, a bundle, and a bargain. If the 4 Ps of marketing are product, price, place, and promotion, this approach ticked three of the four boxes — a unified product, an attractive price, and a unifying concept for promotion.

But very little of that had to do with the actual category. And that's the *Play Bigger* origin story!

Implicit concepts

There is, however, an important insight here. Note just how critical the implicit, unifying concept was in the Macromedia story relative to any specific, generic category. You see this in *Play Bigger's* language with lines like "Macromedia used experience design to invent a new category that it called Rich Internet Applications." That is, there was a much bigger, more implicit umbrella *concept* first ("experience design") and *then* a generic *category* ("Rich Internet Applications") came after.

So, if we do indeed want to play bigger, perhaps the real lesson here is that when it comes to 'category creation,' *the concept comes first*. It's that unifying concept that gives you your memetic fuel, your 'information packet with attitude.' It has attitude because it carves out a new, bigger, more integrated space that helps focus minds, both internally in the company and externally in the market.[9]

This is similar to the other concept-driven approaches of other so-called category creators like HubSpot, Drift, and Gainsight. Why?

- *Concepts* segment markets.
- *Concepts* are the right-brain hook that makes you new.
- *Concepts* wrap up what's happening 'out there' in the environment, especially when you provide a specific playbook to act on it.

As much as I ding *Play Bigger's* category-is-everything approach, I'm a huge fan of unifying concepts that try to bring together disparate elements as a strategic, unified whole. That's what strategy is, in a way — bringing disparate parts together so their impact is greater than the sum of their parts. For Macromedia, the impact of the bundle was, evidently, much greater than the sum of the individual products it consisted of. Huge props to the *Play Bigger* folks for doing the work and pulling off the unifying concept piece for Macromedia.

(But minus points for calling that "category creation," which has been an enduring recipe for confusion ever since.)

Building your own stadium

Again, let's contrast the implicit and explicit here, like we did last chapter. This might seem like a pedantic point, but in contrast to this more implicit concept-first approach, when B2B founders and marketers get ahold of the *Play Bigger* playbook, what do they do? They do what they think they're supposed to do. They start trying to come up with corny new names for *explicit* categories of software because that's what the idea of 'category creation' suggests. (It's in the name!)

That is, they pick up the minor part — like the Rich Internet Applications category name, or, say, Data Security Collaboration (to pick three random words), and miss that there's supposed to be a unifying concept — a right-brain 'whole' — that drives these labels.

Then they think they've got to put all their marketing muscle into 'educating the market' about Data Security Collaboration as a new category of software, of which they hope their actual product, let's call it DataShieldPro (thanks, ChatGPT), will someday, somehow be crowned king.

Define the category, "market the problem," and success will surely follow. Right? That's the 'category creation' narrative we're giving the laser beam treatment here.

Unfortunately, it doesn't really work like that. This kind of *explicit* 'category creation' in a category-of-things sense, with no trend behind it (i.e., unlike Yammer riding the social networking trend in **chapter 32**), is like putting all your effort into inventing a sport that no one is playing and building a stadium that no one asked for just so you have a place to play with home-ground advantage. Sure, it'll be great if anyone ever turns up, but will they? There are a lot of potential points of failure there!

This is the core conceit in the name category *creation*. It suggests, in a subtly ego-flattering way, that you can *create* a *generic* category — and we're talking about generic categories here — by simply willing it into being. But chances are you can't — for a nascent startup, especially, it's next to impossible. And why would you want to? You're just creating a stadium for others to come and compete with you.

In **chapter 30**, when looking at Gainsight, we saw that categories in a 'new movement' sense are usually *co-opted*, not created from scratch. For Gainsight, the customer success movement had been bubbling along for at least *eight years* prior to their arrival on the scene. That is, folks were already playing a new game, and Gainsight's role was (to stretch the sports metaphor) to help the league turn pro.

Category vs. brand

Modern brand theory (which we'll discuss in **chapter 44**) gives us some helpful clues as to why this is so hard.

Branding is fundamentally about building memory structures in folks' brains. That's very hard to do for a single product or company. When folks hear that they need to "educate the market" about the *category* (as opposed to their product), they then need to do this not once but twice:

- First, to associate some situation with the new category ("When you're faced with X, consider the new category Y").
- Then, to associate the category with your product ("When you think

category Y, consider product Z").

It's not that this never happens, but doing it from scratch *as a startup* is brutally hard, slow, and incredibly expensive.

It's far better to jump on an emerging trend instead. That's the "category creation" that *does* work. That's why we've been so focused on secular change — it drives category *emergence*. That's where there *is* tremendous bottoms-up energy that washes over an industry, creating those new category memory structures for you. That gives you the wave to ride so you're not splashing about alone in a lake. And if you can catch the wave, perhaps you'll even become synonymous with it. That's what Yammer, Gainsight, HubSpot, and just about everyone else who is credited as a 'category creator' was able to do.

This approach doesn't sound as cool as being able to create entire globe-spanning market categories as a product of your sheer genius, but it's far more indicative of how the world actually works.

> **The investor-to-founder communication mismatch:** I get why folks like the *idea* of category creation and why investors encourage it. The big change story, the ambitious concept, and the market-defining outcome — it's what we all want. And investors tend to think about that in more implicit, right-brain terms. It's their job to see what's 'out there' and worth investing in. But in a right/left/right jump sense (chapter 26), what happens when the idea of category creation gets pushed down to a founder (or marketer) who has to execute? We get a switch from right-brain to left-brain attention, and that changes the game, as we discussed last chapter. This is another fundamental attention mismatch that brings folks unstuck. Investors use category creation with big winners implicitly in mind. Folks in the trenches end up using it very literally with an explicit category of software in mind. But that wasn't really what the investor meant, and this investor/founder attention mismatch creates crossed wires that can be very difficult to undo.

Category, concept, company

As far as tech goes, so far we've unbundled category creation into:

1. Explicit, generic software *categories*, like Yammer and Enterprise Social Networking, which are usually a product of secular change.
2. Change + playbook *concepts, like* your HubSpots, Drifts, and Gainsights, which are also a product of secular change.
3. Implicit, era-defining *companies* like Facebook, Google, Apple, and

Netflix (among others).

We've covered explicit software categories (Yammer, **chapter 32**) and new concepts (**chapter 30**). But *Play Bigger's* maximalist narrative also lumps in this third kind of company as a 'category creator.' Arguably, these are the most attractive companies from an investor's point of view — generational companies that can dominate markets for decades. Let's unbundle this, err, category of 'category creators,' or 'category kings' to use *Play Bigger's* terminology.

Ideal exemplars

These household names are the source of endless fascination for anyone interested in business and tech. They've been fodder for countless pop business books — and resulting memetic ideas — that breathlessly profess to have found the secret to this kind of world-beating success. These companies *started with why* to *play bigger* in a *blue ocean* as they went *zero to one* to become monopolies, category kings, or whatever term you like for this specific kind of success.

I suspect these kinds of examples tend to tickle our right-brain pattern matching. Here's McGilchrist[10]:

> *The right hemisphere's version [of the world it sees] is more global and holistic, based on the recognition of similarity with an ideal exemplar, and on where this is positioned in the context of other examples, whereas the left hemisphere identifies single features that would place the object in a certain category in the abstract. As a result, where the left hemisphere utilises abstract categories, the right hemisphere operates more effectively using specific exemplars.*

This discussion of your Facebooks, Googles, Netflixes, and so on seems to fit this preference for the 'ideal exemplar,' and investors, founders, and managers who want to develop an intuitive sense of what this kind of business looks like surely find these exemplars relatable and appealing. Who wouldn't want to be the next mega-success on that level? Why *not* aim for that?

Play Bigger uses these sorts of companies over and over again. But what really powered their success? Was it a desire to 'create a category'?

We can... ask

Thankfully, the internet exists, and we can see what famous founders thought at the time. Here's Zuckerberg in 2005, according to a 2012 Business Insider article[11], for example:

> *"I think Facebook is an online directory for colleges," he said then. "If I want to get information about you, I just go to TheFacebook, type in your name, and it hopefully pulls up all the information I'd care to know about you."*
>
> *"I mean, I just really want to see everyone focus on college and make a really cool college directory product," he told the interviewer. "There doesn't necessarily have to be more...Part of making a difference and doing something cool is focusing intensely."*

No problem-marketing, grand market-shaping, category-creating vision there. Just a "really cool college directory product." This is another case of playing *narrower.* This is the story 2 starting point of tight positioning that would help kickstart Facebook to eventually go on to be a market-defining product and company, even if that wasn't clear to Zuckerberg himself at the time. But why did that happen? What drove Facebook, with apologies to The Postal Service, to such great heights?

Well, the life-changing magic of network effects, for one. And a better experience than the competitors of the time. But there's another framework that sheds light on Facebook, and many other industry-defining tech successes, too.

Aggregator, not category creator

It turns out there's a better way of looking at these iconic *internet* companies, and it comes courtesy of Ben Thompson's Aggregation Theory[12].

Thompson began writing about Aggregation Theory as part of his *Stratechery* publication in the mid-2010s, in the same era as *Play Bigger*, and he continues to develop and refine the framework to this day. Here's how he describes it[13]:

> *Aggregation Theory is a completely new way to understand business in the Internet age. Business schools suggest that with the right frameworks, an executive can understand how to manage all kinds of problems: what happens, though, when many of the inputs to those frameworks are zero? Zero distribution costs. Zero marginal costs. Zero transactions. This is what the Internet enables, and it is completely transforming not just technology companies but companies in every single industry. Old moats are gone — and new ones can be built — and Aggregation Theory helps you identify both.*

Thompson has discussed Aggregation Theory in relation to Facebook, Amazon, Netflix, and Google, for example[14] — all examples frequently cited by *Play Bigger.*

Aggregators, in extremely simple terms, are unique in the internet age because the internet creates abundance on the one hand and zero distribution costs to almost everyone on Earth on the other. Therefore consumers have, in theory, the ability to read any web page, watch any movie, order any product, book any room, or connect with any person, at any time, all thanks to the internet.

In this case, what's missing? What's the hard part? The hard part is, as Thompson observes, efficiently connecting consumers to those sources of abundance so they get what they want.

That is, the hard part is making the experience of finding what you want as good as possible — of *aggregating* all that abundance in a meaningful way. History teaches us that those who built the best experience for finding web pages (Google), movies or shows (Netflix, at least in the '10s), products (Amazon), rooms (Airbnb), or other humans (Facebook) dominated their markets.

This means that, to quote Thompson[15]:

> *[T]he most important factor determining success is the user experience: the best distributors/aggregators/market-makers win by providing the best experience, which earns them the most consumers/users, which attracts the most suppliers, which enhances the user experience in a virtuous cycle.*

That is, those that can aggregate all that supply and satisfy demand with the best experience not only capture all the value, they also end up in a virtuous cycle which means they *keep* capturing all the value. It's *this* kind of winner-takes-all effect that investors would dearly love. That's what investing careers are built on. Here's Thompson again from the same piece:

> *What is important to note is that in all of these examples there are strong winner-take-all effects. All of the examples I listed are not only capable of serving all consumers/users, but they also become better services the more consumers/users they serve — and they are all capable of serving every consumer/user on earth. This, above all else, is why consumer technology companies are so highly valued both in the public and private markets.*

When a book like *Play Bigger*, or any other tech-oriented business book for that matter, cites these big players, they're citing this kind of business *implicitly*. They know in a right-brain sense that these companies are special. But they don't have the language or insight to define or explain them explicitly, but they can use them as exemplars and at least point to them. "Be like Mike."

But we *do* have the language to better describe these #3 winners, thanks to Thompson. They're Aggregators, not category creators. And that makes a world of difference in how we understand them. We can see, for example, that there's a world of difference between what a CMO at an enterprise B2B SaaS company might be trying to do vs. a free-to-use Aggregator that happens to strike once-in-a-generation gold.

> **Aggregators and B2C:** Also note how Thompson refers to "consumers/users." The Aggregators we're discussing here are almost entirely B2C in their focus and are heavily entrenched. When these folks get used as examples of "category creation," and the logic of category creation is used in B2B, the result is, again, more confusion. B2C Aggregators make bad exemplars of so-called category creation in B2B. One of the defining features of B2B is, in a sense, the *lack* of aggregation. Business needs are broad and diverse, and there's plenty of money to be found in niches and verticals in a way there simply isn't in the consumer world. That's why story 2 positioning is so important, as we'll pick up in chapter 37.

Category creation unbundled

We can now refine our unbundling of 'category creation' into:

1. Generic categories
2. Owned concepts
3. Aggregators

This isn't exhaustive. If your definition of category creation includes every big winner, then we could stretch our list further to include right-brain, product-as-art winners like Slack, as we discussed back in **chapter 12**, along with similar PLG successes like Squarespace (**chapter 44**), Notion (**chapter 45**), and Figma (**chapter 27**), for example.

But that's a bad definition of 'category creation,' so we won't. But if an investor asks you to 'create a category,' you at least have some idea about what they might mean. They're probably speaking more implicitly along the lines of #2 or #3, and if you rush to a #1 generic category, for example, you might miss the point.

But I could be wrong. Maybe you're in enterprise B2B SaaS, and investors really do care about the #1 generic category label. Now you can ask and find out!

Let me stress again that I'm all for these approaches individually *when they're appropriate*. But they all tend to be story 1 strategies, and not everything is a story 1 problem. We need to keep both the x- and y-axis of clarity strategy in mind, and how you deploy both story 1 and story 2 in the claw game of attention.

Nevertheless, these might be useful tools for stretching your narrative on the x-axis of right-brain attention. And who knows, maybe you *have* found an opportunity to leverage, say, Aggregation Theory in a unique way. If so, more power to you. These can all be a hammer for a particular nail. But not everything is a nail.

> **The *Play Bigger* CMO playbook:** *Play Bigger's* CMO playbook is, for me, very much glass half full/glass half empty. On the one hand, sometimes it's a finger trap for teams who get tied into knots trying to follow the *Play Bigger* jargon (campaigns become "lightening strikes," basic positioning is "category design," and Scientology-style suppressive persons who might dare express doubt about category creation are "Zeds" that need to be rooted out — really!). On the other hand, their separate *Mobilization Kit* encourages a concerted, strategic, company-wide effort to actually engage with the market, and simply *acting in a coordinated way with a single market-oriented message* is a big win for most companies. I don't want to throw the baby out with the bathwater here — if it's helping you coordinate and communicate with your market, that's great. But just realize there are often better ways to think about these things. For example, in chapter 44 we'll learn about the B2C marketing world and its imperfect but much more rigorously studied approach to brand building, where the focus is on building and sustaining memories over time. That's a far more useful mental model than 'category' for most folks in mature companies.

Do category winners take all?

Here's a spicy stat[16]:

> *Our data science analysis of U.S. venture capital–backed tech start-ups founded from 2000 to 2015 shows that category kings earned 76 percent of the market capitalization of their entire market categories.*

We've done a reasonably thorough job of unbundling 'category creation' as a universal strategy, but *Play Bigger* has an ace up its sleeve.

Part of the *Play Bigger* narrative is the idea that, per the quote above, "category kings [earn] 76 percent of the market capitalization of their entire market categories." This is based on research they've published[17] and had written up in *HBR*[18] (which wasn't the first time HBR wrote about 'category creators'[19]).

This stat has — like the idea of category creation itself — stuck. We quoted VC and Yammer founder David Sacks using very similar language in **chapter 32**, too, referring to this outcome as something VCs just "know." (Sacks' wrote, "Over 75% of the market cap in a software category typically goes to the category leader. VCs know this, so funding disproportionately flows to startups that can lay claim to early category leadership.")

I'm not suggesting Sacks plucked this directly from *Play Bigger*, but that's the point — the stat apparently had become a memetic idea in investor circles, even though it's not clear Yammer or Enterprise Social Networking saw those dynamics over time.

This is, dear reader, another superficially appealing idea, and it's funny what happens when you apply the ol' laser beam to these ideas that folks just 'know.'

And yet, it's this stat that people just 'know' that makes category creation matter as a strategy for VC-baked startups and CMOs who have drunk the category creation Kool-Aid. If winners take all, and winners are category creators, then you better be creating a new category where you can take all, rather than fighting over the scraps of an existing category.

I mean, that's what the data says, right?

There's a lot at stake here. If millions of dollars, or perhaps hundreds of millions or billions of dollars, is allocated because of this 'fact' that investors just know, then you'd certainly hope it was true...

Junk science — not even wrong

I'm not going to pull punches on this one, dear reader. This stat, in the context of *Play Bigger*, is just silly. 'Category creation' can't be this universal qualitative vibe from Marx to Drucker to Salesforce on the one hand *and* a rigorous definition for venture-backed tech companies that's appropriate for quantitative analysis (or "data science") on the other.

It must be one or the other, and to suggest otherwise is an incredible example of wanting to have your cake and eat it, too. This is the Schrödinger's cat of 'category creation.' Really, it's junk science, plain and simple.

The stat makes for a great meme — it caught on as something 'everyone knows,' after all — but it crumbles under the slightest bit of scrutiny. The holes are so big you could power slide a truck through them sideways. In fact, let's do so now.

Quantifying categories

We've established that "categories," in the *Play Bigger* world, can be literal software categories, unique concepts, or unique companies (often Aggregators). How do you create a meaningful average of performance across such disparate things? It's averaging out apples, themed restaurants, and golden eggs. Throw in the fact that we're lumping together B2C and B2B companies, and it gets even muddier.

What constitutes a category — especially a new category — is inherently subjective. That's not an insurmountable problem if you come up with a rigorous category definition — people define markets all the time. But *Play Bigger's* study, such as it is, the "category king" criteria includes[20]:

> *[W]hether firms are articulating new problems that can't be solved by existing solutions and whether they are cultivating large and active developer ecosystems, among other criteria.*

"Among other criteria," indeed. The examples offered are Facebook, LinkedIn, and, bafflingly, Tableau, the BI/data visualization tool. BI tools and *Facebook* operate in quite different markets, one imagines. *HBR* notes that this "76%" finding was actually a "qualitative" one, given the researchers could draw category definitions wherever they liked. The category rubber band can, evidently, grow or shrink as much as it needs to.

So much for "data science." We've got an incoherent mix of ingredients with a "qualitative" hand on the scales drawing arbitrary category definitions. It all falls apart very quickly.

I can see why they *suspected* this might be true, though, and that brings us back to Aggregators.

Aggregators do take all

Aggregators like Google, Netflix, Facebook, and LinkedIn (all examples that IPO'd in *Play Bigger's* period of analysis) certainly created the impression that 'big technology company' is synonymous with 'winner that takes all.' These companies *really have* thoroughly dominated their segments. But these companies are all Aggregators.

Perhaps it is, somewhat ironically, *because* those B2C companies *really do* occupy such an outsized place in our minds that their mental gravitational pull biases perceptions and makes it seem like a universal norm that must stretch across B2B, too.

But does it?

B2B vs. B2C categories

Consider how different the hype around once-in-a-generation Aggregators like Facebook or LinkedIn is compared to the less glamorous, in-the-trenches "category creation" work a B2B founder of an early-stage startup has to do.

To use a real B2B SaaS example, let's take Pocus, a sales platform for product-led teams that's raised a Series A round of funding. Two years into their journey, they announced they were evolving from a "Product-Led Sales Platform" to a "Revenue Data Platform" to "democratize data for the entire go-to-market team."[21]

That's all great. This kind of *positioning* work — and finding an explicit label to match — happens all the time in B2B startups, and so it should in these formative early years.

But this is the world of story 2 positioning, not Facebook or LinkedIn aggregation and network effects. In this case, it's about trying to figure out where they fit in the B2B SaaS map, and finding a way to drill down and categorize the *product* rather than naming a *playbook*. (When "platform" is in the name, it's obvious you're trying to describe the product, as opposed to naming a playbook like "conversational marketing," which names the *activity*, i.e., the behavior change, for example.)

And it can be a struggle. From Pocus' blog post:

> *In the early days of Pocus, describing what we built was hard. At the time, every investor wanted to call us a PLG CRM (and some still do), but we eventually landed on the Product-Led Sales (PLS) Platform.*

However, the team determined that:

> *Product-Led Sales is still a use case we enable in addition to a broader set of use cases unrelated to PLS. Our customers have told us this explicitly - Pocus runs their go-to-market, not just their Product-Led Sales motion.*

It's challenging to try and find where to plant your flag in the ambiguous early days of a startup, but credit to them for trying to nail it down.

What does this have to do with 'winner takes all'?

If Pocus takes 76% of the "Revenue Data Platform" category they have coined, that's great, but that may be 76% of a very small pie. What if they bite off 5% of the $60 billion+ CRM market (their investors sometimes call them a "PLG CRM" after all) for a nice $3 billion exit instead? That would be a pretty tidy exit!

I'm not suggesting Pocus' approach is right or wrong. They're trying to work out an explicit, generic category label, and that's fine. Maybe they could use a strong concept instead (in a change + playbook + tool sense), or maybe they already have one. I'm all for exploring those different approaches. Neither am I suggesting that in B2B there's never a winner that takes all. (Figma, for example, certainly is taking all, as we saw in **chapter 27**.)

My point is simply that it makes absolutely no sense to roll these different approaches and companies up into a universal, category-is-everything theory of value creation from either an investor, founder, or marketing point of view.

> **Fat stat:** This 76% stat is another case of *Play Bigger* wrapping dubious ideas up in a very effective narrative. They provided a (supposedly) research-backed simple narrative of all tech markets, which boils down to a single number. That's a seductive thing for investors to believe! Why worry about all the nuance when you've got a (supposedly) universal model for value creation instead? This raises a thorny issue: When we're crafting narratives, do we have to choose between accuracy and simplicity? I believe that's a false dilemma: the best narratives might be brutally reductive, but they endure because they ultimately accord with reality when people put them to the test. And the worst narratives, well... we all know what happened at Theranos. I know which end of that spectrum I'd rather be hanging at.

Broken theory = broken thinking

If the theory is broken, why trust it?

Play Bigger's whole shtick about winner-takes-all was that it meant it wasn't worth investing in anyone but "category kings." They write[22]:

> *It's the dark side of the "long tail" concept pundits have been talking about for more than a decade: In any category of product or service, one entity gets to be the big and valuable dog, while all the rest wind up in the economically challenging tail. The category king dogs unequivocally wag the second-tier tails.*

Is that true? Or is it, again, junk science? Is there just one "big and valuable dog" and a bunch of also-rans in the "economically challenging tail." It *is* often true for Aggregators — Bing is a very distant second to Google, for example, which *Play Bigger* notes.

But what about the rest of, you know, the entire B2B SaaS landscape, where markets are enormous? Would you refuse to invest in HubSpot because Salesforce exists as the CRM category leader? (A category leader with only 19.2% of CRM market cap — winner doesn't take all there, evidently.) What about Drift, who carved out a mere billion-dollar exit in the very same CRM market, as we saw in **chapter 31**? Are you telling me you wouldn't invest in — or found — those companies because they're not CRM 'category kings'? What about Snowflake (<20% market share), for example, that we touched on in **chapter 21**?

This is the vast difference between B2C Aggregators and B2B SaaS, and it obviously makes 0% sense to blend them for a single statistic, or an investment philosophy for that matter. It's like throwing a T-bone steak into your fruit smoothie — it's all food, I guess, but it doesn't exactly mix. It's not something I'd recommend drinking, and I wouldn't suggest drinking from the *Play Bigger* philosophy of markets, either.

> **Pick the parts:** Again, if you find value in *parts* of the *Play Bigger* approach, great, more power to you. I'm all for finding your label (particularly in a story 2 sense) or coining a concept in a story 1 sense. I think it's great when Marketing can focus more on market strategy and play a more strategic role across the org. There's value there. But would I bet my company on it when the core thesis is so obviously not true? No. I'd get serious about meeting — and using — both forms of attention in appropriate ways instead.

88% existing categories

I realize we're getting into "He's already dead!" territory here, and again, dear reader, the next chapter is ready and waiting for you if you'd like to skip ahead. But while we're on this 76% stat, let's also note that *Play Bigger's* research includes data from over twenty years ago. A lot has changed in the two decades since. If the 76% stat isn't true for B2B SaaS, then what is? What do we see if we look at data that just focuses on B2B SaaS?

If we look at more recent data, we could argue that new *generic* categories are, in fact, actually very rare for companies that IPO. For example, in 2020, Greg Kogan, then a B2B SaaS enterprise marketing consultant, looked at IPO activity for B2B SaaS companies in 2018 and 2019.[23] Kogan found that:

> *In 2019, there were 14 B2B software companies that went public. All but two of them marketed their products in existing categories, rather than something unique. [...] For the two exceptions in the list, it's difficult to know how much the unique categories helped them, and at what cost.*

And comparing 2019 to 2018:

> *The proportion was similar in 2018, with 10 out of 11 companies*

> *competing in established, non-unique categories the year before going public.*

24 companies in total, with 3 marketing themselves in a new category. The bottom line?

> *The majority (88%) of B2B startups reach the IPO milestone by marketing themselves within existing categories, not in categories they created.*

We've gone from *Play Bigger's* claim that, over 2000-2015, "category kings earned 76 percent of the market capitalization" to Kogan's claim that, in 2018-2019, 88% of B2B startups IPO *within existing categories.* Let that sink in.

Because 'category' is an elastic term, we can slice and dice this in a variety of ways. As Kogan notes, databases are always duking it out with new *concepts*, but they're all still *databases*. You could argue those concepts are 'categories,' but when category means everything, then category means nothing.

And, in any case, it's a well-worn path for companies to play narrow to begin with, get a foothold in a market, and expand to take a chunk out of an existing parent category, because that's where all the demand is.

That is, a B2B SaaS company like Pocus might, in a broad sense, fit in CRM, but they're still going to come up with a unique name (like "Revenue Data Platform") *for their particular concept* as they get going.

Ironically, picking an obvious, relatable category and offering your unique concept within it (or wedge, as we'll see in **chapter 46**) probably makes you much easier to understand from a buyer's point of view. No 'It's a whole new category!' mental gymnastics required.

This *isn't* the supernova world of the '00s where B2C Aggregators and their ilk could take off and take all in a handful of years. Nor is it the ZIRP world of the late '10s and early '20s for B2B startups where you could raise phenomenal amounts of cash with some grand 'category creation' story from investors with more money to invest than they knew what to do with. The post-ZIRP landscape is, as we've all felt, markedly different, and that calls for new strategies.

But was there ever a time when category creation, in the maximalist *Play Bigger* sense, was the right move in B2B?

Saturated category market

Today, there are 2096 reasons why creating a new generic category isn't what it was.

You could argue, however, that there was a period in the early days of SaaS when new categories (in the generic sense) were being regularly created. For "online accounting software" to exist, someone had to make online accounting software (and teach the market about it), and so on for every category of SaaS, all throughout the '00s. Yammer, the canonical example of B2B generic category creation launched in 2008, for example. Perhaps *Play Bigger*, written largely in the first half of the 10s, was looking back at this period.

But now, in the 2020s, B2B SaaS has matured, and software categories for most business activities across most industries and verticals already exist. That's why IPO activity now largely happens within them.

The starkest demonstration of this fact is software review site G2's list of 2096 categories of tech products, as we touched on in **chapter 32**. G2's list[24] includes these as legitimate B2B software categories:

- Pet Care Software
    - Kennel Software
    - Pet Grooming Software
    - Pet Sitting and Daycare Software

That's great. It's *awesome* that someone created "kennel software" for those kinds of businesses. High five to those folks. And creating a new category in the pet care space is, I suppose, still possible, but we have a name for that: vertical SaaS (and niche vertical SaaS in this case). Going after niche verticals is great; it's just not the market-shaping activity it, err, once may have been. And that's fine — it doesn't have to be. This just means that *generic* category creation is not some earth-shaking strategy that's going to make you the next Facebook. There was a time when there weren't 2000+ generic categories, but that time has passed.

Again, let me be crystal clear that sometimes *generic categories of software* still will emerge as software continues to eat the world, now with a new set of AI dentures. (Don't think too hard about that analogy.) Coming up with generic category names may well help clarify what you're doing. They may well help buyers understand what makes you different in a left-brain, story 2, map-the-market sense.

But don't be misled into thinking that coining a category term is something you *have* to do. Work backward from what's clarifying for your prospects. Often, trying to split hairs over category terms just confuses them. Give them a better metaphor, or playbook, or concept, or attribute (**chapter 46**) instead — something that's about their success, not you and your company — and make yourself as easy to buy as possible.

> **Gall's Law & category creation:** Gall's Law states that "A complex system that works is invariably found to have evolved from a simple system that worked."[25] Category creation, in the maximalist, *Play Bigger* sense, tries to snatch at the big complex story without getting the simple, narrow one right first. But it's figuring out that simple story — that simple system — that enables the complex one. That's why 'category creation' case studies can involve mature companies like Macromedia and Qualtrics. They already had the simple system down and could evolve towards the complex — the idea of not just new tools but new disciplines, as we'll see. But where is your startup? Does it have the simple system down, with the customer success to show for it?

Chucking the bathwater

So far, we've seen that while generic categories *are* created from time to time, particularly in enterprise SaaS, "category creation" as a universal concept is inherently flawed:

- **Investor/founder mismatch:** It conflates an implicit investor message ("Build a big unique company!") with an explicit market message ("We're creating the Data Security Collaboration category").
- **No definition of "category":** "Category," in the *Play Bigger* sense, is so broad as to be meaningless — it's generic categories; it's unique concepts; it's Aggregators; it's Elvis (really). It's everything; it's nothing; it's the vibe, man. It's hard to create what you can't define.
- **It's based on junk science:** The 76% stat and idea of winner-takes-all *as a universal phenomenon* seems extremely dubious, and you could just as easily argue that, for B2B technology companies in more recent years at least, 88% IPO in *existing* categories.
- **Category saturation:** There are 2096 generic categories for tech products anyway, so creating the 2097th isn't the flex it might once have been.

I sincerely hope we can put the everything-is-category era behind us. Concepts are concepts. Positioning is positioning. Playbooks are playbooks. Aggregators are aggregators. Verticals are verticals. Market strategy is market strategy. And, yes, categories, in the literal sense, are categories.

That said...

Keeping the baby

I'm trying to walk the fine line of trying to acknowledge that some of what "category creation" attempts to describe, in its own highly convoluted way, is actually very important; it's just that "category creation" is a redundant way of describing it.

For example:

- **Categories have specific roles to play:** As we saw with Yammer earlier and Pocus above, explicit categories have a role in that left-brain, market-mapping sense. B2B tech analysts will christen them, marketers will try and convince analysts to use their term, and enterprise buyers in particular will use them to locate the aisle they should be shopping in. Secular change can absolutely create new categories, it's just that you don't *have* to create one — there's plenty of existing demand in those 2096 existing categories, after all.
- **Catchy labels matter:** Having a catchy way to describe what you're selling is important, just realize it doesn't have to be a "category." Consider the three T's of 'tech, tool, or transformation' from **chapter 12**, for example. If you've got hot new tech, sell that. If it's a new tool for a specific role, make that clear. If it's a transformation (as we'll see with playbooks in the next chapter), that's great, make it obvious. ("Enterprise Social Networking" was pretty obvious!) If a category label helps folks map you, fine, but don't be afraid of your parent category — that's often what folks want to buy!
- **Being unique is good:** The fundamental idea that *Play Bigger* (and just about any other aren't-the-big-succcesses-amazing pop business book) is trying to channel is that big, unique successes do big, unique things. This is classic business advice popularized by Michael Porter in the '80s and '90s. We'll explore this more in **chapter 49**. It's good advice! It just has nothing to do with creating categories.

The uniqueness angle is particularly important — unique returns come from unique value, and that's created through unique products and activities.

Take something like Notion, for example. "Knowledge management" tools have existed for a long time. (Confluence was launched in 2004, for example.) The Notion folks don't go around trying to "create a category" so they can sell Notion; *they just sell Notion.* They focus on Jobs to be Done (**chapter 36**), their incredible community, and building a powerful, unique product that transcends typical categories. They compete to be unique. "Category" is redundant.

Likewise, HubSpot, an oft-cited "category creator," did far more than create a generic category, they competed to be unique, pioneering a new concept ("inbound marketing") and a generic category (marketing automation), *and* built a big, unique, multi-product business by ultimately moving into CRM. They transcend their category — HubSpot is *HubSpot.*

These folks have a unique *gestalt*, which is an idea we'll pick up on in **chapter 45** when we look at brand. That uniqueness comes from a *combination* of moves — a combination of multiplicative choices — that ultimately builds companies that land you in books like *Play Bigger*, and that's been classic strategy advice for decades.

Trying to bundle all of those disparate elements up as "category creation," as *Play Bigger* did, made for a great narrative and meme, but the result was inherently confusing for those in the trenches. Therefore, it's time to put category creation back in its box and create space for a less myopic view of market strategy. The industry has matured, and our approach to market strategy needs to mature, too. We need to call things what they are. And thanks in part to McGilchrist's work on the hierarchy of attention and how we attend to the world, we now have the language to do just that.

Clarity strategy vs. category creation

It's tempting to say the alternative to this maximalist view of category creation is — *surprise!* — clarity strategy, in the full narrative and positioning (and brand!) sense, but that's not quite right. This chapter hasn't been entirely intended as the business-book equivalent of an American Gladiators-style takedown of a competing idea with those big foam jousting sticks.

It's more about putting things in context and giving you more narrative and positioning tools to use than the big category stick the *Play Bigger* folks want you to hit everything with.

The beauty of clarity strategy and our hierarchy of attention is that it helps us capture the big ambition on the right-brain side while *also* allowing us to go deep and be specific on the left-brain side, using both modes of attention to their fullest.

And, as ever, there's a balance to be struck. You might prefer one mode or the other. For example:

- **Going broad:** Sometimes your 'story 1' approach may well be a new concept and playbook that re-segments an existing category, as David Cancel did with Drift, conversational marketing, and CRM, running straight for the red ocean.
- **Going narrow:** Sometimes your 'story 2' approach might be drilling down to a precise category-ish name, like Pocus above, to try and communicate to folks exactly what you do.

That is, you can do whatever is most appropriate for your company. Maybe it's the 'obvious' pitch of Snowflake from **chapter 21**, or maybe it's the high-concept approach of their competitor, Databricks. And if you've been flying the "category creation" flag and want to use the ideas and techniques of clarity strategy to flesh out your category initiative, go for it. I don't mind what you call it. What matters ultimately is the clarity you create for your audience by matching their mode of attention.

The good news is you can now derive the appropriate strategy from the attention you're trying to hit and the clarity you're trying to create. I'd suggest we don't need speculative theories about category creation anymore, but you do you!

Playbooks are playing bigger

There is one more thing I believe we can take from *Play Bigger's* category creation world.

As much as I enjoy putting the boot up the date of the maximalist, category-is-everything approach *Play Bigger* espouses[26], it would be a mistake to ignore one of the key lessons that emerges if we focus on what the *Play Bigger* folks *do* and less on what they *say*, especially when it comes to playbooks.

In the next chapter, we'll look at Qualtrics, one of Play Bigger's most successful examples of *playbook* creation, and dissect exactly how that works, including how it all started from niche, narrow positioning.

Notes

1. Yes, really. On Drucker: "Sometimes category kings aren't even a business, yet they define and develop something new in our way of life." Ramadan, Al; Peterson, Dave; Lochhead, Christopher; Maney, Kevin. *Play Bigger: How Rebels and Innovators Create New Categories and Dominate Markets* (p. 23). Little, Brown Book Group. Kindle Edition.
2. SAP's brief acquisition of Qualtrics being a case in point, as we'll see next chapter.
3. See: **https://en.wikipedia.org/wiki/The_Invisible_Gorilla**
4. Ramadan, Al; Peterson, Dave; Lochhead, Christopher; Maney, Kevin. *Play Bigger: How Rebels and Innovators Create New Categories and Dominate Markets* (p. 23). Little, Brown Book Group. Kindle Edition.
5. Ramadan, Al; Peterson, Dave; Lochhead, Christopher; Maney, Kevin. *Play Bigger: How Rebels and Innovators Create New Categories and Dominate Markets* (p. 60). Little, Brown Book Group. Kindle Edition.
6. Ramadan, Al; Peterson, Dave; Lochhead, Christopher; Maney, Kevin. *Play Bigger: How Rebels and Innovators Create New Categories and Dominate Markets* (pp. 62-63). Little, Brown Book Group. Kindle Edition.

7. You can see the old Studio MX website from the era here, for example. I don't see much mention of "experience design," but perhaps it was used elsewhere? **https://web.archive.org/web/20031204195927/http://www.macromedia.com/software/studio/**
8. From a 2004 review of the bundle from the sadly now defunct *Digital Web Magazine.* **https://www.digital-web.com/general/macromedia-studio-mx-2004-got-something-to-say-share-your-comments-on-this-topic-with-other-web-professionals-in-reviews-product-reviews-by-jesse-nieminen-published-on-april-14/**
9. The implicit concept can't remain entirely implicit either — it needs a name, and the *Play Bigger* folks gave it one ("experience design"). But in that era, the implicit meaning behind Studio MX meant something much more like "Be a badass web designer/developer with all these fantastic web design tools at your disposal." That was the received implicit meaning.
10. McGilchrist, Iain. *The Master and His Emissary* (p. 52). Yale University Press. Kindle Edition.
11. Shontell, Alyson. "Mark Zuckerberg Never Imagined Facebook Would Become A Big Company," *Business Insider,* May 16, 2012. **https://www.businessinsider.com/mark-zuckerberg-never-imagined-facebook-would-become-a-big-company-2012-5**
12. See Thompson's introduction and related articles here: **https://stratechery.com/aggregation-theory/**
13. Ibid.
14. Thompson, Ben. "The FANG Playbook," *Stratechery,* January 20, 2016. **https://stratechery.com/2016/the-fang-playbook/**
15. Thompson, Ben. "Aggregation Theory," *Stratechery,* July 21, 2015. **https://stratechery.com/2015/aggregation-theory/**
16. Ramadan, Al; Peterson, Dave; Lochhead, Christopher; Maney, Kevin. *Play Bigger: How Rebels and Innovators Create New Categories and Dominate Markets* (p. 22). Little, Brown Book Group. Kindle Edition.
17. See the "IPO Sweet Spot Report" available here: **https://www.playbigger.com/science**
18. "How Unicorns Grow," *Harvard Business Review,* January-February, 2016. **https://hbr.org/2016/01/how-unicorns-grow**
19. Yoon, Eddie; Deeken, Linda. "Why It Pays to Be a Category Creator," *Harvard Business Review,* March, 2013. **https://hbr.org/2013/03/why-it-pays-to-be-a-category-creator** This piece is similarly flawed, in my view, but it perhaps contains the proto ideas that the *Play Bigger* folks would run with, but that's speculation on my part.
20. "How Unicorns Grow," *Harvard Business Review,* January-February, 2016. **https://hbr.org/2016/01/how-unicorns-grow**
21. Grabell, Alexa. "Introducing the Revenue Data Platform," *Pocus,* April 17, 2023. **https://www.pocus.com/blog/introducing-the-revenue-data-platform**
22. Ramadan, Al; Peterson, Dave; Lochhead, Christopher; Maney, Kevin. *Play Bigger: How Rebels and Innovators Create New Categories and Dominate Markets* (p. 16). Little, Brown Book Group. Kindle Edition.
23. Kogan, Greg. "No New Categories," August 7, 2020. **https://www.gkogan.co/blog/category-creation/**

24. See the full list: **https://www.g2.com/categories?view_hierarchy=true**
25. See: **https://en.wikipedia.org/wiki/John_Gall_(author)#Gall's_law**
26. Congratulations, you made it to the most Australian sentence in this book.

35

Unbundling category creation IV — Qualtrics & playbooks

In this final part of our look at story 1 change stories, *Play Bigger,* and category creation, we'll focus less on the theory and more on the outcomes, looking at Qualtrics' journey and the playbook for, err, playbooks.

While we unbundled maximalist category creation at length in the previous chapter and found that the theoretical basis for the need to 'create a category' was flimsy at best, it'd be naive to then conclude there is nothing there. Activities can sometimes achieve valuable outcomes even if the understanding of *why* they work is incorrect. In this case, I believe the thrust of most 'category creation' work is to try and target folks' right-brain radar; it's just that until now, we haven't had the language to understand attention and describe it in this way. In this chapter, we'll see that's exactly what the *Play Bigger* folks were aiming for in their famous work with Qualtrics. We'll also see what lessons we can extract for stretching your own story 1 narrative, and we'll look at Qualtrics' very *narrow* (hint, hint) origin story before we set sail, next chapter, for the world of story 2 positioning.

The concept formula

Let's return to our simple story 1 formula of **concept = change + playbook + tool.** When folks talk about 'creating a category,' usually, in my view, they're actually talking more in terms of this formula — they really want a *concept* that *transcends* their generic category and makes them a unique and highly differentiated choice.

Playbooks are the last piece of the concept puzzle, and so far, we've covered:

- **Concept:** This is the big story 1 narrative we first looked at in **chapter 20** and have further analyzed in this category creation series.
- **Secular change:** We covered the secular change part of the equation in depth in **chapter 30** (Gainsight vs. Hopin) and **chapter 31** (CRM) and introduced the idea that change + playbook went hand in hand.
- **Tool:** I'm taking your tool or platform as a given, but we looked briefly at how you can use the two modes of attention to build for the emerging needs of a new discipline when we looked at Figma in **chapter 27** and their work with modern product teams and Design System Managers.
- **Playbooks:** This is what we're going to explore further here.

Playbooks are the bridge between the story 1 world of secular change — how we understand the world — and the story 2 world of your product — how we manipulate the world.

The idea behind 'creating a category' is, in essence, that you package these elements up into a new conceptual whole that stands alone, i.e., it's a new 'category.' In the last chapter we went to great lengths to unpack what that really meant, but for our purposes here, we can say that at the heart of the idea of category creation is doing something in a fundamentally new way. That means some specific behavior change is required, and that means folks need a playbook for these new activities. We touched on this when looking at CRMs and HubSpot, for example. Digital marketing was the secular change, so "inbound marketing" became the new playbook that bundled up a group of activities (SEO, social, email, etc.) into a new conceptual whole. HubSpot could then own that playbook and bake it into their brand.

> **Should behavior change or stay the same?** Recall that the story 2 pitch of Snowflake back in chapter 21 was grounded in the idea that the prospect *wouldn't* have to change their behavior at all to adopt the tool — they could just keep doing what they were doing, and Snowflake would handle everything in a faster and more reliable way. This is a key difference between a story 1 and story 2 pitch, whether you're pitching a new understanding of the world — with a new playbook — that requires new tools, or whether you're just offering better tools that match a prospect's existing understanding of the world. Either can be a path to success; you just need to know which game you're playing.

Memetic playbooks

But what kind of playbooks work in a story 1 world? In theory, a playbook, in the broadest sense, could be a 1000-page manual and really put the "book" in "playbook." But that's not exactly catchy, story 1 narrative fuel. Likewise, it could be something ChatGPT spits out that's chucked into a poorly formatted PDF, but again, that's not exactly going to set the world on fire.

So what is it about the playbooks that appear in the world of category creation, in particular, that makes them stick?

In a nutshell: they're memetic. They're short, catchy, simple front/to ideas that stick in people's minds. They're the kinds of micro-pitches we covered back in **chapter 10** when we looked at your Hemingway story.

For example, what if:

- The web happened, and *cloud software you rent yourself* became the new playbook?

- Organic digital marketing happened, and *inbound marketing* became the new playbook?
- Messaging happened, and *conversational marketing* became the new playbook?

(That's Salesforce, HubSpot, and Drift, as we covered in **chapter 31**.)
Or what if...

- Churn became an issue to solve, and *customer success* became the new playbook?
- Covid hit, and *virtual events* became — for a very brief moment in time! — the new playbook?
- Soliciting feedback became popular, and *experience management* became the new playbook?

(That's Gainsight and Hopin from **chapter 30** and Qualtrics, which we're covering here.)

As we've seen, the change your playbook addresses has to be enduring (unlike Hopin). Your target, for this *particular* kind of playbook, has to be open to right-brain change stories and have the authority to make it happen, like the C-suite at enterprise companies. Plus, your company has to have the means to make this happen, with the enterprise reps and customer success teams ready to go, backed by a mature product suite.

Those are some very specific conditions. But if you meet them, you have something more powerful than just a tool or an idea or theory. You have a playbook. You have the *conceptual* 'whole product' folks are looking for, not just the feeds-and-speeds tool. The 'whole product' needs to be something folks can exchange money for and get a result — we're not talking about vague thought leadership here. It has to be purchasable, implementable, and ideally measurable in terms of impact.

And, for your playbook to work as part of your clarity strategy, it has to be memetic. We covered memetic playbooks when we looked at *Play Bigger's* own story 1 narrative in **chapter 33**, where we saw that category creation itself was a memetic idea that stuck, for a while at least, because it suggested a simple playbook: create a category!

From tool to transformation

Back in **chapter 12**, when looking at Slack, we touched on the idea of whether you were selling a 'tool, technology, or transformation.'

For many B2B SaaS startups, the journey often starts with a very specific tool sold in a very specific niche to get early traction and then, over time, morphs into selling more of a transformation for bigger businesses, especially if you move upmarket to enterprise sales.

Sometimes that's a planned and deliberate transition, but sometimes the market pulls you there, too. That's because, if you're selling to bigger businesses, you quickly realize that "buying a tool" is the easy part for an enterprise — they have no shortage of cash — but delivering a transformation *around* the tool is where both the challenge and impact lie.

Or, as Matt Wensing, founder of Summit (and speaking from his personal experience), puts it[1]:

> *Complex enterprise sales include consulting that transforms the company enough to adopt the solution. The cost of this transformation is included in the software's price. [...]*
>
> *Once the salesperson is introduced to a roomful of disagreeable people with different ideas, ambitions, and motives (sometimes hidden), and says "yeah, I can fix this", their 6-7 figure price doesn't seem so expensive. "Wait—you're going to fix this* and *we get the software?"*

That is, enterprise customers don't just want a tool — they're a dime a dozen — they want a transformation. They want a company-level outcome to fix a company-level problem. And they want — and will pay — for your help to get there, even if you're ostensibly selling software.

This creates an opportunity for a specific kind of playbook: one that helps businesses stay up to date in a changing world, and this is what underpins the typical story 1 narrative pitch. Sometimes this might be aimed at SMBs, like HubSpot and Drift, with inbound and conversational marketing, respectively, but sometimes it's aimed at enterprises, too — companies that need a more unified way of understanding the world and bringing together the disparate things they're doing.

This creates an opportunity to differentiate your tool and platform from your commodity peers. It creates a branding opportunity as you bake your playbook into the very essence of your brand. And this creates a sales opportunity because you can now access the C-suite with a right-brain, story 1 pitch that matches their attention and opens the door to big enterprise deals.

'Category' as narrative

This is what Qualtrics did, and this is why folks still take 'category creation' seriously, despite the silliness we covered in the previous chapter.

Qualtrics, a client of Play Bigger (the consultancy), went on that journey from selling a tool — survey software — to selling a transformation — "experience management," as they call it.

They started with extremely focused story 2 positioning and these days lean heavily on their story 1 change narrative. Therefore, their story makes a neat bridge, in a reverse kind of way, between wrapping up our story 1 concept development here — this is the top of the mountain, folks — and our look at positioning and getting the early flywheel going in the first place, which we're about to explore in the coming chapters.

Qualtrics is also interesting because they really have baked their big 'story 1' category story into their clarity strategy and brand as a mature company.

When the company went public in 2021 — almost 20 years after they were founded — Yahoo interviewed Qualtrics founder Ryan Smith and CEO Zig Serafin. Smith mentioned their category/concept of "experience management" almost immediately[2]:

> *We founded a company 20 years ago. And I feel like our best days are ahead of us [...] And there's never been a bigger time or a bigger moment for experience management in a space which we've pioneered and created.*

We've been discussing the power of memeing your market and the idea that it's memetic playbooks that catch on. Qualtrics, for their part, does everything they can to seed their meme in their market, mentioning it right up front in everything they do, in this case very deliberately dropping in that "experience management" is the space they "pioneered and created."

Qualtrics, as a company, refers to "experience management" as XM and goes so far as to suggest that "XM" is a generic category like "CRM" or "HRS," but in my view, it's more conceptual and more right-brained than that. This conceptual "space" they've created is what gives them their right-brain change story. It's this owned conceptual space that lets them sell transformation. And it's this owned concept and experience management playbook they build right into their brand.

Playbook as brand

In **chapter 44**, when we discuss modern brand theory, we'll see that brand is fundamentally about the memory structures and associations you build in folks' brains. As we saw above, Qualtrics takes every opportunity to build a connection between their name and their "XM" experience management playbook so folks internalize it as a part of their brand.

Playbooks are, however, just one part of our three-act concept story of change + playbook + tool. Let's start with the secular change, as Qualtrics tells it, and then dive into the playbook for playbooks.

The change

In the same Yahoo interview quoted above, Qualtrics CEO Zig Serafin describes the broad change that "experience management" addresses:

> *[W]e all live in an experience economy today. You take your own personal experience with any business. You know that a business succeeds or they fail based on the experience they deliver to you personally.*
>
> *And the experience management technology space is the process of finding and fixing the experience gap. And it's often the difference between what a business believes is happening and what's actually happening. What Qualtrics does is we enable companies to be able to create, monitor, and manage every experience on a single platform.*

As a secular change story, "we all live in an experience economy" is... pretty vanilla. All kinds of companies could run with that story, and "we live in a [fill in the blanks] economy" is a pretty typical narrative trope.

The secular change is there, but it's not what makes Qualtrics interesting. Instead, it's their playbook, their unifying concept that's very unique to them, that makes them interesting. It's an expression of what only they can do.

Where playbooks compete

Let's remember where your narrative competes — in the big, messy void that is the business world, a world made up of individuals trying to do their job and companies trying to collectively row in the same direction.

On the individual level, people are busy and stressed. They're under-resourced and under-supported. They don't want to reinvent the wheel from first principles in every job they hold. They aren't going to read gigantic business books (oops). They want a simple playbook. They want something that might give them an edge, an advantage, a new way of doing things.

Companies feeling pain, on the other hand, want transformation. They know they need to change, but they don't know how. They have competing priorities, competing departments, and competing personalities. But with everyone on the proverbial hamster wheel focused on the next meeting, the next 1:1, the next sale, and the next task, it's hard to figure out *how* to change, *how* to transform as a company.

Now throw a secular change into the mix. The world has changed, the economy has changed, technology has changed.

Oh, great. Whirlwind + change. WTF do we do?

This is the essence of the right-brain change + playbook concept story. No buyer *wants* a new 'category' in the abstract — who cares? What they want —

and what they need — is a new playbook.

> **Tell them what to do:** This is why Customer Success exists, by the way. Many folks don't need just another tool. What customers often need, as a client once pointed out to me, is to be *told what to do with it*. People sometimes need to be literally told how to succeed. Don't assume they'll figure it out — they may have no idea. And even if they personally know what to do, they may have no idea how to align their team or get their boss on board. A good playbook isn't an instruction manual, it's a transformative guide. Many teams try and pitch their offering as 'not just another tool,' but few companies truly internalize what it means to be in the transformation business and build an end-to-end playbook to match.

The category JTBD

Qualtrics had gone just about as far as they could selling a tool — we'll see how their messaging evolved over the years below — and reached a point where they wanted to take that next step.

To take that next step they teamed up with Play Bigger the consultancy, and here's how Al Ramadan, one of the *Play Bigger* authors, described their work[3] (see if you can spot the job-to-be-done 'category creation' was hired to do):

> *Play Bigger began working with the Qualtrics executive team back in 2016 to design a new category. The objective of the design was to expand beyond the survey category (which they had led) and move into a more strategic / c-level category.*

Let's unpack that. They wanted to:

- **Create a new concept:** "Design a new category" is Play Bigger-speak for coining a new concept.
- **Transcend their category:** They wanted to "expand beyond" their *generic, explicit* software category — survey tools (which had been commoditized) — to move to a more "strategic" category.
- **Engage executives:** Their target was the C-suite.

These are, again, the classic story 1 conditions for a right-brain change story. They wanted to get out of the left-brain commodity world of survey tools and put something new on folks' radars — the C-suite, in particular, who we know tend to be in right-brain radar mode.

This is becoming a familiar pattern at this point. We've seen this story 1 approach with:

- Andy Raskin's "strategic narrative" in **chapter 20**.
- Matt Dixon's "Lift" pitch outline in **chapter 23** (a.k.a. the "Ray of Sunshine" pitch).
- *Play Bigger's* own narrative in **chapter 33**.
- Qualtrics 'category creation' case study here.

Raskin, Dixon, and Ramadan all discuss the benefits of a more C-suite-oriented pitch aimed at bigger deals.

In my view, these folks have all, in their own way, discovered what can work in targeting right-brain radar attention, it's just that, again, it's only now that we've got the language and understanding to appreciate what's happening, thanks to the work of Iain McGilchrist.

This is gold for understanding how you might stretch your narrative on the x-axis, especially in a catchy, memetic way. And the lesson from Qualtrics is not just that it's catchy, it's that for it to catch, you've got to go all in on it and grind it out *for years* until it sticks, which is what they did.

But this wasn't *Play Bigger* folks' first rodeo, and nor was it their last.

Play Bigger's playbook playbook

Here's the secret sauce of *Play Bigger's* playbook for playbooks, or at least some common themes.

When studying Play Bigger's approach, I couldn't help but notice that the Macromedia, Qualtrics, and *Play-Bigger*-the-book "category creation" case studies all shared similar traits. Then I spotted those same traits in their work for the B2B ~~whiteboard~~ visual work platform SaaS Mural, too.

Here are some examples of how the *Play Bigger* folks typically play bigger when it comes to category/concept creation. They might coin:

1. **A simple concept name:** 'Experience design' (Macromedia), 'Category creation' (Play Bigger), 'Experience management' (Qualtrics), 'Collaborative intelligence' (Mural).
2. **'A new discipline':** For Macromedia, "experience design" would be a "new discipline" that would be "part of the fabric of the technology industry"[4]; for Qualtrics, experience management would be a "new business discipline"[5]; for category creation itself "we [Play Bigger] believe it's time to develop a new *discipline* called category design"[6] (their italics); and for Mural, they would introduce "a new discipline" called "Collaboration Design"[7].
3. **A new playbook:** You can't have a new discipline without a new playbook, and Macromedia suggested Rich Internet Applications (or at least a web design suite of tools) was the future; Qualtrics had a specific view on

closing the "experience gap" ("the difference between what a business believes is happening and what's actually happening"[8]); *Play Bigger* was, in part, a literal playbook for market strategy; and Mural had a playbook they acquired (LUMA, and their "LUMA Institute") that was supposed to result in the "Collaboration Design Institute™."

4. **Going all in:** Macromedia didn't just float a concept, they put their money where their mouth was with a "bargain" bundle of tools; Qualtrics made XM a core part of their brand (even their stock ticker was $XM); *Play Bigger* claimed anything and everything was category related, stretching their concept further than perhaps any business concept has been stretched before; and Mural ended up with, well, concept soup: Collaborative Intelligence, Collaboration Design, LUMA, Institutes/CDI, and that's before we get back to the actual product: Mural.

I'm obviously cherry-picking examples here, but you can see what it means for the *Play Bigger* team to try and stretch a narrative out on the story 1 side, especially when it comes to selling transformation and not just a tool.

Does it work?

Let me stress that I'm *describing* and not necessarily *prescribing* this approach for your startup, dear reader. (I don't know that any of those 'new disciplines' ever materialized, for example.) What worked for Qualtrics apparently *didn't* work for Mural — after raising almost $200M in the ZIRP era and announcing their category creation push in April 2022, Mural replaced their CEO months later in December of the same year, and references to category creation were subsequently scrubbed from their website.[9] (And I can see why: new concepts work when they clarify, but Mural's concept soup only confused.)

Let's not forget this is not the only approach to your clarity strategy, either. If you pitch a new discipline along with a company (or function) transformation to rank-and-file folks who really do just want a better tool to do their job, as we saw in **chapter 23**, they will smile politely while backing away as fast as they can.

However, for those with open right-brain attention, it certainly can work, and in Qualtrics' case, it gave them a catchy, unifying, foundational concept they could own to help them *transcend* their category. And no doubt it helped grease the wheels of many C-suite sales, too.

That's great. But we need to remember that's not where they started.

Clarity strategy: Positioning to playbook

While Qualtrics might be the poster child for this kind of playbook/category creation, it's not like the company was born spouting the virtues of "experience management." They started as a single tool with very tight positioning in a very particular niche. Here's how founder, Ryan Smith, describes their journey in *HBR*[10]:

> *Qualtrics started as a single-product survey company used primarily by academics to conduct research. [...] But by the early 2010s we had evolved into a multiproduct company helping organizations manage customer experience, employee insights, and market research. In 2017 we launched the experience management (XM) category and created the first platform on which organizations could oversee the four core experiences of business: customer, employee, product, and brand. That led to our acquisition by SAP for $8 billion in January 2019 and an IPO two years later in which the opening price valued Qualtrics at three times that amount.*

Note their journey from tight positioning to enterprise playbook:

- **Tight positioning:** They started as "a single-product survey company used primarily by academics to conduct research."
- **Enterprise playbook:** They grew into "a multiproduct company" that helped organizations (and enterprises in particular) "oversee the four core experiences of business: customer, employee, product, and brand."

We can see that evolution in more detail thanks to the wonders of the Wayback Machine. Here's how their messaging evolved over the years:

- **2001:** "Qualtrics Labs" offered "Data Gathering Solutions" that could be deployed in your *intranet*.[11]
- **2004:** Qualtrics was, adorably, "Privately Branded Survey Sites For Your Company!" exclamation point and all.[12*]
- **2008:** They were now the 'world's #1 survey software' with the tagline "Easy enough for an intern...Sophisticated enough for a Ph.D."[13]
- **2012:** Their offering became a 'survey research suite' with the tagline "Sophisticated research made simple."[14]
- **2015:** They were an 'insight platform' that let you "Ask Questions. Get Answers. Act Fast."[15]

From survey software to a survey research suite to an 'insight platform.' Then "experience management" emerged:

- **2017:** Qualtrics was now the "Qualtrics Experience Management Platform" with a simplified focus on "Four core experiences of business. Four applications to get them right." (The four core areas were customer, product, employee, and brand.)[16]

- **2018:** They went all in on their "XM" branding — the product was now an "XM platform"[17], not just a bunch of ways to manage different experiences.[18]
- **2022-23:** It's now XM all the things. XM in the logo, XM as a generic category, the product as the "business operating system for XM," and by 2023, it was XM for three different groups: "customer frontlines," "people teams," and "strategy & insights," and I'm sure it will continue to evolve from there.[19]

While we can see their messaging evolve over the years, "experience management" was a clear fork in their evolutionary path that helped them get to where they are today. But even that had years of continual development, and it's that dedication and persistence to developing your story — your clarity strategy — that really matters. If you treat it as a one-shot deal, you'll get one-shot results. But if you treat it as something to aggressively evolve year over year, then your chances of unlocking the growth that comes with a story that resonates are far greater.

And credit where it's due: Play Bigger, the consultancy, was instrumental in that evolution, helping coin "experience management" or "XM," and for all my mixed feelings about the book *Play Bigger*, they deserve massive props for their success with Qualtrics.

Positioning comes first

> *The best way to create a category is to go sell some software.*
>
> – David Kellogg[20]

Qualtrics could launch this level of clarity strategy because they were already *very good* at selling their software. It all started with a niche position, however, and if we dig into their origin story a bit more, we learn just how niche that initial positioning was.

Qualtrics started as a bootstrapped, family-run business. As Smith, Qualtrics' founder, tells it in *HBR*,[21] at first, corporates in the mid-'00s weren't interested in measuring sentiment online at all. So the team focused on a smaller market: academia, one that Smith's father (and Qualtrics co-founder) Scott M. Smith knew from his time as an academic.

They approached CTOs and CIOs at universities, and while "they loved the product," they wouldn't buy it, so they had to get even more niche. Eventually, they went after "marketing professors at business schools, who had almost no budget." (Scott M. Smith was also a marketing professor.)

That is an *extremely* narrow position to start from. But from that niche they could work their way back up the food chain. Their narrow foothold created a "ripple effect" in academia, with the marketing department signing up, then the business school, then other departments, and then eventually "the entire university would sign up."

This is both somewhat predictable as far as product adoption goes (we'll explore niche-to-reach adoption in **chapter 44** and a lesson you'd completely miss if you only knew Qualtrics through the lens of category creation and XM. That's not how they started, and, interestingly, it's not how their software spread, as there was another growth loop in play.

B2E2B

Smith continues in *HBR*:

> *Another benefit was that we created future corporate customers: Students took their software preferences with them to internships and, after graduation, to their full-time jobs. One MBA who went to work for Heineken over the summer used Qualtrics for a field study project. The company wanted to keep using Qualtrics after he returned to school, so it bought a $60,000 license from us. Another student, from the University of Washington, did a project at Expedia, prompting the travel company to sign on as well.*

The college kids that learned to use Qualtrics' software in college didn't stay in college. They went out into the business world and wanted to use it there, too. That creates an interesting kind of organic growth. (No one talks about the spend-years-in-academia B2Edu2B route!)

For all our lofty category/concept/playbook talk focused on enterprise C-suite sales, it's vital to keep in mind just how niche their initial positioning was to get the business going. There was no big talk of enterprise transformation then, just laser-beam focus on the specific needs of their specific cohort:

> *Our tagline in those early days was "Sophisticated enough for a PhD, but easy enough for an intern." It defined our user experience from the start.*

When it comes to clarity strategy, this goes to show how important it is to know where you are with the claw game of attention. Fifteen or twenty-plus years into Qualtrics' life, it's all well and good to be going out on the x-axis and talking about enterprise transformation. But early on, it was, I imagine, much more about that y-axis of positioning, down to the specific needs of a very specific group of customers and designing the best experience for them.

In this case, as David Kellogg says above, 'creating a category' was only possible after they had sold a *lot* of software.

Out-selling the competition

And selling software was something that Qualtrics became particularly adept at. In Evan Armstrong's article "Product-Led Growth's Failure"[22] for his *Napkin Math* newsletter, he describes how Qualtrics "should've lost" to their product-led competitor SurveyMonkey, "but somehow emerged victorious."

Armstrong put their success down to:

> *A 3 pronged strategy of utilizing cheap Utah labor, gutsy product marketing, and aggressive GTM tactics," specifically outbound sales. Was their product better? No. According to Armstrong, "their product capabilities are nearly identical to Surveymonkey.*

Qualtrics isn't just interesting for its clarity strategy around "XM," it's fascinating as a case study in sales-led SaaS beating the product-led competition through superior GTM across the board. The key difference, as Armstrong sees it, is the focus on customer outcomes:

> *Instead, Qualtrics invented something entirely their own, "Experience Management." What is this? Who cares! Sounds a hell of a lot more valuable than "survey software."*
>
> *The core difference between Qualtrics and everyone else is this choice. By doing experience management versus selling software they have shifted from selling tools to selling outcomes. This little idea, of selling impact vs selling technology, is expressed in everything they do. Sales reps (and all Qualtrics employees) are under strict instruction to never make a public-facing statement that uses the word "survey".*

I love that difference between selling tools vs. selling outcomes and *expressing that in everything you do.* A lot of folks think they're selling "benefits" because their value prop on their homepage is about a benefit and not the tool; very few folks have developed a playbook for true customer success that delivers much broader outcomes *and* have centered their company around it.

Qualtrics really did go all-in on "experience management," and I'd suggest it wasn't just a cool-sounding term, either. It had a real meaning of unifying "the four core experiences of business: customer, employee, product, and brand," to quote founder Ryan Smith. They were working with a concrete definition of transformative, measurable customer success around their concept of the experience gap.

Back to Armstrong. What were the implications of Qualtrics' approach? (His emphasis:)

> *By selling experience management, the person they sell to entirely changes. Rather than working through a bottom-up approach [...] Qualtrics can go directly to the C-suite. [...] [The C-suite] don't care if surveys are really easy to make. All they care about is the outcome. [...] Outcomes can have identical product capabilities as competitors, be harder to use than competitors, and have significantly higher prices.* It doesn't matter. *People will pay for it anyway.*

Again, that's the story 1 C-suite pitch we're interested in here: company-level outcomes. And note that "XM" isn't about the product *per se*, in a story 2, map-the-market sense. A lot of folks get caught up trying to describe what their product is — it's *this* kind of platform or *that* kind of suite, but that's not a playbook; that's a generic category label. That's about your tool, not their success. The unifying concept for Qualtrics, on the other hand, was an outcome they could sell — enterprises could now manage (i.e., measure) the core experiences they offered all in one place and close the 'experience gap.'

> **New playbook = land and expand:** This unifying concept and new playbook doesn't just create a point of focus for your company; it creates a point of focus for the organization *you're selling into,* too. I can imagine a Qualtrics pitch that could run something like: *Don't just adopt us as an employee survey tool, roll out "experience management" company-wide! Survey all the things! Customers! Employees! The market! It's not a tool for one department anymore; it's now integral to the DNA of your company!* And, of course, the deeper it's integrated, the less likely customers are to churn.

Fire up the content machine

If you're a mature company and you're going all-in on a concept, what happens? Having a cool-sounding name is great, as is having an enterprise-oriented playbook. But then what?

Well, then you market the hell out of it. That's how you really seed your memetic idea in the marketplace: by brute force.

Enter the content machine.

Your content machine now pumps out ebooks, events, research, and case studies (I'm literally reading from the Qualtrics website here) around your concept and playbook — experience management. You've got to start building those memory associations one well-marketed asset and event at a time.

This is the power of the unifying concept — it creates a point of focus and clarity for your entire organization, and the memory association you're trying to build in prospects' brains. Your organization now serves the outcome-oriented playbook, not just a product or set of tools. With a unifying concept, Product can build towards it, Sales can sell it, Marketing can talk about it, Success can serve and consult on it.

And that's what Qualtrics did. Ramadan, in his Qualtrics case study, writes that in light of their new 'category':

> *[Qualtrics] repackaged their whole tech stack and products, changed their GTM strategy and hosted their annual X4 conferences [...] to audiences of more than 10,000 attendees per event.*

Product, GTM, and conferences realigned around a new narrative is certainly going all in.

A tale of two playbooks

When it comes to 'category creation,' as with all narrative storytelling, it's easy to slip from conviction to delusion, and your left brain is ready to apply all the brain putty necessary to make that happen. That's why it's critical to have clarity about what stage you're at, what resources you have, and what's a natural next step. The 'right' advice applied at the wrong stage is dangerous. It's unnatural, doesn't fit, and doesn't work. Pulling a couple of words out of your hat ("X intelligence!", "Y cloud!") doesn't cut it. What's the playbook? Where's the evidence? What's the driver? What's the customer outcome?

I've seen companies who have spent years spinning their wheels trying to ~create a category~ and have it not stick because it was the wrong play at the wrong stage; it wasn't based on anything.

For example, one founder coined a category term but couldn't get it to stick and ended up splitting the company's attention as they tried to evangelize the category while also selling their product. But I've also heard one founder on a podcast who sounded overjoyed at finally having a concept that put a finger on what they'd been bashing away at for years.

The kicker? Both founders — the overjoyed one and the frustrated wheelspinner — were trying to coin a version of "$noun intelligence" as their concept, their 'category.' (I'm obscuring the names to protect the guilty.)

The difference was that one had earned their label, and one hadn't. Therefore, one had a catchy, memetic playbook, and the other didn't.

The wheelspinner who hadn't earned their label just ended up with two things to sell, two things to prove, two memory associations to build — a product and a made-up category. They were, ironically, the one who had taken the 'category creation' advice literally. They thought if they coined a generic category name ('creating' the category) and marketed the problem, eventually, people would buy their software. But it only muddied the waters and wasted precious resources. That's an easy mistake to make, especially when it's often understood as the category creation playbook.

For the other founder, they had earned their label. Their "$noun intelligence" concept summed up a simple, catchy playbook they had developed, where you could deploy their tool to do the basic job folks were after *and then* use it to level up the efficiency of the business as well. In this case, the playbook and the product were two sides of the same coin. All they needed was a clarifying, unifying concept to tie it all together, one that let them position their product as a transformative tool for a business, not just another point solution. They had earned their transformation story and it helped them transcend their generic category.

Baby steps

It's worth highlighting that for all the talk of transformation, businesses tend to be very conservative and they tend to like the *idea* of transformation much more than the disruptive reality.

Therefore, some of the catchiest playbooks are the ones that don't ask too much. For all I know, the wheelspinner I mentioned above might have been right with their particular "$noun intelligence" take, and it was just too much to chew for their prospects. It may have been too new, too novel, too risky. On the other hand, Qualtrics gave executives a playbook for the kind of work their organizations were probably already doing — surveying customers, employees, and/or consumers and analyzing the results. They just gave them a unifying concept for this work, giving executives a framework for trying to improve the less tangible parts of the business — the experience — in a more holistic way.

That's the Goldilocks dilemma your playbook has to face — too hot and it's too scary; too cold and it's too boring. It has to be *just right*. We'll explore this MAYA ("Most Advanced Yet Acceptable") threshold in **chapter 48**.

Concept = change + playbook + tool

Let's recap. We've covered a massive amount of ground in the last few chapters, looking at:

- **Riding secular change:** In **chapter 30**, we saw that secular change is often the driving force of right-brain change stories. Gainsight and Hopin, who shared the same CMO, illustrated the difference between *secular* and *cyclical* change, with Gainsight riding the secular change of Customer Success all the way to a big exit, and Hopin riding the temporary *cyclical* change of COVID and remote events to an astronomical valuation before falling back to earth.

- **Differentiating in a crowded category (CRM):** In **chapter 31**, we saw our concept = change + playbook + tool pattern play out in the 'red ocean' of CRM, looking at Salesforce (SaaS and the 'end of software'), HubSpot (inbound marketing), and Drift (conversational marketing) deploying their respective concepts with a unique playbook that gave them a rich, owned concept that helped them transcend their crowded category.
- **Creating a new, generic, enterprise-focused category:** In **chapter 32**, we looked at Yammer and the Enterprise Social Networking category they coined. We saw that this kind of enterprise-oriented, market-mapping, generic category creation is distinct from the more conceptual, playbook-driven side we looked at in CRM. We also touched on Yammer's unique, compounding 'architectural' advantages that gave it a leg up over its chief competitor, noting that the best strategy is often multiplicative.
- **Unbundling so-called category creation:** In **chapter 33**, we went deep into the narrative and practice of *Play Bigger* and its, err, idiosyncratic, category-is-everything approach. We unpacked three kinds of 'category creation': explicit, generic category creation (Yammer), of which there are now 2000+ B2B categories alone; concept creation (Gainsight, HubSpot, Drift, Qualtrics, etc.); and implicit, category-of-one products and companies, the latter of which were often just B2C Aggregators (Google, Facebook, Airbnb, etc.) — a vastly different kettle of fish.
- **Going all-in on a branded playbook:** Finally, in this chapter, we looked at Qualtrics and memetic playbooks built into your brand. We discussed the recurring patterns of story 1 change narratives and how a unique playbook can help you connect with executives operating in right-brain contexts looking for a company-level transformation. Despite this success in 'playing bigger' we also saw that Qualtrics started by playing narrower — finding a very tight niche to create a "ripple effect" out in the broader market over two decades.

Play narrower

That wraps up our look at story 1 in clarity strategy. I think of it as the right-brain half of clarity strategy; you can call it whatever you like. Either way, I hope you've picked up a few ideas on how to stretch your narrative out on this x-axis while avoiding some of the pitfalls of 'category creation' in the process.

What comes next? We've seen that Qualtrics's story 1 narrative was built on success with an initial narrow story 2 positioning — positioning that got them a foothold in the market in the first place.

That's what we're going to look at next. We're going to switch our focus and look at stretching your narrative *down* by way of left-brain positioning, down to the specific jobs-to-be-done you tackle and the best-fit customers you serve. This is how you generate the momentum that gets your growth flywheel going, growth that one day may result in a company big enough to sell the C-suite of Fortune 500 companies on an outcome-focused transformation narrative.

But first, you've got to get a foothold by solving a specific problem, a specific *job-to-be-done.* Let's run through a quick JTBD primer and then dive head-first into the magical world of positioning.

Notes

1. Wensing, Matt. April 16, 2023. **https://twitter.com/mattwensing/status/1647402026958204928**
2. La Roche, Julia. "'It's an incredible moment:' Qualtrics founder on IPO," *Yahoo Finance,* January 29, 2021. **https://finance.yahoo.com/video/incredible-moment-qualtrics-founder-ipo-201559668.html**
3. Ramadan, Al. "XM: A Pitch Perfect Example of Category Design," *Play Bigger blog,* January 30, 2021. **https://www.playbigger.com/media/xm-a-pitch-perfect-example-of-category-design-0**
4. Ramadan, Al; Peterson, Dave; Lochhead, Christopher; Maney, Kevin. *Play Bigger: How Rebels and Innovators Create New Categories and Dominate Markets* (p. 55). Little, Brown Book Group. Kindle Edition.
5. Ramadan, Al. "XM: A Pitch Perfect Example of Category Design," *Play Bigger blog,* January 30, 2021. **https://www.playbigger.com/media/xm-a-pitch-perfect-example-of-category-design-0**
6. Ramadan, Al; Peterson, Dave; Lochhead, Christopher; Maney, Kevin. *Play Bigger: How Rebels and Innovators Create New Categories and Dominate Markets* (p. 52). Little, Brown Book Group. Kindle Edition.
7. That quote comes from a since-deleted April 21, 2022 blog post from Mural co-founder and former CEO Mariano Suarez-Battan. **https://web.archive.org/web/20220421171837/https://www.mural.co/blog/introducing-collaborative-intelligence**
8. La Roche, Julia. "'It's an incredible moment:' Qualtrics founder on IPO," *Yahoo Finance,* January 29, 2021. **https://finance.yahoo.com/video/incredible-moment-qualtrics-founder-ipo-201559668.html**
9. As of writing, a quick search of Reddit for Mural vs. Miro (one of their main competitors) turned up a bunch of threads where users frankly hated Mural. Fixing the

product, rather than creating a category (complete with an institute), may have been the play there.

10. Smith, Ryan. "The Founder of Qualtrics on Reinventing an Already Successful Business," *Harvard Business Review,* July–August, 2021. **https://hbr.org/2021/07/the-founder-of-qualtrics-on-reinventing-an-already-successful-business**
11. **https://web.archive.org/web/20011023071932/http://www.qualtrics.com/**
12. **https://web.archive.org/web/20040404201412/http://qualtrics.com/**
13. **https://web.archive.org/web/20080704133622/http://www.qualtrics.com/**
14. **https://web.archive.org/web/20120303122928/http://www.qualtrics.com/**
15. **https://web.archive.org/web/20150311200630/http://www.qualtrics.com/**
16. **https://web.archive.org/web/20170303030601/https://www.qualtrics.com/**
17. **https://web.archive.org/web/20180302163711/https://www.qualtrics.com/what-is-experience-management/**
18. **https://web.archive.org/web/20180301061332/https://www.qualtrics.com/**
19. **https://web.archive.org/web/20231101000943/https://www.qualtrics.com/**
20. Kellogg, David. "Playing Bigger vs. Playing To Win: How Shall We Play the Marketing Strategy Game?," *Kellblog,* August 27, 2022. **https://kellblog.com/2022/08/27/playing-bigger-vs-playing-to-win-how-shall-we-play-the-marketing-strategy-game/**
21. Smith, Ryan. "The Founder of Qualtrics on Reinventing an Already Successful Business," *Harvard Business Review,* July–August, 2021. **https://hbr.org/2021/07/the-founder-of-qualtrics-on-reinventing-an-already-successful-business**
22. Armstrong, Evan. "Product-Led Growth's Failure," *Napkin Math,* June 3, 2021. **https://every.to/napkin-math/product-led-growth-s-failure**

PART VII

CLARITY & JOBS TO BE DONE

36

Jobs to be Done primer

> Jobs-to-be-Done Theory provides a framework for (i) categorizing, defining, capturing, and organizing all your customer's needs, and (ii) tying customer-defined performance metrics (in the form of desired outcome statements) to the Job-to-be-Done.
>
> *- Tony Ulwick*[1]

Here we are, deep in the clarity strategy development mines. So far, we've been looking at high-concept, right-brain story 1 change narratives. These stories tend to be big, broad stories for big, broad concepts — new ways of doing things for whole industries or large companies.

At the heart of those new concepts is, as we've seen, often a new playbook. At the heart of those playbooks, there's a new way of doing a job — the *job to be done.* That might be not just supporting customers but helping them succeed (Gainsight and Customer Success), reaching customers organically online (HubSpot and inbound marketing), or providing high-quality experiences for customers and employees (Qualtrics and experience management). These are big, high-level jobs that a company on the whole or a specific function — either an existing one like Marketing or a new one like Customer Success — might be responsible for. The tools, technology, and times might change, but the need to do the 'job,' broadly speaking, will remain the same.

But there are thousands of smaller opportunities for new, technology-powered products to solve emerging or existing needs, too. What if you want to focus on those? What if you want to play narrower, so to speak, especially if you're an early-stage startup? What if you're focused on a specific niche or problem you've discovered? How do you craft a story that reaches *down* in a 'story 2' sense to hit those specific customer needs?

Well, we'd need a way to describe those needs. And it would stand to reason that the more accurate our understanding of those needs, the more useful a solution we could build, and the more accurate our sales and marketing efforts would be, and that includes your clarity strategy. That is, the closer we could *position* our solution — our supply — to our customer's needs — their demand — the better off we'd all be, just as we discussed back in **chapter 21**.

Put another way, the better we understood the 'jobs to be done,' the closer we could get to product/market fit. We would, in fact, have a way to methodically *describe* product/market fit.

Hold that thought.

The Job to be Done

We're about to switch gears and dig into the narrow, story 2, laser-focused, left-brain, segments-and-attributes positioning side of clarity strategy. Before that, though, I want to give you a brief *(ahem)* primer on the Jobs to be Done framework (a.k.a. JTBD or just 'Jobs') because JTBD and positioning tend to go hand in hand. An overview to JTBD, therefore, also makes an excellent introduction to positioning. It also doubles as a nice bridge between the playbooks we've been discussing — it's easy to say, "You need a playbook," but JTBD helps you drill down into what that playbook should be centered on.

If you're a big Jobs nerd like your humble author, what we're about to discuss should be fairly familiar to you, and if it is, feel free to skip this chapter and dive right into positioning in **chapter 37**. That said, even if you feel like you know Jobs inside-out, you still might like to stick around for this speedrun through the Jobs theory (or theories, as we'll see) as we'll cover:

- **Brand new insights** into the theory, especially as we see how our favorite little framework — the hierarchy of attention — happily explains the different flavors of JTBD to a significant extent (particularly the differences between the Ulwick, Moesta, and Christensen approaches).
- **Five practical JTBD "fits,"** including what JTBD means for language, positioning, sales, innovation, and competitor analysis (and even your homepage).

It's a little theory with big implications.

> **Going deeper with JTBD:** A bunch of books have been written on JTBD, the best of which is, in my view, Jim Kalbach's *Jobs to Be Done Playbook* (especially given Kalbach's B2B SaaS pedigree from his work at MURAL). Want to get your head around the sometimes confusing world of Jobs? Start there. If you want a quick introduction to Jobs to be Done, Kalbach's podcast interviews are also a great place to start[2].

The language of needs

In this chapter, I'm going to give you, dear reader, a high-level lay of the Jobs-to-be-Done land, including the different left- and right-brain approaches to JTBD and then look at how Jobs relates specifically to your clarity strategy

— positioning and messaging in particular.

Let's start with Jobs to be Done as the language of needs. JTBD is the startup world's *lingua franca* for discussing customer needs[3], and that's how I've been using it in this book, too. (I tend to use "Jobs to be Done" for the theory and "job-to-be-done" for the actual job to be done, and JTBD interchangeably.)

But what does "job-to-be-done" actually mean? In my experience, it can be a profound lens for looking at your business, but it can also be, at times, pretty abstract and overwrought, so I hope this chapter provides a practical, clarifying overview.

JTBD has been around for a while now, and it has survived several hype cycles for one very good reason: it stops us from focusing solely on *our* product, *our* features, *our* data, or even what we *think* we know about our customers. Instead, it helps us refocus on our customers and drill down to the actual job they are trying to get done. (And that isn't "sign up," "use feature X," or "submit form Y.")

That is, it gives us a lens for customer success, and that has profound implications across the entire org.

> **Superficial vs serious jobs-to-be-done:** Jobs became popular as a tool for understanding customer needs through research. Over time, however, while the language stuck, the practice — the actual discovery work — has, along with the nuances of the theory, sometimes gone missing. A running theme of this book, however, is that you're only as good as your discovery, because that's where insights come from. A framework like JTBD can help you interrogate reality to generate those insights. But it can be used as window dressing for speculative brain putty, too. I see a *lot* of the latter out there in startup land, dear reader, and it bums me out. But it doesn't have to be that way. If you're serious, every customer call, every sales interaction, and every internal discussion is a chance to understand and share information on your customer's real jobs-to-be-done... but only if you take the theory — and the discovery it depends upon — seriously.

A common unit of analysis

Jobs is an effective common language because, at its core, is a common unit of analysis: the job-to-be-done. Not the product, not the customer, and definitely not the customer's demographics or persona — the customer's job. That's what matters most to the customer, so that's what should matter most to us, too.

Having a common unit of analysis helps keep everyone on the same page, from product innovation to customer research to sales and marketing, all the way down to the words on your homepage.

And while JTBD is a powerful framework, it's also fundamentally just a very helpful *metaphor*.

Metaphor is, as we saw in **chapter 15**, how we understand the world, and in this case, with the metaphoric "job" as a shared unit of analysis, we can then go about methodically identifying them, understanding them, and solving for them. That is, we can precisely investigate what exactly users care most about.

But why stick with "jobs" as the operating metaphor and not just use "use cases" or "requirements" or even just "needs"?

Counterintuitive insights

The big draw of JTBD is the promise of counterintuitive insights — getting beyond the superficial claimed 'need' to the root issue a customer is trying to address. It's a bit like stated preferences vs. revealed preferences, in that it's helpful to observe what people do — how they buy, how they operate at work — and not just what they say or ask for. Then, by taking a more methodical, first-principles approach to understand what 'jobs' customers are trying to 'get done,' we get the revealed underlying job and unlock several advantages. We can:

1. **Zoom in and out from micro to macro** — from little job to big job — which provides more scope for innovating around (and positioning close to) the whole problem, not just one "use case" or "requirement" slice of it. Finding the right level of abstraction is a fundamental part of JTBD[4], and indeed, in clarity strategy, which can go from the big change story (a very high-level job-to-be-done) down to something much more precise and focused, as we'll see.
2. **Think laterally about competitive alternatives,** i.e., how the customer might already be attempting to satisfy the job, be it manually, with generic software (hello Excel), with some solution held together with duct tape, or perhaps not at all ('non-consumption'). That gets us out of the trap of assuming the customer's choice is product A vs. product B when there are other, less obvious alternatives that constitute the 'real' competition.
3. **Focus our entire organization** around the fundamental job(s)-to-be-done we're solving, from product to positioning, sales, success, and support. The entire organization can be oriented around truly and comprehensively solving that job, whatever that takes and whatever that looks like, be that in code, words, expertise, or services. That can be

liberating if we're used to obsessing about features, for example, as it opens up the door to broader innovation and more coherent, cross-functional action.

4. **Interview customers,** as thinking about Jobs gives us a great way to talk to customers, be that zooming out from simple feature requests ("What would feature X help you achieve? Oh, and why is that important to you?"), asking about the real competitive alternatives they're considering ("How do you solve that currently? What else have you tried or considered?"), or going deep into why they switched ("What was going on in your life when you decided to start looking for Y?"), and perhaps exploring Moesta's four forces of push, pull, habit, and anxiety[5].

> **Isn't this just, uh, user research?** User research folks might roll their eyes at the simplicity of this — and not without reason! For you folks, yes, you may have already been doing this for years or decades. But for business folks and those trapped in "Customer asked for THING, we must build THING!" treadmills, it can be quite a revelation to see that there's a deeper way to understand and solve for a customer's real needs.

From innovation to sales and back again

This common unit of analysis — the job — has big implications for finding product/market fit, given that fit comes from either building the right thing or finding the right customer. JTBD gives us a methodical way to do both:

- **Product:** With a better understanding of the jobs to be done, we can more accurately and expertly innovate on the product side, both in the micro and the macro.
- **Market:** On the market side, we can dial in our clarity strategy, focusing our positioning, messaging, and sales process to communicate *that we do in fact* solve your real job-to-be-done, dear customer, and have built a sales and success process that will make that a reality for you.

This product/market split is evident in the way JTBD has developed as a theory, too. JTBD has evolved along several different evolutionary branches over the years, and the key players are also split, to varying degrees, across this innovation (product) vs. sales-and-positioning (market) spectrum.

Here, for example, are three of the biggest proponents of JTBD and the flavor of the theory they represent:

- **Tony Ulwick**, the originator of the JTBD theory. Ulwick has a product engineering background and therefore takes a very precise and, in the view of your humble author, *extremely* left-brained, engineering-y approach to JTBD for customer needs identification.[5] I love the rigor in that approach, however, and Ulwick is likewise one of the most rigorous JTBD thinkers and writers in the space, as you'd expect from the inventor of the process.

- **Clayton Christensen**, the late, great business academic legendary for coining the theory of "disruptive innovation" (as well as the term "job-to-be-done") popularised JTBD as a solution to his 'innovator's dilemma'[6], particularly in his 2016 book *Competing Against Luck*.[7] Christensen takes a broader view of the theory and applies it to, well, everything, from innovation to organizational structure and even day-to-day life itself.[8]
- **Bob Moesta**, innovator and sales expert, popularised JTBD in the analysis of the sales process itself, encouraging you to match your 'supply' (how and what you sell) to the actual 'demand' (the job people want to get done). Moesta has been particularly influential in SaaS circles throughout the 2010s and beyond, with Jason Fried of 37signals and Basecamp fame writing the foreword to Moesta's 2020 book *Demand-Side Sales 101*.[9]

This isn't a definitive list of JTBD approaches; I just want to highlight that success has many parents, to borrow a phrase, and this applies to JTBD, too. And why not? Jobs is an immensely practical way of looking at innovation, sales, customer success, and, of course, your clarity strategy, especially as we start to narrow down.

JTBD 101 — product

JTBD can get a little convoluted at times, and the metaphors can be a little overwrought ('hiring' and 'firing' products, for example). There can be a tendency towards monocle-emoji navel gazing (what is a job, *reeeally?*). But the point here is just to keep it simple. To riff on Ulwick's fundamentals of JTBD[10] focused on product innovation, we can boil JTBD on the product side down to:

1. Customers have needs; those needs are a function of *jobs* they want to *get done.*
2. Jobs are *stable over time, solution agnostic,* and ideally expressed in a brief statement.
3. Customers buy products to help them get their jobs done, ideally on one platform.[11]
4. Customers can express on what *dimensions* they'd like to improve how they do those jobs — faster, cheaper, more often, less often, and so on.

With that in mind, a challenge for you, dear reader:

1. How would you express your customers' jobs-to-be-done? Have you mapped them out?
2. Can you neatly describe what customers are trying to do independent of your product or solution?
3. What percentage of your customer's entire chain of jobs-to-be-done is satisfied by your product or platform?
4. What are the top metrics of improvement your customers care about regarding their jobs-to-be-done (and not your product, per se)?

I'm sure you can imagine how much easier clarity strategy work is when you know what your customer's jobs to be done are, especially those of your best customers.

JTBD 101 — market

That's the product side of JTBD, but what about the market side? Much of our clarity strategy work in this book is focused on how *you* can shape *your* narrative *you* bring to *your* customers. That is, you create the story and broadcast it to your audience. But what if you could unpack the customer's story instead?

Bob Moesta's sales-oriented JTBD approach does just that, using deep customer interviews to essentially reconstruct the buyer's journey — their narrative — as a story, putting it on a timeline that's not dissimilar to a classic movie story arc, with an instigating incident, a struggle to make progress, and satisfying conclusion.[12]

The key to this approach is that when you've got the buyer's narrative mapped out, you can debug your own process and adapt how you sell and what you offer to better serve the folks you're selling to. In Moesta's case, this revolutionised how he felt about sales and the results he got[13]:

> *I thought of sales as a series of techniques, tools, and processes—a trade—and I just had to learn it. So, it's with this mindset that I boldly took a lead sales and marketing role at a home products company manufacturing and selling solid surface countertops for kitchens and bathrooms [...] Immediately, I struggled! [...] To me, the consumer world seemed irrational and random—like falling into a blackhole—very different from engineering. Previously, I'd been responsible for solving problems for people, but in sales, I was trying to get something from people—money in exchange for my product. For some reason, sales felt icky. It was an unnatural "push" to get people to buy my product.*

Moesta then realized two things. One, at the time, there was no real sales academia to learn from, and two, he didn't really know anything about the customers he was trying to sell to. But the more he learned, the more he could adapt the process, even with simple things like offering bigger sample sizes or restricting color options so buyers had an easier time making a decision. Then, he continues:

> *For the first time, I saw the direct impact of solving a struggling moment in the setting of sales. I had flipped the lens on how I saw sales. Sales started to feel more like engineering—solving a problem—and less icky.*

There's a lot to be said for closing the delta between what you discover customers are trying to do and how your sales — and R&D — process is set up, even if that's just helping buyers actually *buy*.

> **Active looking vs. passive looking and SaaS demos:** Moesta runs through six stages of making a purchase, from first thought, passive looking, active looking, deciding, onboarding/purchase, and ongoing use.[14] Back in chapter 17, we discussed the importance of naming things, and here just having names for common steps in the buyer's journey can completely change how you sell. For example, if a prospect is just *passively* looking (a.k.a. window shopping), perhaps you, as Ryan Singer suggests, show them how someone has solved this problem before.[15] That is, you teach the prospect, by way of example, how to shop for whatever it is they're looking for. If they're *actively* looking for a solution, i.e., demonstrably investing time in researching what might be best, then you might spend more time explaining the ins and outs of your solution while moving the prospect towards a decision, helping them understand the tradeoffs that exist based on the options available to them. None of this possible in SaaS if your sales conversations are just rote product demos. Demonstrating how others solved this problem, how your product serves their JTBD, and how to make tradeoffs are all far more valuable 'demos' that serve *buying* JTBD.

JTBD and clarity strategy

You will have noticed that, for a primer on JTBD, we haven't settled on The One True And Only Approach To Jobs™. We've touched on the product side and the market side, but why isn't there one uniform view of JTBD?

I want to suggest the reason there isn't is, in part (beyond people and personalities), due to the fact there are, as we know, two modes of attention and, therefore, two ways of seeing both the world at large and JTBD in particular. That is, if we recall our hierarchy of attention while thinking about jobs-to-be-done, we can see that:

- **Left-brain positioning** is often about drilling *down* and discovering what customer needs — what jobs-to-be-done — you're best at solving and then aligning your narrative around addressing those needs.
- **Right-brain change narratives** can be about sales — the literal change narrative of why a buyer switched — or about broader concepts too, like suggesting entirely new jobs that businesses need to solve for (or entirely new ways of solving existing ones) that they may not have been aware of.

You can see how this roughly maps to the product and market sides of JTBD we covered above. We can use the theory and language for zeroing in on those left-brain product specifics that matter (especially in the Ulwick school of thought), mapping needs, and taking a very analytical approach to the parts and pieces of what we offer, *and* we can use the language of Jobs to discuss a customer's timeline and buyer narrative to unpack how they might struggle well before they even get to a product choice. Or, in a similarly right-brain way, we can look at where Christensen takes the theory, looking for counterintuitive insights 'out there' in the environment that can drive radical innovation by stepping back and thinking about the whole company offering and customer experience.

JTBD and the R/L/R jump

Again, we have to be careful with simple binaries here. Ulwick's approach is useful in sales and marketing, and Moesta's approach is useful in innovation, too. When we talk about "left-brain" or "right-brain" approaches we're talking about dominance, not exclusivity.

The way we actually attend to the world is through the right/left/right jump from **chapter 26**, and Moesta's examples are particularly interesting when he has complete control over the process, which demonstrates the R/L/R jump in action. He can consider the narrative aspects — the right-brain change over time — and then drill down to the specifics (where "Sales started to feel more like engineering—solving a problem") — before reintegrating those insights back into the whole, in terms of the product and offer he could put in front of prospects. Run a few R/L/R cycles like that, and you're going to get a very tight fit between how (and what) you sell and how folks buy.

The milkshake story and disruptive innovation

There are lots of fantastically useful things in both left- and right-brain approaches to JTBD, but I need to pour a little cold water on one of the founding JTBD myths.

Many folks' introduction to JTBD comes via Christensen's milkshake story, which he used in talks and in the first chapter of *Competing Against Luck*. The story, in a nutshell, is that Bob Moesta and then-business partner Rick Pedi did some research at a single McDonald's store, taking notes about customers who bought milkshakes. They observed that customers in the morning had the 'job' of wanting something to eat/do on their boring commute, and customers in the afternoon wanted to treat their kids. They suggested McDonald's reconfigure its milkshakes to serve these 'jobs.' McDonald's rejected the research, and that was, as far as I can tell, that.

The story has since taken on mythic proportions in the way it's been repeated and retold, with varying degrees of spectacular milkshake-sale success, which *Competing Against Luck* makes clear didn't actually happen. (Ulwick, for his part, critiqued the approach as a poor means of market segmentation.[14]) Christensen, who documents his experience in *Competing Against Luck*, however, tries to repeat the (very informal!) McDonald's research approach with margarine and gets a similarly dismissive response, this time from Unilever's executives.

This might seem like extremely inside baseball, dear reader (and it is), but there's potentially quite a profound insight in the *failure* of these approaches.

If you look closely, there's a classic attention mismatch at play. Christensen is fascinated with right-brain, big-picture attention to possibilities 'out there.' The executives are more focused on the left-brain, downward focus on the narrow specifics of the category and the desire for incremental, laser-beam tweaking vs. whole new ways of thinking that Christensen and co were attempting to put on their radar.

Therefore, we could speculate that to avoid disruption — as Christensen advocated — perhaps requires more right-brain attention than most companies operate with, given they're usually content to apply left-brain attention to manage their "success" through analysis, which works, but misses change over time until it's too late.

That is, the narrative *for innovation itself* can fall victim to the narrative mismatches we discussed back in **chapter 6**, especially if you're trying to put novel right-brain ideas on the radar of management that's strictly in left-brain analysis mode. This is, perhaps, a small insight into why companies get disrupted.

From vision to jobs

As helpful as JTBD is, it does have its limitations. It's obviously possible to build great products without trying to formalize everything in JTBD terms, especially for products that are more a product of right-brain vision, ones that are just implicitly *nicer* or more *product-as-art* as we explored back in **chapter 12**. In those cases, the incumbent might superficially tick more of the JTBD boxes by virtue of having more features, but if a challenger comes along with a strong intuitive sense of what the product *experience* should be, then that can create enormous opportunities. Think Linear over Jira, Slack over other chat apps, Stripe over payment solutions that came before, and Figma (and Sketch before it) over, say, Photoshop or Illustrator.

These folks, I imagine, didn't require hyper-analytical breakdowns of dozens of explicit customer needs at the start — there was a much stronger intuitive vision driving them instead, a vision of a far nicer product experience, especially in domains where the main JTBD is well established (issue management for developers, vector drawing for designers, for example).

And that's great. If your vision is strong enough and you can find that magic touch, you don't necessarily need new, hyper-analytical, JTBD-driven ways to understand your market and opportunity. But (a) we should all be so lucky to be as successful as those folks, and (b) they're not mutually exclusive. Vision, even for the best startups, only takes you so far. At some point, you outgrow the original *contexts of use* that informed your product in the early days, and you need a methodical way of understanding where to go next.

And that's where JTBD is so helpful. Especially when it comes to positioning.

> **Contexts of use:** We touched on this idea — that tool makers often *lack* good contexts of use — when discussing Figma and the implications of the left/right-brain hierarchy of attention when it comes to management back in chapter 27. This is a case where close analysis of jobs-to-be-done makes an enormous difference to a company's fortunes, and that's what happened with Figma. I don't know if Figma uses the term "JTBD," but as we saw in that chapter, their ability to understand the real needs of the teams they were serving (design systems and digital product design) *and* their ability to build a platform where teams could perform most of their jobs on one platform helped them soundly beat the competition, which had a become a fragmented 'stack' of solutions.

The five fits of JTBD

That's JTBD as a theory, with a fresh hierarchy-of-attention twist. But what does it mean for your clarity strategy and positioning, in particular? While some folks might be fortunate enough to be doing big Christensen-esque let's-disrupt-ourselves innovation projects, for most folks, JTBD is a tool for innovating in how you go to market. In that sense, I like to think there are at least five "fits" JTBD helps us understand:

1. Language/market fit.
2. Position/outcome fit.
3. Sales process/buyer jobs fit.
4. Innovation/user jobs fit.
5. Analyzing competitor/market fit.

Before we run through each, it's worth noting that JTBD can also be a great tool for driving internal alignment and tracking progress, too.

Jobs theory lets you essentially map the progress of your SaaS (for example) according to what quantity (and how well) you solve a given set of customer jobs-to-be-done. That's what a market is, after all, and JTBD gives you the language to essentially map the problem space and benchmark your progress in satisfying those jobs, which then, of course, fuels your narrative.

What does that look like in practice? Steve Blank, of Customer Discovery and (along with Eric Ries) Lean Startup fame, sees Ulwick-style JTBD as a missing ingredient in the Lean Startup methodology, especially around market definition, writing that[16] (his emphasis):

> *When looking at a market through the Jobs-to-be-Done lens, a market is best defined as:* a group of people and the job they are trying to get done. *[...]*
>
> *When defining markets with a jobs-to-be-done lens, thousands of unique markets exist. They are stable over time, focus on what people are trying to accomplish rather than solutions, offer a focal point for analysis, and form a foundation for deeply understanding customer needs.*

Blank offers the Jobs-to-be-Done Market Definition Canvas at the end of his article as a tool to do just that. Therefore, before we get into our "fits," if you want to think very analytically about who exactly you're serving (or intend to serve) and how exactly you're serving them, then the Market Definition Canvas could be an excellent place to start.[17]

Like I said: little theory, big implications. Let's run through the list above and look at the relevance of JTBD for each.

> **Gitlab and JTBD:** For the canonical example of a company tracking their progress against their customers' jobs-to-be-done, look no

> further than Gitlab. No company (that your humble author is aware of, at least) has been as public and methodical about their use of JTBD as DevOps platform Gitlab. They share their process too; it's no secret,[18] and you can see how they document the JTBD they're building towards and their maturity in serving them.[19] One wonders what would happen if other SaaS companies were this rigorous and methodical in executing against customer JTBD. I suspect it would be a revelation — and perhaps a revolution — for most.

1. JTBD and language/market fit

One of the simplest ways to implement JTBD is to start by getting your customer language right. The B2B SaaS copywriting world, including various home-spun wrappers around JTBD, basically exists to do just that — to achieve 'language/market fit,' as Matt Lerner, consultant and former investor and growth leader, puts it.

Lerner has a great piece in the First Round Review[20] on the conversion rate bumps he's seen once startups figure out the language that resonates with their audience, but it goes deeper than that, as he writes (his bold):

> *[T]he advice I share with founders these days is always consistent:* ***If you do not have language/market fit, put down your tools and go find it.*** *If you're doing anything else, you're wasting precious time. Stop whatever you're doing. Stop chasing product/market fit. Stop pumping money into Facebook and AdWords, stop trying to replace your marketing team. Until you find language/market fit, nothing else will really work. Forget the symptoms. Address the root cause first.*

Lerner shares some impressive conversion rate boosts from B2C/B2SMB startups who nailed their language and multiplied their conversion rate. In cases where your startup lives or dies essentially on direct response marketing, those kinds of gains can be the difference between life and death for the startup.

In complex B2B sales, things are, well, more complex. The language gap is part of a larger Rubik's Cube of pieces that need to line up. But that doesn't mean JTBD is any less valuable, and applying JTBD in your clarity strategy *even just for language/market fit* can be very powerful indeed.

> **Elearning vs. online:** As one small example of language/market fit, I was working with a SaaS and course provider working in the

government employment sector. When interviewing their sales rep, I asked whether they thought of their courses as "eLearning" or not. "Oh no," they said, "eLearning is really old school for us. We just talk about 'online' because everything is 'online' these days, so it's 'online learning'." What sounds like splitting hairs to an outsider can have serious implications for those in a particular industry. Here, 'eLearning' was associated with the previous generation of tools and content they were hoping to replace, not be apart of. Precision in language matters.

2. JTBD and position/outcome fit

Can you confidently say that your current positioning closely matches your customers' specific jobs-to-be-done, especially those of your best-fit customers who are succeeding the most with your product?

When you study the JTBD literature, what you find — or at least what I've found over and over — are actually examples of positioning in action. It's easier to change your position than it is to change your product, after all. And this pops up in all sorts of contexts, in and out of tech, too:

- Snickers, for example, gets *positioned* around the *job* of satisfying hunger, courtesy of Moesta's work with the Snickers product team. (Remember the slogan?) [21]
- Agricultural products for cows (!) get positioned around the outcomes dairy producers care about, which, in a case study Ulwick recounts in his book, involved fresh messaging across the company's product literature and website, along with sellers insisting that the company focus on positioning around outcomes.[22]
- And, of course, SaaS tools like Gitlab (above), Basecamp[23], Autobooks (see the case study[24]) and many, many others get positioned around their specific customer jobs-to-be-done too, often with customer interviews surfacing a number of positioning and messaging insights.

This is similar to the language/market fit idea, just taken all the way to outcomes, thanks to a deeper understanding of what customer success looks like. And it all stems from being able to clearly and accurately articulate the 'outcome a customer is trying to satisfy' in their own language, independent of your product.

Positioning for niche vs. reach: While JTBD is great for dialing in your positioning once you know roughly where you're aiming, one of the things the theory *doesn't* do is tell you which set of jobs-to-be-done

> you should serve in the first place. That's an upstream market strategy choice that informs your positioning at a more fundamental level and often boils down to whether you'll position for *niche* or *reach*. That is, will you serve one set of JTBD really, really well, or will you build a big platform that serves a diverse set of jobs-to-be-done? This is something we'll explore in the coming chapters[25], looking at selling to nerds vs. normies across the tech adoption lifecycle in chapter 40 and building 'mental availability' in mainstream markets in chapter chapter 44. This is a critical choice that folks sometimes miss — they think if they interview enough customers, some key growth insight will emerge, and they'll set their niche on fire. That can happen, but that's a pretty left-brain approach of looking *down*. Sometimes you have to take a more right-brain approach and think about how far *out* your reach can go, instead, because often growth comes from simply serving more needs.

3. JTBD and sales/buying jobs fit

There is, as we've touched on — and hopefully as this book demonstrates — an enormous opportunity in designing your B2B sales process, clarity strategy included, around *buying* jobs-to-be-done, too. Bob Moesta did just that when selling countertops (and many times since) and turned sales into a game of problem-solving, not persuading or order taking.

What if we applied that same approach to the world of complex B2B technology sales? Imagine if your sales process, and indeed your entire commercial organization, was built around the specific buying jobs your customers have to 'get done.' What would change?

In a nice crossover moment, it turns out that Brent Adamson (co-author of the highly influential B2B *Challenger* sales books, whose insights we touched on in **chapter 21** and **chapter 24**) has worked with Bob Moesta in applying JTBD to the B2B sales process.[26] Adamson describes the typical (but not exhaustive — yours may look different!) buying jobs-to-be-done as[27]:

1. Problem identification
2. Solution exploration
3. Requirements building
4. Supplier selection

They're not linear steps, either — buyers might bounce around all of them, snakes-and-ladders style, before they get to a decision, *if* they get to a decision. (If these steps sound familiar, it's because we discussed them back in **chapter 24** when we looked at levels of awareness and the transition from broad right-brain awareness down to left-brain specifics.)

In an article for *HBR*[28], Adamson argues that the modern B2B commercial organization is "hopelessly out of date" and that:

> *In a world like this, simply "aligning" sales to marketing to ensure a seamless "handoff" as a deal "progresses" along a linear buying "process" represents a woefully inadequate solution to a radically new buying reality (similar to a taxi driver hanging an air freshener in the back of their cab and hoping they've somehow "fixed" an otherwise deeply flawed rider experience).*

He goes on to describe how a company called SMART Technologies, feeling the growing gap between how customers wanted to buy and how they were selling, didn't just talk about better 'aligning' sales and marketing. Instead, they decided to "completely dismantled traditional sales, marketing, success, and service altogether and reconfigured them into what SMART calls the "Unified Commercial Engine" (UCE)." The JTBD crux of their approach was:

> *Unlike traditional silos, mapped to internal processes, the UCE is built back from a careful mapping of customers' buying journeys across a range of predictable "jobs to be done" as part of a typical educational technology purchase.*
>
> *Through that initiative the team identified five common buying jobs (Learn, Buy, Order/Install, Adopt, Support) and established an internal team specifically deployed to support each one, reassigning nearly every member of legacy marketing, sales, service, and success staff as a result.*

This apparently led to very impressive growth numbers, especially given this is a mature company founded in 1987. Let's remember, however, this is a case of needing to know the rules before you break them. This company wasn't a startup figuring out sales for the first time (like Segment back in **chapter 8**); this was a mature company with 250 people on the commercial side of the business looking to better align how they sold not with their other commercial functions *per se,* but with how *their specific customers* actually went about discovering, buying, adopting, and using their specific products, and JTBD was the lens that helped get them there.

4. JTBD and innovation/user jobs fit

> *The vast majority of product efforts fail not because of demand, but*

> *because they can't come up with a solution good enough to get people to switch.*
>
> – Marty Cagan[29]

While we're mostly focused on the market side of product/market fit here, it'd be remiss of me not to mention JTBD on the product side, too. And, as product management legend Marty Cagan suggests in the quote above, it's not enough just to discover needs — to find demand — we need to be able to build a solution compelling enough to make people want to actually switch[30]; to go on that buying journey we might serve with our commercial engine.

That solution needs to have that explicit feature set, implicit product magic (on the user side), and/or company-level outcome (on the exec side) that people ultimately want to buy. As we've discussed, this might mean:

- **Zeroing in:** Using JTBD to zero in on and quantify specific needs and desired outcomes in a left-brained, narrow-focused sense.
- **Zooming out:** Using JTBD to zoom out to the whole job, and think more about what might be 'out there' in a right-brain sense and considering the whole as we attempt to innovate across whole workflows and whole solutions, not just the particular piece of work we're narrowly focused on.
- **Consider the whole:** Using JTBD as a lens to consider how the buyer's journey and what we offer need to work hand-in-hand, and to solve the non-obvious or counterintuitive jobs folks might have that may prevent them from buying.

What does it mean to zoom in and out, to go up and down the JTBD ladder of abstraction? Let's take Professor Theodore Levitt's well-worn quip that "People don't want to buy a quarter-inch drill. They want a quarter-inch hole."[31] (A quote that seems to have drilled itself into the minds of everyone that's ever written about JTBD and which I am therefore contractually obliged to mention).

Levitt's quote is trotted out to highlight the need to focus on solutions and outcomes, not the product itself. However, some people, like tradespeople, really do want a quarter-inch drill and take quite a keen interest in what kind of drills and drill bits are the longest lasting (titanium? diamond carbide?), the best value, and the easiest to work with. In this case, you might apply more laser-beam attention to their exact needs, determine which are overserved and underserved, and analyze how you might innovate there. (Along similar lines, Ulwick did just that for tool-maker Bosch, applying his process to their circular saw offering.)

But we can look out at novel innovations that obviate an existing solution entirely, too, more in the right-brain Christensen sense of disruption. Take 3M's hooks that stick to your wall, for example. They solve for a job that goes beyond both the quarter-inch drill and the quarter-inch hole. Many folks' real JTBD wasn't a quarter-inch hole at all, it was hanging something on a wall, and it turns out that doesn't always require a drill. For renters, it often *couldn't* require a drill, and non-permanent hooks tapped a well of unmet customer demand. In that sense, if you're hoping to innovate by entering new markets, as 3M did, you might use JTBD in a much broader way to find an entirely new opportunity, and *then* zero in on exact uses (What weight should be supported? What kind of hooks should we offer? Should there be a velcro layer on top?) and so on.

This works in SaaS, too. Take Drift, for example. As we explored in **chapter 31**, Drift took an existing technology (chat widgets) and applied it to a near-universal business need — talking to prospects — that had traditionally been served by the classic forms-and-email approach. Drift instead solved for the 'job behind the job,' which was communicating efficiently with prospects, not capturing information in a form. Solving *that* JTBD in a new, innovative way drove very considerable success — it was a solution good enough to drive people to switch.

5. JTBD and competitor/market fit

Jobs can be used to analyze your competitors, too. You can use JTBD interview and research techniques with users of a competing or adjacent product to see how well customers are or aren't served by that offering, for example.

Likewise, you can learn a lot from studying the sales and marketing material of competent competitors, too. A simple way to spot the jobs-to-be-done your competitors are focused on (or are at least talking about) is to play a game I call 'spot the verb.'

Most customer needs are usually expressed as verbs (*manage, sell, launch, collaborate, unify,* and so on) with some dimension of improvement (*smarter, more, faster, easier*), and the bigger your platform, the more verbs you generally cover. Spotting the verbs in your competitors' messaging — including their homepage — can be a neat way of seeing how they are positioning themselves and how thoroughly they're covering the range of jobs a given target market might have.

> **How seasoned practitioners talk about JTBD:** What we're discussing here is essentially 'Baby's First JTBD Research,' but I want to give you a sense of how expert practitioners talk about JTBD on the product side, too. For that, check out the Twitter thread from design and product leader Hà Phan, linked in the footnote. Phan describes a process that

includes prototyping two divergent use cases for a new AI feature and recruiting B2C and B2B users for JTBD-style research. She uses behavioral research techniques to understand "if the experience moved the needle for users, [along with] their tolerance and expectations for this AI feature, [including] when and how it was useful, [and] how it compared to a non-AI enabled artifact." She also learned on the B2B side "how they made decisions about purchasing our product [and] what they might pay for the new feature." All of this research "refined the JTBD for me and [made] clear the intersection between B2B and B2C value" which will no doubt help the positioning of this new feature. Phan describes this prototype-driven approach in behavioral research terms, as a "stimulus in [a] test, much like how you block paths in a maze & see what the mouse would do." The net result was that her team now had the "foundation of all the whys to guide the strategy for product and innovation." That's potent JTBD-driven research and learning that can and, I'm sure, will drive strategy. As for the AI in the prototype? Phan adds, "Oh yeah the AI didn't exist yet, but I faked it."[32]

Verbs and quick JTBD wins

We're going to talk a lot about high-level positioning in clarity strategy in the chapters ahead, but an immediate practical outcome you can start on right now is:

- Look at the jobs-related verbs on your own homepage. Where did they come from? Are they accurate?
- Look at your competitors' homepages or product pages. What verbs are they using?
- Look internally at how you talk about customer outcomes. What are the common verbs used across your marketing, sales, and success teams?

Understanding verbs in the context of JTBD helps you understand competitors and provides some context for the language/market gap that may exist between your current messaging and your prospects. I'm not suggesting you blindly copy your competitors, either. Assuming competitors have some special knowledge is a classic mistake. Work from the source — talk to your customers — and see what's real for them first.

And that's our speedrun of JTBD!

Popularized by the mind that gave us disruptive innovation, adopted by the folks behind Lean Startup, and used by startups and mature companies across innovation and go-to-market, JTBD has been and remains an incredibly influential framework in tech and beyond. The hierarchy of attention helps, I hope, shed more light on exactly why, when, and how it works, too.

High five on making it this far. Our exploration of JTBD doesn't stop here, however — as we keep peeling back the clarity strategy onion in the following chapters, you'll see that JTBD is never too far away, and that includes the next chapter when we dive back into positioning and the left-brain laser beam.

Notes

1. Ulwick, Anthony W.. *Jobs to be Done: Theory to Practice* (p. 34). IDEA BITE PRESS. Kindle Edition.
2. For example, see Kalbach on the July 31, 2020, episode of the *UI Breakfast Podcast* with Jane Portman. **https://uibreakfast.com/181-jobs-to-be-done-with-jim-kalbach/**
3. For a couple of particularly relevant articles, see Ben Thompson's 2018 discussion of Aggregators and Jobs-to-be-Done: **https://stratechery.com/2018/aggregators-and-jobs-to-be-done/** or his 2013 take on the iPhone, Facebook, and JTBD, "Apps, People, and Jobs to Be Done": **https://stratechery.com/2013/apps-people-jobs-to-be-done/**.
4. Christensen, Clayton M.; Dillon, Karen; Hall, Taddy; Duncan, David S. . *Competing Against Luck* (p. 205). Harper Business. Kindle Edition.
5. These come from Bob Moesta's JTBD approach, where interviews uncover why folks switched. 'Habits' and 'anxiety' keep folks attached to the status quo (existing behavior, fear of the new), while 'push' and 'pull' take them towards a new solution (the push of their pains, the pull of what's on offer). How do you implement this in a SaaS context? Benyamin Elias, VP of Marketing at Podia, who also happens to have an academic background in qualitative research, gives a very practical rundown here: **https://twitter.com/BenyaminElias/status/1437424048191987717**. When you code your research along these lines, as Elias suggests, "the jobs-to-be-done become blindingly obvious," giving you extremely valuable insights into your market and business, including, I'd suggest, the MAYA sweet spot ('Most Advanced Yet Acceptable') that we'll discuss in **chapter 48**.
6. The innovator's dilemma is the theory of disruption: a company innovates, succeeds, and then becomes complacent as they plod along doing the 'right' thing — they cash in on their success, listen to customer feedback, and tweak their product, and otherwise act conservatively, only to then be blind-sided by a disruptive challenger. For Christensen, JTBD was a theory that could wake companies up to new opportunities. They could stop just doing incremental innovation to their existing product and start them thinking about the broader job they serve instead. Therefore, they could innovate proactively and, theoretically, avoid disruption. For Ulwick's take on introducing his process to Christensen, see his *Medium* article "The Innovator's Dilemma is Solved with Outcome-Driven Innovation" from February 9, 2017. **https://jobs-to-be-done.com/the-innovators-dilemma-is-solved-with-outcome-driven-innovation-**

e5ae588be746

7. Christensen, Clayton M.; Dillon, Karen; Hall, Taddy; Duncan, David S. . *Competing Against Luck* (p. 207). Harper Business. Kindle Edition.
8. For example, on page 207 of *Competing Against Luck*, Christensen muses, "What is the job that our children hire parents to do for them?" perhaps getting a wee bit carried away with the metaphor.
9. Moesta, Bob. *Demand-Side Sales 101: Stop Selling and Help Your Customers Make Progress*. Lioncrest Publishing. Kindle Edition.
10. Ulwick, Tony. "The Core Tenets of Jobs-to-be-Done Theory," November 29, 2017. **https://jobs-to-be-done.com/the-5-tenets-of-jobs-to-be-done-theory-ba58c3a093c1**
11. Ibid.
12. Moesta shared an illustration of his JTBD timeline in August 2020: **https://twitter.com/bmoesta/status/1298324702117666822**
13. Moesta, Bob. *Demand-Side Sales 101: Stop Selling and Help Your Customers Make Progress* (p. 24). Lioncrest Publishing. Kindle Edition.
14. Moesta, Bob. "The 6 Stages of Making a Purchase," *Demand Side Sales*, October 19, 2020. **https://medium.com/demand-side-sales/the-6-stages-of-making-a-purchase-3fdba5322fc3**
15. Singer, Ryan. "By far my most important takeaway from studying @bmoesta's Demand-Side Sales 101 was learning to tell the difference between when I should describe a solution (active looking) vs. when I should tell a story about how someone else solved the problem (passive looking)." November 10, 2022. **https://twitter.com/rjs/status/1590539537012621314**
16. Blank, Steve. "How to Find a Market? Use Jobs-To-Be-Done as the Front End of Customer Discovery," November 4, 2021. **https://steveblank.com/2021/11/04/market-definition-its-the-front-end-of-customer-discovery/**
17. You should end up with a specific group (or groups) of people on the one hand and specific jobs-to-be-done on the other. Combined, that's your market and your opportunity, waiting for your solution. This is, in my view, a much more tangible approach to market definition than hand-wringing over, say, category, whether you define a new category label or not. It's *very* analytical in the left-brain Lean Startup tradition though and, in that sense, it's another fascinating example of the hierarchy of attention in action.
18. For example, see the 2022 blog post "How we use the Jobs-To-Be-Done Framework to rethink user workflow": **https://about.gitlab.com/blog/2022/09/07/jobs-to-be-done-interviews/** and this guide: **https://about.gitlab.com/handbook/product/ux/jobs-to-be-done/**.
19. For example, this page states, "The goal of this page is to document, share and iterate on the Jobs to be Done (JTBD) and their corresponding job statements for the Pipeline Execution group. Using JTBD framework we intend to arrive at the more specific problems to be solved in relation to Continuous Integration workflows." **https://about.gitlab.com/handbook/engineering/development/ops/verify/pipeline-execution/jtbd/**
20. Lerner, Matt. "Finding Language/Market Fit: How to Make Customers Feel Like

You've Read Their Minds," *First Round Review.* (Undated.) **https://review.firstround.com/finding-language-market-fit-how-to-make-customers-feel-like-youve-read-their-minds**

21. Chris Spiek, a JTBD advocate, practitioner, and close collaborator with Moesta, discusses the story on the *Product Mastery Now* podcast with Chad McAllister. "TEI 057: Applying the Jobs-to-be-Done Framework – with Chris Spiek," February 1, 2016. **https://productmasterynow.com/blog/tei-057-applying-the-jobs-to-be-done-framework-with-chris-spiek/**

22. Here's Ulwick quoting a client who says that, after doing significant JTBD research, "We then focused a lot on the redesign of our whole messaging. We took all our product literature and our website and redesigned and focused them on outcomes. Our whole marketing talk and speak turned to outcome based, and that is very prevalent now in our whole business. People talk about outcomes. Now you hear [our] salespeople saying, "We need to position this in terms of an outcome a person is trying to satisfy."" Ulwick, Anthony W.. *Jobs to be Done: Theory to Practice* (p. 120). IDEA BITE PRESS. Kindle Edition.

23. Spiek, Chris. "Jason Fried on using JTBD at Basecamp," *JTBD Radio,* February 2017. **https://jobstobedone.org/radio/jason-fried-on-using-jtbd-at-basecamp/** (Note the discussion of customer language here, too.)

24. Gia Laudi and Claire Suellentrop of Forget the Funnel have a great case study on their JTBD work with SMB-focused SaaS Autobooks. They outlined their results and their 'customer-led growth' framework in Kevan Lee's newsletter (**https://kevanlee.substack.com/p/418-forget-the-funnel**). In a nutshell, Autobooks positioned themselves as an all-in-one finance management platform, but prospects didn't have time to get their heads around all of that. Instead, it was the online invoicing and payment processing that got SMBs into the platform, so they narrowed their focus on that specific job, updated all of their messaging, increased their sign-up rate 64%, and saw a 300% increase in customers accepting payments. Their homepage is a fantastic example of what messaging focused on a specific JTBD looks like: **https://web.archive.org/web/20231127185309/https://www.autobooks.co/**

25. We touched on why enterprise software often takes the 'Cheesecake Factory' approach — awkwardly serving a diverse set of needs in one platform — back in **chapter 12**. Consumers (and SMB) tend to have more universal needs and, therefore, a JTBD insight in that market can be broadly applicable. B2B SaaS, at the enterprise end especially, tends to serve a far more diverse set of customers, and, therefore, a far more diverse set of JTBD. Businesses pick their customers, and deciding which end of the B2B spectrum you'll serve is one of those fundamental choices that changes your orientation to JTBD, including who you interview, how broad you go, and whether you focus on the quality or quantity of jobs served. Doing JTBD blindly without considering your market strategy can be a bit of a dead end otherwise.

26. Adamson discussed this in his interview with Colleen Francis in "The Future of Sales: With Special Guest Brent Adamson," October 2, 2020. **https://fb.watch/p0IC2_ENc2/**

27. Adamson, Brent. "Traditional B2B Sales and Marketing Are Becoming Obsolete," *Harvard Business Review,* February 1, 2022. **https://hbr.org/2022/02/traditional-b2b-sales-and-marketing-are-becoming-obsolete**

28. Ibid.

29. Cagan, Marty. "Discovery – Problem vs. Solution," September 4, 2020. **https://www.svpg.com/discovery-problem-vs-solution/**

30. That quote is from Cagan's essay on why discovery and solution work needs to happen together, which seems relevant to the often discovery-heavy world of JTBD. Research isn't an end in itself; it's an end to getting just enough information to test a solution.

31. For such a famous quote, there is surprisingly little available in terms of an original source. Nevertheless, Levitt's famous "Marketing Myopia" piece for *HBR* from 1960, republished in 2004, provides a strong sense of his philosophy: **https://hbr.org/2004/07/marketing-myopia**. You can draw a straight line between Levitt's arguments there and Christensen's theory of disruption decades later.

32. Phan, Hà. "Here's an example of testing an AI feature and understanding the biz model and defining JTBD. The company recently rolled out this B2B product and at MVP it sold exceedingly well. So now biz leaders want to focus on scaling the program." January 14, 2024. **https://twitter.com/hpdailyrant/status/1746193452386398626**

PART VIII

CLARITY & **POSITIONING**

37

Positioning: The left-brain laser beam & best-fit customers

While most people think of positioning as a marketing concept, a shift in positioning feels more like a shift in business strategy.

- April Dunford[1]

Let's take a moment. Stretch. Move. Breathe. Shake it out. Ready? In this chapter, we're switching gears — and modes of attention — in a big way, and I want to give you a second to loosen up before this next deep dive. Good to go?

We've been looking at stretching your clarity strategy *out* on the x-axis of trends and change, exploring secular change, category creation, playbooks, and the way they all come together to hit target prospects — especially C-suite folks — on their right-brain radars.

That's all great... *if* there's a secular change or mega trend for you to ride. *If.* If there isn't, you end up hitching your product wagon to some made-up or generic trend that no one believes in or cares about, and it's a big dead end. That doesn't help you sell your tech.

But that's just one story type, and clarity strategy is about the two stories for two modes of attention — story 1 *and* story 2, right brain *and* left brain. As we saw back in **chapter 21**, the story 2 approach is the left-brain positioning school, where we use our laser beam of attention to drill down into customer needs and jobs-to-be-done (as we explored in the previous chapter).

Remember, positioning is just a metaphor for taking a "position" close to, or further away from, some other point. Usually, that's closer to your customers and further away from your competition. That means:

1. Our story needs to close the gap between our supply (our product) and our prospects' demand (their needs, or jobs-to-be-done) by moving closer to them in our language and story, and perhaps in our product roadmap and strategy, too.
2. Our product and offer needs to either move closer to the competition in some dimensions if we're trying to catch up (compete to be the same) or, more pertinently, move away from the competition as we try and create meaningful differentiation (compete to be unique).

This is the art of positioning. As April Dunford suggests in the quote above, a shift in positioning isn't just about refining your marketing assets; a shift in positioning can have profound consequences for your whole strategy — how you go to market, what distribution channels you use, what features you build (or cut) — the implications can be far-reaching indeed. And it all starts with that narrow-focused laser beam of attention.

Engaging the laser beam — the chapters ahead

Do you have a positioning problem, and if so, what does it look like? To help answer that question, let's look at what we have covered and will cover when it comes to positioning and messaging.

So far, we've covered:

- **Positioning and language:** We looked at finding a succinct description or metaphor (**chapter 15**) of your product — perhaps as a 3-word Hemingway story (**chapter 10**) — that captures your value. Is your concept clear, catchy, and compelling? We also discussed getting your website messaging right in **chapter 14** and deeper JTBD-driven language/market fit in **chapter 36**. Are you using your customers' language for the jobs and outcomes they care about?
- **Positioning as "story 2":** In **chapter 21**, we discussed Snowflake's story 2 pitch — the "dumbest and most effective sales pitch I've ever heard," as it was memorably dubbed — where the focus is on positioning your product in an extremely obvious way around existing customer behavior. That is, it was a new way to manipulate the world (in a left-brain sense) but not understand the world (in a right-brain sense). We looked at a theoretical example of the Seed Finder 10000®, too, that covered similar ground. Is that the kind of pitch you're making?
- **Positioning-as-story vs. positioning-as-GTM:** In the same chapter, we also touched on the differences between positioning as story, messaging, and pitch vs. positioning as a strategic GTM lever. In the coming chapters, we'll continue exploring both, including deeper looks at positioning-as-GTM and how you can use positioning to get all your GTM ducks in a row, from distribution to brand to target market to competitive strategy.
- **Positioning and your pitch:** In **chapter 23**, we looked at three alternate deck outlines, including the 'positioning headshot' where we go from a targeted right-brain insight to the very focused left-brain specifics of our product or offer. Before we go further, are you comfortable that your pitch is engaging folks in the appropriate mode of attention?
- **Positioning around jobs-to-be-done:** In **chapter 36**, we covered JTBD in all its multi-faceted glory, and that's the kind of core research that

informs what we're about to look at here.

We've covered a lot of ground already, and now we're going to zoom in further, drilling right down to your best-fit customers before zooming back out to GTM-focused discussions about what positioning looks like across the B2B spectrum and, controversially, whether you should do positioning at all.

We're going to cover:

- **Positioning and your best customers:** You've got a bunch of customers, but who are the best-fit ones? It's hard to speak their language without knowing who they are. In this chapter, we'll discuss zeroing in on your most successful (and profitable) customers and how that forms the basis of your customer expertise.
- **Positioning and your early sales pitch:** Getting your early story 2 pitch dialed in requires finding where the enthusiasm in your prospects lies, studying how folks react, and deeply reflecting on what you're hearing. In **chapter 38**, we'll discuss story 2 pitch development, look at an early sales playbook, and touch on the challenge of moving from a story 2-driven narrative to a broader, horizontal, story 1-driven narrative.
- **Positioning and your GTM/distribution strategy:** Positioning and GTM are inextricably linked, and here we'll start to explore positioning and distribution. Niche positioning is great, but you still have to reach those folks efficiently. This is the niche-to-mainstream playbook, and in **chapter 39**, we'll look at how ConvertKit founder Nathan Barry single-handedly bootstrapped his email marketing SaaS through early niche sales.
- **Positioning in new or adjacent markets:** Perhaps you've had initial success — and maybe a lot of success — with early adopters and want to head to the mainstream, perhaps with a new product to 'cross the chasm' (**chapter 40**). In **chapter 41**, we'll look at the highly instructive Pebble case study and their positioning that worked spectacularly well and then failed horribly.
- **Positioning and distribution mismatches:** Going niche can give you a foothold in a market, but it can also be a money bonfire if your positioning strategy doesn't match your distribution approach, as we'll discuss in **chapter 42**.
- **Positioning with tactics to match your (B2B/SMB/Enterprise) market:** Maybe you *shouldn't* use niche positioning at all if you're going B2SMB, and we'll discuss positioning for *reach* instead in **chapter 43** and **chapter 44**.
- **Positioning by attribute as a wedge in a crowded market:** Finally, maybe you need to sharpen your differentiation in a crowded SaaS market. In crowded categories, competitors naturally drift towards a particular

wedge attribute they can lean into, as we'll see in **chapter 46**.

That's a mini-guide to the positioning chapters to come. Positioning can be as simple or sophisticated as you want it to be, from focused language tweaks to fundamental GTM changes. But where should you start?

If only there was an excellent positioning playbook that let you DIY your basic positioning...

> **Simple starting point — your 5W1H snapshot:** Another simple starting point for your positioning journey is your 5W1H snapshot we explored in chapter 11. That snapshot is intended to help you try and capture where you're at *now*, what the market looks like *now*, what that means for your distribution *now*, and whether the pieces you've got add up to a coherent whole. That creates a benchmark for your own from/to process, and with the 'from' down and your current context established, you can start thinking about the 'to' and the kind of future position(s) that you might want to build towards.

The prophet of positioning

Good news! It turns out a how-to positioning guide already exists. Allow me to briefly (re)introduce the prophet of B2B positioning, the doctor of differentiation, and the magician of, err, meticulous market mapping: April Dunford.

No one has done more to push B2B tech companies to be deliberate about their positioning than Dunford, especially with her fantastic book, *Obviously Awesome*, and her countless hours of talks, interviews, podcast appearances, tweets, articles, and more,[2] derived from her many years of in-house experience as a VP of Marketing and work as an independent positioning consultant.

In fact, rumor has it that if you look at your computer and say "April Dunford" three times, a quality positioning resource will instantly appear in front of you. Try it.

Dunford, whom we met briefly back in **chapter 21** when we introduced story 2 and the positioning school, has an expansive, practical, and tightly argued take on B2B positioning. While we're not going to rehash the full approach here (read her book!), there's no better way to set the scene for this deeper dive into the story 2 side of clarity strategy than with a brief look at Dunford's step-by-step approach. We'll also explore Dunford's positioning origin story of sorts and then set off on our own positioning adventure.

So let's crank up the positioning laser beam.

> **Positioning & left-brain pigeonholing:** We're focusing on the left-brain, discovery-driven slice of positioning here, but keep in mind that I'm using "left brain" as more of a metaphor to capture a particular kind of narrow-focused attention. In reality, both our left and right hemispheres are involved in everything; it's the *dominance* of a given hemisphere — and its respective style of attention — that is of interest to us in this book, and in these positioning chapters, we're really studying how we use and engage with what is often the dominant mode of attention: the narrow, left-hemisphere laser beam.

Positioning playbook

In a nutshell, Dunford's approach to positioning is finding a company's differentiated value — the value only they can provide relative to competitors — and the customers that care a lot about that value. In a slightly larger nutshell, Dunford's positioning methodology — a methodology she uses with her own clients — boils down to these five (plus one) components of positioning[3]:

- **Competitive alternatives:** If you didn't exist, what would customers use? (This is the implicit Jobs to be Done question — "competitor" doesn't necessarily mean "competitor in your category," it means: what other way would they solve their JTBD? With a spreadsheet? A manual process?)
- **Key unique attributes:** What features/capabilities do you have that alternatives do not?
- **Value:** What value do the attributes enable for customers?
- **Customers that care:** Who cares a lot about that value?
- **Market you can win:** What context makes the value obvious to your target segments?
- **Trends:** Trends you can tap for increased relevance are, in this methodology, a bonus point. (Contrast this with the fundamental importance of secular change in story 1 approaches, as we looked at **chapter 30**.)

Methodically working through these questions is a fantastic way to start clarifying your positioning, especially when you need to get the leadership team aligned on what your positioning *actually is.*

Dunford has a 10-step process for working through these components as a team. It involves understanding best-fit customers, forming a positioning team, mapping your unique attributes, determining who cares a lot about the value you offer, finding a market frame, and capturing it all in one-page positioning cavas.[4]

Note that this is a very analytical approach. When I talk about the left-brain laser beam, this is what I mean. It's not vision-driven vibes, and it's not a brute force, sell/build/sell orientation to the market. Instead, it's precise analysis with an intentionally narrow focus on *exactly* what matters, *exactly* why it matters, and *exactly* for whom it matters. Some folks will lap that up. Others will probably find that level of left-braininess uncomfortable. I imagine it depends on whether you think in this empirical, inductive, map-it-out-in-a-spreadsheet way or not.

Theories of success

The strength of your positioning is often a function of your *theory of customer success*. That theory is, in turn, a product of what you've learned about the patterns of your best customers, i.e., your customer expertise.

In **chapter 28**, we discussed the difference between operating in the speculation layer vs. the reality layer. The story 1 change narratives people run with *tend* to be more a product of the speculation layer, but your story 2 positioning story is much more a function of the inductive reality layer — the specific, instructive cases you've discovered by doing the hard work of drilling down and becoming an expert in your customers' reality.

To be honest, dear reader, I really love the hold-your-feet-to-the-fire, reality-layer approach of story 2 positioning stories. Positioning, as an inductive, reality-in, reality-out process is a great antidote to a lot of the overly speculative market strategy work that happens out there.

Speculation is fine to a point, but the real magic of technology isn't in the speculative *promise* of what it can do (which is often the world of story 1), but the *reality* of what it *actually does*, and that's what we're trying to position around — to close the gap between our story and that reality.

Finding reality

> *While the ICP starts as an aspiration, over time it turns into a regression.*
>
> – David Kellogg[5]

That quote captures what it means to find reality as part of your positioning process, to drill down to who your ideal customer actually is.

At the start of the startup journey, there's a lot of aspiration about what you *want* to be and who you *want* to serve. So you put your product out there in the market, start selling, and see what happens. Some folks buy; some don't. Eventually, enough of them do that you now have the data to see which customers are best-fit, both in terms of how your product fits their needs and how their profitability and ease-of-reach fits your business goals.

While Kellogg is using "regression" as more shorthand for 'you stop speculating and start analyzing,' what would it mean to run an actual regression?

A regression is — indulge me here — a statistical process of understanding the relationships between variables in a dataset. In a given SaaS business, you have one dependent variable: customers who pay you the most, for example, and a bunch of other variables about those customers, like industry, team size, role, JTBD, product usage, and so on. The question then is: which attributes — and jobs-to-be-done — correlate most strongly with the folks who pay you the most? (Or have the most success, or are the most efficient to serve, etc.) Do you know? Can you find out?

If you know what attributes matter for customers that pay you the most, then you can go out and find more of them, pattern-matching — and positioning — your way to success.

> **One platform, multiple positions:** It's worth keeping in mind that SaaS products, especially at the enterprise end, are often sprawling behemoths that do a million different things, perhaps built on one overarching narrative about a new way of working. But you can still have multiple positioning-oriented story 2 pitches for a variety of contexts, perhaps by industry (hospitality, real estate, finance) or function (marketing, product, sales) based on their specific jobs-to-be-done. So, while we talk about 'positioning' in the sense of finding *the* best position you can take, as you mature and your platform grows, multiple positioning opportunities open up, and positioning becomes more fractal than singular.

Positioning around customer success

Let's say some of your customers are succeeding more than others. You want to learn from the ones that are and have those insights inform your positioning and clarity strategy and, ultimately, the next cohort of customers you find. That's great.

But how do you find those remarkable customers? You can't do it just by averaging out the haystack — that gives you the hay, not the needle. But most startups' mental model of their customer base *is* the average of the haystack. It's perhaps a vague sense of what customer jobs-to-be-done look like, with a fair bit of brain putty (**chapter 28**) patching gaps in your collective understanding with whatever your left hemispheres feel like making up.

If you're looking to dramatically accelerate growth, we're not necessarily interested in what "most" customers do or even what "the majority" of customers say. We want the outliers.

(On the other hand, if you're happy with your growth, then great; maybe you've found a steady stream of best-fit customers already, and perhaps "most" customers *are* ICP customers.)

To grow faster, that growth has to come from doing something different. Different activities, different outcomes. And those activities include finding that something different in your best, most successful, and most profitable customers. If you can do that, that's positioning gold. That's your recipe for customer success that you can reposition around.

But you've then got to have the courage to make that positioning bet.

Pain-driven positioning

When it comes to positioning, the hardest place to be, however, is where things are going *ok-ish*. It's hard to take risks when the current approach *might* work one day, even if it's not particularly likely. The pain to drive change isn't there, and that gets in the way of really putting in the work to understand what customer success looks like for your customers, *as defined by them*. That cuts off opportunities for emergent insight and, ultimately, repositioning.

What if the pain to change *is* there, though? What doors would that open?

Imagine, for a moment, that you could empty your mind of all your assumptions and start with a completely blank slate instead. Imagine that you had unequivocal evidence your product vision was actually dead — the vast majority of customers weren't using your product at all. All your initial assumptions were wrong.

But... you still have a product.

There might be *some* people using it.

What do you do?

100 customer calls

Dunford has a classic story[6] where this wasn't just a thought experiment, it was reality, and it made for a fantastic natural experiment in demonstrating what pure, discovery-driven positioning around best-fit customers — or 'user innovators' as we'll call them here[7] — looks like.

The context was this: the product — a database — was built and launched, but the original low-price, low-touch vision was all but dead, and the company was about to pull life support.

Dunford was tasked to ring around and see if yanking the cord would upset anyone. One hundred calls (!) later, Dunford discovers like *a handful* or so of folks who had a really interesting use case for the database.

It turned out they *were* using the product... but as sales reps out in the field (this predates cellular internet), which was a very different job-to-be-done than the one the company had originally envisaged.

This kind of user-driven innovation is perfect for repositioning around. And this wasn't just a new best-guess theory of customer success, this was folks literally saying (I guess), '*I am succeeding in this interesting way with your product.*'

Sometimes you don't have to speculate.

Long story short, the business repositions the product around this job-to-be-done — including packaging, pricing, GTM (selling to large teams, not individuals, a lesson B2B SaaS folks learn regularly), the whole kit and caboodle — and a couple of acquisitions later, it was a billion-dollar success. All because the company gave their customers their laser-beam attention (through Dunford's research), found the innovation that 'lead users' had developed[8], and positioned around that.

Taking a product from life support to a pathway to billion-dollar success *by fixing its positioning* is a hell of a story.

Finding the signal in the noise

Not to be too pointed, but: could your startup pull that off?

Would you (or your PMs or PMMs) call *one hundred customers* to find out how they're using your product? (Maybe you and your team talk to customers every day, have done for years, and that's nothing — you already know all your 'lead users' on a first-name basis. High fives all 'round if so!)

And if you did, would you be tuned in enough to spot the opportunity in the *single-digit percentage of folks* with an innovative use case, a novel job-to-be-done that, it turned out, your product was a great fit for? Or would hearing "Don't use it, see ya" ninety other times make you tune out? (To be fair, I guess when the contrast is that stark between the don't use it/do use it folks, it's pretty clear which way you should head.)

And even if you *did* spot that pattern with your laser-beam-like discovery, would your organization have enough right-brain bandwidth to put it back on your leadership's radar and have the whole company execute on the insight?[9]

I certainly don't know that I would have caught that (eventual) billion-dollar positioning insight. But Dunford did, and it's a fantastic example of what open-minded, assumption-free learning about how customers are succeeding (or not) with your product can do.

> **Opportunity comes from the edges:** Unique user innovation driving customer success is the ultimate positioning fuel, particularly because

it's emergent, as we saw with Loom back in chapter 10. But picking up on that emergent behavior means being the expert in how your customers really do succeed, especially at the edges, with those weird, unexpected, why-are-you-doing-those-heinous-things-to-our-product cases. Those *contexts of use* are the ones that potentially hold the keys to unique avenues for innovation and growth simply because they represent unsatisfied demand. But only if you can spot them.

Pattern matching and expertise

> *Without pattern matching, there is no intelligence. Without similar scenarios, there is no pattern matching. Without a tight positioning, there are no similar scenarios. Time and time again, you'll come back to this realization, and it will form the basis of everything you do.*
>
> – David C. Baker[10]

The strength of the laser beam is breaking things down into parts so you can find the patterns. (Pattern matching is a right-hemisphere behavior, incidentally, and this is right and left-brain attention working together.)

Developing tight story 2 positioning — as opposed to relying on a broad story 1 concept — lets you create potential leverage through *expertise* and *distribution*. We'll touch on customer expertise here and then we'll discuss distribution in the coming chapters.

David C. Baker, an expert positioning consultant for creative firms quoted above, wrote a fantastic little book on positioning and expertise called *The Business of Expertise*. One of his key insights is the relationship between expertise, pattern matching, and positioning, and it very much applies to selling B2B SaaS, too.

Everyone hammers home the need to 'solve real problems for customers,' but to meaningfully solve them well, you have to see patterns, and there won't be any patterns if your customers are all over the map. That's not to say you can't sell opportunistically — of course you can. But tight positioning comes from a tight ICP and a tight definition of their success.

Therefore, customer expertise and tight positioning go hand in hand — you see the patterns in the challenges your best-fit customers face, and you innovate and solve them better than anyone else, because you know them more intimately than anyone else. That then creates leverage in deals with other folks with those same problems (through point of view, references, features, and/or your playbook). That's the game.

Dunford picks up a similar theme on pattern matching in *Obviously*

Awesome, writing[11]:

> *[A fellow repeat VP of Marketing and I] discovered that our philosophies were strikingly similar—we spent all our time looking for things that work, and then deciphering why. What marketing campaigns brought in the most leads? Which pieces of content were consumed the most? What events had the most attendees? Asking these questions led us to the happiest customers. Once we narrowed in on the happiest customers, we went looking for the reasons they were so satisfied while others were not.*

Let me interrupt the quote here to highlight how Dunford, in her VP of Marketing days, zeroed in on her happiest customers — through a lot of detective work! (Think of this as a more detective-style regression analysis to riff on David Kellogg's earlier quote.) Knowing which campaigns, which content, and which events performed best — *and then developing a working theory of why* — is the kind of radar-level awareness VPs should have. Back to Dunford:

> *What was it about our offering that made them so happy with it, and what was it about those customers that made them such a good fit for us? Answering these questions helped us figure out what our value was, who it was resonating for and why—in other words, it helped us get a start on understanding what our positioning should be.*

I love this because it's so laser-beam driven, so much about drilling down to the *why* — to that deep strategic insight about what drove value. That's the kind of insight — and pattern matching — that drives successful positioning and a successful story 2 in your clarity strategy.

With that knowledge about best-fit customers and how they use your product, you can then start closing gaps in a very productive way — be that language/market fit, sales/buyer journey fit, or product/needs fit, as we discussed **last chapter** in our JTBD primer. That's a powerful way to level up your startup.

This kind of story 2, positioning-by-pattern-matching approach is something of an optimization strategy. It assumes the customer base is there to analyze and optimize in the first place, and the answers are to be found by looking *down* and seeing what you can bring back up to the market.

But what happens in the early, zero-to-one days when you need to build your story 2 sales pitch from scratch? What does your clarity strategy look like in that case? In the next chapter, we'll look at story 2 pitch testing and sales-enablement SaaS Dock.

Notes

1. Dunford, April. *Obviously Awesome: How to Nail Product Positioning so Customers Get It, Buy It, Love It* (p. 175). Ambient Press. Kindle Edition.
2. This book wouldn't be possible without them — thanks April!
3. Dunford, April. *Obviously Awesome: How to Nail Product Positioning so Customers Get It, Buy It, Love It* (p. 45). Ambient Press. Kindle Edition.
4. For a great description of the process in a real-world B2B SaaS setting, see Jane Portman's article "How We Used April Dunford's 10-Step Method to Overhaul Positioning at Userlist," *Userlist blog*, 2019 (updated November 4, 2022). **https://userlist.com/blog/positioning-overhaul/**
5. Kellogg, David. February 19, 2023. **https://twitter.com/kellblog/status/1627114183350157314**
6. You can hear April tell it in an interview with Ben Goodey in "How I Helped Reposition a Database Product That Went on to Make $1B in Revenue," *How the F*ck Podcast*, December 26, 2021. **https://seo.thefxck.com/interviews/product-positioning-april-dunford/**
7. We discussed 'user innovators' in the context of Loom at the end of chapter 10, and touched on it again when discussing Slack in chapter 12, too.
8. 'Lead users' is Eric von Hippel's term, and he was writing about this phenomenon way back in 1986, for example. See, for example, von Hippel's paper "Lead Users: A Source of Novel Product Concepts," *Management Science*, Vol. 32, No. 7 (Jul., 1986). **https://www.jstor.org/stable/2631761**
9. We touched on the need for right-brain bandwidth in leadership to keep innovating and avoid disruption when discussing Clay Christensen and Jobs to be Done in the previous chapter. Further evidence for the fact that *organizational* challenges are often the biggest ones comes from this 2018 article by Tim Harford, discussing Rebecca Henderson's work, the limits of disruption theory, and why companies fail to act on new ideas: **https://timharford.com/2018/10/why-big-companies-squander-brilliant-ideas/**
10. Baker, David C.. *The Business of Expertise: How Entrepreneurial Experts Convert Insight to Impact + Wealth* (p. 26). RockBench Publishing Corp.. Kindle Edition.
11. Dunford, April. *Obviously Awesome: How to Nail Product Positioning so Customers Get It, Buy It, Love It* (p. 66). Ambient Press. Kindle Edition.

38

Lattice, Dock, and testing your pitch

> This is the moment when your audience is viscerally, undeniably excited about what you're talking about. *That* is the reaction you're looking for. Whenever you tell a story, share your value proposition, or test your positioning/messaging...
>
> Eyes lighting up is the ONLY reaction that matters.
>
> - *Wes Kao*[1]

When we talk about clarity strategy — especially if your startup has dreams of big VC returns — there's a constant tension between wanting to tell the big, transformational story that sounds the most exciting to you... and the one that actually lights up the faces of your early-stage prospects — the folks you need to close if you're going to get a foothold in the market.

This is the challenge of balancing radar-level story 1 change stories and story 2 laser-beam positioning. You might have big billion-dollar dreams, but first, you need folks to give you a chance or even the time of day. That means having a story 2 pitch that resonates, and that usually *isn't* something you dream up and roll out, it's something you test and iterate as you go.

That's easy to say, but what does it look like in practice? How do you know if folks actually care about what you're pitching at all, especially in the story 2 world where you're trying to reach down to their exact needs?

How can you tell if your story actually resonates?

It's all in the reactions.

In this chapter, we'll return to pitching and the idea that when it comes to high-touch sales, your pitch *is* customer development. We'll look at pitch reactions first and then move on to the approaches of the founders of B2B SaaS companies Lattice and Dock, respectively.

Affirmation isn't allocated budget

Startups are fundamentally about behavior change. Using your product means, by definition, that folks change their behavior from what they were (or weren't) previously doing. And when it comes to sales, we want to see behavior that indicates their interest in making that bigger change. Not just talk, behavior. The advice to not follow what folks say but what they do is cliche for a reason. Talk is often aspirational — 'Yes, I'd absolutely *love* to use a product that did X!' but behavior is reality — 'Err, no, I'm not going to pay $/month for *that.*'

When it comes to your story, we're not after empty agreement. We're not after polite enthusiasm. The ultimate behavior change we want is people making some kind of financial commitment. But before that, we need those teeny tiny bits of proxy behavior that indicate we're on a promising path. We need to see — and feel — reactions. That's why testing your message IRL is an integral part of refining your positioning and your pitch.

> **User testing your pitch:** I like to think of pitch testing like user testing but for your narrative. Software has a long history of qualitative user testing, and selling in early-stage B2B — or in new market segments, e.g., targeting the C-suite — is essentially one long series of qualitative tests, i.e., whether buyers will respond positively to your questions and your pitch. But how deeply do you study those reactions? How much analysis do you give them? That's the kind of learning we're exploring in this chapter.

Meaningful reactions

In actual sales calls, those reactions can be subtle.

Back in **chapter 22**, we were discussing the natural left/right-brain pitch experiment where both inbound and outbound folks were getting 'radar' pitches. The inbound folks apparently didn't react very well to those pitches, but the outbound reactions were different and a surprise to your humble author.

Some of those prospects would listen to the pitch, saying very little, then, when asked what they thought, they'd express some vague interest... and then (much to my surprise) they'd agree they'd like a demo of the product. These *weren't* super enthusiastic reactions, but these were marginal prospects being pursued by a mature company. And while I'm not a seller, I've watched enough calls to know that's often the sales life — trying to find that thin sliver of interest, getting the prospect to volunteer to take the next steps, and hoping you capture their attention in a way that generates *just enough* interest to create some momentum, however small.

That's what can happen in mature startups.

But what about the early days when you're trying out new ideas, particularly core parts of messaging around story 1 change stories and/or story 2 positioning? If you're trying to find that story to orient your company around, you need strong reactions, not subtle cues. What does it look like when someone actually catches on to what you're telling them?

At the simplest level, their eyes light up.

Eyes light up

I love this "Eyes light up (ELU)" concept from Wes Kao, marketing veteran and co-founder of education startup Maven. Here's how Kao sets it up[2]:

> *Positioning and messaging is exciting because you're picking your angle. Of all the ways you could talk about your product, which way should you talk about it?*

That is: "How do we know if our story resonates?" Kao's response:

> *You already know when people are excited to hear a story. They perk up. Their energy level shifts. All of a sudden they look alive—and they want to hear what you have to say. I call this the Eyes Light Up moment. Or ELU for short.*

ELU is, as Kao points out, the opposite of bored, polite, matter-of-fact answers. The more folks you can talk to, the more you can try out variations and see what gets the 'eyes light up' reaction — questions, interruptions, *energy* — and what puts people to sleep.

"ELU" is particularly interesting to us for a couple of reasons:

1. It's a neat example of very simple forms of behavior change. It's not (just) what people say, it's the physical reaction that comes with it.
2. Kao named the concept and made it catchy! That's a theme we've been discussing throughout this book (e.g., **chapter 17**). Putting your finger on something, giving it a catchy name, and making it memorable so people can talk about it helps ideas spread (this being a case in point).

Eyes, err, not lighting up

There's an opposite of the ELU reaction, too — the complete flop. A great example of this comes from Francisco T. Bram and The *Adventures in Messaging* podcast.[3] Bram was working on the pitch for a new ultrasound medical product for doctors treating folks who were clinically obese. Traditional ultrasound devices don't work well with these patients, with 25% of them (!) being misdiagnosed.

New tech, obvious problem, clear benefit... must make for an easy pitch, right? Bram took the pitch to doctors about this serious and (evidently) solvable problem. Their reaction?

I don't misdiagnose.

Ouch.

(You can imagine the 'Of course *other* doctors might misdiagnose, but I most certainly *do not* misdiagnose 25% of these patients' conversations that would follow.)

Back to the drawing board, but, as Bram noted, catastrophic launch averted, thanks to testing your pitch and seeing how folks react.

> **Panel testing for what customers actually value:** Real-life reactions are great for that visceral, qualitative feedback. Another way of sanity-checking your pitch in a more quantitative sense to understand what's important to your ICP is to use paid surveys and, well, ask. This tip comes from copywriter extraordinaire Momoko Price, who recounted one example of working with a point of sale company in a talk.[4] Price, thanks to paid panel surveys, discovered (a) that what her client was investing in wasn't what their customers actually wanted (!), and (b) what customers *did* want was actually much more obvious: stability and speed. However, it was so obvious everyone overlooked it until the surveys made it crystal clear. Sanity checks can pay off, but so often, startups raise millions and just YOLO it instead. A little bit of research can go a long way!

Product/market fit reactions

Back to 'eyes light up.' What does that reaction look like in early-stage B2B SaaS?

First, let's note that at this stage, sales often *is* customer development, and vice versa. For example, here's how former VP of Marketing Alex Kracov talks about his early days at HR startup Lattice (valued at $3B, pre-tech market meltdown)[5]:

> *I remember watching [Lattice co-founder] Jack [Altman] do sales in the early days of Lattice. It felt like he basically networked his way into making Lattice happen. He was always talking to a potential customer or meeting with someone who could introduce him to a potential customer. He would then take the insights from these conversations back to the Lattice team. He would talk to the product team about what we would need to build next. And then I'd work with him on how these insights would inform our positioning and go-to-market strategy. This formative experience was extremely helpful when I started to do sales for Dock.*

Every startup doing sales starts at the bottom of the sales learning curve, and your journey up that S-shaped learning curve is how you get to product/market fit and beyond, riding customer reactions as you go. Those reactions — and the insights they generate — feed back into your GTM approach, and over time you zero in on a pitch that works, which in turn enables you to scale. But it requires deliberate, laser-beam attention to the sort of reactions you're getting.

Kracov moved on from Lattice to found sales enablement startup Dock, and took a similar approach of selling his way up the learning curve, writing[6]:

> *When it comes to early stage sales, you're still figuring everything out. You have a barebones process, a pitch you're not sure will work, a product that's still being developed and likely no formal pricing proposal. This makes the sales process very different. It needs to be a creative process that you're constantly iterating on the fly. After every call, you need to look and think about how you can evolve the pitch. This is one of the many reasons why a founder needs to do sales themselves.*

At this stage, the pitch *is* the product, in a sense, and needs just as much care, (laser-beam) attention, and iteration, otherwise you're going to struggle to get the plane off the ground.

> **"Feedback" vs. "sales" calls:** How you position your initial calls matters too. They don't have to be "sales calls." Kracov called them "feedback calls" instead and mentioned they weren't focused on revenue, they were focused on making the product better. A lot of folks were happy to jump on a feedback call with a founder in a way they might not have been if it was a rep doing sales.

Pitch and test

During and after the calls, Kracov would try and find the patterns in use cases (picking up on our theme of positioning through pattern matching we discussed in the previous chapter), thinking hard about how the product and the market, well, fit. He would ask himself whether his vision for Dock was correct, what features they would need, who the ICP was, and what words resonated with which customer profiles.

That is, he would do a lot of careful, laser-beam-y analysis of what he was seeing, with product, pitch, and customer development all going hand in hand. But he was also just looking for 'eyes light up' moments too (my bold):

> *While the tactical features were important,* ***what I was really***

> ***looking for on all of these calls (especially in the super early days) was the level of enthusiasm*** *the person I was talking to had for the idea and our vision.*

The ELU folks became the ICP, that's who they built for, and that would in turn generate another round of even more enthusiastic calls, and so on.

It was an ELU flywheel using sales to drive the product — and pitch — toward product/market fit. (This is not dissimilar to what we'll see Nathan Barry do with ConvertKit next in **chapter 39** — relentlessly iterating on the early pitch to generate that all-important momentum.)

> **Listening *after* the conversation:** Everyone says 'talk to customers,' but what's implied is *listen* to your customers. Listening is a skill, and it goes beyond the conversation — it's what happens after the conversation that lets you really process and integrate what you've heard. Asking questions and listening to responses in the moment is important, but it's the deliberate, laser-beam thinking and reflecting after the conversation that's the real 'listening.' That's why, I imagine, Krakow says, "After every call, you need to look and think about how you can evolve the pitch." That's where deep listening happens.

Less telling, more learning

Kracov's experience is unique in that he'd seen up close what the development of a robust sales process looked like at Lattice and then went back to square one to do it again at Dock. He says that when he started his Dock demo calls, he had a "grand vision" of running the conversations using the process Lattice landed on — a mature pitch deck and a lot of telling. But that was too awkward for his very early-stage startup, so he made it much more conversational instead: a brief intro on the founding story and problem they were tackling, then into questions, and then a tailored demo.

It's fascinating to see how different late and early-stage selling is, especially along explore/exploit lines.

Your late-stage pitch is the product of hundreds or thousands of customer conversations, and you're able to exploit everything you've learned prior. But just because you learned that in one context doesn't mean it works in another context. That is, even if you've seen what a mature sales process — and a mature clarity strategy — looks like (and in Kracov's case, helped create it), when you've got a new product or market, it's back to square one: new context, new customers, new learning curve.

> **Older companies and unlearning:** This approach can be useful for older companies struggling with growth, too – you may need to *unlearn* some of what you think you know, allocate some opportunities to fresh exploration, and see how your pitch develops *without* the baggage of all the iterations that have gone before.

Horizontal vs. vertical positioning

Dock's story is still unfolding as I write, and it's fascinating watching them try to find the balance between a tight niche and something broader. Kracov tried to go broad but ended up coming back to their niche, writing[7]:

> *One of the mistakes I made early on at @dock_us was trying to build for too many uses cases. I wanted to create a horizontal product for any type of business to business interaction from sales to recruiting to finance.*
>
> *But this proved to be an enormous challenge on many levels.*

Customer feedback was "all over the place," the product was "too open-ended," their positioning was "hard to define," and it was "difficult to gain GTM momentum." However:

> *Everything changed in our business when we decided to focus.*
>
> *Today, Dock focuses on helping revenue teams collaborate with customers.*
>
> *We have three core uses: sales deal room, customer onboarding, and client portals. This niche works for Dock since all these use cases are interconnected.*

Many startups try and build initial momentum by focusing on a particular vertical, a particular problem, or a particular niche to start with, going down the story 2 y-axis of attention to solve a very specific job-to-be-done, hoping to discover an initial customer base that way.

For Dock, Kracov focused on the pains he had felt around sales enablement, specifically the lack of a central place to manage collaboration and communication between seller and prospect. That's the *vertical* positioning and story 2 pitch around a specific job-to-be-done.

That's great to get started, but the temptation is to then chase the broader story 1, radar-friendly narrative that will give you a broad horizontal position. Kracov flirted with that horizontal positioning as he mentions, at one stage describing Dock in category-creation terms, tweeting that they were building "the first external collaboration platform."[8] They were casting a wide net for use cases, too, with their homepage featuring client portal and design project uses. But with no secular change to ride, that was always going to be a tall order for a small startup.

It wasn't to be, at this stage in their journey at least, with Dock doubling down on their original niche positioning instead. In any case, we're *enormously* fortunate to have a talented founder and gifted communicator like Alex sharing the ups and downs of the startup journey openly and transparently.

> **Get the full playbook:** Alex Kracov generously shares his experience on his blog, and that includes his whole early stage, founder-led sales process, from emails to the call, the pilot ("a classic tactic for B2B SaaS companies in order to prove out their product"), to manual migration and onboarding, and how he got the flywheel started (years of building a network in anticipation of becoming a founder!).[9]

The story 2 → story 1 transition

Kracov's experience highlights the difficulty of the transition from a narrow story 2 to a broad story 1. It's difficult not only for the reasons he describes, but because it's a shift from one mode of attention to quite a different mode of attention, too.

Those initial 'eyes light up' reactions were to the narrow story 2 pitch, not necessarily the broader story 1 pitch. The story 2 work was a product of reality-layer observations — felt pains from doing the work, while story 1 narratives tend to be more speculative and aspirational, as we've discussed (which is why riding a secular change is so important, as we discussed in **chapter 30**).

The transition from story 2's narrow jobs-to-be-done world to a broader story 1 concept-and-playbook approach can be a treacherous one. It's a chasm of sorts that most folks don't realize exists, and it's one many startups face as they try and grow beyond their initial positioning.

Next, we'll look at ConvertKit, an email marketing SaaS that *did* successfully pull off the niche-to-mainstream, story 2-to-story 1 playbook, all while bootstrapped, thanks to a founder discovering — and persevering with — niche sales.

Notes

1. Kao, Wes. "Eyes light up (ELU)," August 14, 2020. **https://www.weskao.com/blog/eyes-light-up-elu**
2. Ibid.
3. Bram, Francisco T. (guest). Stratton, Emma (host). "E01 with Francisco T. Bram, Head of Global Product Marketing, Uber," *Adventures in Messaging,* March 4, 2021. **https://www.youtube.com/watch?v=dzOJ52T4L4A**
4. Price, Momoko. "Data-Driven Copywriting for Brand-Spanking New Products," *CXL,* April 16, 2020. **https://www.youtube.com/watch?v=XVa7VQUSXmQ**
5. Kracov, Alex. "How I've approached founder-led sales," January 18, 2022. **https://www.kracov.co/writing/founder-sales**
6. Ibid.
7. Kracov, Alex. December 6, 2023. **https://twitter.com/kracov/status/1732073573253329286**
8. Kracov, Alex. February 7, 2023. **https://twitter.com/kracov/status/1622633943278227457**
9. See the full post here: **https://www.kracov.co/writing/founder-sales**

39

Market strategy, ConvertKit, and the niche-to-mainstream flywheel

Most startups don't die because they don't get their tech to work or the competition out executes them. They fail to launch because of indifference. Nobody cares. Not potential customers, employees, investors or the media. It's death by a thousand shrugs.

– D'Arcy Coolican[1]

How do you craft your narrative to attract those first few people that care? When it comes to story 2 in clarity strategy, it's all about reaching down to find them (and their needs). So far, we've looked at:

- **Existing customers:** Reaching down to study your best-fit customers, their jobs-to-be-done, and the way they're using your product (**chapter 37**).
- **New prospects:** Reaching down to folks' real-time reactions to see how they respond to your pitch (as we saw previously in **chapter 38**) and doing the deep listening and reflection required to integrate the insights you discover.

And now, in this chapter, we're going to look at:

- **Niche targeting:** Reaching down to a very specific niche to get the growth flywheel going, as demonstrated by Nathan Barry taking ConvertKit from zero to one through niche sales and finding folks who cared *just enough* to switch (or be switched, as we'll see).

Positioning × distribution

In this chapter, we're looking at how picking a tight target niche upfront — as ConvertKit did — can drive early sales and distribution and get the growth flywheel going. This approach to *sequencing* your positioning (and your approach to sales) can generate momentum in your market in a way that simply taking a random sales walk does not.

To understand why this matters, we need to understand the basics of market strategy, particularly the relationship between your positioning and your distribution, which is what we're going to explore over the next few chapters. It's critical to get this right for your broader clarity strategy to land, and getting it wrong can be an incredibly expensive mistake (as we'll see later in **chapter 42**).

Market strategy 101

Every startup has an implicit market strategy, and your clarity strategy (what you say) has to work hand-in-hand with your market strategy (who you say it to).

B2B tech market strategy is unique because we're selling something new, innovative, and often very expensive relative to the classic B2C market*ing* world of selling toothbrushes or shampoo. We'll explore the implications of this further in subsequent chapters, but for now, here's a super simplified version of tech market strategy.

Let's imagine we can put your potential customers in three concentric circles.[2]

- Core
- Mainstream
- Marginal

In a perfect world, you'd build your initial customer base with your core early adopters (perhaps including "design partners" in your seed days in the enterprise world), use that as a base to go after the mainstream, and then having conquered a given mainstream market, you'd go after increasingly marginal customers. Reality is, of course, much messier, but that's a simple representation of the classic B2B market strategy model. The point here isn't that this is a perfect model by any means; it's simply that *market strategy is a thing* and that market strategy changes by stage.

The sales learning curve

Universal theories of tech markets are fine to give you the outline of what *might* be happening in your market, but ultimately, you want to develop or discover your own theory of your specific market so you can make wise choices about how you might get a foothold in that market.

Two of those fundamental choices are:

1. How you position your product, in a customers-you-choose sense.
2. How you go to market, in a sales-and-marketing sense.

Those might seem obvious, but what's less obvious is just how deeply intertwined those choices are *and* how highly dependent they are on the stage of your startup.

This is where we hit a bit of a left-brain/right-brain conflict. In a left-brain sense, we have the different static parts and pieces: positioning over here, some sales and marketing tactics over there, and a rough sense of who you want to sell to that underpins both.

These are fine as parts, but the challenge is to integrate them as a right-brain whole *and* locate that whole on a timeline, so you know the pieces you have make sense for the stage you're at.

Fortunately, a couple of smart folks — Mark Leslie and Charles A. Holloway[3] specifically — have given us a handy framework to think about your sales progress over time: the sales learning curve[4], a concept we very briefly touched on in the last chapter. Founders and operators who can shape *both* their positioning *and* their GTM motion in tandem as they scale this curve have a huge advantage relative to those folks just taking a random walk in the market instead.

In this chapter, we're particularly interested in the very start of that curve — lighting the fire that gets you started and gets that sales momentum going. This is where founder sales is so important. One person — perhaps that's you — with the support of their team, refining both the pitch *and* your ICP simultaneously, shaping supply and choosing the demand to serve all at once. We saw how that worked for Dock in the previous chapter, and in this chapter, we'll see how this worked for ConvertKit, too.

> **The sales learning curve in a nutshell:** One of the most famous ideas in strategy consulting history is Boston Consulting Group's "experience curve," which is the idea that the more product you manufacture, the more you learn, the more efficient you get, the lower the cost of your product, and therefore the more market share you can win by being the low-cost leader in a segment. The sales learning curve adapts this idea for the sales process, creating an experience curve for sales itself, which includes early-stage "initiation," mid-stage "transitional," and late-stage "execution" phases.[5] The point is to apply the right sales strategy at the appropriate stage, and not rush into formalized "execution" sales with an army of reps before the early team — and, I'd suggest, the founder in particular — has done the early exploratory work to learn how to sell efficiently in the first place. That is, you've got to explore before you exploit, and rushing to exploit too soon is a classic mistake that startups make over and over again, one that frameworks like the sales learning curve exist to prevent in the first place.

Early vs. late stage

How you position — and sell — your product can change dramatically over time as you ride the sales learning curve. We saw this with mature Lattice vs. early Dock in the previous chapter, for example. Alex Kracov, Dock's founder, had experienced Lattice's journey up the sales learning curve, helping them become a well-oiled sales and marketing machine. He tried to apply those same late-stage principles to Dock in the early days but quickly realized that didn't work, and therefore he had to go back to the start of the curve and begin again.

We also saw this with Qualtrics, for example, in **chapter 35**, which went from playing *very* narrow and establishing a foothold with a core audience (marketing professors at business schools) that let their product ripple out into the market, to playing bigger as a mature company and going after very large enterprise accounts. They are obviously very different techniques for very different stages of the curve, and they're only found through the grind of learning and trying ideas.

The journey from rickety point solution to robust enterprise platform is, in some ways, the typical B2B SaaS playbook. Start by solving a very specific, very painful problem, grind out sales, build out in concentric circles pulling in broader and broader use cases as your product and organization becomes more capable, close more deals, gain more references, generate more word of mouth, and ride the sales learning curve all the way to being an established mainstream player — perhaps *the* established player.

These are market strategy playbooks everyone "knows" in a know-it-exists sense, but far fewer people really *know* in a know-what-it-takes sense. What it takes is execution that's efficient, methodical, relentless, and relentlessly curious, as we're about to see in Nathan Barry's journey through the early days of ConvertKit's sales learning curve.

Laser-beam positioning, laser-beam distribution

How do you get started at the very bottom of the sales learning curve?

Niche positioning often works because it enables niche distribution, and when you're early, niche distribution is all you've got. The whole point of the niche positioning playbook is to create that early leverage *through* niche distribution.

If you could just sell The Universal Thing to The Universal Market, then obviously, that would be the more lucrative choice, and you should do just that. But if you can't — and almost all early-stage B2B startups can't — then you have to generate momentum from a niche.

Why does this work? Why would you start by positioning your product in one specific niche when your ambitions are much bigger? The simple answer is: the magic of compounding returns.

Early on, if you land, say, five customers in five different verticals on the back of spray-and-pray marketing, it's very hard for compounding effects to kick in. Each sale is, in effect, very expensive because your entire sales and marketing budget is being spent to generate these ad hoc customers. Word of mouth can't kick off if each customer is essentially a lone voice in their own room.

Most founders and sales orgs learn pretty quickly that this approach won't take them up the sales learning curve, and they need to be more efficient. What if all five customers were in a single room instead, along with a bunch of other potential customers? That would generate that much more word of mouth and, therefore, momentum. So you learn how to sell to a niche, sales becomes more efficient, and you progress up the sales learning curve.

Now, in this more efficient, niche-focused sales world, potential customers don't hear about your new product once and forget about it, they start hearing about it again and again, all without requiring your marketing spend to generate those touches. The references you build go a lot further, too — peer references are obviously more impactful than ones from companies folks have never heard of.

This matters for product development, too. Trying to satisfy the different needs of folks who aren't in the same proverbial room is a lot harder than satisfying the needs of the ones who are. A stronger product for one set of core users again creates more momentum than an okay product for a variety of users, simply because the happy core users will *tell other people* in a way the so-so folks won't. These are the 'Eyes Light Up' folks we discussed in the previous chapter.

These compounding effects make each subsequent sale easier. You need a tight enough pool of early customers to make that happen, but if you can keep selling to folks in the same customer pool, each sale gets easier and easier, and eventually, after much rubbing of sticks, your fire lights and stays lit. You progress up the curve.

Again, this is not new advice — I'm drawing, in part, on *Crossing the Chasm,* first published in 1991, which we'll explore in the next chapter. But it's one way to avoid dying a death of a thousand shrugs, to borrow from this chapter's opening quote. Nor is it universal advice — some products really do go viral and waltz into the mainstream market without breaking a sweat. In the B2B world of complex sales, however, that's exceptionally rare, almost by definition. It's far more common to grind out sales instead, learning as you go, becoming more efficient, and driving your way up the sales learning curve almost through sheer force of will.

This is perhaps not the sexy, be-the-next-iPhone/Facebook/megahit, create-the-category narrative of how startup success is supposed to happen. But this is what separates B2B from B2C: grinding out sales through the sales learning curve. It might not be pretty, but it sure can be effective.

Early stage positioning — open or focused? There's an argument that early-stage folks should keep their positioning open to start with so they can collect enough users to then apply the left-brain laser beam and see which segments are worth targeting, i.e., which customers are 'best fit.' Closed betas, for example, for product-led folks, can help in that regard. For sales-led folks, if you've got enough muscle to get out into the market and accept GTM inefficiency upfront to increase your learning surface area, then you can certainly give that a go; just don't confuse this capital-fueled exploration phase for having a genuine exploitable market insight. Burning loads of capital can be a net negative in that sense if it lulls you into a false sense of progress up the sales learning curve when really you're just learning how to light a bigger and bigger money bonfire. In any case, especially in a post-ZIRP world, most sales-led folks need efficiency from the get-go, and that's where tight positioning and tight distribution loops come in.

ConvertKit and niche sales

ConvertKit is a fantastic example of just what it takes in those very early days to get that momentum going and start making progress up the sales learning curve.

ConvertKit is a bootstrapped, B2SMB-focused email marketing platform targeting creators. It's a classic example of what happens when a founder deploys the niche-to-mainstream positioning playbook, and because its early success was a product of a founder making it happen, we get an excellent end-to-end case study of how the process worked.

The founder in question, Nathan Barry, was no stranger to selling products online. He'd had success selling ebooks and courses, and he had also tried his hand at several SaaS products with... not much to show for it.

In January 2013, he founded ConvertKit because, as the company tells it, he "needed a better way to reach his audience as a blogger and creator to sell books and courses through email."[6] His aim was to reach $5,000 in MRR (monthly recurring revenue) within six months. He made it to $2,000 MRR by marketing primarily to his existing audience, which then dwindled to below $1,500 MRR. (A strong personal brand only goes so far.)

Then he discovered sales.[7]

Niching down down down

Barry described his early sales journey of trying things and learning from his mistakes in an excellent Twitter thread, which I'm drawing from and quoting below.[8]

First, Barry made the classic mistake of having super broad positioning that gives you no leverage in the market:

> *I was trying to target everyone who needed email marketing. That's a sure way to lose.*

Picking a huge TAM and thinking you can just peel off some random 1% of the whole is a classic mistake because you have 0% reason to be on *anyone's* radar relative to the incumbent solutions. You need a reason to get on folks' radar, and in this case, that meant niching down hard:

> *Instead I narrowed in on "email marketing for professional bloggers." which allowed me to write much more focused sales page copy. When I tried to raise capital all the VCs said our market was far too narrow. But I actually niched down further. I'd go to: "email marketing for professional paleo recipe bloggers who are women" or "Men's fashion bloggers in NYC".*

He then built online relationships with those folks:

> *This allowed me to use Twitter, Google, & 'top blogs in x category' posts to compile lists of who to contact. I subscribed to their newsletters, followed them on twitter, & interacted with their content so when I reached out they'd recognize me.*

The niche positioning unlocked niche distribution.

That is the key, and that's what a lot of folks forget when encouraging niche positioning — *it must create meaningful leverage in sales and distribution.*

Listening before pitching

In Barry's case, he could focus his positioning laser beam on a very specific crowd and get to know them. And then what, pitch them? Not directly. Instead, he'd work his way onto their radar by asking about frustrations with the dominant email marketing platforms like Mailchimp, dropping a mention that he runs a competing service with customers they'd know. As Barry puts it:

> *Then I'd email them directly. Rather than asking for them to buy or demo my product, I'd ask what was frustrating them about Mailchimp, AWeber, or whatever they were using.*
>
> *The email was short, personal, and name dropped the most relevant*

customers.

The name-dropping is important — the point of the niche positioning playbook is to build like-for-like references you can use to convince the next tier of customers to consider you. It de-risks the switch in a "You go first" sense. If your friend has already jumped in and the water's fine, then it's a lot less scary for you to jump in, too.

But they still need a reason to switch, and Barry could tap directly into that:

> *The responses poured in. These bloggers were frustrated with not being able to segment their list, create opt-ins to give away a free guide or incentive, creating automated email courses, & more.*
>
> *The exact things I was frustrated with that got me to start ConvertKit.*

Note the left-brain narrow focus here, both in terms of his prospects and his own frustrations. There's no story 1 'Email is the future of X' change narrative here or high-concept talk about new categories or new disciplines. That would completely miss the mark. Instead, we're focused right down on very specific pains and frustrations — classic story 2 territory.

Enabling things that do scale

For Barry, more selling meant more learning — he had validation that the frustration was there. But that didn't magically create rivers of cash. Barry says he did hundreds of calls (!) but kept hitting a wall with prospects. They were reluctant to actually switch from their current tool. So, out of desperation, he offers to do it for free. That becomes his "concierge migrations program," and he does *everything* to migrate folks... all to land a $50/month customer.

> *My hourly rate was less than $5/hr for those early customers. Far from the $150/hr I'd been charging for design services.*
>
> *But it was working to get momentum.*

This is why Paul Graham's oft-repeated "do things that don't scale" quip is so popular — you have to get the momentum going somehow. Here's Graham in his famous essay (published, incidentally, six months after ConvertKit launched)[9]:

> *One of the most common types of advice we give at Y Combinator*

> *is to do things that don't scale. A lot of would-be founders believe that startups either take off or don't. You build something, make it available, and if you've made a better mousetrap, people beat a path to your door as promised. Or they don't, in which case the market must not exist. Actually startups take off because the founders make them take off. There may be a handful that just grew by themselves, but usually it takes some sort of push to get them going. A good metaphor would be the cranks that car engines had before they got electric starters. Once the engine was going, it would keep going, but there was a separate and laborious process to get it going.*

Barry cites Graham's advice in his thread and offers (my bold):

> *Do things that don't scale,* ***because that enables the channels that do scale****.*

That is, it gets the flywheel going. It helps you learn how to sell *at all*, even for the $5/hour or less Barry was netting at the time. It's that flywheel of *learning* that matters, as that's what drives you up the sales learning curve.

For Barry, ConverKit's early momentum was built one customer at a time. By targeting micro-niches, Barry says some bloggers thought "everyone" was switching to ConvertKit when, in reality, it was everyone *in their little itty bitty niche*, and the product was still only doing $5K MRR.

Inflection point

Sales got easier as time went on, with Barry asking for — and getting — warm referrals to other bloggers after each successful customer onboarding. Referrals grew and grew, and then they hit an inflection point:

- **Mid 2015:** After six months of doing direct sales ConvertKit hits $15k MRR in July 2015.
- **Late 2015:** They roll out an affiliate program and hit $98k MRR in December 2015.
- **2023:** A decade after founding, ConvertKit crosses $3M in MRR (an annual run rate of more than $36M), all while profitable and bootstrapped.

But it all hinged on that inflection point. It's amazing to see that 2015 change in trajectory and the (distinctly linear) growth that followed in subsequent years in their MRR chart[10]:

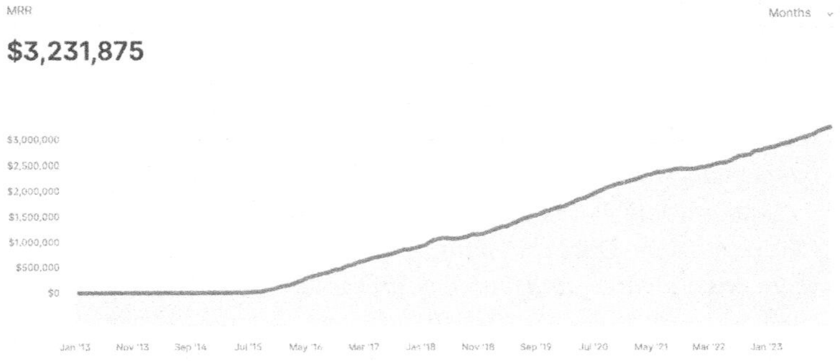

Figure 39.1: ConvertKit's revenue growth.

All because one very clued-in founder went as niche as possible in their early positioning, discovered sales, and did whatever it took to get that initial flywheel going after marketing alone failed.

That meant closing those early sales and migrating folks by hand, building that momentum one customer at a time, and having the confidence to do so because, I imagine, Barry knew the real pains his product solved that the incumbents didn't.

> **Who owns your positioning?** In B2B VC land, bigger companies might attempt aspects of this approach. Maybe they'll call it Account Based Marketing, meaning the deliberate, specific targeting of prospects that meet your ICP criteria and building 'relationships' with them, in some form. But is your positioning muscle strong enough to make that work? Consider the core ingredients in ConvertKit's story — ultra-niche positioning and one person iterating and learning end-to-end across product, positioning, pains, objections, and buying success. It's a whole package — right and left brain, radar and laser beam — not an atomized slice of marketing that better-funded folks are often reduced to. Where is your company on the sales learning curve? Who in your organization is responsible for your positioning/distribution fit? Who oversees the end-to-end process and assesses what works? Often, I suspect, the answer is no one, really, and perhaps that's one issue that's holding you back.

The attention model

Let's run this case study through our trusty radar-and-laser-beams model of attention and see why it's the founder's attention in particular that's so important here:

- **Laser beam insight:** Barry first applies his laser-beam attention down to the specific email-related pains and jobs-to-be-done he faced as a blogger and marketer. He's not speculating; he's lived in the same reality layer as his prospects. He builds a product to match and then looks out to see how he can find customers.
- **Deliberate and indirect:** After failed marketing attempts, he switches to direct sales, builds some rapport, and (in true sales fashion) manually puts the product on the radar of those he thought were a good fit with similar pains, albeit somewhat indirectly.
- **Their pains:** When engaging, he invokes the prospect's own pains and frustrations, which is the subtle art of getting folks to sell themselves by bubbling up their specific narrow-focus issues.
- **Niche solution — story 2 in action:** ConvertKit is then positioned as the solution to these very specific problems, in true story 2 form, going well down the y-axis of attention.
- **Obstacles:** But it's not enough! Switching friction needs to be overcome. Barry does manual migrations and eventually starts getting successful switchers to put the product on the radar of folks in their network, either directly or through affiliate sales.
- **Ignition:** Then the flywheel kicks in. By this time, Barry had solved for the entire sales loop — from pains to product to outreach to migration to referrals and beyond. He could apply his full spectrum of attention — and effort — to the whole process and make it work.

Again, per Paul Graham, "startups take off because the founders make them take off."

From niche to mainstream

ConvertKit, as a product and offer, had *just enough* magic to get those early customers to switch. It still took a heroic effort from Barry, but the combination of product, targeting, and hand-held migration gave folks a compelling reason to switch.

ConvertKit didn't stay targeting tiny niches, either, and it's interesting to compare this niche origin story to their mainstream clarity strategy today, which has a much stronger story 1 flavor. They're now positioned as the "creator marketing platform" — a unique category definition with a secular trend to ride (the rise of the 'creator'), providing tools for folks to sell their creative output entirely through the ConvertKit platform.

That is — at least thus far — the mainstream destination they've landed on in their niche-to-mainstream journey.

That's a compelling clarity strategy, and it all started with niche positioning, niche distribution, and tight pattern matching that drove the learning needed to get the flywheel going. The result is a fantastic story and an incredible achievement. (Credit to Barry, too, for generally being a dude with a lot of integrity and doing the right thing by his employees as well, something that doesn't always happen!)

In the next chapter, we're going to keep our focus on market strategy and the relationship between positioning and distribution, but we're going to wind the clock back a quarter of a century from the time Barry discovered sales and rediscover some of the classic lessons — with a few modern twists — from B2B tech market strategy bible *Crossing the Chasm*.

Notes

1. Coolican, D'Arcy. "Product Zeitgeist Fit: A Cheat Code for Spotting and Building the Next Big Thing," *a16z Consumer*, December 9, 2019. **https://a16z.com/2019/12/09/product-zeitgeist-fit/**
2. I'm simplifying the Diffusion of Innovations model here, which we'll revisit in the next chapter when we explore nerds vs. normies and *Crossing the Chasm*.
3. Leslie, Mark and Holloway, Charles A.. "The Sales Learning Curve," *Harvard Business Review*, July–August 2006. **https://hbr.org/2006/07/the-sales-learning-curve**
4. Also see Leslie's 2015 article for Sequoia: **https://www.sequoiacap.com/article/the-sales-learning-curve/**
5. Ibid.
6. "Our Story," ConvertKit Handbook. **https://convertkit.com/handbook/our-story**
7. Barry, Nathan. "My SaaS business hit $1,500 MRR and then got stuck. Until I learned an important trick... Here's the exact method I used to scale to $100k MRR in the next 12 months:" January 16, 2022. **https://twitter.com/nathanbarry/status/1482374302414426116**
8. Ibid.
9. Graham, Paul. "Do Things that Don't Scale," July 2013. **https://paulgraham.com/ds.html**
10. You can see the live chart for yourself thanks to their public revenue data: **https://convertkit.baremetrics.com** (set the date range to 'All time').

40

Crossing the Chasm: Nerds vs. normies; cast & catch vs. segment & sell

[There] is indeed a chasm, and into this chasm many an unwary start-up venture has fallen. Despite repeated instances of the chasm effect, however, high-tech marketing still struggles to get this problem properly in focus.

– *Geoffrey Moore*[1]

In the previous chapter, we looked at ConvertKit, where founder Nathan Barry struggled until he, as he put it, "discovered the key." That key? "Direct sales."[2] Barry's niche positioning & niche distribution (direct sales to bloggers) worked hand-in-hand for ConvertKit to build early momentum, help them climb the sales learning curve and, ultimately, unlock scalable channels, enabling a successful, momentum-driven niche-to-mainstream journey.

Alex Kracov, likewise, built momentum for his early-stage startup Dock by growing a personal network pre-founding, focusing on the sales enablement niche, and learning from the 'eyes light up' reactions to his pitch, as we saw in **chapter 38**. And prior to that, in **chapter 35**, we saw how Qualtrics started by playing *narrower* for years before they had any inkling they could play bigger and transcend the survey category.

These case studies all reflect, to varying degrees, what we discussed in **chapter 37**, where we looked at positioning in the April Dunford mold of zeroing in on your best-fit customers who care a lot about the value you uniquely offer. This is the left-brain-leaning world of positioning by analysis, drilling down to find who loves you and why.

All these approaches are about building early momentum by focusing down the y-axis of left-brain, laser-beam attention with tight positioning. In this chapter, we're going to give this approach some more context and switch from positioning-as-story (i.e., how and who you pitch) to more of a positioning-as-GTM approach around broader market strategy.

The shape of demand

This part of *Clarity*, along with the forthcoming chapters, is particularly focused on *the shape of demand* in your market and how folks commonly shape their positioning to meet that demand.

I'm a strong believer in your positioning being a function of your understanding of the shape of demand that exists in your market. The best clue for where that demand lies is usually found in the customers you've already got — hence the ICP-as-regression and customer expertise ideas we looked at in **chapter 37** — but it's not the only approach.

Eventually, you might want to:

- Make the jump from a nice to the mainstream and 'cross the chasm', which we'll explore in this chapter and the next.
- Go broad from the get-go to reach a mainstream market, which we'll discuss in **chapter 44**.
- Compete in the crowded mid-market to try and peel off a slice of the market, which we'll explore in **chapter 46**.
- Go after enterprise accounts, as we'll see in **chapter 47**.

That is, there's tight positioning to *exploit* (i.e., maximize) the demand you're currently tapping (even if it's just in the fringes of your user base), and there's broader positioning to *explore* demand that might exist in adjacent niches, up or down market, or in mainstream markets. This is the shape of demand we'll be exploring.

> **Clarity strategy and the shape of demand:** You can think of the work we've done on clarity strategy thus far — dialing in your story 1 and/or story 2 pitch in particular — as understanding the shape of demand that exists in your market in terms of attention. This is the true nature of product/market fit in many ways — finding what product (and pitch) will fit the *shape* of the demand of the market. And because that shape will vary by industry, market maturity, up and down market, and so on, product/market fit will always look different. But the skill of *understanding* the shape of the demand that exists in a given market is a muscle you can strengthen, and that's what the heavy lifting over the next few chapters is about.

The classic approach

If you're going to build towards a *new* position in a market, there are two ways you can do it:

- Reactively, responding to whoever shows up, and selling your way to wherever that leads.
- Proactively, by doing some analysis and strategizing about where you

want to go.

The vast majority of folks find themselves in reactive situations, where they're responding to the demand they happen to be seeing, while remaining blissfully unaware of the demand they're not seeing. If that works, great, and for 99% of typical businesses, it does. (That includes most bootstrapped, a.k.a. self-funded startups, too.)

But venture-backed startups are not 99% of normal businesses. The whole point of being a venture-backed startup is to have the ambition for that mega exit, and that requires plotting a path to get there. That is, it requires understanding the shape of demand that will unlock that opportunity.

Some folks think they need to go as big as possible straight out of the gate, and while that does work occasionally — market opportunities come in many different forms — mostly we get a lot of half-baked 'category creation' efforts that are all hype and no substance and fall flat on their face.

But for decades there's been a more sophisticated playbook for high-tech startups that involved starting narrow first, building momentum, and then attempting a very focused attack on a broader market.

This classic approach is called *crossing the chasm*, derived from the book of the same name that we'll explore in this chapter. The core idea is that there's a, well, chasm between your early niche position you might have and the more mainstream market position you probably want, especially if you're venture-backed.

The *Chasm* narrative

Before we get into that, though, I want to make a broader point about what we're really talking about here. In the '90s, *Crossing the Chasm* was seen as *the* way that startup tech markets play out. The *Chasm* playbook both described the shape of demand for you (visionary customers look like *this*, early adopters look like *this*, mainstream pragmatists look like *this*, etc.) and prescribed the solution you should follow (build a leadership position in a niche with a strong reference base, then make a focused attack to establish a beachhead in an adjacent market).

We're going to explore that playbook, but I can't resist touching on *Chasm* as narrative. For one, by simply naming concepts (as we discussed in **chapter 17**), *Chasm's* author, Geoffrey Moore, did the industry a huge service in helping folks identify and proactively address dynamics that may have otherwise destroyed their companies. That is, he gave folks the narrative that helped — and still helps — folks see things and think thoughts they otherwise wouldn't have. The *Chasm* narrative gave founders and investors new categories to think in, and particulars of the approach aside, *that* is a gift. That's ultimately what S-tier clarity strategy does, too: it gives folks a powerful flashlight they can use in an otherwise dark room and (to mix metaphors) it helps folks see around corners and avoid predictable mistakes.

But three decades on, the *Chasm* approach may be *an* approach, but it's not necessarily *the* approach for every startup, in part because a lot of tech is less disruptive than it once was (folks are used to buying SaaS, for example), and therefore adoption isn't such a battle like it once was. Likewise, as we'll see, markets are far bigger — incomprehensibly bigger relative to the '80s and '90s — and that changes the shape of demand again.

Moore himself, as we'll see below, happily admitted things were far more messy than the elegant narrative of *Chasm* suggested, but that doesn't diminish its usefulness in helping us understand how markets work and what the shape of demand can look like.

Therefore, while you might want to position your company up or down market, left or right in adjacent niches, or just *out* into the mainstream, keep in mind that what we're about to discuss is just one way of seeing the shape of demand in a given market. But looking out to see the shape of *any* demand in your market will put you ahead of 99% of folks who are just reactively responding to the leads that knock on their proverbial door on any given day.

Going niche to mainstream

For most startups, just getting traction in their initial niche is difficult enough, and if that's where you're at, focus on solving that problem before you worry about crossing over anywhere else. But once you've got the flywheel spinning, you need to look out at the shape of your market to know not just where you fit now but where you *can* fit — where the gaps are, where the occupied spots are, and where the traps are.

One trap in particular — the chasm that can exist between early adopters and mainstream pragmatists, or *nerds* and *normies* as we'll call them — is what we'll be looking at here. We'll also take a deeper dive into the world of market strategy, returning to tradition, as it were, as we pick out some of the forgotten momentum-building lessons of 30 years of B2B tech market strategy, which also come courtesy of *Crossing the Chasm*. We'll look at what's changed too, especially given that startups tend to start with a 'cast and catch' mental model, not the traditional 'segment and sell' approach of the *Chasm* world, which, in turn, has profound effects on positioning and GTM motions.

Then, in the next chapter, we'll look at a modern, textbook case of the company-killing chasm in action, examining how Pebble, by their own admission, got their positioning, distribution, and narrative horribly wrong when they attempted to cross over to the mainstream.

Internet first vs. business first

To set the scene, let's first look at a weird — and not uncommon — B2B startup phenomenon: not knowing sales exists and having to discover it from scratch. That was true for ConvertKit, as we've seen, as well as Segment, as we saw back in **chapter 8**, and even for the proudly bottoms-up, category-creating Yammer, as we touched on in **chapter 32** (see footnote #11).

That said, plenty of folks in B2B SaaS world obviously do know sales is the name of the game — take Dock and Lattice in **chapter 38** as typical, early and late-stage examples. And that's especially true in enterprise, where sales is everything. (We've looked at how right-brain stories for executives can grease the enterprise sales wheels several times, most recently in **chapter 35** on Qualtrics, for example.)

The strange thing then is that in the world of B2B SaaS, saying "sales is how business gets done" can be both (a) as revelatory as saying the sky is blue and (b) like saying "here's a secret key to success you may have absolutely no idea about" with that being 100% the case.

That's pretty weird! And even for the sales-aware folks, the fact there's decades of wisdom and process to draw on — sales is not just repeating the same pitch every day until you die — can be a revelation.

Why is there such a split in understanding? Why is it a revelation for some and as obvious as water being wet for others? I suspect this split represents something of an evolutionary fork, with the effects of that fork persisting to this day. That fork was the internet itself, and it has created a situation where folks think in:

- **Internet-first terms:** You sell to folks via the internet by default, and sales is a secondary tool to convert buyers who find you.
- **Business-first terms:** You sell to businesses directly, and your internet presence is a secondary tool for information that supports the primary sales effort.

I'm sure you're seeing shades of product-led vs. sales-led approaches here, and there is plenty of overlap there, but that doesn't quite capture it. Instead, this is more about a deeper distinction between the fundamental mental models of how the B2B world works at all, and what you might see as your primary approach to the market.

I suspect the vast majority of folks reading this fall into the internet-first camp, even if they're sales-led, simply because the internet is now the air we breathe in tech, especially SaaS, which is, by definition, software delivered over the internet. Throw in buyers who can do much of their research online without ever talking to a vendor, and you've got an incredibly internet-centric buying culture.

But it wasn't always like this, and for the sales-driven upmarket end of the B2B world, the business-first approach is still the name of the game. That's why we're about to dig into the traditional business technology world for insights we in the internet-first world may have missed out on.

Why does this matter?

While this is an imperfect binary, can you see the impact on positioning, for example, in each kind of B2B startup? Some internet-first folks are going to lean very horizontal, waiting for the entire internet to arrive, while some business-first folks are going to lean very vertical, focusing on a particular challenge a specific industry might have.

But which approach makes the most sense in B2B? Traditionally, it's been the latter, but we all secretly dream that B2C-style internet success is just around the corner, so the temptation to *not* nail down your positioning and ICP is very real, but that's a mistake.

For SaaS startups serious about selling, it's worth instead learning the lessons of the past — and present — and understanding how these two mental models fundamentally shape both positioning and GTM motions.

How is software sold, really?

In the view of your humble author, these two mental models have led to two broad approaches to distribution in SaaS which have profound implications for your positioning:

- **Cast and catch:** Casting a wide net and trying to grab a slice of a broad market.
- **Segment and sell:** Zeroing in on specific niche prospects with a specific burning business pain.

Cast and catch is what everyone tends to do, but segment and sell is the domain of focused, left-brain, story 2 positioning, and that's what we're going to focus on here, shining a light on some forgotten fundamentals. To do that, we're going to have to get a bit zen and empty our minds.

> **Which one best describes you?** What's your mental model for how B2B software is sold in your market? Does it lean more towards the business-first, segment-and-sell approach or the internet-first, cast-and-catch approach? Do you go spearfishing for very specific accounts, or do you hope your broad net catches a bigger quantity of smaller accounts?

Let's get zen

Let's think about the typical internet-first mental model for how software is sold online, particularly from a marketing perspective: search ads, social ads, SEO and blog traffic, social media, online communities, review sites, podcasts, webinars, videos, etc.

Now consider all the ways word-of-mouth happens online: Twitter/X mentions, Facebook groups, Discord channels, community discussions (Reddit and Hacker News), blog posts and comments, LinkedIn posts, etc. I'm sure you can think of many more examples.

Now, imagine none of those things existed.

Imagine the internet itself didn't exist.

What's your mental model for how software is sold now? What's your go-to-market playbook? How do you target a specific B2B market? What does a market for 'high tech' even look like without the internet? No web, no email, no search, no pinging a colleague with a quick text message, no downloads, and definitely no Zoom.

This is a clarifying exercise intended to strip away all the distractions of modern tactics and help us think about the fundamentals of how *communication of innovation* happens. And that communication happens *socially*, with distinct stages of adoption, stages that require distinct strategies to match.

Tech before the internet

It's 1991. *Seinfeld* is in its second season, as is *The Simpsons*. "Everybody Dance Now" is top of the charts. Your humble author is but a child, reading computer magazines on sick days (Macworld), picking shareware games from black and white catalogs, going to annual conventions, and, now and then, on very special days, going to a retail store to pick some boxed software off the shelf — floppy disks and chunky manual rattling around inside.

This is the year Geoffrey Moore would publish his landmark work on tech market strategy, *Crossing the Chasm*. *Chasm*, written in 1990 reflecting on the lessons of the 80s, would introduce both the Tech Adoption Life Cycle (TALC) and the concept of the 'chasm' — the gap between early adopters and the pragmatic mainstream in that adoption cycle — to a tech industry growing at breakneck speed. Those concepts have been updated and expanded upon by Moore in subsequent books, and Moore is now a living legend, still out there giving talks[3] and posting on LinkedIn[4] discussing his ideas.

This is the business-first world of technology — and technology sales — that has become something of a lost art for early startups.

The lost art of segment selling

Chasm has been around for more than three decades, and folks still love it.

It's still the market strategy bible for many VCs (especially the business-first ones), even if the lessons are increasingly lost on today's internet-first founders. It's still popular, and for good reason: it's about getting big and making a fortune selling technology.

Chasm's playbook to get there is all about building momentum in a very deliberate way: going niche to get started, finding those first visionary early customers, getting word-of-mouth going, and positioning your product and company in such a way that you can make that leap from your niche to the early mainstream. Investors obviously want to back big mainstream winners, and before disruptive innovation or category creation were hot concepts, it was *Chasm* that gave folks the playbook to get there.

The trick was understanding that before you attack the mainstream, you've got to generate momentum in your niche.

Visionary customers

Imagine you are selling new B2B tech in the 80s or 90s. How would you get the word out? You'd aim to reach the nerds first — the folks who were reading the enthusiast magazines, going to the conventions, and trying the hot new thing. (They're the ones who try new stuff, after all.)

If you were particularly lucky, you might be able to locate the nerds lurking in the executive ranks — folks spending real money who were game to take a risk and bet on your tech to improve the productivity of their business. Moore calls these folks "visionaries," and it's a nice acknowledgment that it's not just visionary founders that matter, it's visionary early customers, too.

> **Goldcast and selling to a role (and a visionary buyer)**: Remember Drift's CEO David Cancel, whose ingenious market strategy we discussed back in chapter 31? Cancel was the visionary early adopter for B2B event marketing platform Goldcast. Palash Soni, Goldcast's CEO, in a discussion with B2B SaaS veteran, founder, and educator TK Kader,[5] mentioned the "massive bet" Cancel took on their platform, and that's exactly what an early, visionary customer does. Soni also hit on a bunch of the themes we've been discussing, like VC's right-brain obsession and the mismatch with customer's left-brain focus, with Soni saying (lightly paraphrased) "VCs only care about visionary stuff, right? [...] So we only talked about 'the message' and 'category creation' [...] but we were so wrong about it [...] because the customer is buying features ultimately." So they focused much more on story 2-style product marketing instead of the story 1 'message' and were much better off for it. This is very familiar to the *Chasm* playbook,

> finding those visionary customers, selling to a social group (event marketers), and establishing leadership in a niche with a strong, capable product. Hats off to Soni (and Kader and his program) for growing Goldcast with $38M raised and hopefully plenty of success ahead.

Print and people

In the *Chasm* playbook, you'd spend a lot of time building momentum in the early adopter niche before you'd attempt to make the perilous crossing into the mainstream, being equally deliberate about the mainstream niche you'd land in, too. Remember, this is the pre-internet era of print and people. Print isn't especially targeted, sellers are expensive, and print communication is slow (magazines only come out monthly).

Therefore you, as a B2B technology company, need to be very deliberate in how you position your product and deploy your resources as you go to market. You can't just barge your way into a market by hoovering up leads from Google ads or by aggressively targeting folks on LinkedIn. It's a much more fragmented world and you need a niche GTM strategy to match.

Targeted is better

The striking thing about reading *Chasm* (and Moore's later work from that era) today is that it assumes a degree of phased targeting and GTM flexibility that we rarely do anymore. You're going to reach *this group*, then *that group*, then pivot to a horizontal mainstream approach to capture all the demand, then, then, then...

It's like a rhythmic gymnastics routine — there's a fixed sequence of complex steps, with Moore telling you when to throw the ribbon, what leaps and flips to do, and when to catch it again.

Moore developed his own universe of metaphors to describe each stage of market adoption, too. (Metaphors are great for narrative!) There are enthusiasts and visionaries, then the chasm, D-Day and beachheads, then a bowling alley, then a tornado, then main street. Gorillas and chimps make an appearance as market winners and runners-up, as well.[6] It's... a lot, but hey, more power to the dude; he knows his way around a metaphor.

Where does momentum come from?

Compare the traditional *Chasm* approach to the market adoption mental model that more internet-first, cast-and-catch B2B startups have, both for PLG and sales-led motions.

In internet-first PLG:

1. We start doing marketing (a blog, some SEO, and some ads, maybe) to drive folks to our product and hope they try and then buy.
2. We do more of it.

With a sales motion, but still internet-first:

1. We do the above, but with a demo call to action, not a product trial.
2. We use sellers ("sales development representatives" and "account executives") to do outreach, qualify, and sell, turning MQLs (marketing qualified leads) into SQLs (sales qualified leads) and then opportunities, selling to more or less anyone we think might buy.
3. We do more of it.

That is, we cast a big net by putting information (or our product) on the internet, and then we hope if we keep doing more and more of that, then the internet will do its magic and more and more people will buy.

That's HubSpot's classic inbound playbook (we discussed HubSpot in **chapter 31**) and it's been so successful it's how plenty of folks (not all!) assume just about all SaaS gets sold. That's the GTM strategy. And if that works for you, great. If that approach, which obviously may grow in complexity (if not sophistication) as the startup grows, delivers the growth you want, fantastic.

Market thinking before marketing doing

But often there's a missing ingredient: where the early focus and momentum will come from. You need focus to prevent you from getting stuck doing random acts of marketing (a little social, a little SEO, a little email...) and you need momentum to build accelerating, compounding growth.

This requires thinking about the market (i.e., market strategy) before doing activities to reach the market (i.e., marketing and sales itself), which is why positioning exists.

Internet-first logic tends to make us believe that products just magically catch on at some point if you persist with enough 'marketing' (in whatever form), and sometimes that does happen, in a right-time-right-place sense.

But often, founders *make* it happen through a very targeted, very deliberate market and/or sales strategy, as we saw with ConvertKit last chapter, and what's required there is the sequencing, the sophistication, and the targeting that *creates* momentum through compounding returns, rather than just doing 'marketing' and waiting for growth to arrive.

Positioning and 'creating the competition': Moore wrote that "Positioning is the most discussed and least understood component of high-tech marketing,"[7] arguing that the goal of positioning was to "to create a space inside the target customer's head called "best buy for this type of situation" and to attain sole, undisputed occupancy of that space." That's a neat Trout and Ries-esque summary of what positioning can do, but Moore's positioning-statement process has long since been superseded by April Dunford's more regression-oriented approach, as we looked at in chapter 37. That said, one positioning trick Moore had up his sleeve is worth revisiting: creating the competition. Moore wrote that this "involves using two competitors as beacons so that the market can locate your company's unique value proposition." One is the incumbent (or your buyer's default). The second is a disruptive innovator they know. You then position your product as (a) one that solves the same problem as the incumbent (highlighting you're targeting the same budget) but with (b) a disruptive innovation that addresses some core fault with the incumbent. An example Moore gives is Box, which was positioned as a Microsoft Sharepoint replacement on the one hand (the incumbent) but with the ease-of-use of the consumer-oriented Dropbox (the disruptive innovator) on the other. This technique is not for everyone, but if you can triangulate your value using two other competitors — an incumbent and a disruptor — it makes for a fantastic Hemingway story micro pitch (chapter 10).

The two market strategies

Before we get into where that momentum comes from, let's further flesh out these two business-first and internet-first mental models that B2B startups tend to opt for:

- **Narrow segment and sell:** Focusing on specific verticals, for example, with a very targeted outbound motion. You might be hunting for whales that justify the large sales force you put in the field, like the visionary exec to get you started or the pragmatic company that will give you a key reference for other mainstream buyers. Or you might just be trying to find those early users who will put your product through its paces.
- **Wide cast and catch:** The classic internet-first, funnel-based approach, casting a wide net to grab as much as you can at the top of the funnel, be that through inbound marketing (organic social marketing, SEO, and/or broad Google Ads-based approaches, for example), or untargeted outbound, where you attempt to drive *any* slice of the broader mass

market to you. You cast wide, catch what you can, sort what's left by the time you've reeled the net in, and, for startups with a sales motion, offer the pick of the bunch to sales to close.

This has massive implications for positioning and your clarity strategy:

- To sell efficiently with narrow distribution, you have to really lean into your differences in product, positioning, and story.
- To sell efficiently with broad distribution, you have to be adept at serving all comers to make the math work.

They are *incredibly* different goals. Can you see how this impacts your clarity strategy, too? For one, it creates more potential strategy mismatches.

Product × distribution mismatches

Consider the following mismatches:

- **Broad product × narrow distribution:** If you had a broad, horizontal product like Notion (an incredibly internet-first product), for example, and tried to restrict it to a vertical or role, like education (students and teachers), say, then you'd obviously miss a lot of broad market upside. This is rare, though, because it's a nonsensical idea — market pull and team ambition would make it clear that broad distribution was the right play.
- **Narrow product × broad distribution:** If you have a narrow product and throw it to the mainstream, your acquisition costs (as we'll see in **chapter 42**) are going to burn your capital faster than you can say "bridging round." This is far more common, as everyone wants their crappy MVP to take over the world, but realistically you need to focus your scarce resources on doing one thing really well for one specific group of people.

Ideally, instead, you would go for a product and market strategy that matched:

- **Broad product × broad distribution:** If you have The Universal Solution, then sure, start selling it to The Universal Market and hold on tight. You're headed straight for the tornado, as Moore calls it. But very few B2B startups go straight there.
- **Narrow product × narrow distribution:** Instead, most folks realize that trying to be everything to everyone straight away is a fool's errand, and therefore they aim to (or at least try to) start by seeding their particular innovation in a given segment or niche and build out from there, as we've been discussing in recent chapters.

Your GTM preferences

Which way do you lean — segment and sell or cast and catch?

- Do you lean on marketing to drive a certain quantity of leads, and if so, what's your story and positioning in the broader market?
- Do you lean on sales to go niche, and if so, what story and positioning are they taking to that niche?
- Do you expect sales to serve broad leads, and if so, can they qualify based on your niche, or are they expected to serve all comers?

This is the art of clarity strategy as it relates to your distribution. There's no right or wrong, but there are more and less plausible ways to build momentum based on your stage. The more fundamental point is simply that the type of distribution you use — broad or narrow — has massive implications for your positioning, product, and ultimately your narrative, because the demand you see will vary greatly on the distribution approach you use.

The positioning question then is whether you meet that demand or choose new demand to serve. The next question for you, then, dear reader, is: what's feasible? Because that ultimately will determine who you can reach, the demand you see, how you position, and therefore what you say in your clarity strategy. That is, to achieve 'fit,' you need to let the shape of demand you serve shape you, too.

That's a lot of ducks to get in a row, and it doesn't stop there.

Nerds vs. normies: two theories of success

We've discussed the internet-first cast-and-catch approach and the more traditional, business-first segment-and-sell approach to going to market. But when you do go to market, what kind of customer do you find?

We've talked a lot about buyers with left- or right-brain attention and the need for your clarity strategy to account for both. Now, we're going to introduce another dimension: appetite for risk.

When we kicked off this deeper, left-brain look at positioning in **chapter 37**, we discussed how drilling down to discover your *theory of customer success* was vital for your positioning. To return to *Crossing the Chasm*, in the *Chasm* world there are (to simplify greatly) essentially two different contexts for customer success in the early stages, one for nerds and one for normies:

- **Nerds:** Innovators and early adopters *welcome* the new, risky, somewhat unproven technology because it gives them an edge. That's what success means to them.
- **Normies:** Mainstream folks *avoid* the risky and new and aim to throw

their lot in with the winner once the winner has become clear, so they can take advantage of the ecosystem and other benefits that accrue to the winning player. That's their very different definition of success.

The gap between these two groups is the chasm.

The chasm exists between early adopters and the early mainstream users when selling a breakthrough, *discontinuous* innovation. On one side, there are the nerds who will take the raw, unpolished version of your product for the cutting-edge productivity boost it will give them. On the other side, there are the much more cautious pragmatic normies who will wait for a 'whole product' solution to appear from the emerging segment winner.

Some innovations cross the chasm from nerds to normies with ease; others stay in early-adopter land forever; and some innovations — and the companies behind them — fall into the pit of snakes that is the chasm, never to return.

Smartphones, for example, did a triple backflip over the chasm and into the mainstream (but were really PDAs for nerds before that); VR is perpetually an early adopter innovation no matter what Meta does (strapping on a bulky face computer will never not be dorky, though we'll see if Apple can cross the chasm with its Vision Pro headset); and Pebble, as we'll see in the next chapter, fell right into the chasm.

Back in the days of *Chasm* — the '80s and '90s — being an early adopter was a big deal. Technology was more foreign, more unwieldy, less polished, more expensive, and just harder and slower to adopt and roll out. But it could be massively transformative compared to the status quo at the time, so the bet was often worth it, if you had the stomach for it.

The folks placing those bets were the true nerds (or 'visionaries'), and the gulf between them and the mainstream could be large indeed. Those differences still exist today, but the cast-and-catch world can make them difficult to spot.

Finding HR early adopters: To pick a corporate B2B example, imagine you're selling into HR. If you're fielding cast-and-catch inbound sales inquiries, how can you spot the HR early-adopter nerd who's into Agile HR (it's a thing!) from the normie 9-5er? Usually there's a bit of a secret handshake that happens in the call, but it's not like folks rock up to demos with NERD or NORMIE tattooed on their forehead. This is why you need to be clear on where leverage is going to come from — are you trying to build long-term momentum by focusing on a niche or are you trying to serve all comers? Those are two very different games.

Peer references

Nevertheless, how you position your offer and pitch can change significantly depending on whether you're aiming for the nerds or the normies. Here's a tweet from a DTC ecommerce founder that neatly illustrates the pragmatist (normie) vs. early adopter (nerd) divide in B2B:

> *If you're a SaaS platform and you want to sell me your software - the ONLY way you can do so is by showing me how other DTC founders are using it to win Nobody cares about whitepapers, PDF's, and Webinars. All we care about is how we can win as big as other peers*

The whitepapers, one-pagers, and other tech-specific things were, in this case, for the nerds and early adopters — folks interested in the technology itself.

For normie buyers[8], especially in this case with a *very* tight ROI criteria, they don't care about any technical information or demonstration you could give them, they *only* care about peer references. And fair enough — they're watching every dollar and don't want to speculate on something that *might* offer an advantage; they want to buy a sure thing that will help them 'win big.'

This is the classic normie mindset, and it demonstrates why building peer references is so important and why it is, as we'll see, the crux of the *Chasm* strategy.

> **What's the point of a demo?** What do nerds and normies want out of a demo? And what is a demo for, anyway? A demonstration of your *technology* implies that the technology itself is where the value is — that someone will be so wowed by what they see that the next obvious step is to buy, shut-up-and-take-my-money style. That may well be true, especially if you're selling to visionaries and early adopters who care about that specifically. But is it true for pragmatist customers looking for a whole solution to meet their needs? Paul Wiefels, a collaborator of Moore's, made this point in his book *The Chasm Companion*, where he bemoaned the prevalence of "sell the demo" sales behavior *in 2002* (!) and suggested instead a "highly consultative [approach] based on an understanding of the customer's needs—overt and latent" that is "validated with the customer" before moving on to the inevitable demo. Today, folks often want a demo because they've done their own research and are already at the end of their buying process.[9] Either way, it's worth considering what the actual point of your demo is and what a nerd vs. normie demo process might look like in your organization, especially in terms of raw tech vs. outcomes and

> references.

The chasm vs. radars and laser beams

Let's bring our hierarchy of attention to the challenge of the chasm. If we think about radars and laser beams, what's interesting here is that different customer segments *in the same market* can have their radars calibrated for opposite things. That is:

- **Nerds want an edge:** Nerds have their mental radars calibrated to look for the success and the early adopter advantage that *comes from* the risk of the new. That might mean a short trip on the x-axis — *this cool new thing exists!* — before a long trip down the y-axis — *and here are the very specific technical things it can do.*
- **Normies want a sure thing:** Normies have their mental radars calibrated to look for the success that comes from jumping on board with the low-risk, established winner. That might require a much more comprehensive x-axis, right-brain, reference-driven story — *here's a complete solution that many other people have had success with!* — before heading down the y-axis — *and here's why it's an easy, safe bet.*

You can see how this has serious implications for both your clarity strategy (from your pitch to your positioning) on the one hand and your broader GTM strategy on the other.

On the clarity strategy side, are you pitching new (risk) or completeness (low risk), for example? Are you pitching your 10x product wedge that makes up for all its other deficiencies, or are you pitching your mature 'whole product' that has accrued the network effects of success with a healthy ecosystem and community around it?

And on the GTM strategy side, can you zero in with narrow distribution on nerds alone, or are you casting a wide net and picking up a random mix of normies too? For example:

- **Segment and sell** folks try and focus on one group or the other — they're trying to *prove* their case with the early adopters and then *demonstrate* their success with mainstream folks.
- **Catch and cast** folks may have a mix of folks showing up and are at the mercy, to a degree, of who happens to find them.

Either way, this can create confusion for early-stage startups. Perhaps the startup gets very positive signals initially — maybe their would-be champion in the organization is a visionary early adopter — only to have their hopes dashed by the normie department head that says you're too new, too unproven, too small, too whatever.

Or maybe the CEO or VP is the ambitious, visionary nerd who wants to try the new thing for the potential organizational benefit, but to their normie team, it just screams risk and painful change, and they torpedo the idea.

Navigating these social dynamics and understanding the shape of the demand you're seeing requires a degree of flexibility in how you sell and the story you tell, a degree of flexibility that's not always available when folks are stuck with the same old SaaS cast-and-catch, hello-to-demo GTM playbook with minimal awareness of how segment-and-sell works.

Creating momentum

Chasm, along with Moore's later work, offered a very specific niche-by-niche playbook for jumping from the nerds to the normies. But as Moore himself found out, writing in the foreword to 2002's *The Chasm Companion*, "far from being as simple and elegant as specified [in *Crossing the Chasm*], things were in actual fact a mess."[10] So he wrote more books to address the "mess," however, he only found more mess:

> *Once again, the reality and exigencies of execution began to teach and inform the idealized forms of strategy, leading to more messes and a couple of more books, and eventually to the realization that this is all a dance, and that the job of the strategy consultant, at the end of the day, is simply to keep dancing.*

I appreciate Moore's candidness here, and it's a good reminder that *Chasm* wasn't set in stone in 2002, much less in the 2020s and beyond. Strategy ideals can only ever be just that, and the quest for a universal theory of market adoption will always be something of a white whale, simply because business opportunity is often emergent and the shape of demand tech startups see can vary wildly.

Nevertheless, there is one clear universal theme that comes through *Chasm* and Moore's subsequent work, and that's the mechanics for building and capitalizing on momentum in a given market. For example:

- Focusing on early adopters helps you build initial momentum.
- To cross the chasm to the mainstream, in Moore's playbook, you need to build a very deliberate kind of momentum, with a normie-friendly 'whole product' or solution (think resources, support, integration, an ecosystem, community, etc.) and references to match.
- You keep building momentum by targeting adjacent niches.
- And then *(in Soviet Russia...)* momentum starts driving you — you have more demand than you can handle, and you try and capture as much of it

as you can.

This is a kind of spring-and-release strategy, or to use Moore's analogy, like getting a fire going.

> **Does it always work out this way?** B2B startups can of course succeed without reaching some kind of market-pull nirvana.[11] And on the chasm-crossing side, "early adopter" niches and verticals VCs assumed were small markets can, in fact, turn out to be huge multi-billion dollar opportunities in their own right, precluding any need to pivot to the mainstream, as MongoDB and pharma-focused Veeva have demonstrated.[12]

Momentum is *built*

By definition, fast-growing startups generate momentum from somewhere; that's how they grow. That might be through an internet-first product breakthrough with bottoms-up growth (Notion, Slack, Figma, etc.), capitalizing on an emerging secular trend (Gainsight et al., as we discussed in **chapter 30**), or going niche-to-mainstream (like ConvertKit, Qualtrics, and most sales-led startups), assuming your niche doesn't turn out to be a billion-dollar opportunity in its own right.

And clarity strategy has its role to play in building momentum too. That might be:

- **Internal momentum:** Your clarity strategy might be about that one idea that unifies the company so you can build momentum in one clear direction.
- **Market momentum:** Perhaps your narrative is about a broad change narrative for the market — or enterprise deals — so you can build momentum in the broader market and/or in deals.
- **Niche momentum:** Or, perhaps you want that specific, narrow-focused story for a niche so they can build momentum and get the flywheel going in the first place.

Either way, momentum drives growth. But momentum is a pretty abstract idea. What does it mean in practice? In the segment-and-sell world, it means picking social groups you can sell to. Which social groups can you reach?

> **How Brex built momentum with startup founders:** VC and former CRO Sam Blond has some fantastic in-the-trenches stories about

running effective sales programs. His Twitter thread[13] about launching Brex — a corporate card for startups — by targeting founders in San Francisco through marketing and sales is a classic example, especially in terms of showing what's possible when you think creatively and smash the tired standard GTM playbook. Brex's launch involved billboards, bottles of champagne with "handwritten" notes sent to founders (with CEO follow-up via email [14]), podcast sponsorship, and a very visible event presence (magic shows, burritos, and a Brex credit card), all focused in one concentrated, highly targeted way to build that initial momentum.

Cycles and social groups

The core idea of *Chasm* was, well, the chasm — the gap between nerds and normies in market adoption.

But Moore's work was essentially him riffing on the classic *Diffusion of Innovations*[15] curve (popularized by Everett Rogers back in 1962 in his book of the same name), which he dubbed the Tech Adoption Life Cycle. Here's the classic illustration of the adoption curve with Moore's chasm[16]:

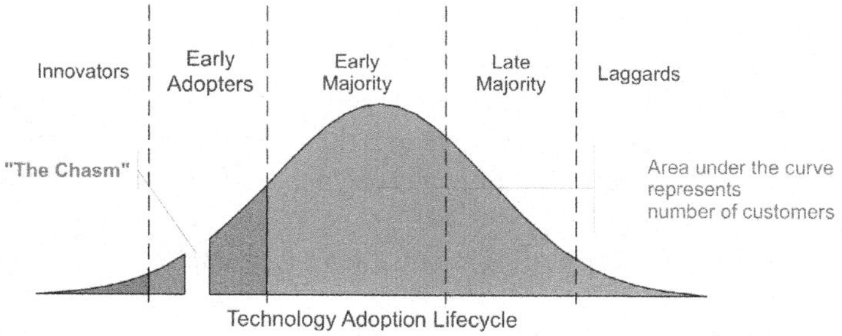

Figure 40.1: Moore's Technology Adoption Life Cycle via Wikipedia.

This is generally understood as how innovations that require new behavior are adopted in the market. Moore was ostensibly fascinated by the behavior and differences of each of these segments for tech markets, and that was the basis of his books.

But Moore picked up on the deeper insight from the Diffusion of Innovations and that's that *momentum comes from social groups who communicate with each other.*

This may sound either abstract or obvious, but when we talk about niche positioning and finding best-fit customers, this is the crux of what we're trying to do. We're trying to connect and sell to like-for-like folks (nerds in the early stages, normies later) so we build momentum that's bigger than our sales and marketing efforts alone.

We touched on the importance of peer references earlier, for example, and that highlights the need for *self-referencing groups* you can sell to. DTC ecommerce founders want to hear about the success of other DTC ecommerce founders. Developers want to hear about — and from — other developers, so Dev Rel is a thing. Even startups themselves are often their own community, especially around the San Francisco Bay Area, so many startups start selling to other startups, for example, because that's the early adopter social group they can reach.

All of this is, in a sense, facilitated word-of-mouth.

In traditional B2B tech market strategy, as *Crossing the Chasm* describes, this is the whole playbook — finding self-referencing groups you can sell to. You were simply dead in the water otherwise. Your only path to survival was to light a word-of-mouth fire in the social groups you were targeting.

Light a fire

> *Trying to cross the chasm without taking a niche market approach is like trying to light a fire without kindling.*
>
> – Geoffrey Moore[17]

The chasm exists between nerds and normies. You want to cross the chasm and get to the normies because (in theory) that's where the bulk of the market and money is. To make that jump, you have to build momentum with your early adopters — that's where the fire starts. To Moore, "enthusiasts are like kindling." They get the fire going, and it's your job to help that fire catch on in the broader market.

When we discuss the way niche positioning unlocks niche distribution (and vice versa), this is why: we're trying to light that fire in a given market. For Moore, the very definition of a market *is one that can sustain a word-of-mouth fire*.

This is fundamental to your growth prospects: the hotter you get that fire, the more momentum you'll build, the faster you'll grow (assuming you can cross the chasm).

The key to these early markets is, Moore says, customers "who reference each other when making a buying decision." Whether that's enthusiast nerds pre-chasm or pragmatic normies post-chasm, the point is that these people can reference each other. These are the discrete social groups you segment and sell to.

Chasm assumes no company could afford every touch for every customer

— you *needed* word of mouth to do much of the work for you. You *must* get the fire going. Therefore, you really do need to spend a lot of concentrated time in one narrow market rubbing two sticks together, otherwise you're SOL.

That's the classic, business-first, segment-and-sell approach. Remember, this predates the internet. There's no Google, no web, no online communities.[18] In the age of the internet, with search, social media, ad targeting, and digital products (SaaS, specifically), things are obviously a *tad* different, and getting a fire going will look different, too.

But you still need some way to get the fire going. Clarity strategy, in that sense, exists in part to give you the story to discuss that's not your product with social groups who reference each other, perhaps folks in a specific role or industry.

But which subset of those folks will you reach? Which one will you speak to — nerds or normies?

Perhaps your answer is "both." Maybe you'll have no choice but to try and reach both — balancing two different stories for two different types of radar attention (new and edgy vs. safe and proven).

This creates a dilemma of discipline.

> **Dock's horizontal positioning mistake:** Previously, in chapter 38, we looked at Dock's niche positioning and founder-led sales approach to get their growth flywheel going. We also looked at what the founder described as one of their early mistakes: going too broad and too horizontal in an attempt to expand beyond their initial niche of revenue teams. Beyond the lack of product or positioning clarity this created, we can see that trying to light a word-of-mouth fire for disparate social groups like sellers, recruiters, and, say, designers is *incredibly* hard. That's why early-stage startup advice is so often focused on sticking with a niche — lighting one fire is hard enough, lighting several in groups that don't otherwise reference each other is *exceedingly* difficult.

Strategic discipline

For Moore, in the business-first era, early on, you are either:

1. A disciplined, market-driven (that's 'market' as in market *strategy*, not 'market' as in market*ing*) company that knows *and executes* on a niche-to-mainstream strategy (using sales).
2. An undisciplined "sales-driven" company and the truth is that you "do not have, nor are [you] willing to adopt, any discipline that would ever require [you] to stop pursuing any sale at any time for any reason."[19] You know

about niche strategies, of course, but you think you just don't have the time — if someone's buying, you're selling.

Note that a sales motion is assumed in both "market-driven" and "sales-driven" companies; the question is whether you have the discipline to sell *only* to your niche and say no to everyone else or whether you'll sell willy-nilly to anyone and everyone, no matter the fit.

Moore paints the undisciplined "sales-driven" approach in a very dim light, and you can understand why. In the business-first, pre-internet, print-and-people era, if you pursued sales with folks who didn't match your ICP and they had a not-amazing experience, then you couldn't use them as references for the next sale, *and sales-by-references is how you get into the mainstream market.* Everything hinges on those references. (See our DTC ecommerce buyer quoted above, for example.) That is your GTM approach *in its entirety.* That's how the word-of-mouth fire gets lit and deals get done. It's how you establish leadership and start building an ecosystem. And if you don't have references from overjoyed customers, then what have you got? A random bunch of semi-satisfied customers pulling you in different directions and a slow sales trajectory based on a random walk through the market.

That's... not ideal.

The more things change

But, note the difference in eras. Today, it's more a case of balancing tradeoffs. If you sell opportunistically to folks who come via SEO, for example, you might unlock a well of demand you were otherwise unaware of... or it may distract you from building those core references in your chosen niches. It's a classic explore/exploit tradeoff, and in our modern cast-and-catch world, it's highly likely you'll see a broad cross-section of a given market eventually (given folks can seek you out and throw money at you despite being poor-fit customers) so that discipline and strategy is extra important if you're running the niche playbook.

Nevertheless, what does this approach to lighting a fire mean in an internet-first, cast-and-catch world?

First, the challenge from Moore remains: can you light a fire? Can you grow in a compounding way, a way that's faster than your 1:1 effort-to-outcome or current CAC to LTV, and improve your compounding coefficient (if I can put it that way[20])?

This is where the more things change, the more they stay the same. The tactics and landscape may be vastly different, but that need to light a fire in your B2B market — a world where B2C growth techniques don't always apply — still very much exists.

We get so caught up in the world of tactics these days that sometimes we forget to ask whether our combined actions are working together to light a fire in a given market — getting the kindling going with early adopters and compounding from there.

And in terms of things staying the same, it may be a revelation to some folks that segment-and-sell is still a thing. Sales exists! Fortunes are still made on the back of it! It's still how a lot of business gets done!

> **Selling vs. spamming:** Let's be clear that segment and sell isn't mindlessly spamming every executive in a given industry. For example, Calendly CMO Jessica Gilmartin described on LinkedIn how she receives, on average, "100 emails, 40 calls, and 50 LinkedIn connection requests from salespeople" *every week*.[21] Gilmartin is a special case in that she's incredibly empathetic to sellers — she was one! — but now she doesn't respond to any sales outreach and had to change her phone settings because she was so sick of cold calls. The kicker, as she describes in her post, is that she's not even the right person to talk to. She allocates budget, but she doesn't direct how it's spent. That happens lower down the org chat, and she suggests sellers start there instead. This is also why it's not the case that cast and catch always means inbound (organic or ads) and segment and sell always means outbound (email, call, or LinkedIn message). There's a lot of lazy outbound that's essentially cast-and-catch spamming, and there's a lot of opportunity for targeted segment and selling that uses more inbound or organic means. But it starts with knowing who your buyer actually is.

Positioning vs. ICP confusion

At the core of all this is your positioning, and your positioning rests on your "Ideal Customer Profile" or ICP. These days, we talk a lot about ICPs. What's your ICP? Have you nailed your ICP? Does your positioning reflect your ICP? Do you have multiple ICPs?

Analytical, left-brain positioning exercises (like April Dunford's that we touched on in **chapter 37**, for example) are founded on the idea that you can isolate your best-fit customers, i.e., your ICP, and position around them specifically. This is the discrete group you can sell to or, perhaps, you're already selling to, but they're getting lost in the noise.

This is where the positioning-as-GTM rubber hits the road.

Will you have the discipline to be "market-driven" and sell in a targeted way, or will you be purely "sales-driven" (in the *Chasm* sense) and opportunistically sell to all comers, whether they're an ICP match or not?

That is, even if you do a positioning exercise to work out your ICP, will you retool your GTM approach to focus on *just* that ICP, potentially switching from cast and catch to something more akin to segment and sell?

This issue of matching your positioning to your distribution is a fundamental one for all startups. Everyone knows having a clear ICP is something you 'should' do, and back in 1991, everyone 'knew' having a niche strategy was the best way to go to market, but as Moore says, there's a big difference between knowing what you should do and actually doing it.

What makes that so hard? I suspect that the thought of cutting off potential avenues for revenue when you fundamentally have a cast-and-catch mental model can be very difficult to do and feel very unnatural.

What also makes it hard is that this advice is very stage dependent. This is about early-stage B2B high-tech startups. Sometimes, you *should* go broad, however, and sell opportunistically if that's what the shape of the demand you're targeting looks like. This is where a little strategy — and knowing your modes of attention — can go a long way.

A little strategy goes a long way

We've talked a lot about matching your message with your (or your audience's) preferred mode of attention in previous chapters. Left-brain story 2 pitches for left-brain audiences, and so on.

Here, it's not that segment and sell is necessarily right in every instance and cast and catch is wrong. Ultimately, through his books, Moore would go on to advocate a kind of dynamic positioning that, as we'll see in **chapter 44**, flipped from *reach* to *niche* and back again as you moved through the stages of adoption. In Moore's world, you need to cast a net to find an initial niche to mine; you need to get established in that niche, and then you might branch out again, tackling an adjacent niche or perhaps getting pulled into a more mainstream position by word-of-mouth.

The point here isn't that there's an exact playbook you need to follow; it's that having some basic strategic GTM awareness and understanding the shape of the demand you're targeting can go a long way so you can create your *own* framework and playbook that matches your own unique circumstances, especially if you're the founder and you're the one responsible for knowing when you need to pivot your GTM motion.

To start with, being more strategic simply might mean taking a more market-driven approach to sales, for example, where you use that one super-successful customer in one industry you landed as a reference for the nearest adjacent peer (or peers) in the same industry. That's focusing on the clear demand you can see. It doesn't have to be a huge shift in approach, it just might be a case of finding your existing best buyers and mapping out adjacent peer companies in an opportunity tree. Pick a branch of that tree, do some recon, see what the real frustrations are, and then see where you can put your existing references to work. (This is, I suspect, what most folks have been trying to achieve with various 'Account Based Marketing' approaches.)

That's the more left-brain approach, where a bit of analysis and strategy can give you a *far* higher chance of building momentum and getting to a meaningful exit. Tackle the obvious shape of demand you see before speculating about what might lay beyond it.

In a right-brain sense, however, what matters is having a greater sense of the whole — your whole journey, knowing where you are in terms of stage, and knowing what positioning and GTM options are available to you, which we'll continue to explore in the chapters ahead.

> **Generating broad sales *and* broad word-of-mouth:** Unlike the print-and-people B2B era of *Chasm*, these days B2C tactics *can* work in B2C-ish business markets. That is, your GTM doesn't *have* to be built around this high-tech-for-big-bosses, niche-first perspective, as we'll begin to explore in chapter 43 when we look at the B2B spectrum. If your marketing game is strong enough, you can both advertise broadly and get a broad word-of-mouth growth machine going at the same time. Monday.com and ClickUp, for example, used B2C tactics like social media advertising to drive SMB B2B growth with great success, and that generates its own broad word-of-mouth feedback loops, too. This, however, requires a broad distribution × broad positioning approach (not to mention mountains of capital), and we'll explore that further when we discuss Squarespace in chapter 44.

Category concepts in social groups

We've been focused on left-brain, story 2, drill-down positioning here, but Moore's observations — and Rogers before him — about how social groups communicate about new innovations apply to story 1 concepts, too.

Think about successful so-called "category creation" attempts, especially in marketing, like HubSpot's inbound marketing, Drift's conversational marketing, or even just Goldcast's event marketing focus we touched on above. What's unique about them? They targeted digital marketers — a highly online, internet-first, idea-spreading, meme-able group who will reference each other, with high potential for self-sustaining ideas.

Highly connected groups with budgets will *always* make attractive sales targets, which is probably why there will soon be more martech tools than there are marketers in the industry. (The same is true about connected groups like developers, for example, and ideas like NoSQL[22], or even Agile, for that matter, too. And let's not forget Gainsight and customer success.)

You don't *have* to sell to a particular role — if you can sell to whole organizations either bottoms-up through mass PLG, or top-down through executives, great. But for business-to-business tools, specific business roles will always be obvious distinct groups that you can sell to. Likewise, if you don't sell to a highly connected group, you're going to have a tough time getting your narrative and product to catch on.

Compare those successful approaches with vague attempts at new concepts or categories, one of which I came across was "conversational commerce." This was a riff on Drift's "conversational marketing" concept, but it was more of an amorphous trend term. Drift targeted marketers — a self-referencing group of buyers if ever there was one — with a clear, novel from/to, a chat-not-forms Hemingway story, and the idea took off *and led back to Drift*.

Ideas like "conversational commerce," for example, lack all those elements — there's no clear group of buyers who reference each other (is it for shoppers? Store owners? Store marketers?), and no clear from/to, apart from considering chat in its various forms.

Likewise, consider something like Mural's since-departed "Collaboration Intelligence" category concept we touched on in **chapter 35**. 'Collaborators' aren't a coherent group who can spread a meme; they're just any white-collar worker anywhere, really. There's no self-referencing group there (and there's no real secular change to drive it, either), and that's often true for other 'X Intelligence' category concepts, too.

Does that make those concepts worthless? Not necessarily — they might be helpful for closing deals or getting your foot in the door with a C-suite prospect. Or, perhaps they give you some direction in terms of the 'whole product' playbook you're building that mainstream normies want. That's great, but there was little plausible reason why something like "Collaboration Intelligence" would catch on in a market-adoption sense, because how can it? What social group will adopt and spread it? What does it even mean?

The point here isn't to ding companies like Mural; it's simply to state that when we talk about the shape of demand, we're often talking about the *social shape of markets*, and those social dynamics are integral to how new technology — and new concepts and narratives — are adopted, from nerds to normies. This has been known for decades, but it's often neglected when folks try and work out their GTM and clarity strategy.

Therefore, when you're thinking about the shape of demand you serve when targeting a niche, consider starting with what's working and then think about the shape of the social groups your best-fit customers — especially the early adopter nerds — represent. How will you reach them: through cast and catch or segment and sell?

Next, we'll see a canonical example of the chasm between the classic nerds and normies social groups when we look at Pebble and how they failed to make the jump from their initial nerdy, hacker, early adopter crowd to a mainstream fitness-oriented normie one. Get your grimace emoji ready; you're gonna need it.

Notes

1. Moore, Geoffrey A.. *Crossing the Chasm, 3rd Edition: Marketing and Selling Disruptive Products to Mainstream Customers.*
2. Barry, Nathan. January 16, 2022. **https://twitter.com/nathanbarry/status/1482374304704565251**
3. There are plenty on YouTube if you want to get the gist of his ideas during a lunch break.
4. He has ~650k followers on LinkedIn: **https://www.linkedin.com/in/geoffreyamoore/**
5. TK Kader interviewed the Goldcast founders as part of his *SaaS & Scotch* series. "Go-To-Market Strategy: Raised $40M, 100+ Employees, 150+ Customers (Goldcast)," January 6, 2023. **https://youtu.be/q74axHUARrs**
6. This comes from Moore's other '90s books, as summed up in 2002's *The Chasm Companion. Play Bigger* would resurrect the same dubious winner-takes-all idea of market 'gorillas' with "category kings" more than a decade later.
7. Moore, Geoffrey A.. *Crossing the Chasm, 3rd Edition: Marketing and Selling Disruptive Products to Mainstream Customers.*
8. Normie buyers in one area might be early adopters in another — we're talking about specific buyer behavior and not overall personality here.
9. We discussed this back in **chapter 21** when we looked at how small the window of opportunity was becoming for B2B sellers. In a 2022 blog post (**https://chasmgroup.com/from-on-prem-to-saas-how-has-enterprise-sales-changed/**), The Chasm Group also makes similar observations in their potted history of enterprise tech sales ("From On-Prem to SaaS – How Has Enterprise Sales Changed?"), writing that "one key change [...] is in *how* enterprise customers choose to buy these days. Overall, when they engage with vendors they are already much further along in their selection process than in the old days. The pervasive availability of online vendor/product information, analysis, and reviews has significantly compressed the initial evaluation phase of every customer's decision process." There's also a great quote from sales leader Rich Nieset, writing that "There is a tremendous pressure to go from hello to demo almost immediately and SaaS makes committing this sin much easier," and goes on to bemoan a lot of pointless demoing showing too much functionality to the wrong stakeholders who then get turned off further sales engagement, all of which tends to be perpetuated by younger founders with little sales experience. That sounds grim, but the article goes on to say that, for bigger

businesses, the need to "match the product to the prospect's business problem" is the same as it ever was. And in nothing else, the low bar gives you, dear reader, plenty of room to stand out.

10. Wiefels, Paul. *The Chasm Companion: A Fieldbook to Crossing the Chasm and Inside the Tornado.* HarperCollins.

11. Martin Casado, in a blog post for *a16z Enterprise* ("The Unending Chasm, and How to Survive It"), writes that "Sometimes even very large companies spend their entire lifecycle gutting out a push-based sale cycle all the way through IPO and beyond. Relatively young companies have been forced to introduce multiple products to navigate market shifts; others pivot into service companies; and still others go through low-margin, long, and complex sales cycles for years... and yet, can still be successful. Even in my own experience, we had to push hard into the market up to half a billion dollars in revenue and beyond" and offers advice for different market scenarios: **https://a16z.com/the-unending-chasm-and-how-to-survive-it/**

12. That's according to VC Jeff Bussgang, in a piece on Medium titled "After 30 years, Crossing The Chasm Is Due for a Refresh — Why Markets Are Larger Than They Appear." Bussgang writes that, "Market sizes have proven to be much, much larger than any of us had ever dreamed. [...] A glaring example in our investment portfolio of market sizes is database software company MongoDB. [...] [W]e boldly predicted the company had the opportunity [...] successfully take a piece of a market that could grow as large as $8 billion in annual revenue in future years. Today, we now realize that the company's product appeals to the vast majority of the market, one that is forecast to be $68 billion in 2020 and growing to approximately $106 billion in 2024." The same was true, Bussgang wrote, for pharma-focused Veeva, a pioneer of 'vertical SaaS,' writing that "[The] founders concluded they could at least achieve a few hundred million in revenue on the backs of pharma and then expand, as the company's initial name indicated, to other vertical industries from there. Boy, were we wrong! The company filed their S1 after that fifth year showing $130 million in revenue and today the company is projected to hit a $2 billion in revenue run rate next year, all while still remaining focused on just the pharma industry."

13. Blond, Sam. "1/ How Brex went from an unknown company to a household name among startups (our target market) in a matter of weeks [...]," January 31, 2023. **https://twitter.com/samdblond/status/1620123634353451008**

14. Blond, Sam. "1/ The best outbound campaign we ever ran at Brex (75% demo rate, 75% demo to close):[...]," January 10, 2023. **https://twitter.com/samdblond/status/1612518983726424080**

15. Rogers, for his part, would write in the final edition of his book that "Past research shows no support for this claim of a chasm between certain adopter categories," which is a bit of an academic burn. But in the tech world, when you see the chasm in practice, it's hard to unsee it, and I'm surprised Rogers didn't investigate further.

16. Image by Craig Chelius, CC BY 3.0. **https://en.wikipedia.org/wiki/Crossing_the_Chasm**.

17. Moore, Geoffrey A.. *Crossing the Chasm, 3rd Edition: Marketing and Selling Disruptive Products to Mainstream Customers.*

18. I'll note for the nerds that the internet did indeed exist at this point, with email being popularised in the late 80s with the release of Eudora in 1988 and Lotus Notes in 1989,

the same year Sir Tim Berners-Lee would float his initial idea for what would become the web.

19. Moore, Geoffrey A.. *Crossing the Chasm, 3rd Edition: Marketing and Selling Disruptive Products to Mainstream Customers.*

20. In terms of virality, this would be your "K-factor" (**https://en.wikipedia.org/wiki/K-factor_(marketing)**), but we're talking about old-fashioned word-of-mouth here, or even just sales efficiency through references rather than viral loops.

21. Gilmartin, Jessica. May, 2023. **https://www.linkedin.com/posts/jessicagilmartin_prospecting-sales-leadership-activity-7066759701056675840-5iqD/?trk=public_profile_like_view**

22. For better or, perhaps, worse, as Nemil Dalal describes in his July 6, 2017, blog post "Why Did So Many Startups Choose MongoDB?" **https://nemil.com/2017/07/06/why-did-so-many-startups-choose-mongodb/**

41

Pebble's failure to cross the chasm

Yes, in the end we failed. We failed to build a long term successful and profitable company. We certainly had successes on the way (the company was over 8.5 years old when we sold!) but the end result was definitive. We ran out of money and sold what we could to preserve the product experience for the millions of people out there with Pebbles.

– Eric Migicovsky[1]

Let's recap what we looked at in the last chapter in terms of who you're trying to reach and how you're reaching them before we dive into the story of Pebble. We covered:

- **Segment and sell vs. cast and catch:** Have you gone niche? If not, can you reach niche, best-fit prospects directly? Or are you trying to let them come to you, casting a wide net and gathering them up that way? What's the coherent positioning × distribution play for your specific product and market?
- **Nerds vs. normies:** Are you selling to nerds or normies? Are you trying to build momentum in one group or the other? Have you observed any chasm between the two? Are you targeting visionary early adopters to get the flywheel going?
- **Self-referencing groups vs. discrete buyers:** Are you targeting highly online, internet-first groups (as others have, targeting marketers, developers, CS folks, etc.) who perhaps can be pulled together offline (as Gainsight did with their conferences)? Or are you just aiming for a broad, well-articulated story 1, perhaps with an enterprise playbook, to close deals? Either approach is fine, just understand the momentum you're developing relative to the stage you're at.
- **Tech vs. concept early adopters:** The adoption cycle applies to innovative ideas, not just innovative technology, and in clarity strategy, that's what we're often selling — the concept, not just the tool. Understand how much behavior change your concept requires, what that means for crossing the chasm for mainstream adoption, and what kind of social momentum you need to generate to get there.

Know thy game

Let's recall the core idea in *Crossing the Chasm* — the, well, chasm. We've got nerds on the one side and pragmatist normies on the other (folks who want a 'whole product' solution from a clear winner).

If you intend to attack the mainstream market, then it helps to understand the typical phases of market adoption and ensure you're playing a game you can win — converting early adopters and not banging your head against the wall of the conservative, pragmatist mainstream with something that's too new for them. Just knowing there can be a chasm between these two groups is half the battle. Crossing that chasm remains a challenge because it's not necessarily obvious you're going to hit it. Even if it does appear, it's just a little iceberg that surely won't sink our titanic-sized dreams, will it?

In theory, the bigger the behavior change your product requires (that is, the more "discontinuous" it is as an innovation), the greater the potential chasm between early adopters and the mainstream market. As a founder or operator *with* early traction (there's no point worrying about chasms if you haven't cracked the early market!), you need to be aware of that risk. Your mental radar needs to be calibrated to look outward and see where the risks of the chasm lie, and then your laser beam needs to focus down on what you need to do to get to the other side.

And while the original insight around the chasm is decades old at this point, it can still trip up otherwise successful startups, just as Pebble learned the hard way.

Pebble and positioning

Pebble, one of the original smartwatch companies and Kickstarter sensations, sadly turned out to be a case study in getting their positioning wrong and failing to cross the chasm.

This book is about B2B software, and while this is a B2C hardware story, it's *such* a fantastic illustration of nerd vs. normie positioning, misplaced attention, and the chasm itself that I couldn't resist including it. It also acts as a cautionary tale of dropping the ball due to brain putty (assuming you know what the mainstream market wants), failing to direct your laser-beam attention at discovery (talking to customers), and failing to articulate the big picture, radar attention (narrative and vision).

It's got it all, and I have infinite respect for Pebble's founder, Migicovsky, for sharing it.

Millions of happy customers

That intro might sound a bit grim, but it's worth highlighting that Pebble built and sold millions of a really cool product: a fun line of smartwatches that some folks still use to this day. And it was indeed a *discontinuous* innovation, the kind that splits nerds and normies. Slapping a computer on your wrist was a very new behavior — this predates Apple's entrance into the arena. The first Pebble watches debuted in 2012 and shipped in January 2013, more than two years before the first Apple Watch hit the shelves in April 2015.

Unfortunately, it wasn't to last: the company was dead by the end of 2016. What happened?

A *smart* watch? Why would I need *that*?

It's 2013. Smartwatches are a brand new thing.

What are they for? Should I get one? What would I do with it? What's the point?

New product, new behavior. Pebble was trying to prove the concept of a smartwatch itself — there was no direct predecessor. Perfect for nerds and intriguing to normies. Or, to quote from the story we're about to unpack:

> *In 2013, Pebble was the cool nerdy colourful watch that your coworker kept looking at during lunch.*

Pebble wasn't just a new innovation; it was the poster child for new, early-adopter distribution, too: crowdfunding. (It's striking how innovations that reach some breakout success also leverage some new distribution channel.)

Crowdfunding gave direct access to early adopters in a way traditional retail never could, and that gave Pebble a fantastic place to start, raising $10.3M in 2012, making it the most funded project in Kickstarter history at the time.

This is a double hit of early adoptery-ness:

1. They had a **product** just for early adopters.
2. They had **distribution** just for early adopters.

They were leaning hard into the nerd segment of the market. They *were* the segment. Their product was built for folks just like them and sold to folks just like them — buyers who reference each other! — and that got them off to a fantastic start.

So why did it fail just four years later?

Migicovsky, in a Medium piece[2] I'll quote from below, describes how Pebble tried to cross the chasm from their "initial geeky/hacker user base" (i.e., nerds) to the fitness/productivity world (i.e., normies) and, sadly, didn't make it.

They had a product that was supposed to take them mainstream — the Pebble Time. Apple Watch had hit the scene around the same time and, to quote Migicovsky, "We thought the smartwatch market was about to explode." That was an entirely reasonable assumption — in the classic Diffusion of Innovations theory, a large mainstream majority *does* follow the nerdy early adopters.

But they didn't account for the chasm — the fact mainstream folks want different things from early-adopter nerds. And when it came to launching that mainstream-focused product, it flopped. It flopped for a number of reasons, but perhaps the main one was that there was no story (story 1 or story 2), no discovery, and, therefore, no narrative or meaningful positioning for the mainstream. As Migicovsky puts it:

> *[I]t was exceedingly difficult to explain why [mainstream] customers should care. [...] We didn't do our product research, we didn't talk to enough of our users."*

> **Pebble, Slack, and loveable products:** Pebble is, like Slack, a great study in creating a loveable product where the character of the product, the *implicit idea*, really shines through. We discussed software as art when looking at Slack in chapter 12, and Pebble was a case of product as art — people really saw it in human terms. Strong implicit ideas can be double-edged swords, though — they're subtle in the way the team itself can forget they weren't just building 'smartwatches,' they were building *Pebble* watches, or perhaps just *Pebbles.* And that's perhaps another reason why the initial vision didn't generalize — the implicit idea behind the whole project was so specific and nerdy. How do you turn that into a mainstream product without losing the magic?

No discovery, no positioning

> *The underlying problem was that we shifted from making something we knew people wanted, to making an ill-defined product that we hoped people wanted.*

Pebble, as Migicovsky writes above, tried to make the jump from the reality layer — what they knew people wanted — to the speculation layer — what they hoped people wanted.

The problem for the Pebble folks was that pre- and post-chasm represent what are essentially two separate learning curves. They'd mastered one and thought that naturally continued into the mainstream market. Obviously, it didn't.

The learning curve for early adopters included their choice of early adopter distribution: Kickstarter. Pebble were *fantastic* at that, as their crowdfunding results demonstrated. You are what you learn, and Pebble had learned to use crowdfunding to great effect.

But the learning curve for mainstream normies also meant mainstream distribution: retail. And Pebble struggled mightily there, ending 2015 with, according to Migicovsky, "over $15m of inventory sitting in our warehouses." Success on one learning curve masked the need to start from scratch on another, which is something SaaS folks have to learn all the time when they try and expand into new channels.

Double the sales curves, double the learning

> *We didn't know if there was actually a market for a more 'productivity' smartwatch and we weren't a fitness company at the core.*

That's the trap of the chasm. It masks the fact that there's a whole new set of things to learn, and in Pebble's case, it's not just that they didn't do the learning, it's that it appears they didn't know they *needed* to do the learning — it was a killer 'unknown unknown.'

And to be fair, they built the original product more or less for themselves. They *were* the geeky hackers that got a kick out of their OG smartwatches. They were the known knowns. You don't need to do as much discovery when you are your target market. But that also means never developing the learning muscle to discover *exactly* what normie customers want and *exactly* what their normie "fitness/productivity" jobs-to-be-done were and how they might solve them.

And if we think about this in terms of social groups that reference each other, the nerdy buyers obviously took note of what other nerdy buyers were doing, but, as *Crossing the Chasm* taught years earlier, pragmatic normies don't necessarily value references from nerdy early adopters. They want peer references. Without them, why should they care?

> **Brain putty in action:** Back in chapter 28, we discussed the brain putty trap, where perfectly earnest founders can still get tripped up by their own assumptions. That's the danger of left-brain brain putty — your brain patches your reality and doesn't tip you off that you don't know

> what you don't know. Instead, your brain tells you what you do know — you're a big Kickstarter success! You're a fun company with a cool hit product! What could go wrong?

You don't *have* to cross the chasm

> *We forgot the cardinal YC rule: talk to customers. Build something people want. In 2015, we had the option to narrow our focus to a smaller but more distinct market positioning. Looking back, this is one of my biggest regrets. We could have been* the *smartwatch for hackers but we tried to grow our volumes and market share (and failed).*

If you truly are a discontinuous innovation with strong early adopter positioning, then you don't *have* to cross the chasm. You can stay serving the hackers, the nerds, the innovative folks on the edge — the social group you know. (That's where modern, PC-oriented VR has landed, for example.)

What's particularly interesting about Pebble's story is that they struggled with the right-brain, big-picture, radar side of their story, too.

> **Are early adopters always nerds?** I'm using "nerds" as shorthand for early adopters here as it tends to describe the kind of folks who will go out of their way to use technology that might not quite be ready for the mainstream. But sometimes nerd adoption is actually a *negative* signal to a different kind of early adopter — the fashionable early adopter of culture. Think about the reaction to Google Glass when it launched in 2013 — it looked incredibly dorky and became an object of derision. Tech companies have since tried to launch wearable consumer products in partnership with the fashion industry to avoid this fate, particularly Apple with their fashion-centric launch of the Apple Watch[3], which contrasted dramatically with Pebble's nerd-focused approach. More recently, Meta's smart glasses were launched in partnership with Ray-Ban, and Humane's ill-fated Ai Pin debuted on the runway at Paris Fashion Week. There will still always be a place for new gadgets launched at nerd-friendly places like CES too, of course — not all consumer tech is worn — but early adopters can look very different depending on the market you're going after.

What's on the radar?

Migicovsky says that getting Pebble's product positioning wrong was one of his biggest regrets, and so was his difficulty in articulating his long-term vision. As he tells it, he struggled to step back from the day-to-day grind and actually make a plan around that vision:

> *Startup founder lesson learned — never forget to define and talk about your long term vision for the future. When things are going well, it's easy to get caught up in growth. But you need this to carry your company through hard times.*

Leading through the two axes of attention

This is a great example of the two axes of attention in action — Pebble sadly missed on the y-axis of story 2 positioning where you drill down to what (normie, post-chasm) customers want, and they didn't articulate their x-axis, story 1 narrative for where they were headed, either.

There's a subtle aspect to this, too — the proactive and reactive modes of learning.

Companies often start with that right-brain story 1 vision of what's 'out there' that they are going to build, like the nerdy, fun, technology-powered future of watches. This right-brain vision can drive the company forward for a long time. But when it comes time to launch new products in new markets, that same intuitive vision doesn't work — you've already caught up to it, so what comes next?

Here, the company — and leadership— has to switch modes of attention. They need to switch gears to an after-the-fact kind of learning of drilling down to observe what works. That laser-beam learning is very different from the initial right-brain radar narrative that gets them going. That's a point that's often missed, perhaps because it's such a different muscle — or hemisphere — that needs to be engaged.

That's what clarity strategy seeks to address — our two modes of attention do very different things and need to be used in very particular ways, and it pays to know your radar from your laser beam.

Game over man, game over

In Pebble's case, this failure of laser beam positioning and discovery on the one hand and big-picture radar on the other — beyond their initial, highly successful product — meant that, ultimately, Migicovsky and the Pebble team had no clarity about what market actually existed on the other side of the chasm. Nor did they know what market beyond smartwatches *could* exist if they were to develop their vision for it and repeat the early adopter playbook that very much did work for them in the beginning. Sadly, like many startups before them, they paid the price, winding up in insolvency in December 2016, with Fitbit acquiring their IP and team. (Fitbit themselves would then be acquired by Google in November 2019 for $2.1 billion.)

Migicovsky's insightful and painfully honest story of Pebble's failure is well worth studying. And while Pebble didn't work out in the long run, it was still cool as hell while it lasted, and thousands of people are still keeping their Pebble watches going to this day. Better to have succeeded for a while than never succeeded at all. Migicovsky has since thrown his hat back in the ring with his startup Beeper, and it will be interesting to see how the lessons of Pebble shape that new project.

For us, we again see the power of niche positioning and niche distribution working hand in hand — Pebble's nerdy, fun, early-adopter product matched its early, fun, early-adopter distribution in the form of crowdfunding.

And we see what happens when folks try to make the leap to mainstream positioning and mainstream distribution without understanding the risk the chasm poses and how it can end in tears.

Matching positioning to distribution is vital, and as we tack back to B2B SaaS next, we'll see what happens if you try to use niche positioning with mainstream distribution. In the previous chapter, we talked about lighting a word-of-mouth fire, and in the next chapter, we'll talk about lighting a different kind of fire: a marketing money bonfire.

Notes

1. Migicovsky, Eric. "Success and Failure at Pebble," April 12, 2022. **https://medium.com/@ericmigi/why-pebble-failed-d7be937c6232**
2. Ibid.
3. Remember the solid-gold Apple Watch Edition, priced between $10,000 and $17,000, that Beyoncé and other celebs briefly wore? Hard to believe that didn't catch on.

42

Distribution mismatches & the $100k CAC trap; L-factor vs. left-brain marketing

> Failure to take into account context, [...] an inappropriate demand for precision where none can be found, an ignorance of process, which becomes a never-ending series of static moments: these are signs of left-hemisphere predominance.
>
> *- Dr Iain McGilchrist*[1]

Over the past few chapters, we've looked at niche positioning driving niche distribution and vice versa — direct sales for ConvertKit, a vertical focus for Dock, and early-adopter crowdfunding for Pebble. Those examples are fascinating studies in positioning and distribution working hand in hand (until, for Pebble, it didn't).

In Pebble's case, we looked at what happens when you get your positioning and distribution wrong when attempting to cross over to the mainstream — to cross the proverbial chasm.

In this chapter, we're going to return to the world of B2B SaaS and get a little mathematical as we look at another classic positioning and distribution mistake — trying to use niche positioning with paid mainstream distribution. This is a bit of a transitional chapter as we go from looking at niche sales to broad marketing distribution.

We'll also explore some of the messy reality of marketing, especially when left-brain attention runs amok, as tends to be the case all too often in the very left-brained tech world. The symptoms of left-brain dominance in marketing in particular echo McGilchrist's general description of left-hemisphere predominance quoted above — a lack of context, faux precision, and a fixation on static snapshots, along with a failure to understand the whole... it all rings a little too true.

Thankfully, some folks are doing their best to elevate what B2B SaaS marketing can be, and we'll give them (and their resources) a shoutout, too.

You call that a knife?

If you try and run the niche positioning playbook but don't have access to niche distribution, you end up bringing monopoly money to a CAC fight.

Let's say you're trying to sell a niche SaaS product through high-intent, bottom-of-the-funnel Google Ads keywords. I've seen this up close. A client of mine once sold, let's say, a CRM for dentists (I'm making up the category and customer to protect their identity, but the details are accurate). They hired a reputable agency to handle their Google Ads account, and the agency burned close to half a million dollars for fewer new accounts than you could count on one hand. This resulted in a customer acquisition cost of about $100,000 per new account. $100k CAC was not, shall we say, something this company could afford.

You'd think Google was a needy charity the way folks throw money at it, and Google is, of course, more than happy to take your donations. I've certainly made a few small donations over the years myself. In the tech world, burning cash on Google Ads is a tale as old as time, and as far as tech company money fires go, this was but a single, solitary lit match. Many far bigger companies have burned several orders of magnitude more cash on pointless ads and other activities over their lifetimes, I'm sure.

But why does this happen?

> **Paid drives brand, and vice versa:** You could argue that the paid ads I investigated might not have been driving direct sales, but indirect sales, with folks seeing the brand and doing their own research before converting. There is truth to this — 'performance' (i.e., direct response) and 'brand' marketing are often two sides of the same coin.[2] Paid ads are often how people discover a brand in the first place, and there's often secondary (supposedly) 'organic' activity that wouldn't exist if it wasn't for paid spend driving these incremental conversions. In the above case, however, the PPC spend was so egregious that there would have to be perhaps 10:1 incremental conversions for every known conversion for the math to work, and that both defied logic given the strategy and was essentially impossible in light of available data in any case.

Niche positioning × mainstream distribution

I mention this example not to ding my poor client — they were good people who trusted a large agency, an agency that said all the right things, had all the right credentials, showed all the right "directional" data, and still burned a big pile of their money. And if my client's experience was anything to go by, if you multiply that outcome by all of this agency's other clients, you could probably warm a small village with heat generated by the money they were burning.

Instead, I mention this example for two reasons:

1. As a stark illustration of the difference between positioning in the 'cast and catch' world vs. the 'segment and sell' approaches that we discussed in **chapter 40**.
2. To highlight what happens when left-brain predominance becomes the norm, as McGilchrist describes in the opening quote.

Casting the net

On the first point, the pyromaniac agency was trying to cast a big net in the Google Ads ocean for folks searching "CRM software," hoping to catch a few dentists that might pop up. (They advertised on "dentist CRM software," too, but for the real keywords, the volume was all in the parent category.) There is a certain logic to this — if we advertise broadly but make our ad clearly just for dentists, then only dentists will click, right? Well, no, and you're bidding against vastly bigger companies with vastly deeper pockets who can serve the whole market.

On the face of it, it was never going to work. Imagine if ConvertKit, our go-to example of niche direct sales from **chapter 39**, had tried this approach to enter the email marketing software market by bidding on "email marketing software" — it would have been a blood bath.

This is why niche positioning and niche distribution must work together. (Unless, of course, you're feeling particularly charitable to Google and its shareholders.)

> **Geographical niche × mainstream distribution:** There is an exception I've seen work, and that's where you leverage a different kind of niche: a geographic niche. If you are in a non-English speaking country, for example, and the dominant global players have not started with internationalization, then there's something to be said for using mainstream distribution to sell your way to a mainstream position simply by (quite literally) speaking your buyer's language. You can work your way up the sales learning curve while flying relatively under the radar relative to competitors who are busy duking it out in the American market. But making the jump to the US requires crossing a chasm of sorts, too[3], where you may have to rediscover product/market fit for a new, significantly more competitive market.[4]

Cast and catch vs. math

If we zoom out a bit, we can see, in simple terms, how niche positioning works — or doesn't work — with mainstream distribution. Let's do a little bit of back-of-the-napkin math to understand the mechanics involved, looking at the cost-per-click math of mainstream search advertising and how it can destroy any efficiency gained from being niche.

If you're paying $30 a click (not uncommon) for a mainstream, high-intent, B2B software search term, and your strategy is to *only covert those who match your niche criteria*, the math will not be kind to you. Here are some ballpark numbers for differentiated-but-not-totally-niche folks targeting a given market:

- $30/click.
- 1% conversion rate to Sales Qualified Lead (SQL), i.e., someone in your ICP who requests a demo.
- 20% close rate from demo to customer (a.k.a. your win rate).
- $30 / 0.01 / 0.2 = $15,000 customer acquisition cost (CAC), which is not at all uncommon in B2B.

If your niche positioning effectively says, "We're only going to serve half the market" (which is not very niche!), but you stick with mainstream distribution, you effectively halve your conversion rate or opportunities, and your CAC doubles to $30,000.

You might counter and say yes, but it's only by positioning around that half of the market that lets us win any deals at all. That is, the point of targeting that one-half of the market is to drive efficiency in converting those folks specifically.

That can work if you can target them directly in a segment-and-sell sense. But if you're trying to find them in a broader search context, cast-and-catch style, you need to be at least 2x as efficient in winning deals to make up for the smaller market. That is, you need to double your SQL conversion rates or your win rates for your niche ideal customer profile (ICP) just to match what a broader positioning would have given you. That's just to get back to where you were.

This is a vast simplification of course, but now we can start to see that there's another variable at work — let's call it L for leverage, or your L-factor, to give it a corny name, and riff on the K-factor of virality.

The L-factor

This L-factor can be the difference between whether a distribution channel is viable for you or not. If your niche positioning creates a very strong L-factor — you only serve half the market, but you serve them *unbelievably* well —

then that might more than offset any loss in efficiency in not targeting the broad market. But if your niche L-factor is weak, then you lose mainstream distribution options and will struggle to generate the momentum that helps propel you into the mainstream.

How do you develop your L-factor? That's a question that folks have been trying to answer for decades. It's what *Crossing the Chasm* is about, it's what *Play Bigger* is about, and it's what this book is about. On the product side, you can focus on specific jobs-to-be-done, or on core 'disruptive' innovation that creates technological leverage by doing something so well it doesn't matter that it does everything else much worse than the incumbents.

And on the GTM side, broadly speaking, you can:

- Develop the right distribution muscle for your stage, and try and segment and sell to generate market leverage that way, as we've been discussing.
- Pick appropriate positioning strategies (we'll look at positioning by *attribute*, not *segment*, in **chapter 46**, for example, where folks split the difference between the mainstream and a niche by trying to serve all comers in a particular way).
- Develop your clarity strategy by going further out on the x-axis, building a stronger right-brain narrative for executives (secular change strengthens your L-factor there, as we saw in **chapter 30**), for example, perhaps landing a visionary early customer that way, or going further down the y-axis and focusing narrowly on your best customers, and pattern matching against them, as we saw in **chapter 37**.

The point is to find market leverage — that elusive but all-important L-factor which is, in some ways, what this entire book is about. It doesn't matter where it comes from; it just matters that you can methodically work towards finding and developing it. And if you can stack multiple forms of leverage and reap the compounding rewards, all the better.

Nothing wrong with a niche

Let me be clear that I'm not saying niche positioning is bad or that there was anything wrong with my client's niche position. Niche positioning can work spectacularly well as we've seen, especially at early stages when you're building momentum *and have the distribution to match*. It's often essential to get started. (And sometimes software markets are so big your "niche" turns out to be a multi-billion dollar opportunity in its own right.)

Nor am I saying it's bad to accept some inefficiency upfront to get the flywheel going. We discussed 'doing things that don't scale,' for example, back in **chapter 39**.

But niche positioning *and* a mainstream distribution strategy *is* bad. It's mixing oil and water — the point of niche positioning is to drive niche distribution and open doors that you wouldn't otherwise have access to. The point of positioning is, really, to create that L-factor, that leverage in the market, and because companies and markets evolve over time, the kind of positioning you choose, and the leverage you create, needs to evolve, too.

> **Growth at all costs & your L-factor:** In the early 2020s 'foie gras' era of VC-backed startups it was assumed that you should grow at all costs to reach scale, and once at scale, efficiencies would kick in, and you'd find the pot of gold at the end of the rainbow. Unfortunately, and quite ironically, one of the costs of growing at all costs is knowing how to grow at, err, not all costs, i.e., to grow efficiently. For this cohort of startups, markets eventually realized that customer acquisition costs for these businesses often went *up*, not down, as they grew because they had to try and win increasingly marginal customers.[5] This is what happens when you try and spend your way up the adoption curve. You can do it, but if you never develop the muscle to create genuine momentum through unpaid sales and marketing touches and develop an L-factor that isn't just "we have loads of capital to burn," then the only move you have available is to spend more and more and more, until the music stops.

Positioning evolves

Remember that positioning, in the broadest sense, is a metaphor for moving your 'position' from one point to another point, usually picking up your 'supply' to drop it closer to a given point of demand. In this book, I tend to use "positioning" to describe that left-brain, narrow-focus, snapshot-in-time approach to the demand you serve. And they really are static snapshots in time.

If that happens on the y-axis, what's the x-axis? It's growth over time, usually represented as some kind of curve — adoption curves like the sales learning curve, the chasm-containing Tech Adoption Life Cycle that Geoffrey Moore espoused, or the Diffusion of Innovations theory more broadly that it was based on.

These curves exist to capture the idea that the position you take — and the leverage you generate — should evolve over time. The leverage you need to get started is different from the leverage you might apply once you're a unicorn at scale. At scale, perhaps the strength of your brand and the breadth of your product *does* make broad advertising work, and you can drive customer acquisition that way.

But if these growth-over-time models are well known, what stops folks from following this well-trodden, well-documented, fairly straightforward path of trying to build the right leverage at the right time as they ride the adoption curve?

Part of the problem is that modern B2B marketing — and perhaps most functions, including product, engineering, and sales, too — is often broken, and broken in an overly left-brained way.

Left-brain attention is great for plenty of things, especially applying the laser beam and manipulating the world, as we've seen. I'm not arguing that we should act as though we've had left-hemisphere strokes. (In fact, there's often not enough left-brain attention applied to a startup's narrative and strategy, which is why we're going so deep into the left-brain positioning world in these chapters.) But more broadly, an unconstrained left-hemisphere approach to marketing in general (and product, sales, and engineering) leads us down some strange paths — paths that don't lend themselves to good, holistic positioning and clarity strategy.

What overly left-brained marketing looks like

Let's discuss four of those extreme left-brain paths:

1. The left-brained failure to join the dots.
2. Tooling being insanely left-brained.
3. Being data-rich with faux precision but insight-poor.
4. Marketing theatre.

> **Choose your own adventure:** While marketing isn't remotely unique in committing these extreme left-brain sins — other functions are, as mentioned, equally guilty — given the close connection between positioning and marketing, marketing is what we're going to look at. If that's of interest to you, great, keep reading. If not, skip ahead to chapter 43, where we'll kick off our look at the difference in positioning strategies across SMB, the mid-market, and the enterprise.

1. Failure to join the dots

In the right/left/right-hemisphere jump we discussed way back in **chapter 26**, we looked at how, in simple terms, our brains process information from the right (radar) to the left (the parts) back to the right (a reformed whole). Well-functioning orgs do this too, as we saw with Figma in **chapter 27**.

In marketing, from what I've seen, it's rare for folks to do this. It's much more common to get stuck on the parts, with, at best, a vague appreciation for the whole. That is, there's a lot of activity but not much coherent, end-to-end strategy. This is a failure of management, not individual contributors, who will often burn themselves out working hard on their particular specialty or against some poorly considered metric, all without a strong overarching strategy to guide their efforts. And when there *is* a strategy introduced, it's often the silly, overly ambitious, borderline-delusional category creation stuff, which we dissected in depth starting at **chapter 32**.

In a right-brain/left-brain sense, the right-brain whole is often absent, while the left-brain narrow focus is running amok, leaving marketers flying blind with neither much by way of macro strategy or, ironically, meaningful feedback loops on the narrow-focus side either. That is, there's neither:

- Coherent strategy, in terms of a shared mental model for how their specific B2B buyers actually *buy* and any beyond-the-superficial understanding of the jobs-to-be-done their customers want to, well, get done.
- Coherent feedback, in terms of laser-beam attention to what *attributable* results they're driving, even the sense of the visitor-to-customer journey captured in your CRM.

Nevertheless, people do what's expected of them or what they think will be good for their career[6], and most leaders expect marketing folks to churn stuff out at a frenetic pace, which they do. It's no wonder, then, that folks get very good at *quantity* of actions without ever having time to stop and contemplate the whole and the *quality* of their actions. Leaders use overwhelming activity to try and mask their underwhelming understanding, and that's why you end up with agencies burning hundreds of thousands of dollars on pointless advertising — no one is connecting the dots.

This gets worse as startups grow, with CMOs coming in with a noose already around their neck, making some rushed bets based on arbitrary growth goals with their budget, squeezing folks to make sure as many activities are done as possible, and then wondering why it all didn't seem to add up to the growth they wanted.

The need to appreciate the right-brain whole goes beyond marketing leadership, of course. It's the job of the executive team to coordinate as a whole, not pull levers at random and hire and fire when it doesn't work out. That creates a negative feedback loop where people are even less inclined to think seriously about the whole because what's the point if you're not going to stick around?

That's not to say good folks aren't out there — they are, and if you've got them, hold on to them! They're gold and should be compensated accordingly. And when I say folks need to "join the dots," this doesn't have to mean grand strategy either; it might just mean simple steps like:

- **A bit of spreadsheet detective work:** In my client's case, back when I wore my analyst hat from time to time, all I did was what the agency never felt the need to do (or perhaps was afraid to do) — ask how many of these PPC-sourced "leads" actually became customers (answer: very few). A few spreadsheets and a little bit of right/left/right-brain detective work was all it took to figure out what was going on. Likewise, a bit of time in your CRM actually understanding what content your actual buyers consume — and I do mean buyers, not visitors — can go a long way.
- **Pattern matching from existing customers:** If the data is there, you can work backward from your best customers and pattern match on what brought them to you and what made them happy, as April Dunford wrote about, as we saw back in **chapter 37**. Positioning exercises are a great chance to ensure you know where your customers — again, not your visitors, not your leads, your customers — are actually coming from.
- **Locating yourself on the curve:** That might be the adoption curve that we discussed in **chapter 40** or the sales learning curve that we mentioned in **chapter 39**. Either way, as we've discussed, positioning evolves as markets evolve, and understanding where you are on the x-axis helps you understand the position you take on the y-axis.

In the cast-and-catch world, it's very easy to do a lot of casting, catch a few random things, and then think, 'Hey, I guess this is how it works!' But it really doesn't take much to develop a mental model that's much more anchored in reality; you just need to concentrate your attention down on that reality layer and see if the dots really do lead from what you're doing to what you're catching. Don't be afraid to ask what seems like the most obvious questions in the world, like: Do these activities actually work? How do we know? What will we do if it doesn't?

You might be surprised by what you find, and you might free up loads of budget that could be deployed much more fruitfully elsewhere.

2. The lies of martech tools

There's an old saying that the number of opinions people have about a given topic is inversely proportional to the amount we know about it. You could say the same about martech tools, where the number of tools available (about seventy billion at last count) is inversely proportional to how well they actually work, or at least measure up to their hype.

When we talk about clarity strategy and your GTM plan in a holistic, left-*and* right-brain sense, it's important to stress what a poor job marketing tools do of reflecting that complete reality. Instead, they tend to represent an extremely left-brain take on the world — all faux-precision and independent slices of activity.

For example, if positioning and distribution are as intertwined as I've been suggesting, where do we see this relationship articulated in martech tooling? Absolutely nowhere, and this is after B2B market strategy has been a thing for 30+ years. These aren't new ideas!

As one very simple example of the way tooling can mislead, I recall one story from SaaS bootstrapping legend Ruben Gamez[7] of Bidsketch and SignWell fame that demonstrates why you need to understand how customers actually buy vs. what you see in your tools, especially from a Jobs to be Done point of view.

Gamez's Bidsketch proposal tool requires users provide their email address to get a sample proposal in a classic 'lead magnet' sense. Bidsketch's landing page focuses *very* heavily on making that the one-and-only call to action — there's not even 'pricing' or 'features' links in the top nav until you add your email. So, lots of folks give their email address and then receive a value-focused nurture sequence along with regular blog posts.

Here's where it gets interesting. Gamez also did a bunch of switch-focused Jobs to be Done interviews where they would "focus on the story, capturing, "What happened when you bought the product? What was the first thought you had?" [The customer would then] walk us through every step" of their buying journey.[8]

What was really going on? Here's Gamez:

> *One of the interesting things that came up a lot during those Jobs to be Done interviews was [that] they got an email that reminded that we were there. So when they had that need, they signed up, or they got an email way before that point in time.*
>
> *In some way, shape or form, that story came up, over, and over, and over. I knew that happened. I didn't know that it happened that much. [It was] super surprising.*

JTBD switch interviews are about trying to understand the causal factors that drive buyer behavior. Here, prospects that had expressed some interest in the tool were often signing up *not* because a specific email had a killer offer or an amazing insight, but *simply because the email reminded the buyer that Bidsketch existed.*

And it's that *memory* that was often all-important, as Gamez found:

> *Another variation on that story was that [...] sometimes they would remember our name, sometimes they wouldn't even remember our*

> *name. They remember* something, *and they search their email, and find us that way.*

Building these kinds of product <> need memory associations is what branding is all about, as we'll see in the next chapter. Here, however, the customer presumably had some degree of need for a proposal tool that took them to Bidsketch in the first place, but it wasn't until that need became acute enough *and they were reminded that Bidsketch existed* that they felt compelled to switch to the tool.

In light of that, what should conversion tracking say, especially for folks who had a memory *and went back and found an email* to click through to the product site? I'm sure Gamez's newsletter tool would have reported that conversion as coming from that specific email, as though it was that email that "converted" the user. While that's true in a very narrow sense, the conversion was really a product of the strategy — of building memory connections and maintaining *mental availability* (a concept we'll begin exploring in the next chapter) through the blog-post-as-newsletter emails over a long enough period of time such that when the need arose in the customer's life, Bidsketch was, as they say, top of mind.

That kind of attribution simply doesn't show up in your marketing reports, which is darkly ironic when you think about it. For a discipline that's supposed to be grounded in market strategy, most martech tools (and their reports) have next to no concept of strategy whatsoever, or brand, for that matter. There's no way they could attribute it to a strategy in the first place, as the "strategy" in most orgs often exists in some disconnected doc or deck with, again, no one actually joining the dots.

But it does go to show what happens if you get customers to walk through their purchases with Jobs to be Done switch interviews. You get to see a world that your typical martech tools never reveal, including a much broader sense of the 'whole' of buying and a real feel for the actual shape of demand you're serving, not the vastly diminished left-brain view that our tools often re-present to us in context-less numbers and metrics.

3. Data-rich; insight-poor

Left-brain dominance in B2B SaaS marketing tends to really come to the fore in the world of testing and metrics. McGilchrist says left-brain predominance results in a lack of context, "inappropriate demand for precision where none can be found," and a series of static moments, and that's especially true when it comes to marketing data.

And this, dear reader, is coming from someone who really likes analytics data and A/B testing! I cut my teeth in the digital marketing world by getting into analytics and A/B testing way back in the day[9], so don't take this the wrong way, but it's bonkers how many teams out there are (seemingly) data-rich but insight-poor. I've seen this in extremely well-funded teams too, with folks *drowning* in metrics, often to two decimal places for every page, every ad, every campaign... but with minimal understanding of attribution, what they can meaningfully test, or what their overarching strategy even is. It's not great.

This is particularly bad in spreadsheet-driven, later-stage startups trying to make their numbers every quarter. Strategy gets substituted for squeezing — squeeze more out of your channels, your content, your customers, your people... while basic questions like "What does a buyer's journey actually look like?" both in terms of JTBD and actual engagement with the company's sales and marketing resources go unanswered, and the net result is a lot of frantic wheel spinning. Squeezing has diminishing returns, after all.

This kind of faux optimization is particularly pronounced in low-volume B2B contexts. The agency in our example above that was burning PPC budget was operating in a low-volume world with little data to work with. Therefore, the "testing" they were doing — routinely trying different landing pages and ad copy — was entirely meaningless. It was statistically impossible for them to run real A/B tests, and they may have well been rolling out worse performers for all they knew.[10] The volume was just never there, and in complex B2B sales, it often isn't.[11] The only thing being optimized here was the agency's bottom line.

This is not to say you shouldn't experiment — far from it. You should *experiment* far more boldly; you just shouldn't waste your time trying to *optimize* the unoptimizable. The reality is that the low-volume B2B world is one where physicist Ernest Rutherford's quip that "If your experiment needs statistics, you ought to have done a better experiment" tends to hold true.

In low-volume scenarios, you need to prove what's physically true before worrying about what the statistics say. And there are a lot of obvious things that need to be physically true in SaaS marketing — you do need a website with strong messaging (**chapter 14**), you do need a strong search presence with the ability to ladder people up towards your offer (**chapter 24**), you do need to build a brand in the memory-association sense over time (**chapter 45**), and you may well need to use advertising and/or outreach to make that happen.

But making those programs effective doesn't happen through scrutinizing metrics to two decimal places. It's about knowing your growth rate, sure, but your growth is only as strong as your inputs, as your discovery and insights, and that discovery means rolling up your sleeves and getting to ground truth with individual successful customers. Think of it like understanding the physics — i.e., the actual mechanics[12] — of the buyer's journey and their JTBD. We tend to call this "qualitative" data, but that's not quite right — that's still the framing of statistics. I'm much more interested in *cases* and the mechanics of real, tangible, engineered success and what that looks like for a given buyer. We're ultimately in the business of helping folks engineer their own success after all, not just flogging cheap commodities to whoever is buying.

In any case, the lack of meaningful quantitative data in B2B doesn't, alas, stop folks from filling the gaps with what they wish was true instead. This is the land of brain putty (**chapter 28**) — our ChatGPT-esque left brains patching reality with whatever seems plausible, not what's actually true. The vacuum a lack of right-brain thinking creates needs to be filled, and it's often filled with left-brain tea-leaf reading. In B2B, you'll never find a strategy by staring at low-volume, low-powered test results. And indeed, for founders, it's not enough to simply demand more and more metrics from your marketers. You're never going to see the answers you need in those tea leaves.

Setting the strategy — and using marketing as a strategic discipline — is your job as a leader. Marketing can absolutely be a strategic discipline, but it needs to be saved from all the hyper left-brain marketing theatre that adds up to bupkis, and I hope clarity strategy, in some small ways, helps folks do that. Marketing leaders, in particular, need to drag folks out of metrics myopia so they can think about the bigger story they're trying to tell and help set everyone's sights on either the bigger picture out there and/or the narrower picture down in the patterns of your best customers. It's leverage in markets — markets that are made of people, people with two brain hemispheres and two modes of attention — that should be the focus of your marketing team, and those markets aren't won by fretting over meaningless metric fluctuations.

Metrics are good, actually: Dear reader, don't confuse my disdain for pointless testing in low-volume B2B contexts as disdain for measuring and testing on the whole. When you *do* have the data, and you're doing essentially direct response advertising at scale, then you *can* build a very efficient marketing machine because you can indeed model the whole (or at least enough of it) to make sound decisions with meaningful feedback loops. That's what project-management-turned-"Work OS" platform monday.com did, for example. They built an internal tool they called BigBrain[13], which, in part, allowed them to generate an incredibly efficient $5 in ARR for every $1 in cash burn[14] on their way to $550M+ in ARR, with their customer acquisition costs *decreasing* at scale.[15] But the fact they had to build their own tool is telling — off-the-shelf stuff is unlikely to cut it, and in Google's case

> with its Analytics 4 product, for example, only getting worse. This is another example where self-serve SMB and enterprise or complex sales, for example, despite both being "B2B," are really nothing alike. (For operators serious about being truly data driven in business, see Cedric Chin's fantastic writing on the topic, including his discussion of XmR charts.[16])

4. Marketing theatre

The testing, the reports, the meetings, the revisions — the work the agency was doing for my client that we've been discussing was all marketing theatre. And that's a drop in the ocean compared to what routinely happens in poor-performing marketing departments out there, and again, I've seen folks with a lot of capital doing a lot of marketing theatre, too.

Part of the problem is not knowing whether you're in explore or exploit mode. If you follow lean startup co-creator Steve Blank's view that a startup isn't "a smaller version of a large company,"[17] but a "temporary organization designed to search for a repeatable and scalable business model"[18], then that will require specific strategies and tactics focused more on exploration in that temporary state, as we've been discussing. Folks just bringing in the standard, post-product/market fit marketing playbook from a mature company are likely to be a bad match, simply because the exploration required to find leverage looks very different to the exploitation of leverage that has already been long established.

The way forward

All that said, hope is not lost! Things are improving in leaps and bounds in terms of knowledge sharing and education in the B2B SaaS marketing world, and there's never been a better time to learning from practitioners who get it. That includes:

- Emily Kramer & Kathleen Estreich of MKT1 fame who have an excellent newsletter, resources, and a course for B2B startup marketing: **https://newsletter.mkt1.co/**
- TK Kader, who is educating founders and marketers on B2B GTM strategy through videos and courses: **https://www.youtube.com/c/TKKader**
- Executives like Kevan Lee who are sharing startup marketing frameworks for just about everything through their newsletter: **https://kevanlee.substack.com**

And, of course, April Dunford and other folks we've met along the way in this book who have been instrumental in helping marketing be the strategic discipline it can be.

> **Fuel and engine:** Speaking of Emily Kramer and MKT1, Kramer has a fantastic metaphor — and, dear reader, you know I love a good metaphor — that helps us think about these sorts of marketing problems: the fuel and the engine.[19] A lot of the clarity strategy we've been talking about results in *fuel* for your go-to-market engine: a new position or story with new messaging, new assets, a new pitch, and so on. But the *engine* matters too — the channels, tools, and data you use to get the message out there is a critical part of the equation. Startups, Kramer finds, often over-index on one or the other. This book indexes very heavily on the fuel, but in this chapter, we've been discussing the broken engine folks try and feed fresh fuel into — poor data, poor tooling, and poor distribution choices. If you can get both working well together, however, you'll be rewarded with compounding returns that very few folks achieve.

To go up or to go down...

Over the last few chapters, we've been discussing the niche-to-mainstream approach. We've discussed what it means to get momentum going in a niche and where that can go wrong, e.g., getting caught in the chasm trying to cross over to the mainstream or picking mainstream distribution tactics that don't match your niche positioning.

That's the niche side, but what's the "mainstream" side? That is, when we talk about trying to go "mainstream," what do we mean? Or, more specifically, *which part* of the mainstream market are we going after — upmarket or downmarket?

This is a perennial question for B2B folks. Do you double down on sales and go after the enterprise, polishing your story 1 C-suite pitch, or do you embrace product-led growth with a strong story 2 on how the tool works, focusing on end users and generating leads to upsell that way?

There's obviously no right answer there — the money is in the enterprise, and the volume is in SMB, and they are two quite separate markets, as we touched on in **chapter 24**. The bigger question is whether you have a coherent market strategy and can construct a clarity strategy, including your positioning, to match.

In the next chapter, we'll pick up this topic as we look at SMB vs. enterprise positioning and distribution approaches. For now, though, I hope you have a much deeper sense of how distribution and positioning can work together or against each other. In combination, they can create great go-to-market leverage, but if you're not careful, they can diminish it, too. Next, we'll begin our look at the full B2B spectrum by starting with mainstream, SMB-targeted distribution and one of the most misunderstood topics in business: brand.

Notes

1. McGilchrist, Iain. *The Master and His Emissary* (p. 139). Yale University Press. Kindle Edition.
2. The "performance" vs. "brand" marketing debate is one of those endlessly recycled topics that tends to reflect whatever's in fashion at a given point in time. In a December 2021 Substack post, Michael Lorenzos sums up a recent iteration of the debate and hits on the key point that "The truth is, a brand drives performance and performance marketing informs brand perception." **https://michaellorenzos.com/p/brand-vs-performance-lets-end-this** We'll explore what "brand" means in marketing terms from **chapter 44**. If you're interested in diving down this rabbit hole, Mack Grenfell has a detailed blog post on the topic, too: **https://mackgrenfell.com/blog/what-if-performance-advertising-isnt-just-an-analytics-scam**
3. See David Kellogg's June 2022 blog post "The Top 5 Mistakes European Startups Make in US Expansion", for example: **https://kellblog.com/2022/06/27/the-top-5-mistakes-european-startups-make-in-us-expansion/**
4. Jen Abel, a sales consultant who specializes in helping B2B startups crack the US market, has some thoughts on this, too, writing on Twitter that "The longer a startup waits to expand into US market the harder it is to find product/market fit. Local market traction is often a false positive for US entry. Why? Founder mistakingly think US is 'similar enough' for GTM and product to translate well; Founders delegate US entry/expansion to team [and] many layers now [exist] between Founder and customer; [there is] unrealistic expectations for how long it takes to deeply understand market and establish US PMF; [and it just] seems harder, [so] many execs want to retreat back to 'known market'" (lightly reformatted). **https://twitter.com/jjen_abel/status/1670044863620952064**
5. I'm butchering an argument Ben Thompson made on his Sharp Tech podcast here.
6. Tom Kerwin has an insightful take on this in his September 23, 2024 post, "OKRs sound good but they don't work (Part 1)." **https://tomkerwin.substack.com/p/okrs-sound-good-but-they-dont-work** The core idea is that people will rationally want to avoid risk, including the explicit objective and result setting in OKRs because it exposes them to risk, and no one wants to be seen to not be achieving results (especially when layoffs may be around the corner). Instead, Kerwin writes that he "saw [that] most people [...] were choosing an idea that felt right [which included] "the kinds of things that successful people with my role do"; were personally interesting, with enough-but-not-too-much challenge; could plausibly-but-vaguely be justified as helping the team and company." Kerwin says he found this frustrating at the time but

eventually recognized that "These team members resisted specifying an explicit objective [in the OKR sense] because they subconsciously felt like they would be putting themselves at risk." In the tech marketing world, deploying the standard marketing playbook is the risk-free option. That might get some results, but it's unlikely to create breakaway growth. Ironically, for what are inherently risky ventures (i.e., startups), far too much B2B startup marketing is massively risk-averse, because we don't make it safe for people to take risks. Instead, we have an overwhelming left-brain focus on trying to derisk it further by scrutinizing impoverished data sets that only makes it even more risk-averse. Even doing something like Bidsketch's email-centric lead gen approach (discussed later in this chapter) would be seen as too risky for most outfits just because it doesn't look like everyone else.

7. The story is from a since-removed podcast, *Starting & Sustaining*, by Garrett Dimon. The transcript from the February 26, 2017 episode with Gamez is available here: **https://www.listennotes.com/podcasts/starting-sustaining/ruben-gamez-founder-of-iMVY03I1KWN/#transcript**.
8. Ibid.
9. I wrote, but thankfully never published, a book on each topic back in the late '00s (they were bad). My curiosity about what drives behavior is apparently much the same, though, as that's the theme we're tackling here, a decade-and-a-half later.
10. Probability tends to break people's brains, and this is especially true with marketing and growth experiments where folks assume anything that looks "directionally" (ugh) like it might be better probably is. But it only takes one bad test to erase *any* gains that may have been previously made, as Lars Lofgren notes in one of my favorite articles on A/B testing: ""Don't Build a Growth Team?": 9 Reasons Why They Might Fail," *CXL*, September 10, 2019. **https://cxl.com/blog/dont-build-growth-teams/**
11. If the data *is* there, though, i.e., you do have substantial volume, then great, test away! And let me introduce you to another one of my favorite pieces on meaningful A/B testing, Andrew Anderson's "The Discipline Based Testing Methodology," *CXL*, May 12, 2015. **https://cxl.com/blog/the-discipline-based-testing-methodology/**
12. I'm drawing in part from Sergey Roussakow's Quora answer here, which elaborates on Rutherford's quip: **https://www.quora.com/What-did-Ernest-Rutherford-mean-by-If-your-experiment-needs-statistics-you-ought-to-have-done-a-better-experiment**
13. "Meet the geniuses behind our BI tool BigBrain," Monday.com blog, April 2, 2017. **https://engineering.monday.com/meet-the-geniuses-behind-our-bi-tool-bigbrain/**
14. Per their Q4 2022 earnings call: **https://finance.yahoo.com/news/monday-com-ltd-nasdaq-mndy-214331304.html**
15. Lemkin, Jason. "5 Interesting Learnings from Monday.com at $550,000,000 in ARR," *SaaStr*, November 16, 2022. **https://www.saastr.com/5-interesting-learnings-from-monday-com-at-550000000-in-arr/**
16. Chin, Cedric. "Becoming Data Driven, From First Principles," *Commoncog*, 28 January, 2024. **https://commoncog.com/becoming-data-driven-first-principles/**
17. Blank, Steve. "A Startup is Not a Smaller Version of a Large Company," January 14, 2010. **https://steveblank.com/2010/01/14/a-startup-is-not-a-smaller-version-of-a-large-company/**
18. Blank, Steve. "Why the Lean Start-Up Changes Everything," *Harvard Business Review*,

May 2013. **https://hbr.org/2013/05/why-the-lean-start-up-changes-everything**

19. Kramer, Emily. "Building an efficient marketing machine: the fuel & the engine," *MKT1 Newsletter,* November 19, 2021. **https://newsletter.mkt1.co/p/fuel-engine**

43

The B2B spectrum I — Squarespace, How Brands Grow, & passive vs. active attention

> We almost have two businesses — a-mid market and 'corporate and above' business and a 'corporate and below' business. They have vastly distinct characteristics — one has relatively high churn [...] the other is a lot more repeatable. You have to optimize each of them separately.
>
> *- Manny Medina, CEO of Outreach*[1]

In the past few chapters, we've been exploring the relationship between positioning and distribution, with a particular focus on 'segmenting and selling' to get that initial momentum going.

We then looked at what happens when you use mainstream distribution tactics with niche positioning — you burn a whole lotta cash.

In the next few chapters, we're going to kick off our B2B spectrum mini-series and look at what happens when you get mainstream positioning and mainstream distribution right. In contrast to the niche positioning approach we've been discussing, we'll look at the opposite — going for reach instead, as we introduce the *niche vs reach* positioning split.

I'll also weave in some spicy takes on the B2C world of brand, positioning, and marketing 'science,' looking at popular B2C brand marketing concepts like mental availability and physical availability, and we'll ask whether positioning matters at all. It turns out there's a very popular school of B2C thought that thinks it doesn't (!) and that it's essentially a mistake, which is... well, you'll see.

The B2B spectrum positioning mini-series

The next five chapters will make up our little mini-series on mainstream positioning across the B2B SaaS spectrum, and we'll cover:

1. **SMB:** We'll introduce legendary B2SMB SaaS Squarespace in this chapter, along with arguably the most influential voice in the B2C brand marketing world, their fundamental ideas of mental and physical availability, and then finish with a look at passive vs. active attention.

2. **Positioning in SMB:** We'll continue exploring the brand marketing narrative in **chapter 44** as we look at whether positioning is an old-school mistake, the dynamic nature of positioning, and the niche-to-reach positioning framework.
3. **Brand & attributes in SMB:** In **chapter 45**, we'll look at marketing as a battle of memories vs. clicks, where brand attributes do and don't make sense, the brand positioning journey of Patreon, and we'll touch on how Notion combines differentiation, distinctiveness, and positioning for reach to great effect.
4. **Mid-market:** In **chapter 46**, we'll take an extended run through positioning around wedge attributes and why positioning can be such a challenge in the hyper-competitive B2B SaaS mid-market. We'll look at this through the lens of the corporate LMS market, and I'll describe why attempting to balance niche & reach results in wedge positioning around a particular attribute.
5. **Enterprise:** Finally, in **chapter 47**, we'll discuss the enterprise chasm, 10x'ing (or 32x'ing) your pricing as Whatfix did, plus we'll look at the 'hyper-enterprise' UiPath and how they position for reach *and* niche, and the fundamental role narrative differentiation plays at this end of the market.

Is it completely bonkers to kick off another mini-series this late in this book? Yes. Are we going to do it anyway? Also yes, as this will give us the final piece of our positioning puzzle.

Two markets, two businesses

In **chapter 40**, we discussed the internet-first and business-first approaches to selling software, with the arrival of the internet creating something of an evolutionary fork — no longer was selling "high tech" B2B software to large businesses a matter of targeted top-down sales, you could now end-run IT and sell to end users directly, in self-serve, bottoms-up or "product-led" fashion, as Salesforce originally pioneered. What's more, buyers could now do their own research and search out exactly who they might want to evaluate, and businesses could cast a wide net to try and catch them. Thus, the cast-and-catch GTM approach was born.

This cast-and-catch/segment-and-sell binary is, of course, a gross simplification, as all binaries are. But it illustrates that there isn't a homogenous way to go to market for B2B SaaS companies, with B2B startups now going to market in a variety of ways, with plenty of hybridization and cross-pollination of approaches, too. You might have cast-and-catch style marketing to generate inbound leads that are converted with a sales assist on the one hand, with SDRs doing targeted, segment-and-sell outbound outreach on the other. Or, maybe you start with PLG and then go upmarket, having a free plan and enterprise sales living side by side, as analytics provider Heap does, which moved to a sales motion instead of just relying on self-serve and started closing $10,000 deals rather than $500 ones.[2]

Distinct GTM approaches exist because distinct market segments and market *shapes* exist. For SaaS companies that serve both ends of the B2B spectrum, the difference between the two can feel almost like running two separate businesses, as the opening quote from Manny Medina says. This has significant implications for how you position your product and company. In the coming chapters, we'll explore those extremes of the B2B spectrum, think about what "B2B" actually means, and look at positioning for reach, not niche.

The B2B spectrum

What does B2B actually mean? As a term, it's almost too broad to be useful. Imagine if you were trying to do positioning for the term itself...

"So, let me get this straight... it means you're a business... and you sell to other business... and that could be one solo operator, a handful of people in a dingy basement... or, err, enterprises with tens of thousands of employees in the Fortune 100?"

If we're trying to find a market segment to position your product and narrative around, we obviously need to be a teeny weeny bit more specific than just "B2B," which, at the extremes, could mean either:

- **Tiny outfits & B2C motions:** At one end of the spectrum, prosumer software is sold to business users in a very consumer-like fashion, where the lessons of B2C brand marketing are more applicable.
- **Gigantic enterprises & complex sales:** At the other end of the spectrum are complex sales to large businesses that take months to close *at best* and require a lot of story and positioning finesse.
- **The messy middle:** There is, of course, a broad mid-market between these two extremes, which can skew towards either end.

That gives us our classic SMB, mid-market, and enterprise market segments that we're going to explore over the next few chapters.

To put those segments in rough numeric terms, looking at US businesses[3] by employee count[4], for businesses *with at least one paid employee*, the spectrum runs from:

- **Small (1-99 employees):** ~8m businesses.
- **Mid (100-999 employees):** ~177,000 businesses.
- **Large (1000+ employees):** ~8,000 businesses.

Those cutoff points are somewhat arbitrary — if you raised the threshold for large businesses to a very enterprise-y 5000 employees, then you're going to get even fewer businesses in that bracket[5]. Either way, it's incredible to think that "B2B," as a label, covers (in the US at least) both 8,000,000 SMBs *and* 8,000 large and enterprise businesses. That's quite the spectrum!

> **What's best — SMB, mid-market, or enterprise?** If SMB and enterprise are wildly different, which one is best? VC Tomasz Tunguz suggested in 2017 that there was no "best" segment to pursue[6]. He argued that what matters more is team/segment fit in terms of strengths, capital, and insight. Depending on how you slice up the market, each third ends up with about the same number of employees within it anyway[7], and therefore, according to Tunguz, "The size of customer does not materially impact the opportunity to generate revenue" on a seat-selling basis. That is, given there are roughly equal revenue opportunities in each segment, what matters more is that your strategy matches your target customer base (assuming you've picked the right segment), which is what we are exploring here.

Mini-series: Positioning across the B2B spectrum

At the extremes of that spectrum, GTM and positioning are obviously going to look different. We explored the differences in pitching back in **chapter 23**, for example, when we looked at levels of awareness and different pitches for, say, a C-suite exec who may be unaware of the right-brain, story 1 trend you're discussing. And we compared that to a highly aware SMB end user who just wants to know how they'll use the tool in a story 2 left-brain sense.

In those chapters, we didn't spend much time on the SMB folks because we were developing our ideas about narrative, and that tends to be more applicable in the mid-market and above. We're going to flip that in this and the next couple of chapters, though, and use the SMB folks as a lens for positioning for reach, not niche, before touching on how some of those elements do and don't apply at the mid-market and enterprise end of the B2B spectrum, too. Let's dive into the world of mainstream positioning by starting with a look at Squarespace.

B2SMB and mainstream positioning

Squarespace, launched in 2004, does over $1B in revenue as of writing and has millions of customers. They're a great example of a very mainstream B2B-ish player selling to those 8M SMBs in the US (and many more outside the US). What's Squarespace's narrative for the market, in so far as there is one? It's very simple: Get started with a website! It's easy! It will look nice!

And then they plaster that message everywhere.

When it comes to Squarespace's positioning, here's a spicy question: Does it even exist? Does positioning itself matter when your story is that simple and your audience is that broad? Is the niche-to-mainstream playbook relevant here?

Surfing the web

Squarespace had to start somewhere, and in an interview with *Bloomberg*, Squarespace founder Anthony Casalena said[8]:

> *We really anchored the company around blogging in the early days and very quickly realized that people were creating blogs [and] websites for all kinds of reasons, and it became an interesting problem.*

Casalena says "anchored," but we might say "positioned." So there's a bit of a niche position to start with, but it didn't last long. As Casalena says, they "very quickly realized" the opportunity — the market pull — was much broader than that. Then, skipping ahead a few years, from the same *Bloomberg* interview, Casalena explained that:

> *Squarespace differentiated itself by providing a technical solution that also took design into consideration. During the platform's relaunch in 2012, the company intentionally sought to attract a more creative customer base. Back then, the Web was changing and the visual presentation of a portfolio was a challenge, but that style of design could be used for a variety of purposes. "We said, if we can solve for the websites that are hardest to make and the most visually difficult, we can solve that for other people and apply those techniques elsewhere," said Casalena.*

This was summed up in their mission statement from their 2021 IPO[9]:

> *Squarespace exists to help people with creative ideas stand out and succeed. We enable millions to build a brand and transact with their customers in an impactful and beautiful online presence.*

That's pretty broad!

Squarespace's run is a quintessential example of internet-first B2B — low cost, low touch, growing with the web, serving just about anyone, anywhere, and *staying* internet-first, for the most part. And what's particularly interesting in positioning terms is Squarespace's *lack* of niche focus. Blogging was a part of their value prop, but only a part[10], and the market quickly pulled a much broader product out of them for general content management.

Squarespace started out riding a very powerful wave, to be sure. We covered the importance of secular trends back in **chapter 30**. When Squarespace launched, the web was exploding, blogging was all the rage, and people wanted simple ways to manage their content. That's one heck of a wave to catch, and that is a large part of why what was originally a scratch-an-itch product turned into a mainstream phenomenon. That wave of energy was there, and Casalena, as founder, was able to use mainstream tactics like Adwords (remember, this is the mid-'00s) to drive early sales, too. The mainstream wave of the web — and some incredibly savvy marketing — would ultimately take them all the way from dorm room to IPO.

In light of all this, does positioning even matter?

The B2C brand marketing school

There's a very influential school of thought in the B2C marketing world that says, essentially, no, it doesn't matter — and they insist that goes for B2B, too. Niche positioning and targeting is a bad idea and for your brand to grow (and that's brand, not product), you should target *all* buyers in a given segment. You should try and maximize your reach, not limit it by focusing on a niche. That is (so the theory goes) if you want to be a big brand, study what big brands do, and then do likewise — serve the whole market, don't fret too much about differentiation, and instead worry more about brand *distinctiveness* so you get embedded in consumers' brains and they remember you when it matters — when they want to buy. Roughly speaking, the greater the number of people who *can* buy you (by going broad, being memorable, and being available), the greater the number of people who *will* buy you, and therefore you'll grow.

In other words, broad market penetration trumps narrow market segmentation. Marketing, in this world, is mostly advertising, and advertising becomes mostly 'creative publicity,' not persuasion.[11]

This school of thought is championed by Professor Byron Sharp and the Ehrenberg-Bass Institute (EBI). Sharp is the author of the popular marketing book *How Brands Grow,* first published in 2010. It's hard to overstate Sharp's influence in the B2C brand marketing world, especially in FMCG (fast-moving consumer goods), and he's certainly not afraid of trumpeting his approach as the empirical, research-driven, scientific approach to marketing.

In the B2C marketing world, the ideas of Sharp and the EBI are debated ad nauseam. In the B2B world, they often don't come up at all. But if Sharp and the EBI have the scientific keys to the marketing kingdom, isn't that a problem for us B2B folks?

Twin memes

> *Decades of research into the patterns in buying behaviour and marketing metrics has led to the surprising conclusion that brands compete for custom primarily in terms of mental and physical availability. Brands that are easier to buy, for more people in more occasions, get bought more often.*
>
> – Byron Sharp[12]

Before we get into the nitty-gritty of the EBI's point of view and the issues of maintaining market share vs. getting into the market in the first place, let's explore two of the EBI's most memetic ideas, ones that are at the core of the Sharp and EBI worldview of marketing. These are the foundational ideas of positioning for mainstream reach, and they shed light on Squarespace's strategy, too.

The gist is that when you're selling commodities, you're essentially in a never-ending battle to see which brand can achieve and maintain:

- **Mental availability** by embedding themselves foremost in the minds of prospects *in buying situations* (i.e., being the most salient, not just well-known).[13]
- **Physical availability**, i.e., who can make themselves most available *to be bought* (e.g., by literally being on the supermarket shelf or having a store near you).

Sharp draws on a variety of 'laws' beyond these two ideas that fill out the broader brand marketing picture, but it's the brass-tacks logic of these twin concepts that's at the heart of the EBI approach. It's a very uncompromising view, too, with Sharp writing that "No marketing activity, including innovation, should be seen as a goal in itself, its goal is to hold on to or improve mental and physical availability."[14]

That's because:

1. People who don't remember you when they're thinking about buying won't buy you.
2. People who can't physically access you won't buy you even if they *do* remember you.

Therefore, in B2C terms, brands like Starbucks, for example, might try and maximize the number of people who can buy them by opening as many stores — which are all distinctively Starbucks — as possible. That makes them both mentally and physically available when someone wants a coffee. That is, in part, what big brands do; that's why they're big: they're very available to sell to a lot of people.

In B2B terms, plenty of folks may recognize "Salesforce" as a brand (i.e., high brand awareness), but if you live in the marketing world and want a CRM, your first thought may be "HubSpot" instead (high brand salience). In that case, HubSpot wins the mental availability game, where it's the thought that comes to mind when considering a purchase that matters (that is, when customers shop from memory). That's helped by physical availability, too, and in the digital world, that might mean showing up in the search results for "marketing CRM," for example.

> **"{Product} vs" strategy:** A popular challenger approach in B2B SaaS has been to create pages for "{product} vs" or "{product} alternative" pages in the hope of capturing this specific search traffic, and these rules of thumb help explain why this can work. You're trading off the high mental availability the incumbents have created, and because they can't be all things to all people (try as they might), this creates a wedge opportunity (chapter 46) for you to exploit. Sometimes, it's better to orbit around an incumbent with a strong center of gravity — given they have the strongest mental availability — than it is to try and create your own gravitational pull from scratch.

Being easy to buy in B2B

How you get to that point of high availability is, of course, another thing (we discussed Salesforce and HubSpot and the secular changes they rode in **chapter 31** and we'll touch on adoption below). Still, there is a lot to be said for simply being easy to buy, B2B or B2C.

This much more nuts-and-bolts approach to brand marketing stands in stark contrast to the (often misplaced) attention brand marketers apply to their brand in terms of values and purpose and so on. Some brands *do* have strong symbolic value[15], but at the same time, some folks just want a coffee with a twist, so why not just focus on being the easiest to buy, i.e., the most mentally and physically available?

This approach is particularly relevant at the SMB-focused B2B end of the spectrum, where SaaS brands are closer to Starbucks than they are to Salesforce. Here, this logic of mental availability really starts to matter.[16] Squarespace, to pick up our running example, tried to go as wide and broad as possible, and this is where concepts like mental availability and brand salience are particularly useful.

To pull this off, you need an imprint-your-brand-on-their-brain strategy to build that mental availability — to be remembered, which the EBI sees as the point of advertising — and Squarespace famously found an untapped way to do just that, as we'll explore in the next chapter.

> **Canva's mainstream positioning and physical availability in the SERPs:** Canva, also on the B2C end of B2B SaaS, built a great brand and an incredible distribution engine by doggedly pursuing mass 'physical' availability online via SEO. If you want to design a Christmas card, certificate, invitation, or whatever it might be and search with those keywords, you better believe Canva is going to be at or near the top of those search engine result pages (SERPs), as this case study breaks down.[17] Canva has niche roots — they started as a yearbook design and printing service — but staying niche obviously wasn't the path to becoming a business worth billions; going relentlessly mainstream was, reaching into every SERP for every mainstream design need, being as 'physically available' as they possibly could be.

Marketing... science?

The twin ideas of physical and mental availability are a core part of Sharp and the EBI's broader project of marketing science. Some folks think "marketing" and "science" are about as compatible as minced meat and ice cream. That's why it's often a surprise for folks to learn (including yours truly at one point) that, despite having spent years in marketing or marketing-adjacent roles, there are actually decades of empirical research into consumer behavior, including predictable 'laws' of marketing, that we could be drawing on.

On the face of it, that sounds great. Who doesn't %^&*ing love science? And science applied to marketing no less, a discipline with a long history of feelpinions and anecdotes (or straight-up theatre, as we looked at in the previous chapter). The EBI, part of the University of South Australia and where Sharp calls home, certainly loves science, and they have made that a core part of their brand.

But we have to ask: the science of which part of marketing, exactly? For getting into the market or staying in the market? Because it turns out they are two quite different things. Now, dear reader, I need to give you a little bit of backstory, but it perfectly sets up what we're about to discuss, so strap in (or skip ahead — your choice!).

> **Choose your own adventure:** We are, dear reader, about to explore passive vs. active attention, Sharp and the EBI's narrative, and various criticism of the EBI's approach. If you want to nerd out on this stuff, keep reading, but if you'd rather stay focused on Squarespace and some of the practical SaaS implications of these theories around positioning for reach, skip ahead to chapter 44.

The missing "B"

The EBI was originally called the Marketing Science Centre. In 2005, they changed their name to the Ehrenberg-Bass Institute for Marketing Science as a "tribute" to two academics who made major contributions to the field, German-born British academic Andrew Ehrenberg (1926-2010) and American academic Frank Bass (1926-2006).

Why am I telling you this?

- **Bass** was famous for his work on new product adoption, creating a mathematical model of product adoption, and contributing to Everett Rogers' *Diffusion of Innovations* model. That's the very same model we've been working with in recent chapters, as it's the one *Crossing the Chasm* is based on.[18]
- **Ehrenberg** was famous for his studies of consumer behavior and the models he developed, deriving from these models of buyer behavior "a view on advertising for established brands."[19]

That last part — "for established brands" — is important, and this split between new product adoption on the one hand, and models of buyer behavior for established brands on the other, is a fundamental distinction we'll explore in this chapter.

One more bit of trivia. The EBI is a continuation of an organization Ehrenberg founded called the Research and Development Initiative, which was "funded by businesses internationally to pursue and disseminate quantitative marketing knowledge." That eventually became the EBI in South Australia, and Ehrenberg was succeeded by Byron Sharp as its Director.[20] Today, the EBI, per its LinkedIn description, "provides customised marketing research solutions, marketing and advertising consultancy to large corporate clients," that is, enterprises. (That the EBI sells to enterprise clients will become particularly relevant in a moment.)

So, to recap:

- The EBI is named after two dudes — Ehrenberg and Bass.
- Ehrenberg ran an organization funded by big, multinational brands to do market research, which the EBI carries on, doing independent research

and consulting for these large corporate clients.
- Sharp succeeded Ehrenberg and carries on his legacy.

That is, despite Bass being the "B" in the EBI's name, ol' mate's work doesn't seem to come up much in what the EBI publishes, which is very much focused on "a view on advertising for established brands" derived from Ehrenberg's (and others') models.

This may seem highly arcane, dear reader — and it is — but it's important to understand the Ehrenberg-to-Sharp lineage of ideas simply because Byron Sharp really is one of the most influential voices in modern marketing and *what* he preaches is very much of the Ehrenberg big brand staying-there school, not Bass and the getting-there school of adoption. (*How* Sharp preaches, however, is straight out of our clarity strategy playbook, which is particularly fascinating, as we'll soon see.)

That's the backstory, but who cares what marketing school he advocates for?

The difficulty is, dear reader, that Sharp and his many followers like to stress that marketing — all marketing — is about hard scientific laws, like the laws of physics[21] or evolution[22]; laws that apply universally across the B2C *and* B2B world.

There is some truth to that, as we'll see. When the EBI are right, they're really right, but when they're wrong, the baby often goes flying out with the bathwater. Therefore, it pays to unpick what's reasonable from what isn't.

Getting there vs. staying there

The absence of the influence of Bass on the EBI's work — the missing "B" as I like to put it — highlights the "getting there vs. staying there" gap we've between discussing, especially when it comes to positioning, where new product adoption looks very different to maintaining mature product market share and dominance.

This has big implications. On the one hand, we have one of the most prominent voices in the field of marketing declaring universal scientific laws that all marketers should follow. In this world, positioning is out, salience[23] is in. That's important to know.

On the other hand, we have someone — Frank Bass — whose empirical work on new product adoption was apparently worthy enough to be the "B" in "EBI," but gets scant mention in the EBI's actual output, which is, as mentioned, tailored to their big-brand "sponsors" (i.e., clients). That's also important to know.

That puts us in a strange place. When we're discussing startups, we're acting as descendants of the line of research Bass contributed to — new product adoption. And when we're discussing sales-led B2B SaaS, especially for complex sales, we're usually in the segment-and-sell world, not the cast-and-catch world of mass advertising for established brands.

Nevertheless, it's not surprising that different market stages and segments have different GTM approaches and different methodologies of marketing. There are things we can — and will! — learn from the Ehrenberg/Sharp school; we just need to keep them in context.

Some B2B startups *do* leverage B2C-style cast-and-catch, and not all B2B SaaS companies are startups, either, with the OGs now a quarter of a century old. Knowing where you are on the adoption curve matters a lot, and the point of the *Diffusion of Innovations* model is that you want to get to the mainstream sooner or later, and that mainstream approach is what we're exploring over the next few chapters.

Let's put that on pause for a moment, though, and understand why Sharp and the EBI's work lands in the marketing world in the first place and think about how branding works at all.

How branding works: passive vs. active attention

To start to understand how branding works, let's zoom out a bit and think about the core theme of this book: attention. Our pursuit of clarity strategy has been about trying to derive appropriate lenses and activities from our simplified version of McGilchrist's hierarchy of attention with our story 1/story 2 radar-and-laser-beams model. But that's not all there is to attention. In more general terms, there's also the question of the *quantity* of attention, not just the *mode* of attention. (We touched on this way back in **chapter 5**, when we looked at clarity strategy as narrative, positioning, *and* brand.)

We can think about quantity of attention in buyer terms as:

- **Passive attention:** Where someone is paying next to no attention to the thing being broadcast at them, which is the case for most consumer advertising.
- **Active attention:** Where someone is very actively engaged in understanding a product with a view to purchasing, which is common in complex B2B (and some B2C) sales.

If you're familiar with Daniel Kahneman's famous 2011 book *Thinking, Fast and Slow*[24], you will no doubt see shades of System 1 (automatic — fast and instinctive) and System 2 (effortful — slow and deliberate) modes of thought here. (We touched on *Thinking, Fast and Slow,* from which we derived 'story 1' and 'story 2,' back in **chapter 5**.)

Brand advertising, in a sense, exists primarily to train folks' System 1, heuristic-driven mode of thought to deliver a certain memory association when they encounter a given situation. (E.g., feel thirsty, see Coke vending machine as a cue, brain searches memory, retrieves brand association that Coke = fun to drink, consider purchase.)

The world of complex B2B sales, however, is far more deliberative — much more System 2 oriented – and that appears to be outside the purview of the EBI.[25] While the System 1 and System 2 modes of thinking are a fantastic insight (and one that McGilchrist comments on, for what it's worth[26]), let's just use "passive" vs. "active" attention to think about attention quantity here. While it might be obvious to anyone outside of marketing that consumers don't pay much attention to brands or advertising, marketers, especially in the B2C brand world, do pay a vast quantity of attention to these things, and this creates a classic attention mismatch.

Brand/consumer attention mismatch

In the B2C world, Sharp and the EBI's main argument is that consumers apply extremely limited, passive attention to brands. They're just not thinking about them very much. Brand marketers, however, apply a great deal of attention to the meaning they want to imbue in the brands they're advertising — who they're targeted at, what they stand for, how they're positioned, what purpose they represent, and so on.

The problem is consumers, for the most part, just don't care. They do perceive some differences, but they're just not that into brands as a whole. They don't want to love them, have a relationship with them, or otherwise spend more than fleeting amounts of attention on them. They're operating under the "law of least effort".[27]

As an example, I spotted a Chief Brand Officer on LinkedIn bemoaning the death of a consumer-focused delivery company. But they weren't just sad about the demise of the service (though they did mention that); they were sad about the death of the company's logo and brand persona they had worked on. And not just because they worked on it, but because the brand was a person — a 'best friend' you'd 'want to hang out with,' someone who was 'never afraid to be who they were,' and that, apparently, was what this executive would miss 'most of all.' What was this BFF brand? It wasn't actually a person, character, or even a mascot. It was, as far as I could tell, a logo with a non-descript outline of an animal.

I suspect most consumers, on the other hand, if asked if they'd miss this brand, would probably respond with a confused, "Who was this?" (They might miss their deliveries, though.)

This is another fundamental attention mismatch.

At the very start of this book, we looked at the attention mismatch between right-brain, story 1 change narratives, and left-brain, story 2 narrow-focused positioning. Startups sometimes like talking about investor-oriented, dent-in-the-universe, big-picture change narratives; buyers may just want to know where you fit and how you're different. That's the mismatch.

Here, we have another attention mismatch, this time between overly focused brand marketers and largely indifferent consumers. This cuts to the heart of what branding is and how it works, and Sharp & co have been very successful in pushing the brand marketing pendulum back towards acknowledging customer reality and not brand marketer fantasy.

A marketer's job, in this world, is mostly to embed and maintain relevant brand memories — i.e., create brand salience — in the brains of as many consumers as possible.

B2B brand buying

This can be true in some B2B contexts, too. Businesses buy lots of commodities — fuel, toilet paper, insurance, etc., and buyers shop much like B2C consumers shop in those contexts. They're just not thinking that hard about what a toilet paper brand means and would prefer to buy whatever's available — something they've heard of and can see on the shelf or access at a store close to them. That's it.

And this can be true, to a point, with technology as well, as we'll continue to see with Squarespace and other SaaS brands in the coming chapters.

But it's plainly not always true.

Marketer vs. buyer attention

The EBI has done the world a service in pointing out this fundamental attention mismatch and providing plenty of evidence for how consumers actually behave. I like to sum up this mismatch as:

- **Active *marketer* attention:** Marketers care a lot about what brands mean.
- **Passive *consumer* attention:** Consumers just want to buy something they recognize and get on with their day.

That's a great insight. The problem, however, is that Sharp and the EBI restrict *their* attention to this B2C and B2C-adjacent marketing and buying context, deduce supposedly universal laws of marketing physics from this specific scenario, and then insist these laws apply everywhere, including all B2B buying. They argue we're all human, whether we're in B2B or B2C contexts, so the rules still apply. That would be true if our attention remained the same in all contexts, but not all buyer attention is passive. There is also active System 2 buyer attention:

- **Active *buyer* attention:** In the considered world of B2B sales, especially with "we" not "me" buying, buyer attention is far more engaged in the buying process.

  This active buyer attention, in turn, shapes marketer attention:

- **Active *product* marketer attention:** Marketing, in this world, engages more from the perspective of product marketing, not brand marketing, where innovation, differentiation, and positioning matter far more, as we have seen and will see.

  That gives us, at the extremes:

- **Brand** marketing targeting **passive** consumer attention in low consideration, low innovation sales.
- **Product** marketing targeting **active** consumer attention in high consideration, high innovation sales.

That's *not* a universal description of B2C vs. B2B marketing — mature B2B companies will end up doing both because they'll need to engage folks who are in *and* out of market, capturing the attention of those actively looking while building memory associations for those that aren't about to buy (as we saw the bootstrapped Bidsketch did in **chapter 43**). But B2B marketers further up the B2B spectrum shouldn't rely on marketing science that treats buyer attention as though it's always passive when, in complex sales, it clearly isn't.

Tech vs. TP

Active attention tends to exist more in the B2B technology world simply because:

- Businesses have the need and budgets to pay for very specific, complex tools.
- Those tools can have far-reaching impacts on organizational performance.
- Buyers are often betting their reputation and perhaps even their career on those tools.
- Therefore, you better believe folks are going to apply more active attention and be more deliberative than when they buy, say, toilet paper.

To meet that demand, marketers and sellers use segment-and-sell GTM techniques, as we've been discussing, often skipping the whole brand marketing playbook.

But I want to strike a careful balance here. There can be meaningful overlap, too. It would be a mistake for us to imagine that more brand-style, passive, heuristic-driven attention doesn't ever exist in B2B buying, either. People can't buy what they've never heard of. Memory building can still work over the long term. There are important lessons we can learn from the B2C world, which is what we'll unpack in this mini-series. Likewise, the logic of market penetration is something we need to keep in mind so we don't become overly fixated on finding a niche and staying in a niche.

So let's not throw out the baby with the bathwater. If the consumer-side "science" has something to say about positioning and market strategy in general, then we really should understand that to see if they influence our B2B narrative and positioning work, too, and in the coming chapters we'll see how that plays out.

Fiiight!

Your humble author is not, as you might imagine, the first person to take issue with the EBI's view of marketing. And I do say the EBI, which draws on decades of research and includes a number of researchers, including Professor Jenni Romaniuk, who has branding books of her own and came up with the idea of "mental availability" that we touched on earlier[28]. (That is, it's not all Byron Sharp, even if "Byron Sharp" has become a proxy for their work, and I'm sure he'd be the first to admit that.)

Nevertheless, given his outspokenness, Sharp — the "Dark Lord" of marketing, as former marketing professor Mark Ritson humorously puts it, tongue firmly in cheek — attracts quite a range of responses from within the B2C world, too. Ritson is perhaps Sharp's most vocal sparring partner, offering both effusive praise and biting criticism[29], suggesting Sharp is broadly right about brand marketing (in the brand advertising sense) but too extreme.[30]

Despite the criticism, many marketers, including executives at big brands that Sharp targets with a highly differentiated message — **ahem** — take Sharp's claims about brand advertising as gospel.

Why?

Well, he wrote the book, and he dominates the narrative.

> **Sharp with a grain of salt:** Others offer quite pointed criticism about Sharp's methodology[31], with Marketing professor Felipe Thomaz, in an interview for *Contagious*[32], saying the EBI's approach is a simplification that only works when you're already large.[33] I would also suggest that *How Brands Grow* is overly left-brained, suffering from the same symptoms we saw in the previous chapter — faux precision, a lack of context, a need for certainty, absolutely no room for ambiguity,

> and a view of reality as snapshots in time, not a dynamic, evolving whole. That's not to say it's wrong — it may well describe the world it purports to (big established brands), but it's striking how different this Ehrenbergian slice-of-time approach is compared to the change-over-time nature of Bass and product adoption. We need the E and the B.

I sunk your battleship

To a certain segment of B2C marketers, finding Sharp's work — a world of black-and-white answers and absolute certainty — is like being rescued by a battleship while drowning at sea. Rather than being tossed about by the waves of opinion and flash-in-the-pan trends (*"What's our metaverse brand strategy, maaan?"*), not to mention endless hand-wringing about 'brand purpose,' these folks could now blast holes in the antiquated, backward, fad-driven approach to marketing they had found themselves in and ride the good ship Science into a bright, modern future, leaving the archaic, pre-scientific, "medieval" views of their unenlightened brethren behind.

I say that somewhat tongue in cheek, but only somewhat. Sharp, in *How Brands Grow*, writes that just as medical professionals "enthusiastically and universally taught and practised bloodletting" for 2,500 years despite it being useless/harmful, today "marketing managers operate a bit like medieval doctors – working on anecdotal experience, impressions and myth-based explanations."

But why is it that Sharp owns the 'evidence-based' position — **ahem again** — in the marketing landscape? Why is Sharp and the Ehrenberg-Bass Institute the bastion of capital-S Science when it comes to market research? They're hardly unique in doing academic research on brand performance. So why did their work capture the imagination of a significant chunk of a profession?

Well, narrative, for one.

> **Taking evidence seriously:** To be fair to Sharp, it's not as though there isn't precedent for this kind of thing, either. Ignaz Semmelweis famously discovered in the mid-1800s that the lives of many mothers could be saved if doctors washed their hands, but he was rejected by the medical establishment of his day and ended up dying in an asylum. Sometimes 'serious' professional people will carry on unwittingly causing a great deal of harm because they don't like what the evidence suggests.

Narrative selling vs. brand marketing

There's an unexplored aspect of Sharp's work that's particularly relevant to us, and that's the power of narrative. Sharp is *incredibly* good at marketing and *selling* his ideas through narrative — a highly differentiated one at that — and his approach has all the hallmarks of classic story 1/story 2 clarity strategy.

First, we've got a great big change story: everything you know about marketing is wrong. *How Brands Grow* lines up common beliefs and knocks 'em down (or at least tries to). The work itself might be overly left-brained, but the meta, right-brain change narrative wrapped around it is as stark as they come. There's even a table in the book comparing the tenets of this "New world view" vs. the "Past world view." Andy Raskin, from **chapter 20**, would be proud. And hey, you don't want to be one of those medieval blood-letters, do you?

Sharp's organization, let's remember, works with big brands spending big bucks — i.e., enterprises. I imagine this story 1 narrative helps persuade right-brain executives much along the lines of the C-suite-focused sales narrative outline we explored back in **chapter 23**.

Then we've got a highly memeable, self-referencing group of buyers — our poor misguided marketers, who constitute the buyers for these ideas — that we discussed in **chapter 40**. (Again, not that marketers are remotely unique in this regard!)

And finally, we've got the actual memes (we looked at memes back in **chapter 15**), where Sharp & co boil down decades of academic thinking into two potent 'information packets with attitude' that we discussed above.

All of this makes for incredibly effective communication and narrative framing. Everyone knows what Sharp stands for. There's a strong right-brain sales narrative on the one hand and very transmissible ideas for folks in the trenches on the other. And, of course, there's the nitty gritty left-brain substance, too, which contributes to Sharp's positively evangelical zeal for 'educating' his peers. We've talked a lot about stretching your narrative both out and down, and Sharp and the EBI do both.

This is classic enterprise sales and marketing, relying on a strong, differentiated narrative. That narrative, ironically enough, is that this kind of enterprise sales and marketing doesn't exist, because marketing "science" is all passive-attention brand advertising, all the time! Despite the irony, Sharp's narrative is nevertheless instructive in a "do as I do" sense when thinking about complex sales and "do as I say" sense when thinking about the B2C end of the B2B spectrum.

> **Confidence vs. cynicism & industry research:** I don't want to suggest

> this use of narrative is in any way an exercise in cynicism, either. I'm sure Sharp believes it and as far as the B2C brand world goes, there's the historical research, EBI big-name clients, their popular book(s), and even a textbook behind their work. That's where the confidence comes from, and while it certainly does stray into dogmatism at times, there's a depth of substance there that many B2B outfits could only *dream* of when it comes to their own narratives. If you're struggling for confidence in your narrative, consider beefing up the substance, not just the form, and look at what you can draw from. Sharp didn't invent all of this; he and his colleagues tapped and continued the research that had been going on for decades, research that was funded, in part, by industry partners. What research can you tap that drills down to that reality layer of your market and does some serious myth-busting?

Positioning for reach

With all that said, there is a strong internal logic in Sharp's work that sheds light on an approach to positioning that we've yet to fully unpack — reach. That's particularly useful at the consumer-like, B2SMB end of the B2B spectrum for companies like Squarespace and also plays a role in the enterprise and mid-market, too, *at the right time*, as we'll see.

There's one sticking point, however. The EBI's view is that positioning is inherently a mistake. We must therefore ask: Is everything we've discussed about niche positioning wrong? Was all that hard work of the last few chapters for nothing? Let's find out in the next chapter.

Notes

1. Medina, Manny. In conversation at Cloud 100, "The Go-to-Market Learning Curve," posted September 17, 2020. Edited for clarity. **https://www.youtube.com/watch?v=Hv6Ngj4Vjaw**
2. Tan, Mitchell. "How adding sales turned $500 into $10,000 deals for Heap," HeadsUp blog, November 2021. **https://www.headsup.ai/blog/self-serve-sales/**
3. These numbers are drawn from 2021 data available here: **https://www.census.gov/newsroom/stories/small-business-week.html**
4. I'm using simple small/mid/large labels here because segment thresholds vary. That is, I'm not painting this as the one true enterprise segmentation, for example.
5. As Jen Abel does in her groupings, which inspired this section. See Abel's 2020 tweet here, for example: **https://twitter.com/jjen_abel/status/1339564916534075393**

6. Tunguz, Tomasz. "SMB or Enterprise - Which is the Better Go To Market in SaaS?," April 13, 2017. **https://tomtunguz.com/optimal-path-smb-mid-ent/**
7. Tunguz, Tomasz. "What Average Contract Value is Best for a SaaS Company," May 24, 2016. **https://tomtunguz.com/market-cap-saas-by-acv/**
8. "How an Idea for a Website Turned into Squarespace," *Bloomberg*, November 27, 2018. **https://www.bloomberg.com/company/stories/idea-website-turned-squarespace/**
9. Squarespace, Inc S-1. May 3, 2021. **https://www.sec.gov/Archives/edgar/data/1496963/000110465921059918/tm213918-11_s1a.htm**
10. Check out their early homepage from 2004, for example: **https://web.archive.org/web/20040701013348/http://www.squarespace.com/do/display/external/Frontpage**
11. To borrow the title of Ehrenberg, Barnard, Kennedy, and Bloom's 2002 paper "Brand Advertising As Creative Publicity." **https://www.researchgate.net/publication/260555133_Brand_Advertising_As_Creative_Publicity**
12. Sharp, Byron. *How Brands Grow.*
13. Sharp describes the difference between this and brand 'awareness' in his March 2011 blog post, "Mental availability is not awareness, brand salience is not awareness." **https://byronsharp.wordpress.com/2011/03/26/mental-availability-is-not-awareness-brand-salience-is-not-awareness/**
14. Sharp, Byron. "How Mental And Physical Availability Drives Brand Growth," *Branding Strategy Insider*, May 30, 2023. **https://brandingstrategyinsider.com/how-mental-and-physical-availability-drives-brand-growth/**
15. Stephen Chen's criticism of *How Brands Grow* and its treatment of Harley-Davidson is particularly relevant here: **https://www.linkedin.com/pulse/pseudo-science-how-brands-grow-part-2-stephen-chen/**
16. Salesforce tries to play the mental availability game with mascots these days, including its "Astro" character. Characters are a common recommendation of the EBI simply because they're memorable. Western companies still have a long way to go to match the deranged wonder of Japan's legion of truly memorable mascots, though, who you may have seen on the magnificent Mondo Mascots Twitter account: **https://twitter.com/mondomascots**
17. Simmonds, Ross. "The Canva Backlink Empire: How SEO, Outreach & Content Led To A $6B Valuation," June 24, 2020. **https://foundationinc.co/lab/canva-seo**
18. Bass, Wikipedia says, contributed to Rogers' *Diffusion of Innovations* (**https://en.wikipedia.org/wiki/Bass_diffusion_model**), which Geoffrey Moore ran with as the Technology Adoption Life Cycle, introducing the chasm concept.
19. From: **https://en.wikipedia.org/wiki/Andrew_S._C._Ehrenberg**
20. Ibid.
21. Here's Sharp in June 2022, describing his book: "[*How Brands Grow*] deals with the basic rules of how brands compete and how customers buy. It's like the laws of physics that don't change even after the advent of digital technologies." **https://www.marketingscience.info/byron-sharp-marketers-still-dont-understand-how-advertising-works/**
22. Sharp, in a September 24, 2020 blog post titled "The importance of descriptive research," writes that science "needs lots of descriptive research upon which to build

discover and document laws, and then build theories that fit the laws," uses evolution as an example, and finishes with "I urge you to be proud of descriptive research. Let's work to educate our peers." **https://byronsharp.wordpress.com/2020/09/24/the-importance-of-descriptive-research/**

23. The idea behind brand "salience" is that it's not enough for buyers just to be *aware* of your brand — I might be very aware of McDonald's but never eat there. What matters instead is how 'salient,' i.e., important or relevant your brand is when someone is about to buy. That is, I might be equally 'aware' of both McDonalds and Starbucks as brands, but if I always think of Starbucks when I want a quick coffee but never think of McDonalds when I want a quick meal, Starbucks has far higher brand salience. That, in turn, becomes a far better measure of how our brand is doing in the market (or so the theory goes) rather than just trying to measure who has heard of us, and the simple math is that the more people who think of your brand when they want to buy = the more people who will buy you = the more you'll grow.
24. Kahneman's System 1 and System 2 in *Thinking, Fast and Slow* was the inspiration for my 'story 1' and 'story 2' labels thanks to an observation a client once made.
25. To be fair to Sharp, he was right to be critical of behavioral economics early on, a much-diminished field that's been plagued by failure and fraud. Sharp wrote in a July 2014 *Marketing Mag* piece that "What really worries me is that this infatuation with theatrical psychology lab experiments – that show participants in lab experiments can be nudged without knowing it – is bringing back a discredited theory of brand image of besotted/manipulated consumers." **https://www.marketingmag.com.au/tech-data/byron-sharp-behavioural-economics-is-not-the-biggest-story-in-marketing/**
26. McGilchrist, for his part, comments on *Thinking, Fast and Slow* in his 2021 book *The Matter with Things,* writing that System 1 and System 2 *do not* map to the left or right hemisphere *per se*, and that Kahneman agrees that both hemispheres are involved in heuristics (System 1) and deliberation (System 2). McGilchrist, however, suggests it is "the left hemisphere that is considerably more prone to settle for its heuristics, [while] the right hemisphere [is] more inclined to be cautious." In the B2B world, we still deal with snap judgments, but often with an additional degree of caution that comes with highly considered purchases, too.
27. From *Thinking, Fast and Slow*: "A general "law of least effort" applies to cognitive as well as physical exertion. The law asserts that if there are several ways of achieving the same goal, people will eventually gravitate to the least demanding course of action. In the economy of action, effort is a cost, and the acquisition of skill is driven by the balance of benefits and costs. Laziness is built deep into our nature."
28. Or get it straight from the source. Romaniuk, for example, was interviewed by Samuel Scott in January 2022 for *The Drum* in an article titled "In 2022, should B2B content marketers focus more on advertising?" **https://www.thedrum.com/opinion/2022/01/14/2022-should-b2b-content-marketers-focus-more-advertising**
29. Ritson, Mark. "Mark Ritson: Byron Sharp is wrong – of course brand perceptions influence sales," *Marketing Week,* May 25, 2021. **https://phvntom.com/mark-ritson-byron-sharp-is-wrong-of-course-brand-perceptions-influence-sales/**
30. See Ritson's comments in this 2021 podcast episode of *Brand Tuned,* for example: **https://www.brandtuned.com/Mark Ritson on Differentiation and Positioning**
31. Stephen Chen, for example, raises serious questions about Sharp's methodology and

use of data, accusing him of pseudoscience (!): **https://www.linkedin.com/pulse/pseudo-science-how-brands-grow-part-1-stephen-chen/**

32. Thomaz was interviewed by James Swift in "'Byron Sharp is ignoring 60 years of published work," *Contagious,* September 30, 2022. **https://www.contagious.com/news-and-views/byron-sharp-how-brands-grow-ignores-60-years-of-published-work-felipe-thomaz**

33. Note that, per our discussion of the missing "B" in the EBI, Thomaz only talks about 'Sharp and Ehrenberg' not 'Sharp and Bass,' saying, "Could a firm replicate Sharp's and Ehrenberg's results? Sure! Provided they are already large, operating in a static marketplace with other large competitors, and everyone is largely identical. But for everyone else, this will break down. Altogether, the work highlights the advantages of being large, but cannot address how to grow."

44

The B2B spectrum II — Squarespace cont., dynamic positioning, & niche vs. reach

> Mr. Casalena said that since he founded the company, he has spent over $1 billion promoting his brand across all platforms. The revenue from those ads, plus all that attention, helped the company go public in May [2021], landing Mr. Casalena himself $2.4 billion in stock in the $6.6 billion company.
>
> – *The New York Times*[1]

In the last chapter, we:

- Introduced the B2B SaaS spectrum of SMB, mid-market, and enterprise.
- Started at the SMB end by discussing Squarespace.
- Looked at the ideas (and narrative) of one of the most influential voices in B2C brand marketing, Byron Sharp of the Ehrenberg-Bass Institute (EBI).
- Discussed the theory of passive vs. active attention and the attention mismatch that can exist between (busy, indifferent) consumers and (invested, focused) brand marketers.

In this chapter, we're going to keep exploring these ideas, looking at how mainstream advertising works from the EBI's point of view, and what that means for positioning as a dynamic tool that can go from niche to reach and back again.

Channel innovation

Squarespace's great claim to fame, apart from their product, is their path *to* fame: their adoption of podcast sponsorship, discovering a gold mine of ROI — and mainstream mental availability, as we described last chapter — before just about anyone else.

In 2009, according to the *New York Times*[2], Squarespace founder Anthony Casalena took a big bet, dropping $20,000 on sponsorship of a podcast called "This Week in Tech." Squarespace used post-purchase surveys for attribution, and lo and behold, a third of their new sign-ups heard of Squarespace from their podcast ad.[3]

Squarespace, the *NYT* wrote, then started advertising on hundreds of podcasts a month. Their small podcast ads team struggled to spend their budget — there weren't enough podcasts to sponsor. They even thought about promoting *the idea of podcasts themselves* just so they'd have more podcasts to advertise on. The fact Squarespace was so ubiquitous in podcasts became something of a meme — Squarespace *was* the podcast ad company. And all that investment — more than $1 billion across all platforms —paid off, as the opening quote above makes clear.

> **Casper and the ghosts of DTC past:** While we're talking about Squarespace's success here, it's worth noting that while Squarespace may have pioneered the approach, they were hardly unique in building their brand through podcast advertising. Casper, for example, the tech-flavored, direct-to-consumer, mattress-in-a-box innovator, also plowed mountains of cash into podcast advertising. As a brand, Casper aspired to be the "Nike of sleep." They had, from my vantage point at least, strong mental availability through both earned and paid media around their innovative approach to mattresses and strong physical availability through their online DTC model. It turns out, however, that slinging slabs of compressed foam isn't a particularly defensible model, and the market quickly became saturated. Casper made it to IPO in early 2020, but that was labeled a "disaster."[4] They never turned a profit, their share price sunk, and they were acquired by a PE firm less than two years later. The new CEO rejected the lifestyle brand approach, declaring they were "Not in the business of not making money anymore" and were instead focused on just being a mattress retailer.[5] Many B2B startups are, in a post-ZIRP world, also no longer in the business of not making any more, and Casper's experience is a sage warning that a successful brand doesn't always equate to a successful business, and distribution alone isn't enough.

Pavlov's bell

> *"The most important search engine is still the one in your mind." This statement makes a profound point all marketers should internalize about buyer behavior: most purchases start not by searching Google, but by searching our memory. If you believe most buyer behavior starts with memory, it then follows that the primary job of marketing is not to generate clicks, but instead to generate memories.*
>
> – The B2B Institute[6]

In the Ehrenberg-Bass Institute's marketing world, traction is built and maintained, as Squarespace demonstrated over time, by generating enough consistent and ongoing marketing touches to create sufficient memory structures in consumers' brains that link your brand to a given need. This is the world of 'mental availability,' and it might pertain to needs like creating a website, buying underwear, or posting a job ad (to borrow from other podcast advertisers). When consumers want to do *the thing*, that is, when they hit a 'category entry point' as the EBI puts it, if your brand is the one that comes to mind, you get the sale — *physical* availability permitting — and that's how you grow.

To maximize marketing efficiency in this world, you need to be selling a mass-market commodity to make the math work. Squarespace commodified small business websites, poured $1B+ over its history into winning the mental availability race through innovative mass-market distribution, and created conditions where, when any SMB owner, anywhere, ever, wants a website (or needs to recommend a solution to others), the first thought they conjure in their minds is "Square...space? I think that's what it was called?"

The memory comes first, the Googling second.

That is mental availability as scale — sponsoring your way to brand fame. Next time you hear a "Sponsored by..." podcast ad, that's what's going on. It's very Pavlovian. Hear Squarespace, think website, build memory.

This is perhaps the strongest B2B SaaS case we can make for the EBI view of marketing. It's by no means the only example — plenty of B2SMB SaaS companies do mass advertising these days and have done so for many years[7], but none have become synonymous with a medium in the way Squarespace has.

Where does positioning fit in this world?

Brand name & need vs. visual identity: The EBI's view helps us understand what "brand" actually is. We often think about it as a visual representation in terms of logo, wordmark, and perhaps colors — the classic "visual identity" designers work on. That visual identity is particularly relevant when competing products sit on supermarket shelves, for example, and you have to choose from the brands you can see. These are, in EBI-speak, a brand's 'distinctive assets,' but they aren't limited to visual assets, either. Squarespace, for example, demonstrates that 'brand' can be a name-and-need memory association encountered only in audio, well before a prospect ever encounters a visual identity. That's interesting to consider, given "brand" and "logo" are usually treated as synonymous. But in podcasting, "brand" is hearing a name; in Google's search ads,

> "brand" is seeing a name in plain text; in review sites and word-of-mouth, it's about the name and how customers describe you. No logos there necessarily, just a name and need association that hopefully sticks in a prospect's brain. What do your customers encounter first — a name or a visual identity? What does that mean for your marketing strategy in creating that name and need connection?

The EBI vs. positioning

It doesn't, according to the EBI.

Let's think about the shape of the market and demand we're serving here. This is the world of mainstream distribution, where the dynamics of brand-driven B2C marketing and B2SMB marketing overlap. To become a big B2C brand, EBI style, you need big distribution channels to create memories across vast populations, and that means mainstream print, TV, radio, and, these days, online advertising (especially YouTube) in all its forms.

This is the cast-and-catch world of marketing — you need to cast your net as wide as possible to catch as many people as you can. The breadth of the net you cast determines the number of people you catch, or so the *How Brands Grow* logic runs.

In this world, brand *salience* — those memories that appear at buying time — is in, and positioning is out. As the EBI's Professor Jenni Romaniuk puts it in a piece for LinkedIn's The B2B Institute, their approach is in stark contrast to what they see as the old "positioning" view of the world[8]:

> *Old-school views of marketing claim that to grow, you need to attract a special type of customer - one whose needs that matches the product or service you offer. By making sure you have the right offer for that group, you can then lock in their loyalty for the long term and insulate yourself from the competition. This thinking underpins the segmentation, targeting, differentiation, positioning view of the world.*

That's a view of the world they staunchly disagree with, and here we get to the nub of the positioning-is-dead argument.

Reach vs. loyalty

This is a big claim! And if that claim was restricted to the B2C world, that'd be fine. I would neither know nor care how true it was, but the EBI insists it applies universally in the B2B world too, and again, *these are some of the most influential folks in marketing.*

First, to try and do justice to their argument, let's illustrate the difference between the EBI's world and (what they see as) the "positioning" world in a B2C context, using a B2C example of theirs — McDonald's vs. Burger King.[9]

McDonald's targets the whole market — breakfast, lunch, dinner, cafe, salads, snacks, kids, teens, adults, you name it — while Burger King tries to focus on their best customers — the bros, young men in particular who eat at fast-food burger chains almost ten times a month (or, at least, that was the strategy at the time). By targeting those buyers, Burger King theoretically 'locks in' their loyalty and generates more sales that way.

The merits of these respective strategies get debated endlessly in ad land, but in the EBI view of the world, this belief in loyalty is a myth unsupported by research; McDonald's is a much bigger brand than Burger King and their ability to serve the whole market is part of the reason why. So much for bro loyalty.

That is what Romaniuk means when she writes about the "segmentation, targeting, differentiation, positioning view of the world" that they reject. They reject doing this *in the mass market*. In their view, sacrificing reach for niche is always a mistake. (Note that, in this instance, it's *demographic* segmentation they're rejecting, which we'll discuss in the box below.)

This is the same logic we covered in **chapter 42**, where we discussed niche targeting using mainstream distribution, i.e., trying to find niche dentist CRM prospects by advertising for all "CRM software" searches. If you're only going to focus on a certain segment of the market, you need to generate a lot of efficiency to make up for the potential buyers you're excluding, the ones you're giving reasons *not* to buy you. Some marketers believe that efficiency comes from loyalty; in the EBI view, drawing on decades of research, that kind of excess loyalty doesn't exist and, if anything, the reverse is true — smaller brands have slightly *less* loyal buyers.[10]

Is all of that true in B2B as well? Do burgers and business software share the same market dynamics?

In Squarespace's case, quite possibly, and if they focused only on a demographic niche or even a vertical like, say, dentist websites, perhaps those buyers would have actually been *less* loyal than what they achieved by going for reach. There would certainly be fewer of them, too. There *is* a time and place in B2B where the EBI view matters, which is why we're exploring it here.

But in other cases, as we've seen in the previous chapters, plainly not, because segmenting, targeting, differentiating, and positioning *does* create that efficiency — efficiency that gets you into the market in the first place. There's a whole world of dentist-specific B2B products that are marketed and sold, segment-and-sell style, directly to dentists. If you cut off that niche positioning in B2B, you cut off your ability to efficiently enter a market, and that's suicide. You're dead on arrival, and that's why we have to be so careful when parsing "universal" marketing advice when looking at the similarities and differences between the B2B vs. B2C, active vs. passive attention world of buying. The critical difference is that distribution must match your positioning, and vice versa.

This positioning-is-bad view is another example of overly left-brained thinking — taking a static snapshot of a particular kind of market (mature mass markets) and confidently extrapolating it as a universal truth, ignoring the more right-brain aspects of change over time, nuance and ambiguity, and a broader sense of the whole, like the whole world of B2B startups and complex sales.

> **JTBD and why demographics are a dead scene, man:** Part of the EBI's antipathy towards targeting is based on *demographic* targeting — focusing on a given gender and age group, for example, like Burger King's bros above. Interestingly, Clayton Christensen expressed the same criticism and saw the Jobs to be Done theory as the answer to a misguided focus on demographic segmentation when it came to marketing.[11] Think about McDonald's and Burger King for a moment in JTBD terms — what's the difference between the two brands in solving for the fast food JTBD? Not much. Therefore, if we advertise targeting young men, will that somehow cause them to use us for their JTBD over others that target the same market more broadly? Well, why would it? Both companies are serving similar products at similar prices in similar locations — there's no functional difference in the way they serve the JTBD. Therefore, if some bro has a job-to-be-done of getting a quick, cheap burger, what reason would they have to *not* solve that in the most convenient way possible, regardless of what advertising was targeted at them? Sure, they still need to know BK and Mickie D's exist, but that's a matter of frequency, not targeting. This JTBD view is a much more customer-centric view vs. the brand-centric view too many B2C folks buy into and renders discussions of *demographic* targeting somewhat moot. As we discussed in chapter 36, it's the *job* that defines the shape of the market, not demographics. Those jobs might be broad and mainstream (fast food, SMB software) or specialized and niche (high-end restaurants, dental tools, enterprise B2B software), but the fundamental unit of analysis is the JTBD, not the brand.

The niche to reach framework

In the EBI world, marketing's job is to maximize brand sales by maximizing reach. This is a fascinating counterpoint to the segment-and-sell approach we explored when discussing *Crossing the Chasm* in **chapter 40**, where (product) marketing and sales work together to build momentum in a niche. We've focused a lot on what it means to go niche, to light that word-of-mouth fire, and zero in on your best customers, but it's important to know when you should position for reach, too.

In the B2C world, brand is input — evoke the category[12], create memory structures, get sales. But for B2B folks (and perhaps plenty of B2C folks, too), brand is outcome — get sales, create memory structures.[13] The memory structures are a secondary flywheel built from the sum of all activities, including performance marketing, sales experience, product experience (which then drives recommendations), support experience, and so on. There's just far more that goes into the 'brand' — and associated memories — than a catchy TV ad for candy and the (prompted) impulse buy at the checkout for a cheeky snack.

This is also a much more right-brain view of brand as a whole — the whole set of memories that people associate with your company — and it goes well beyond the "weak force" (as the EBI put it) of mainstream advertising.

This is where we in the B2B world need to draw a line between what we do and the myopic view of marketing as static, big-brand B2C marketing. Especially when, in the B2B world (and perhaps in the B2C world too), you often have to *go niche to get to reach*.

Going niche to get reach

This is the niche-to-reach framework, and it's a fundamental part of the Technology Adoption Life Cycle, as folks like Geoffrey Moore saw it. Moore's *Chasm* was the first of a series of books, and later in the '90s he would ultimately prescribe going niche *or* for reach, depending on your stage:

- At the very start? No segmenting.[14]
- Crossing the chasm? Niche all the way.
- Past the chasm and in the "tornado"? No segmenting — you need to capture that mass demand.
- Past the tornado and in "Main St"? It's back to segmenting and specific customer focus.

Positioning and product/market fit

In product/market fit terms, we can take this model of product adoption and positioning and give it a modern update with a bit of a radar-and-laser-beams spin. For example, the stages of a theoretical adoption curve might look like:

- **Reach (pre PMF):** Target innovators as you try to put your product on *anyone's* radar to get started.
- **Niche (pre PMF):** Switch to niche (crossing the chasm by finding a beachhead in the early mainstream) as you apply your narrow-focus laser beam to see where you're getting niche traction, perhaps around a specific JTBD.

  Then, assuming you find fit:

- **Reach (post PMF):** Capture mainstream demand by looking out again, as you attempt to go mainstream by putting your discovery on more folks' radar.
- **Reach + niche (post PMF):** Switch to niche across the spectrum (specifics across that mainstream demand), refining fit for different segments at scale, as we'll see with UiPath in **chapter 47**.

That is, you're basically running explore/exploit loops either side of PMF. Explore to find initial traction, exploit it to create a strong reference base; explore adjacent market segments or more mainstream positioning, exploit it when you're mature enough to drill down again into different segments and verticals, getting your product and marketing tentacles deeply embedded in the whole market.

Dynamic positioning — out or down

This is just a theoretical model. Reality will, of course, punch you in the face several times throughout this journey, and in practice, your positioning and attention might be focused far more at one end or the other.

The problem with universal models is they all assume a uniform starting point and a universal market shape, but your starting point and shape of opportunity might look considerably different, as it's obviously highly contingent on your insight, experience, and resources. You might, for example, spend a lot of time focused down with laser-beam attention solving a very specific problem in a niche, or you might go straight to the mainstream radar phase, the 'radar rush' that only a lucky few experience.

The point I want you to think about, however, is how you might diagnose your own position on the adoption curve and the shape of opportunity — and jobs to be done — that appears before you. How you generate momentum to get to the next stage of your business will, I imagine, both be highly unique to your situation *and* more predictable than you'd otherwise imagine. Fundamentally, you've got to go either out (reach & radar) or down (niche & laser beam), or in rare cases both, and make a bet on generating more growth and momentum one way or the other.

We'll discuss this further in the next chapter. For now, this dynamic approach to positioning gives us a nice high-level framework. Positioning, in B2B at least, doesn't have to be *only* about going niche or *only* ever being a mistake because you have to go for reach, always, as the EBI suggests. Markets are dynamic, adoption changes over time, and your strategy has to account for that.

Why not start with reach?

> *"More traffic/conversions/customers please."*
>
> – Every founder ever

The temptation, however, is to just want to skip all this and go straight to the mainstream. And, sometimes, you *can* start with a mainstream product, take a mainstream position, ride the 'radar rush' with mainstream distribution and adoption, and swan dive into a pile of money. Slack, for example, acted as their own early adopters, in a sense. They could dogfood their tool internally as they spun it out of their MMO — and then they could (I imagine) go after other startups as early adopters too, but as we saw when exploring their founder narrative in **chapter 12**, there wasn't a niche *vertical* focus or anything like that. They were always aiming at broad workplace communications, i.e., aiming for reach.

If that's possible, and if, in the EBI view of the world, going for broad market penetration and not segmentation is 'how brands grow,' then why not just stick with that strategy instead?

Well, if you can, great! Do that and prosper. What founder doesn't wish they could just tap some magical source of mainstream demand and have customers rushing in left and right... and sometimes that does happen!

The fly in the ointment, however, is that for most B2B founders, going for "reach" is just not that easy. If you could just do more broad advertising to generate more sales — if marketing and advertising were that simple — then surely everyone would.

This was particularly apparent in the ZIRP era. B2B SaaS startups had more money than God, and if they could have just plowed all that into ads, going for as much reach as possible and getting $2 out for every $1 they put in, then we would have seen a wave of glorious IPOs from unicorns, decacorns, and other mystical creatures, and not wave after wave of layoffs and cutbacks instead.

And on the bootstrapping side, I'm sure Nathan Barry (**chapter 39**), all things being equal, would have liked his initial, marketing-driven approach to ConvertKit to have taken him to the moon, but he *had* to go niche with direct sales to *individual bloggers* because the broader marketing approach simply didn't work — there was no leverage there in mainstream distribution. He had no reach. Likewise, I'm sure the initially bootstrapped Qualtrics (**chapter 35**) would have liked it if mainstream corporations picked up their tool straight away so they could reach the whole corporate market from the get-go. That's what they tried. But it didn't work. Therefore, they had to go very niche to start, zeroing in on marketing professors (perhaps like Byron Sharp!) first.

But that's sales...

You might counter and say, well, that's sales, not marketing. But in B2B tech, guess what? Marketing exists to make sales easier, as David Kellogg reminds us[15]. Marketing is, as Kellogg says, the help. Customers need to be understood, niches need to be found to create leverage, positioning needs to be defined, and messaging needs to be written, all to help sales (and perhaps even sell without sales if your ads and targeting is good enough).

This is the segment-and-sell product marketing world of finding and exploiting niches. Theoretically, you might be sacrificing reach for niche, but in practice, you never had any mainstream reach to begin with. It's only by segmenting and doing something unique that you generate any reach — any penetration into the market — at all. (We'll explore how SaaS companies try and split the difference, with an eye towards maintaining reach, when we look at wedge positioning in **chapter 46**.)

All of this is why your fundamental clarity strategy is so important — it's vital to know what you're saying, who you're saying it to, and the stage-appropriate market strategy you're using to connect those two things. That's where product/market fit is found. But if a disciplined approach to clarity strategy is what's really required in the modern B2B marketing world, what do we see instead?

An overwhelming focus on leads.

Feed the lead machine

We need leads. Where can we get more leads? Are they quality leads? Marketing qualified leads? Sales qualified leads? WE NEED LEADS!

This happens when marketing is reduced to mere promotion, which was just one of marketing's traditional 4 Ps of product, price, place, and promotion. When the marketing department is really just the promotions department, founders and CEOs tend to treat their marketing org as a leads piñata, thinking if they hit it with a big enough stick (*we need 30% more leads this quarter!*), more leads will fall out. Marketing will happily oblige, shoving any poor soul who submitted an email address into the leads bucket, and you get a funnel filled with garbage leads.

But that's what you asked for: more leads.

Likewise, on the sales side, SDRs will send follow-up emails over and over trying to "activate" a lead (to borrow the B2C term), when really they're just trying to catch someone in active attention, essentially contacting them over and over saying "How about now?" ... "How about now?" ... "How about now?" ... "How about now?" ... which is about as annoying as it sounds.

Having a short-term focus isn't always bad — sometimes you do need to get on with it, contact people, and see who will actually buy — but doing it mindlessly or over and over is an exercise in futility, whether it's in marketing, sales, or product for that matter, too (see: the dreaded 'feature factory').

On the brand side, founders and marketers will often pour money into pointless content, ads, and high-visibility, one-and-done vanity projects in the hope that it will 'build the brand' which, in turn, is supposed to magically generate leads, without stopping to ask how that's supposed to work, or if they can point to a single instance where it has. (This was painfully true in the ZIRP era).

In these cases, the assumption is that marketing = promotion (including advertising), and if we just promote our stuff more, more people will buy. Maybe, but what are you giving your people to actually say? What exactly are they promoting? This is where your clarity strategy comes in. I love the impulse to be creative and/or persistent and put something out there, but it needs to be a part of a bigger system with a much more grounded approach to understanding the first principles of attention with meaningful feedback loops (not just fake lead counts) around success.

How marketing became a leads piñata: Sales often annoys Product with feature requests, as they're the ones face-to-face with customers and prospects all day, discussing needs and what it would take to get a deal done. But how often does Marketing offer meaningful input into product direction these days? Or, more pointedly, what founder or CEO would *trust* Marketing to offer meaningful product input these days? Remember, 'product' really was one of Marketing's traditional responsibilities, and if traditional product/market fit was getting the 4 Ps right (product, price, place, and promotion), you can see Marketing was uniquely placed to make that happen. Unfortunately, in both the B2C and B2B world, Marketing, as a department, has, as mentioned, often been reduced to just promotion instead. (John James has an excellent potted history of why this happened in the B2C world of brands and agencies.[16]) To be fair, promotion these days is massively

> complex due to channel fragmentation, especially in B2B, so it's easy to see how that would consume a whole department. Nevertheless, my hope is that clarity strategy and a focus on attention — especially a more right-brain wholistic view of the message, market, product, and customer JTBD — provides a more unified approach to the core concepts driving your company, rather than this very siloed approach we have now. Who leads that process in the C-suite doesn't matter much to me; but *someone* needs to be overseeing how the GTM pieces fit together as a whole, and clarity strategy is a great lens to do that.

Attention-aware marketing

While a smaller startup can't necessarily measure brand performance like a giant B2C brand — and nor should they — we can at least work from first principles of attention, both active and passive, and left and right-brained, and try to build a coherent system of attention-aware marketing.

- **Laser focus for active attention:** If you can reach folks through ads, outreach, or content who are in-market, have a problem to solve, are perhaps early adopters, and would pay for your solution, then obviously you should exploit that as much as you can. This is where product differentiation is vital for giving folks a reason to try you.
- **Radar focus for passive attention:** For those not in-market (95% of folks), the point is to get on their radar and stay on the radar over the long term. That might be on LinkedIn, email, a podcast, or paid advertising if it's economical. This is where the depth and substance of your narrative is vital for sustaining interest over the long term.

That might be as simple as Bidsketch's email-gated approach we touched on in **chapter 42** where leads who want to convert today *and* folks who were just curious were captured at the same time. Active leads could buy, and the more passive folks got memory-association-building emails over time. Or it might be as sophisticated as monday.com's BigBrain system for highly efficient targeted advertising, as we also saw in **chapter 42**.

If you're really lucky and you can do both activation and brand/memory building at the same time (so-called 'double duty' advertising) *and* you can do that at scale, then that's a golden ticket. That's what Squarespace was able to do with their podcast advertising. Easier said than done, though: If advertising is essentially an auction for attention, in most digital channels, bidding is fierce and the channel owner (Google, Facebook, LinkedIn) will happily take all the margin in a market while competitors bid themselves into oblivion. That's why Squarespace's story is so interesting — there was a strong right time/right place element to it, sure, but they took a bet on a new, emerging, unexplored channel rather than just forever trying to wring more juice out of the existing, highly competitive ones. They innovated at the channel level and were about to achieve mainstream reach through mainstream distribution.

Creative publicity and narrative differentiation

Those channel-level opportunities only come along every so often, and in the meantime, you can generate considerable opportunity by knowing exactly what game you're playing and what you're doing when it comes to engaging the market. That might include one or all of:

- **Creative publicity** in the Ehrenberg sense, where you're just trying to find creative ways to remind people you exist at scale.
- **Narrative differentiation,** where you're genuinely trying to get a function or role in business to do things differently, which is what the EBI themselves use, in a story 1, old world/new world, change-story-for-executives sense.
- **Stark reminders,** if you're #1, perhaps you can get away with simply stating you are, in fact, #1. (Or if you're Apple, you can just run iPhone ads that say iPhone [whatever] with a product shot — the reminder is most of what matters.) Much B2C advertising is designed to *maintain* memory connections, and the bigger you get, the more you need to do because there are more folks whose memories you need to keep fresh.

For example, Salesforce's famous turn-of-the-millennium "end of software" campaign that we discussed back in **chapter 31** was both high on narrative differentiation ('use web-delivered not on-prem software') *and* creative publicity (fake protests, big launch parties). If you can tap a secular change *and* milk it for publicity, you're in for a good time.

Even if there's less of a secular change, if you can nail creative publicity around a launch, as Sam Blond and Brex did (as we touched on in **chapter 40**), you can still get your flywheel going in a very tightly targeted way.

And once you've made it, you can literally just remind folks you're #1. That's the luxury of being a successful mature company targeting mainstream normies, and that's part of what Salesforce does today with campaigns that literally say "Salesforce. #1 CRM."[17] (along with Dreamforce and much more, of course[18]).

Marketing reality — priests vs. pragmatists

Let's wrap up. For the EBI, in terms of what they preach, most marketing is advertising, most advertising is big B2C brand advertising, most big brand advertising is a relatively weak force, and that weak force is largely about reminding mass audiences that products that have existed for years or decades still exist, and you might like them. Growth comes not from building loyalty with core, heavy buyers but instead by trying to remind the (far more numerous) light buyers to have that occasional can of Coke or make that monthly trip to McDonald's. The broader you go, the further you reach, the more you can sell. That is, as they say, *how brands grow.*

At least in theory. In practice, despite the dogmatic certainty with which that message is preached and despite the fact there are not insignificant number of devoted adherents to the EBI's way, the B2C marketing world is nothing if not syncretistic. Marketers in the trenches tend to go for a little Sharp-inspired reach here, while also indulging in the heretical practice of a bit of targeting and activation there, with perhaps some (downright satanic) brand purpose sprinkled on top. That is, plenty of marketing academics, practitioners, and academic practitioners are far more pragmatic than what our focus just on the EBI's sacred texts might suggest.[19]

Nevertheless, the logic of positioning for *reach* is incredibly important to understand at the B2C-esque B2SMB end of the B2B spectrum, as companies like Squarespace have demonstrated. We just have to remember that the SMB and enterprise ends of the B2B spectrum are, as we've seen, almost nothing alike.

For the EBI, when they want to pitch big businesses and meme the marketing industry, turning marketers into message-carrying devotees — that is, when they go *really* B2B — it's narrative time, baby. Few folks have meme'd harder with absolute laws and rules than they have. Here, it's do as I do, not as I say. There's still much we can learn from them, but there's much to be wary of, too.

For B2B SaaS folks, the elements of your clarity strategy — narrative, positioning, and brand — have to go hand in hand. These aren't static elements, either. Strategy and positioning are dynamic because market adoption is dynamic, especially if you're attempting to go from niche to reach. But even if you're going for reach, surely your product or brand still has to be about *something*, doesn't it? In the next chapter, we'll look at positioning by *attribute* while tackling another of the EBI's contrarian views: that differentiation doesn't matter at all.

Notes

1. Stein, Joel. "How Did Squarespace Know Podcasts Would Get This Big?," *The New York Times*, February 13, 2022. **https://www.nytimes.com/2022/02/13/business/**

how-did-squarespace-know-podcasts-would-get-this-big.html

2. Ibid.
3. Podcasts and post-sale surveys have become a hyped tactic in some SaaS marketing circles in the 2020s, this time with vendor-owned podcasts, in an 'everything old is new again' sense.
4. Valinsky, Jordan. "Casper's IPO is officially a disaster," *CNN Business*, February 5, 2020. **https://edition.cnn.com/2020/02/05/investing/casper-ipo-share-price/index.html**
5. Jansen, Caroline. "Casper refinances debt with new $80M credit facility," *Retail Dive*, November 17, 2022. **https://www.retaildive.com/news/casper-refinances-debt-credit-facility/636829/**
6. From the anonymous foreword to the EBI's Professor Jenni Romaniuk's "Category Entry Points In A B2B World Linking Buying Situations To Brand Sales," *The B2B institute*, June 2022. **https://business.linkedin.com/content/dam/me/business/en-us/amp/marketing-solutions/images/lms-b2b-institute/pdf/b2bi-cepinb2b-final.pdf**
7. Gail Goodman's famous 2012 Business of Software talk, "The Long, Slow, SaaS Ramp of Death," discussed how Constant Contact discovered radio ads well over a decade ago, for example. **https://businessofsoftware.org/talk/how-to-negotiate-the-long-slow-saas-ramp-of-death/**
8. Romaniuk, Jenni. "How do Business-to- Business (B2B) brands compete?," *The B2B Institute*. **https://business.linkedin.com/content/dam/me/business/en-us/marketing-solutions/resources/pdfs/b2b-duplication-of-purchase.pdf**
9. Sharp, Byron. "Marketing: Theory, Evidence, Practice," 2012.
10. In *How Brands Grow*, Sharp writes, "Loyalty definitely exists, but it is a characteristic of consumer behaviour, rather than being driven by brand differentiation. This means that all brands should benefit from loyalty." However, niche brands with "small but highly loyal customer bases" are rare. That is, going niche doesn't generate *excess* loyalty that would justify the strategy, and the reverse is actually true. This is Ehrenberg's Double Jeopardy law, where brands with less market share have fewer buyers *and* those buyers are slightly less loyal.
11. "We believe, instead, that some of the fundamental paradigms of marketing—the methods that most of us learned to segment markets, build brands, and understand customers—are broken. [...] The problem is that customers don't conform their desires to match those of the average consumer in their demographic segment. [...] When people find themselves needing to get a job done, they essentially hire products to do that job for them. The marketer's task is therefore to understand what jobs periodically arise in customers' lives for which they might hire products the company could make. If a marketer can understand the job, design a product and associated experiences in purchase and use to do that job, and deliver it in a way that reinforces its intended use, then when customers find themselves needing to get that job done, they will hire that product." Christensen, Clayton M.. et al. "Marketing Malpractice: The Cause and the Cure," *Harvard Business Review*, December 2005. **https://hbr.org/2005/12/marketing-malpractice-the-cause-and-the-cure**
12. This is why brands in the same category all tend to look the same — they're trying to

tap customer memory associations around the category, not one specific brand, which is an act of positioning in its own right.

13. I'm channeling Dave Kellogg here, who writes, "I think branding is a potential marketing rathole for startups, particularly early-stage ones. [...] Want to build a brand? Go sell some software. Want to improve your brand perception? Go sell some software. Want to have a distinctive brand visual territory? Go sell some software. You see the pattern. Some startups put the cart in front of the horse. Don't do that." **https://kellblog.com/2022/07/17/practical-thoughts-on-branding-for-software-startups/**

14. April Dunford advises this too — better to see what sort of demand you catch early on, so you can find out who gets the most value out of your product and then niche down on them.

15. Kellogg, David. "Marketing Exists to Make Sales Easier," *Kellblog*, April 26, 2020. **https://kellblog.com/2020/04/26/marketing-exists-to-make-sales-easier/**

16. James, John. "Brand Campaigns, Part 2: Where Did They Come From?," *Brandingmag*, October 3, 2023. **https://www.brandingmag.com/john-james/brand-campaigns-part-2-where-did-they-come-from/**

17. See, for example, Salesforce's CEO Marc Benioff tweeting an image of a Salesforce billboard with the "Salesforce. #1 CRM." headline next to a market share chart proving their point: **https://twitter.com/Benioff/status/1034357868424650753**

18. I don't want to suggest saying "We're #1" is all Salesforce does. Drew Beechler has a nice writeup of mid-00s Salesforce's attention to narrative and themes as well: **https://www.highalpha.com/blog/the-second-greatest-sales-deck-and-pitch-ive-ever-seen**. In that sense, Salesforce tries to tick all the marketing boxes — narrative differentiation (annual themes), creative publicity (Dreamforce), and memory associations using characters (Astro), too, as we touched on in a footnote in the previous chapter.

19. Mark Ritson, for example, who enjoys commentating on these inter-industry spats as both academic and practitioner, often champions 'Bothism,' which he defines as "the rare capacity to not only see the value of both sides of the marketing story, but actively consider and then co-opt them into any subsequent marketing endeavour in an appropriate mix." Perhaps that's a little too right-brained, though, with the consideration of the whole and the ability to sustain a degree of ambiguity for the left-brained folks out there! "'Bothism' is the cure for marketers' fascination with pointless conflict." *Marketing Week*, September 3, 2020. **https://www.marketingweek.com/ritson-bothism-cure-marketers-fascination-conflict/**

45

The B2B spectrum III — SMB cont., brand positioning, distinctiveness, Patreon, & Notion

One of the more durable generalisations about the hemispheres has been the finding that the left hemisphere tends to deal more with pieces of information in isolation, and the right hemisphere with the entity as a whole, the so-called *Gestalt*.

- Dr Iain McGilchrist[1]

You could argue positioning, in a niche sense, was irrelevant to Squarespace (unless "SMB" is a niche), as we explored in the previous chapter, and you'd be correct, to a point. There is little by way of positioning *by segment*. But there's still positioning *by attributes*.

In this chapter, dear reader, we're going to explore the world of brand positioning, where reach and positioning meet, finishing with a look at creator platform Patreon and, briefly, Notion. Let's dive right in by returning to the idea of brand fundamentally being about memory associations.

Brand marketing — memories vs. clicks

Here's a quote from the EBI and LinkedIn's B2B Institute that we discussed in the last chapter:

> *"The most important search engine is still the one in your mind." This statement makes a profound point all marketers should internalize about buyer behavior: most purchases start not by searching Google, but by searching our memory. If you believe most buyer behavior starts with memory, it then follows that the primary job of marketing is not to generate clicks, but instead to generate memories.*

This gives us a succinct definition of brand marketing from the EBI perspective, with the author going on to say that "moving from clicks to memories is a massive paradigm shift for marketers."

This idea of clicks *vs.* memories (as opposed to clicks *and* memories) is a bit of a silly dichotomy, but it contains an important insight.

The silly part is the idea that *most* purchases in B2B start "not by searching Google, but by searching our memory." That is both highly debatable and largely irrelevant when we consider:

- How complex most B2B sales journeys are, which often involve a significant amount of research.
- The type of buyers you, dear reader, might be targeting (e.g., early adopters looking for an alternative to the memory-dominating incumbent).
- The fact that most marketers understand they want to be *both* memorable and generate clicks from their advertising and marketing efforts.

Nevertheless, the role of memory, and memory *associations* in particular, is a crucial point that's worth unpacking, especially as it relates to our broader concept of clarity strategy.

"Brand" has come back into fashion in B2B SaaS as the pendulum swings back from the hyper-analytical — and hyper-efficient — digital advertising of the '00s and '10s, especially as privacy laws bite, tracking becomes much harder, and the competition bids those clicks up to ever more irrational heights.

Likewise, the rise of "zero-click" content[2] and unattributable social traffic has created interest in more brand-centric media that's more focused on seeding ideas than it is calls-to-action to buy now/sign up/download/etc.

This renewed interest in brand and being known for your point of view on LinkedIn and beyond makes it very important to understand the mechanics of what we're trying to do here.

In this case, the EBI's approach is actually very helpful.

When the EBI folks talk about memories vs. clicks, you've got to imagine the kind of products they have in mind — cola, shampoo, whatever. These aren't click-to-purchase products; they're remember-to-purchase products while out shopping. To build that memory association, you need to deliver your 'information pack with attitude' (i.e., your memetic message, per **chapter 15**) within a single ad. For example:

*Pop... pfffiiiizzz... gulp... ahhh... *product & logo*.*

That's a soft drink commercial in a nutshell. That's a message that's trying to build and maintain a memory association — remember sugary soft drinks? Remember how you like to drink it? Remember our brand? Remember how you like to drink it when out with friends, perhaps? (Or whatever social scenario is relevant.)

Expose enough people to that association frequently enough over a long enough period of time, and you will have built and maintained that all-important mental availability so that when someone wants a drink, they opt for your brand when scanning the shelves. (Or, in Squarespace's case, if they want a website, they remember "Squarespace," as we discussed.)

In B2B, however, particularly at the more innovative or complex end, our messages tend to be far more complex than '*pop, fizz, gulp, ah*' and, therefore, we often *do* need the click, or at least the view, scroll, or whatever form of engagement it is to communicate *the need you should associate with us* in the first place.

Of course, the more succinctly we can do that, the better, and that's why your Hemingway story is so important, as we saw in **chapter 10**, as is having a tight ~10-slide deck like the one we looked at in **chapter 9**.[3] This is also why clicks vs. memories is a bit of a false dichotomy in B2B — it's not either or, it's both.

The SMB brand battle

We also need to remember that B2B is a broad spectrum, and we're discussing the SMB end here. The enterprise end might be complex, for example, but the more universal, understood, and commoditized the category, the more we'll see B2C-style brand effects. Consider the project management category and tools like Asana, monday.com, and ClickUp, for example. If you're a small business with basic project management needs, any of these tools would do the job. The question is which tool — which brand — comes first to mind.

In these cases, the memetic message you'd use to try and build associations between your brand and the typical SMB project management need may well be expressed in a simple ad. For example, in the most basic, boring terms: "Outgrowing spreadsheets? Struggling to manage all your projects? Try [brand] for free." How do you make that name/need connection memorable in a way that can be sustained over time? Therein lies the challenge of creative brand work.

In B2B marketing, we tend to evaluate those ads just on clicks — which converted best? But there's a memory building aspect there that matters too, so that when people search their mind, and not just Google, our brand is the one that comes to mind.

Remember that we're not interested in brand *awareness* per se, but brand *salience.* It's not about folks being 'aware' of your brand for its own sake, it's about folks being aware of *what buying need they connect you to.* Consider our largely (but not exclusively) SMB project management trio of Asana, monday.com, and ClickUp — the challenge is (a) to be thought of at all and (b) to be thought of first when specific JTBDs arrises. That's the race for mental availability that marketing teams are competing in.

In light of this, it's worth thinking hard about what exact memory association you're building and when you expect it to be triggered. In the EBI world, the idea is to come to mind at 'category entry points' (CEPs), which for your SMB SaaS project management brands might indeed be when they outgrow spreadsheets or the use of some other generic tool. This is where JTBD switch interviews shed light on the customer's journey and help us identify these 'category entry points,' i.e., when folks started passive looking, and then active looking, and so on, as we discussed in **chapter 36**.

What need do you want to be associated with?

If brand is about memory associations, stopping and asking exactly what need (or set of needs) you want to be associated with, in a name/need, brand/memory sense, can be a powerful, clarifying question. It tends to become pretty obvious pretty quickly whether you're trying to be associated with a broad, existing need or a new, novel need or application. And that brings us back to your clarity strategy. Knowing what game you're playing and the kind of attention you're targeting is fundamental because everything else flows out of those choices, including your market *and* market*ing* strategy.

Whether you've got a strong left-brain story 2 for a very specific group of people or whether you've got a broader story 1 you're trying to put on folks' radar, those fundamentals dictate your position and how you go to market, the memory associations you're trying to create and sustain, and ultimately the brand you're trying to build.

Advertising vs. brand memories

In the consumer world, marketing is often seen as just mass advertising of largely fungible commodities, and that drives a lot of these 'laws' and rules. But in B2B SaaS marketing, especially in the early days, paid direct-response advertising is actually pretty low on the agenda for how you might start building those name/need brand memory associations with your ICP.

Instead, you might be:

- Producing content
- Hosting a podcast
- Running events
- Participating in or hosting conferences
- Sending a newsletter
- Posting on socials

All of those activities have brand-building potential insofar as they:

- Help create and sustain memory links (decay is real!) between your brand

and the problem you solve in the minds of your prospects.
- Do that with sufficient frequency for those associations to stick.
- Contain a memetic 'information packet with attitude' so the message lands.
- Are not focused on expecting folks who read/watch/listen to buy today, necessarily.[4]

(Plenty of marketers doing these activities have a 'shoot first/figure out the narrative later' approach, but in my view, if you can test your message as we discussed in **chapter 38** so you know what sort of 'eyes light up' reaction you'll get ahead of time, you'll be much better off.)

And sure, if you've got the cash, many B2B startups often want to — and need to — capture folks who are in-market and are ready to buy today, too (or at least get a deal going, even if it takes months to close), which might involve:

- Bidding on search keywords ("CRM software").
- Targeting folks on LinkedIn or social networks.
- Even doing SDR outreach.

This mix of brand activities and "performance" advertising is fairly typical, and as we've been discussing, *doing* the activities is not the hard part, it's having something to say — having your memory-building clarity strategy dialed in — that's the unique part.

What you say, why you say it, how you say it[5], and how frequently you say it all adds up to a certain impression, a certain *gestalt* as McGilchrist mentions in the opening quote. This, in other words, is what we might call your brand.

Brand fundamentals

Brand can mean a lot of things to a lot of people, so let me make a few points about brand fundamentals that are worth keeping in mind in our B2B context:

- **Brand is memory, not visuals per se:** We discussed brand as memory associations in the last chapter, noting that often that has very little to do with a specific logo or aesthetic in B2B. Buyers can hear about you, read about you, and click to learn more about you without having ever seen, or at least noticed, your logo or "brand" in a visual sense.
- **Brand is driven by product:** "Brand" and "product" are generally two sides of the same coin in the B2B technology world, especially for startups (the company is the product is the brand), so I'll use them interchangeably, but obviously, in the B2C sources I'm drawing from,

"brand" might mean something quite different. (A 150-year-old French luxury brand and a 5-year-old B2B SaaS app don't exactly share the same brand significance, but the rules of engaging attention still apply.)
- **Brand salience vs. associations:** Brand *associations* are what we're working with here once you're well and truly in the market. That is, if people don't think about you at all — you have no salience — fix that first. But in doing so, I imagine you'll want to ~~stand out~~ fit in as a coherent whole, or *gestalt*, which is the form of positioning we're exploring in this chapter.

Those fundamentals are important, but so is acknowledging that our specific B2B context is a vastly different world to consumers in a department store, for example, looking at appliances or electronics and shopping price and brand. "Oh, I heard XYZ is a good brand." Or, "Who is ABC brand? I've never heard of them!" Here, as we've all experienced, simply having heard of, and having a broadly positive view of a given brand is enough of a reputational signal to consider a purchase from that brand if the price is right. Conversely, seeing a brand you've never heard of raises suspicions — who are they? Can they be trusted? What's the quality like? In these cases, the positioning of the brand is an almost non-existent concern relative to familiarity, quality, and price.

But is that always true?

Positioning by attributes

When it comes to innovative or more complex products, however, most folks have a general sense that brand is a means of communicating *something*, perhaps that we're the cutting-edge new player, not the old fuddy-duddy incumbent (especially in tech). Or perhaps it's something more 'intangible.' And even in the most mainstream circumstances, there's still a case to be made for positioning your product to give it a certain kind of *gestalt* as we've been discussing, which means an "organized whole that is perceived as more than the sum of its parts" (thanks, Google).

This, however, is controversial.

Some folks, like the Chief Brand Officer we touched on at the start of this mini-series in **chapter 42**, imagine a brand communicates something very specific: a whole BFF personality for a startup's brand... for what is essentially the outline of a generic animal. Is it really possible for a startup brand like that to "own" a whole personality in that sense? Not really, but that's an obscure example. What about famous, successful, mainstream brands? Can they own a specific attribute, at least?

Here, Byron Sharp and the EBI would say no again — consider our consumer appliance shopping example above.

But ultra-mainstream brands aren't the only game in town. Shortly, we'll look at a tech example where brand positioning was vital to turning the business around. For now, though, it's worth acknowledging that the EBI isn't the only authority in town when it comes to brand positioning, either.

A quick lesson in 'brand' history

Mark Ritson, PhD, brand consultant, former marketing professor, and sweary podcast guest who I've cited a few times in this mini-series, is the more balanced ying to the EBI's dogmatic yang. He takes a more nuanced view, writing that (my bold) [6]:

> *I do not think a brand can own an attribute or association either. The perceptual map never reveals a special place where only one brand can covet a particular association.* ***But a brand can differentiate itself by being more of one thing than its competitors.*** *And in combining the right associations with the right execution, ideally powered by the distinctiveness that comes from clear and well applied brand codes, a brand can present itself differently to the market from its competitors.*

That is, you may not be able to *own* an attribute, but you can be *more of* one thing than the competition. To understand where this fits in brand positioning approaches that have been tried and failed, we need a quick history lesson from the former professor. To start with, this is *not* the classic Ries & Trout positioning approach of the 70s and 80s where, Ritson writes[7]:

> *By the start of the 1970s Al Ries and Jack Trout had revolutionised branding with their concept of positioning. And at the heart of this new approach was the search for a single attribute or association that a brand could own in the mind of the target consumer. [...] Volvo positioned on safety, FedEx positioned itself around 'overnight'. According to Ries and Trout a brand won when it owned the most important potential benefit in the mind of the prospect.*

However:

> *But there was a snag. Whenever marketers left the conference circuit or classroom and actually tried to do the differentiation thing it rarely worked. When they actually looked at data it rarely showed a brand as radically different from the other competitors in the market. In fact, more often than not, the perceptions were incredibly similar.*

That is, marketers ran the experiment. They tried to do what the textbooks and gurus of the time suggested. But then the data came in and contradicted the whole approach. Enter the EBI, where Ritson writes:

> *It was at this point that the scholars at Ehrenberg-Bass made a vital and ultimately discipline-shredding intervention. Led by the dark lord of penetration himself, Professor Byron Sharp, Ehrenberg-Bass began to challenge the power of differentiation. Using data from a range of different markets these scholars showed a distinct lack of perceptual differences between brands and queried whether differentiation really was such a golden branding objective.*

Far from 'differentiate or die,' the EBI would go on to declare differentiation itself dead, as we'll see.

That, however, didn't stop many marketers from wading deeper into the world of so-called 'brand purpose,' where brands didn't just stand for a specific attribute, they stood for virtues and values so lofty it made you wonder why brands like Pepsi hadn't solved social inequality years ago, given this was now apparently within their purview.

Ritson, as one of the world's foremost authorities on brand positioning, loves to make fun of that sort of stuff, and ultimately agrees with the EBI to a point while still making a case for some degree of positioning and differentiation by attribute. Take Volvo and the attribute of safety, for example. In an interview, Ritson says:

> *[R]elative to the competition, you are going to have a relative strength if you're Volvo on safety, [but] you're not going to own it, it's not going to be unique to you. [...] The most important thing for Volvo [however] is to come to mind when people think about cars driving new cars [at all]. And that is much more important than safety, even though brand managers might struggle with that thought.*

That is, the first job of a brand is to be thought of at all. Volvo, first and foremost, needs to be thought of when someone wants to buy a new car. Once someone has that thought, *then* the relative attributes of the competitors come into play — perhaps safety is more important to some set of buys in that case. But that's a comparatively minor concern vs. getting into the consideration set at all. But it's not irrelevant, either, as Ritson says (my emphasis):

> *[For Volvo,] it's not just about safety. It's about the three or four associations that together we're trying to link to create [a]* gestalt, *that all [those attributes]* together *create something valuable in a*

> *category. So Swedish-ness* and *safety* and *a little bit of conservatism perhaps,* and *environmental concern [...] I'm making these up, but you get my point. [W]hen you position a brand, usually there are three or four things going into that overall position, and they all track off each other.*

This is a much less reductive view of B2C brand positioning where you go for salience *and* create that *gestalt* that helps distinguish you from your category competitors. Again, that brings us back to McGilchrist's quote that opened this chapter: "One of the more durable generalisations about the hemispheres has been the finding that [...] the right hemisphere [deals] with the entity as a whole, the so-called *Gestalt.*"

This requires a balanced use of attention. Get too left-brained, and you start trying to zero in on a specific part of the puzzle to the exclusion of all else (that one single attribute you wish you owned). But get too right-brained in creating a *gestalt,* a whole that's greater than its parts, and you risk constructing elaborate fantasies of personality and purpose for a brand when actual buyers just see colors and shapes and a name that matches their need, i.e., a solution to their JTBD.

Salience vs. differentiation

To return to Squarespace as our running example, riffing on Ritson and using entirely made-up numbers:

- 80% of the reason you might choose Squarespace is because you heard about it and remembered it (i.e., salience).
- But the remaining 20% might be because it doesn't seem overly technical and turn you off, the templates it offers have a design quality that you aspire to for your site, and perhaps it just seems easier and more fun and friendly than what you thought this kind of software would be. (These are things you could survey for, by the way.)

In Squarespace's case, they positioned around very simple attributes — beautiful, friendly, simple, and so on — but they were the things that made Squarespace, *Squarespace.* They made it a coherent *thing.* That was their *gestalt.* These attributes are captured to a degree on their brand page[8], which echoes their mission of "helping people with creative ideas stand out":

> *Squarespace is one of very few technology companies that can truly call NYC home. The city has inspired our attitude, our aesthetic, and our mission to democratize good design for every ambitious entrepreneur, artist, or visionary with a dream.*

This is where it becomes tricky to tease out the substance from the fluff. Squarespace's brand page goes on and on about its NYC roots, opening with "IcoNYC" (ugh) and "Creating a new identity for a tech brand rooted in New York City."

On the one hand, there is a strong, right-brain, *implicit* idea that drives Squarespace's product, brand, and positioning, which is driven by the attributes we mentioned, not a niche or vertical.

On the other, I have never associated NYC with Squarespace before and thought, "Oh yeah, this is such a New York City aesthetic" (or attitude, or vision, or product, or whatever).

But just because that's not relevant to me on the outside doesn't mean that it's not relevant to the Squarespace team internally, who may all get the implicit cultural expression the company is going for, which then informs those thousands of product and marketing choices that add up the whole that is *Squarespace.*

> **Brand and making the implicit explicit:** Whether we think it's fluff or not, there still is a job-to-be-done here for the brand guidelines, and that's to share an *implicit* idea through *explicit* metaphor, which is tough, and why we get strange-sounding associations to a city, or personality for that matter. It's just hard to capture implicit concepts in words, and while that can invite (sometimes entirely deserved) ridicule, that doesn't mean it doesn't exist, either.

Building a brand, product up

That is, in a nutshell, the B2C brand marketing world of positioning (with a little bit of B2SMB thrown in). There, product plays a role, but you then mostly have to impart your chosen attributes through advertising alone, and that often has very weak effects, which is why perceptual maps only show relatively minor degrees of difference between brands.

In the market for family cars, for example, differentiation is relatively minor as brands like Volvo and everyone else have been competing to be the same (**chapter 49**) for decades. Certain brands *do* take on certain highly differentiated positions — being more offroad, or sporty, or high-end, and so on, but in the mainstream, if you want to be a big brand, you're going to converge on much the same point as everyone else in the category.

In the innovation-driven world of B2B technology, however, the whole product can live out your particular attributes. You literally build your brand by how you build your product. The degree of complexity Squarespace chooses to offer (or not offer), the types of templates they do (or don't) make available, and the types of problems they choose to solve all drive the brand. That includes taking design seriously and solving for creative portfolios when they relaunched in 2012 (as discussed in **chapter 43**[9]), and these days it very much includes their focus on ecommerce.

All of these decisions add up to a certain *gestalt* that goes beyond simple adjectives — and beyond the vanilla world of mass convergence that happens in, say, the mainstream automotive market. (This need to literally build out your brand from the product up is especially true for mid-market B2B SaaS products looking for a wedge, as we'll explore next in **chapter 46**.)

Shared attributes

When trying to discuss this idea of *gestalt,* we run into the limits of language, or at least English, given we're borrowing a German word to even name the concept. We talked about software as art back in **chapter 12**, and here we run into the same issues where words don't fully capture what a given product or brand is really about, in the same way it's hard to describe music or a joke.

For very left-brain folks like the EBI, this is proof that the brand positioning emperor has no clothes. If you can't fully and explicitly describe it — and if consumers don't mention it in brand surveys — then it must not be there. Brands in a category are all effectively the same, save some meaningless distinctiveness, so just focus on that distinctiveness to drive salience — to be the first to mind when a buyer hits a 'category entry point.'

In some cases, I'm sure they're right. Attempts to imbue established brands with new attributes — like "sustainability" for Shell or BP — are, as marketing attempts go, about as naked as emperors come.

But when positioning is more than ideas in an ad campaign and instead underpins the product, the brand, and the company's point of view of the market, then it's possible to build that more implicit, attribute-driven *gestalt* that makes you, you. And that's something the left-brain folks miss.

Nevertheless, on a macro level, the EBI folks like to point out that brand attributes exist, but they're often shared by all brands in the category. Byron Sharp writes in *How Brands Grow* that "Nike is routinely held up as a super brand, but its sales and style are similar to Adidas," for example. We see this as well for both tech companies and luxury French brands especially, who all echo the same traits to a degree, or at least the same sans-serif, character-free logos[10].

This is a very useful insight as it sheds light on what many folks actually want out of their brand: to fit into a category by 'standing out' like everyone else.

Cool brand, bro

> *When a founder focuses too much on making the brand cool, you know that the core of their ambition comes from their personal desire to be culturally relevant. I made this mistake myself so many times it's now impossible to not see how many other people are doing it.*
>
> – Toby Shorin[11]

It's easy to miss the fact that, on the more superficial aesthetic side of visual identities at least, most branding exercises are actually exercises in *fitting in*, not standing out. (If you really are trying too hard to make a cool brand in B2B and truly stand out in cultural terms, you're probably in the wrong business, as Toby Shorin mentions above.) This is a hidden JTBD of brand positioning. And it's a predictable one if, as the EBI finds, most brands share similar attributes in a given category.

If we want to position ourselves in a given category — and I'm using category in the broadest terms, like "B2B technology" vs. "European luxury" vs. "sports," for example — then we first have to discover what those attributes are.

This is why a lot of the branding work marketers do in a given category, in terms of voice, aesthetics, or "personality," especially in the homogenous world of B2B, all appears fairly interchangeable. That's why a B2B brand's voice is *professional — but never stuffy*. Or a DTC ecommerce brand's voice is *fun and slightly silly — but never judgey*. These are exercises in finding the right associations to *shared* cultural memories so we 'stand out' in much the same way as everyone else in our category.

This should be a relief! Making a B2B visual identity that fits in is relatively easy, even if the point of the brand is ostensibly to 'stand out' (again, like everyone else). It's a low bar to clear. What you see in branding, in terms of visual identity, often commands a disproportionate amount of internal attention, but beyond a certain threshold of competence, for 90% of buyers, it's arbitrary. Ask ten talented designers and you'll get ten different-but-similar treatments. Pick any of them; they're fine.

And if your B2B brand visuals are familiar to buyers in a category, fitting in like this is actually a service to the customer. Buyers need branding to be legible and to share traits so they know they're in the right place. You might think this won't make your brand memorable, and there are tricks to being *distinct*, but for the most part, your brand isn't memorable because of how it looks. It becomes memorable because you sell a lot of software. That's the way causation actually runs. Not the color you pick.

This brings us to another of the EBI's hotly-contested views on marketing, and that's the difference between "meaningless" *distinctiveness* (like a recognizable color, logo, or characters — 'distinctive brand assets' as the EBI's Romaniuk dubs them) and *differentiation* (benefits, reasons to buy, unique selling points) for a given brand.[12]

> **Brands that actually stand out**: There are, of course, some folks who do visually stand out in their category — your Stripes, Linears, and Notions of the world, as we've discussed (e.g., chapter 14). But they had a 10x product, and their brand acted as a fame multiplier, if I can put it that way. If you have a 10x product and want a brand to match, awesome, go for it. I love great branding that expresses the nature of the product — simple branding for simple products, hot marketing UX for hot product UX, and so on. But without product/market fit, 0 x a brand multiple of 1.5, say, is still 0, and 0 is where a bunch of pretty dead startups have ended up over the years.[13] And if you're just a normie product, that's cool, just run with normie branding and worry more about product differentiation instead. But if you *can* bring everything together — product, brand, market insight — you'll have a compelling, multiplicative combination.

Distinctiveness vs. differentiation

> *Rather than striving for meaningful, perceived differentiation, marketers should seek meaningless distinctiveness. Branding lasts, differentiation doesn't.*
>
> – Byron Sharp[14]

That one of the world's most influential voices on marketing doesn't believe in differentiation is both insane and true. Imagine a tech CMO walking into the CEO's office and regurgitating that line to their boss. The CEO would do a spit-take. Then they'd immediately start looking for a new CMO while booking a brief vacation in a padded cell for their existing one.

(This is also more than a little ironic coming from Sharp, given the brutal differentiation Sharp and the EBI highlight between their approach and the old, backward, 'medieval' way of doing marketing. But that's narrative for the enterprise in action!)

Again, we have to remember the EBI's context is largely brand marketing at the end of the adoption curve. Yes, differentiation in toothpaste is going to be short-lived — other brands can roll out their Super Spearmint Ultra Whitening Paste too. But being recognizably, distinctively *Colgate*, on the other hand, is something you can profit from for decades.

For Sharp and Romaniuk, it's the latter distinctiveness, not the former differentiation, that matters, and they are very proud of that fact, taking to task everyone from Theodore Levitt to Philip Kotler to Jack Trout for their misguided support for differentiation, a view that's "taught more like religion than hard-nosed business."

That's understandable for brands at scale, but as a universal truth, it's a bit rich. Remember poor old Frank Bass, the forgotten "B" in the EBI, the guy that was focused on adoption? In the B2B startup world, differentiation through innovation is obviously how you grow because it gets you on the adoption curve, starting with innovators, then early adopters, and so on, as we saw in **chapter 40**. It's what puts you on folks' right-brain radars to begin with.

This is a failure on the EBI's part to understand the power of uniqueness. Trying to confect some unique selling point for a commodity *is* something of a futile endeavor, granted[15], but the reason differentiation doesn't last is precisely because it is so powerful. Having some genuinely unique product advantage is what fortunes are made from — that's why everyone is quick to copy them, especially incumbents who perceive threat to their dominance. This is the battle of competing to be unique vs. competing to be the same that we'll look at in **chapter 49**.

(This also highlights the limits of brand as a unit of analysis — if you myopically see the world through the lens of mainstream brands, then *brand* differentiation probably does seem somewhat futile. But differentiation at a fundamental product, GTM, and business strategy level is what great companies are built on, as we've seen in **chapter 32** and will see in, as mentioned, **chapter 49**.)

Nevertheless, as with all things EBI, that doesn't mean there isn't a kernel of truth there or that their view of *distinctiveness* is irrelevant.

Differentiation vs. *gestalt*

In sales-driven B2B contexts, you need to keep developing differentiation because you'll be asked point-blank why you're different. You're engaging with folks' active, left-brain, laser-beam attention, and that's a different kettle of fish from the EBI folks. (We're not selling the same old chocolates or candy we have been for decades, like mainstream B2C brands.)

But differentiation isn't just that One Cool Feature; it's the sum total of what you're about — differentiation and distinctiveness are two sides of the same *gestalt* coin, and that's where much of B2C marketing has landed, too.

At the B2C-ish end of B2B, your gestalt — what you're about — can run out of steam. The market can move on, growth can slow, and you may decide you need to revisit your positioning to rethink what it is you're about.

That was the position Patreon found themselves in. Patreon just scrapes into the B2B spectrum here insofar as they provide tools for folks with small businesses — creators — but they make a fantastic case study in refining your *gestalt*, positioning included, from a brand perspective.

> **Brand as a coordination point:** Back in chapter 5, I described how clarity strategy wasn't just a framework for execution, but could also function as a *coordination point* that transcended functional silos." Whether it's narrative, positioning, or in this case, brand, having a meaningful, shared coordination point helps everyone build towards a cohesive whole. Some founders might see that coordination point through the narrative lens, while others might use the brand lens, especially if the founder understands that brand isn't just the shiny visual identity Marketing is responsible for. Take Linear's CEO Karri Saarinen, for example, who writes that he sees brand "as much internal as it is [an] external thing. The brand is essentially what you are as a company, and internally it affects the culture and the outputs."[16] That is, it's a coordination point that drives alignment and consistency where everyone knows what you're doing, why, and the level of quality you're expecting. Saarinen offers a simple way to maintain that consistency: "The question you ask yourself is [...] "Is this on brand". To answer that you need to know what your brand is and then have the integrity to answer that question honestly. If it's not on brand, you don't do it." Easy to say, hard to do, and very few folks have done it as well as Linear, from the product out.

Patreon & diagnosing positioning problems

Adam Fishman, former growth and product leader at Patreon, has a great writeup on addressing Patreon's positioning and messaging problems through the lens of brand or, as he puts it, "When Branding is a Growth Strategy,"[17] which I'll quote from below. As Fishman tells it, he arrived as Patreon's growth was slowing and had to fix it:

> *My opinion on brand investment has changed. And it's because of my time at Patreon. This is a story of when it was of paramount importance to invest in branding. In fact, without this investment our company would have failed. This was my initiative and the stakes were very high.*

Fishman frames this as a brand story, but it could just as easily be framed as a positioning story, and indeed, it essentially is one long positioning story, as we'll see. We use terms like "branding," "positioning," and even "jobs to be done," but we're all trying to zero in on the same fundamentals — clarity around the concept at the heart of our business, clarity in how that's communicated internally and externally, and clarity in how that plugs directly into growth and revenue.

What I love about Fishman's story, however, isn't that he just magically thinks up the correct answer like those too-perfect LinkedIn stories we've all seen. Instead, he does what any responsible executive does (or should do), and that's properly diagnose the situation through research first. The Patreon team was contemplating building more vertical-specific features, deploying the proverbial left-brain laser beam by focusing on musicians, podcasters, YouTubers, etc. Maybe they needed to better serve their niches? Instead, they commissioned a "comprehensive research project" which came back with "existentially scary" results:

> *What we were observing was that creators kept launching, but it seemed as if their earnings were capped. We weren't seeing the type of high-value creator launch on the platform.*
>
> *We commissioned a comprehensive research project to understand why. Thinking that maybe we needed to build more vertical specific features for musicians, podcasters, YouTubers, etc. It was called Project Patronaut. What we found was eye-opening and existentially scary for the company.*

Losing product/market fit

It turned out that their early operating metaphor and category anchor of being "recurring Kickstarter" and a home for crowdfunding was running out of stream:

> *The problem was not the efficacy of our growth loops or lack of certain features; it was actually the perception of the brand. We were starting to lose the product-market fit we had achieved since launch. Why?!?*

Kickstarter had stopped growing, and Patreon's research found that creators, A-listers especially, didn't want to be associated with crowdfunding. They didn't want to look like they were begging for money:

> *We knew that we needed to change. We needed to drop the crowdfunding perception and reignite interest amongst the "hot shit" creators. The problem was that we didn't know what we needed to change to.*

In the end, they had to redefine the *gestalt* of the brand. That meant agreeing on new attributes to position around — attributes that were much more fundamental than generic adjectives. These were hard strategic choices that were hashed out in a *year-long* branding process, one that started with a classic positioning exercise (and also included brand personality exercises, fresh assets, PR, and more[18]).

Fishman points the finger at brand *perception* here, but perception was a symptom of something deeper, which included the changing JTBD creators had, and that's what Patreon ultimately had to address.

The dreaded positioning statement

The positioning statement has been a staple of branding and marketing exercises for decades. The one Patreon used, working with consultant Arielle Jackson, looked like this, according to Fishman:

- **For** (target customer)
- **Who** (statement of need or opportunity),
- (Product name) **is a** (product category) **That** (statement of key benefit).
- **Unlike** (competing alternative) (Product name)(statement of primary differentiation).

These exercises are often highly contrived, and it's worth noting that April Dunford, with all her experience in B2B positioning, passionately hates them. (To whit: "The Positioning Statement is Harmful and You Should Never Do One."[19]) Dunford's criticism is that the positioning statement exercise:

- "Assumes great positioning is magically locked inside your head and you simply have to write it down."
- That the resulting dog's breakfast of a statement is so convoluted no one will share or remember it.
- Completing the statement gets confused with completing your actual positioning.

I'm sure that's true most of the time when folks think they can just dream up a position by filling in the blanks in the positioning statement and, hey presto, positioning done. But that's a wish that won't come true if ever there was one.

Dunford's approach for finding and articulating differentiated value (which we discussed in **chapter 37**) is focused on proper analysis, not wish fulfillment, and is therefore aimed at distilling what's working for your best customers.

In this case, Patreon's approach wasn't actually that far off Dunford's, given they were indeed analyzing their customer base. That's the research and discovery that matters, and Patreon was operating from a proper, research-driven diagnosis. The positioning statement exercise was used as a coordination point as they tried to forge alignment on where they needed to go. That meant, per Fishman, whittling dozens of different statements of purpose down to 'one distinct key benefit' and having hard conversations about why they existed at all. What were they really trying to do?

Fishman quotes Jackson, Patreon's positioning consultant, in a way that sums up Patreon's situation neatly (my bold):

> *Patreon was seen as a tip jar both by the creatives who used it and their fans. This was 2016, before 'creator economy' was a term, and before we had all of the subscriptions we do today.* ***It was clear that the most successful creators on the platform and their biggest fans were using it more for premium content subscription and less as a tip jar.*** *That's where the company was going. So we repositioned the company to reflect that membership model. You didn't throw in a few bucks to support an artist or musician whose work you liked; you subscribed to them and through that got access to exclusive content and experiences.*

That is, just as Dunford advocates, Jackson and Patreon looked at what their most successful customers were doing and how they were getting value out of the platform, and ultimately positioned around that. That's not something a positioning statement or branding exercise (or narrative deck for that matter) will ever give you in and of itself. But if it prompts the appropriate analysis of what success looks like for your best-fit customers, then that's a ticket to valuable insight and customer expertise you can reposition around.

Back in **chapter 21**, we looked at positioning as, in the broadest sense, a metaphor for finding a "position" in some abstract space. In B2B sales-driven contexts, it's usually around a specific job-to-be-done (or set of JTBDs) as well as the language around that job that makes your differentiated value shine.

That was true for Patreon as well. Tip jar or subscriptions? Casual fans or members and patrons? Crowdfunding or membership? Supporting existing work or new work? Different JTBDs, different language, different positions. The market was moving on from Patreon's initial vision, and while Jackson says where they landed seems obvious in hindsight, it wasn't at the time, and the company — including the leadership — had to go through a big shift to get there.

But why look at this through the lens of brand at all? This brings us back to memory associations — the name/need impression people have of you.

Those associations in the market are what we're ultimately concerned with, but they exist internally, too.

Internal repositioning

The shift Patreon had to go through internally included the repositioning of the CEO and executive team's mental model from the old way to the new way. That was a big part of the battle. A positioning statement on its own doesn't do that, but having conversations around it can. It's hard to get the public to form new memory associations about your brand or product without having first formed them internally and that can mean 'breaking up' with those old associations, which can be hard to do.

> *So fast forward to the decision around the Membership Platform category language. I did what any "decider" would do – I barricaded myself, Jack, our brand/copywriter (Taryn), and our head of communications (Mollie) in the hottest, least ventilated conference room I could find and said we weren't leaving without a decision. [...] I doubled down on Membership Platform using the evidence we had gathered from the qual and quant research and experimentation. Things got a little weird (probably because of the heat) and we started to riff on combinations of fan community and membership platform (nope). And finally, after hours of conversation, although Jack still weakly disagreed with Membership Platform I got him to commit.*

To his credit, Patreon's CEO Jack Conte deferred to Fishman as the 'decider' for the project, and Fishman, to his credit, got everyone in a room, including the CEO, and hashed it out until they could all commit to the new direction. (I'm sure having the quantitative and qualitative evidence the move was the right one certainly helped his case, too.) The result was new positioning that stuck, with new branding, new messaging, new products, and new growth. As Fishman tells it (his bold):

> *In short, we changed* ***everything****. The positioning, the personality, and the product features. It wasn't a pivot, but we repackaged what we offered and how we talked about it. And it worked. Rebranding and repositioning Patreon was the catalyst to 10x-ing the business over the next several years, watching our first creators earn over $100,000/month on the platform and millions per year, and leapfrogging competition that didn't know what to do when they heard "we don't want to ask our fans for money."*

That's an incredible result, and it all started with getting the initial diagnosis right, getting alignment around the deep attributes that drove what Patreon was about, and then leaving no stone unturned in ensuring the execution was on point, too (which Fishman covered in a second article[20]).

> **Positioning as messaging vs. product:** People often talk about positioning as PositioningAndMessaging. That's not a typo, that's just how people often describe it — joined at the hip with messaging. But what about product? On the messaging side, Patreon's repositioning was grounded in messaging choices: "fan clubs," "crowdfunding," and "tip jars" vs. "memberships" and "subscriptions," for example. Closing the language/market gap, as we explored back in our look at Jobs to be Done in chapter 36, can be a powerful result of new positioning. But in Patreon's case, they needed proof in the *product* that this was a real change, too, and not just a cosmetic one. Therefore, Fishman drove product initiatives to make that happen in time for the brand launch. This is, again, marketing being leveraged strategically, driving the whole *gestalt* of the business forward, positioning, messaging, and product, too. So it seems it is possible for Marketing to have input into the product and reclaim at least some input into one of the 4 Ps after all!

Distinctiveness beyond product

In the world of technology, it's hard to imagine any brand being meaninglessly distinct in the EBI sense and that being it, job done for marketing. There's certainly some truth to meaningless distinctiveness in the most superficial sense of colors or aesthetics, but any sensible approach to product positioning is going to go much deeper. For example:

- With Squarespace, we saw meaningful differentiation and distinctiveness come from attributes that underpin the brand and get expressed in the product and its marketing.
- With Patreon, we see in their rebranding/repositioning a whole raft of changes from marketing to product to the fundamental JTBD they addressed that unlocked their next phase of growth.
- In the next chapter, we'll look at wedge attributes by way of product differentiation that are used to compete in incredibly crowded markets.

There's so much opportunity for both differentiation *and* distinctiveness that splitting the difference is moot. In the segment-and-sell, corporate-and-up world that can happen in the sales experience, in your playbook, in your product trial, in your references, in your customer success experience, and of course, in your narrative — there's plenty of room to differentiate and use that to drive your *gestalt*.

Caring more

Sometimes, even just caring more about the customer and selling with integrity can make you highly distinct when everyone else is competing through feature bingo or being, shall we say, not-super-high integrity. That's a point Lattice's former CRO, Dini Mehta, made in an interview with Dock's Alex Kracov:[21]

> *Instead, we just focus on the things we do really well and really sell with integrity, really care for the customer. Because, ultimately, my core driving principle as a revenue leader is: the number one thing we sell is the fact that we care.*
>
> *There's always going to be a competitor with a better brand, or with more money, or a better feature set. But sales and go-to-market overall can be a differentiator. I think we really embrace that at Lattice, where we were like 'we're going to do things the right way.' We're going to focus on the customer. We're going to really care about the customer and just build from there.*

All told, this combination of a *gestalt* that captures your differentiation, distinctiveness, and the attributes you live out — even the fact that you just care more *and customers recognize that* — gives us a much broader view of brand than we started with. If you can build your brand on coherent positioning choices — multiplicative ones especially — then you'll get both distinctiveness in a narrow left-brain sense and a much richer, coherent, right-brain whole, too.

Notion & B2C2B

Before we wrap up our SMB discussion, let's go even further down the consumer end of the B2B spectrum with Notion, who has one of the best *gestalt* combinations of *differentiated* product, *distinctive* brand, and broad, JTBD-aware mainstream positioning of its generation.

"B2C2B" was how Notion's COO Akshay Kothari described their approach in an interview with *Protocol*,[22] where, at the time, Notion wasn't targeting enterprise CIOs, they were instead going very broad and very mainstream. They were running mass awareness campaigns encouraging people to use Notion for free to organize their personal life with the hope they'll become product champions and bring Notion with them into the workplace. That means billboards, bus stops, TikTok, and pop-ups with free coffee — a big mainstream *mental availability* push.

They employed their distinctive brand for these campaigns, with their black-and-white illustrative style and bold typography. What underpinned it all was, however, their highly differentiated, highly distinctive, versatile, once-in-a-decade product that reinvented some of the core primitives of documents and workspaces.[23]

And that's why these things aren't mutually exclusive. It's multiplicative choices that drive multiplicative growth — using distinctiveness *and* differentiation *and* positioning around jobs-to-be-done *and* horizontal reach to grow.

That's true reach — perhaps even more than the EBI folks would admit — and at this SMB end of the B2B spectrum, it makes for fantastic market strategy.

Not everyone is fortunate enough to have cracked the mainstream quite like Patreon or Notion, however. Plenty of B2B SaaS folks are instead fighting it out in highly competitive, mid-market-focused categories where there's enormous pressure to both be like everyone else *and* have something that makes you unique, some kind of wedge, which is what we're going to explore next.

Notes

1. McGilchrist, Iain. *The Master and His Emissary* (p. 4). Yale University Press. Kindle Edition.
2. As coined by Amanda Natividad, VP of Marketing at SparkToro. See her July, 2022 piece on the SparkToro blog, for example: "Zero-Click Content: The Counterintuitive Way to Succeed in a Platform-Native World." **https://sparktoro.com/blog/zero-click-content-the-counterintuitive-way-to-succeed-in-a-platform-native-world/**
3. Granted, this is a rather hypocritical point for your humble author to make this far into this book!
4. These brand-building activities might be sustained by performance advertising, however, where the focus is on the click, download, sign up, etc. Creating an effective system where both performance advertising and brand-building activities come together effectively is hard work — the more links in the chain, the more likely something breaks. But because so few folks can do it well, it becomes a competitive advantage for those that can.
5. No one does actually unique brand voice like the funny, brutal, customer-roasting robots at dbrand: **https://dbrand.com**. Google and see.
6. Ritson, Mark. "Distinctiveness doesn't need to come at the cost of differentiation,"

Marketing Week, March 3, 2021. **https://phvntom.com/mark-ritson-distinctiveness-doesnt-need-to-come-at-the-cost-of-differentiation/**

7. Ibid.
8. **https://brand.squarespace.com**
9. To revisit the Bloomberg quote from that chapter (my bold): "Squarespace differentiated itself by providing a technical solution that also **took design into consideration**. During the platform's relaunch in 2012, the company **intentionally sought to attract a more creative customer base**. Back then, the Web was changing and the visual presentation of a portfolio was a challenge, but that style of design could be used for a variety of purposes. "We said, if we can solve for the websites that are hardest to make and the most visually difficult, we can solve that for other people and apply those techniques elsewhere," said Casalena."
10. Sienkiewicz, Radek. "Why do so many brands change their logos and look like everyone else?", *VelvetShark*, November 18th, 2020. **https://velvetshark.com/articles/why-do-brands-change-their-logos-and-look-like-everyone-else**
11. Shorin, Toby. Nov 24, 2019. **https://twitter.com/tobyshorin/status/1198463296980103168**
12. See, for example: Romanuik, Jenni, Byron Sharp, and Andrew Ehrenberg (2007), "Evidence concerning the importance of perceived brand differentiation," Australasian Marketing Journal, Vol.15 (2), pages 42-54. **https://byronsharp.files.wordpress.com/2008/04/different.pdf**
13. A few favorites from the archives include Quill (**https://web.archive.org/web/20210625045456/https://quill.chat/**), Maybe (**https://web.archive.org/web/20230618022849/https://maybe.co/**), and Mighty (**https://web.archive.org/web/20210428013448/https://www.mightyapp.com/**). RIP.
14. Sharp, Byron. *How Brands Grow.*
15. On differentiation vs. distinctiveness, Ritson, in his potted history of brand differentiation, also explains where unique selling points, or "USPs" came from. Ritson points up Rosser Reeves, a 1960s adman, who believed advertising needed a unique selling point or brand benefit to differentiate from the competition. "The prototypical USP was M&M's," Ritson writes, "which used a patented sugar coating to justify the claim that its chocolate 'melts in your mouth, not in your hand'." That became an incredibly memorable slogan repeated for decades. These days, however, Ritson thinks USPs are largely worthless: "I do not believe in USPs – they are usually bullshit and even when they are not, they can be quickly replicated." What's the alternative? At the other extreme, we have *distinctiveness*. Consider, for example, the modern M&M's ads with the animated chocolate characters — very distinct, no pretense of product differentiation, and no real meaning to it, other than that they're fun, and it's the literal product talking. This, in B2C brand marketing terms, is the difference between classic differentiation and distinctiveness. Can you have both? Sure, if you're realistic, with Ritson writing, "I think the [EBI's] work on distinctiveness is a revelation and an important course correction for those of us who were sold on the dreams of USPs and attribute ownership. But a smart marketer can also enjoy the benefits of a distinctive brand and a realistically differentiated one too."
16. Saarinen, Karri. January 26, 2024. **https://twitter.com/karrisaarinen/status/1750592508651016256**

17. Fishman, Adam. "When Branding is a Growth Strategy," June 23, 2023. **https://www.fishmanafnewsletter.com/p/when-branding-is-a-growth-strategy-part-1**

18. I'm focusing on the positioning aspects here as I feel they were the least fungible and most impactful parts of the process Patreon went through (that, and I really dislike brand personality exercises), but you can read both pieces for the full story, which helpfully includes briefs, work-in-progress, and more: **https://www.fishmanafnewsletter.com/p/when-branding-is-a-growth-strategy-part-1** & **https://www.fishmanafnewsletter.com/p/when-branding-is-a-growth-strategy-part-2**

19. Dunford, April. "A Product Positioning Exercise," October 17, 2019. **https://www.aprildunford.com/post/a-product-positioning-exercise**

20. Fishman, Adam. "When Branding is a Growth Strategy (Part 2)," July 18, 2023. **https://www.fishmanafnewsletter.com/p/when-branding-is-a-growth-strategy-part-2**

21. "Crawling Upmarket: How Dini Mehta grew sales to $100M at Lattice," June 20, 2023, **https://www.dock.us/grow-and-tell/ep-1-dini-mehta**

22. Lawrence, Lizzy. "Notion still wants to become 'the next Microsoft'," *Protocol*, October 3, 2022. **https://www.protocol.com/workplace/notion-cfo-rama-katkar-ipo**

23. That visionary product also — and by design — unleashed a wave of user innovation with the tool, fuelling an incredibly enthusiastic community of users sharing tips, templates, and customized workspaces. A visionary product that *enables* 'user innovators' to apply the tool in myriad innovative ways is another powerful, multiplicative combination.

46

The B2B spectrum IV — the mid-market & wedge attributes

> Fortunately, there's advice for this stage: to break into the market we need a wedge. But what is a wedge, exactly? [...] The wedge is not your company, your product, or your strategy. The wedge is a shape that's able to enter the market because the market recognizes it, allows it in, and (and very importantly) because it fits.
>
> – *Matt Wensing*[1]

In our journey across the SMB end of the B2B spectrum, we've discussed passive B2C attention vs. active B2B sales-driven attention, positioning for niche vs. reach, *How Brands Grow,* positioning statements, differentiation vs. distinctiveness, attributes, and *gestalt*. Phew! We've covered a lot of positioning and marketing theory, and in this chapter, we're going to cop a dose of mid-market reality.

Unlocking the mid-market can be tough, but we've been building up an array of tools we can use on the market side of product/market fit, including concepts like:

- **The B2B spectrum:** If you're serving SMB, the mid-market, and/or enterprises, you need to account for the fact the upper and lower ends of that spectrum behave very differently.
- **Niche-to-reach framework:** You're on an adoption curve, either going for reach or niche.
- **Adopter segments:** You might be targeting nerds vs. normies (a.k.a innovators, early adopters, the early and late mainstream, and laggards).
- **Distribution style:** You're using either 'catch and cast' or 'segment and sell' for distribution (or some combination of both).
- **Positioning choices:** Your distribution style informs whether you position by segment and go niche or position by attribute for reach.
- **Brand:** The collection of memory associations centered around a name/need connection that you need to build and sustain over time.
- **Attention quantity:** The folks you're trying to reach might be using active or passive attention, and you might differentiate for clicks and use distinctiveness for memories.
- **Attention mode:** The buyers you engage with are either using left-brain

narrow focus attention (how it works and how to manipulate the world) vs. right-brain broad attention (change and what's 'out there').

The point of all this work is that if you can chain some of these concepts together, along with your fundamental market or product insight, you'll end up with a very compelling positioning and narrative — perhaps as a brief deck (**chapter 9**) and Hemingway story micro-pitch (**chapter 10**) — that unlocks a wedge in your market.

But what is a wedge, really?

Wedge-shaped gaps

We really are looking for a wedge of *opportunity* here, as 2x founder Matt Wensing says in the quote above. Most folks think of their 'wedge' as a product, feature, or strategy, but those are things that *exploit* a wedge-shaped hole in the market, as Wensing reminds us.

Not everyone needs to find their wedge of opportunity by getting this far into the positioning weeds. This is pretty analytical, left-brain stuff. If your initial product concept and narrative captures lightning in a bottle — like Slack did, back in **chapter 12**, or if you find a goldmine of attention as Squarespace did, as we saw in **chapter 44** — then great. You're off to the races and unlikely to require this level of positioning scrutiny (assuming your growth doesn't top out, as it did for Patreon, as we saw in the previous chapter). You're more likely just holding on for dear life.

But in highly competitive B2B SaaS markets — which is just about all of them — if you're looking for that wedge to exploit through positioning, you're in the right place, especially if you're battling it out in the messy middle of the market.

In this chapter, we're going to look at a typical, early stage startup that you've never heard of and see how positioning can play out for B2B SaaS teams focused on the corporate mainstream mid-market. It's here that finding an attribute to be 'more about' can work wonders for your positioning and direction. This is not a sexy tale of breakout success like Slack, Salesforce, or Squarespace, but I imagine it will be far more relatable for the majority of B2B SaaS folks out there, and it does end in a nice exit. Let's dive in.

Attacking the (mid) market

One SaaS client (well, two, briefly, if we include their acquirer) I consulted for sold an LMS, or Learning Management System. The target market was L&D (Learning & Development) professionals, folks who usually live under the corporate HR umbrella and usually handle things like compliance, training, onboarding, and change management.

L&D is an interesting SaaS opportunity for two reasons:

- **Every company:** Almost every company of a certain size on the face of the planet needs some degree of L&D involvement, even if it's just onboarding or compliance.
- **Every employee:** L&D, as a function, often needs to communicate with every employee in the company — every employee needs to be onboarded, every employee may need to complete compliance training, every employee needs HR-specific communication, and so on.

Think about that market potential: per-employee licensing with a TAM of every company on earth (in theory, at least). And that, dear reader, is why there are 500+ corporate Learning Management Systems to choose from on G2. You're welcome to try LMS #512.

This is classic, mid-market, corporate, B2B SaaS. But how do you craft a narrative and position for a solution in a market that broad with competition that fierce?

First, let's talk about the buyers.

The buyers and their playbooks

LMS buyers range from self-described office admins to learning professionals, instructional designers, and C-suite HR folks (the "CHRO"). That's quite a cross-section of people to sell to.

That is, there's a mix of:

- **Adoption:** There are both very progressive and very conservative folks, i.e., nerds and normies, in the target market.
- **Attention:** At one end, you've got folks with limited responsibilities and very laser beam-focused needs, and on the other, you've got C-suite folks wanting that radar-level narrative or playbook. (These are some of the types of buyers we saw in our natural narrative experiment in **chapter 22**.)

And what do these buyers want?

It turns out there are roughly two broad schools of thought, or two playbooks, in L&D (at the time, at least):

- **The traditional playbook:** Regular L&D folks cared about *engagement* with the courses they'd created, despite the fact employees often hate these courses and just want to get their work done. (If you've worked in corporate land, you've probably had the displeasure of 'engaging' by clicking through a cheesy compliance course of one kind or another. Everyone — including many folks in L&D — hates those, which is part of the reason why people keep trying to invent new tools to solve the underlying job-to-be-done.)

- **The progressive playbook:** Progressive L&D thought leaders — whom I was fortunate enough to work with — felt these problems with the profession acutely. In the broader L&D world, progressive L&D folks had been pushing for 20+ years for the function to *solve business problems* using just-in-time resources and not long courses. (I know it's bonkers that 'solve business problems' was — and is — a revelatory idea in the space, but it really was. And still is.)

So we've got a mix of buyers and a mix of playbooks. My client was very much anchored on the progressive end. However, note the first clue here regarding the market's adoption of these ideas (or lack thereof). Those progressive ideals, as sound as they were (and are), had evidently failed to cross the chasm to the mainstream — innovators and early adopters were keeping them alive, but the fire wasn't catching on in the mainstream.

Let's set up the challenge of selling to these folks — and there were a lot of challenges — and then I'll explain the positioning solution of sorts that we landed on.

> **New playbooks, new risks:** A key positioning decision when you're pitching a new tool is whether you aim to support or replace your buyer's current playbook. Do you support it and just provide better UX? Do you aim to replace it and create a new playbook with a new UX, pitching the behavior change that comes with that? They are very different sales learning curves to be on, and it's again worth knowing which game you're playing so you can be clear-eyed about the challenge you've set yourself. Pitching a new playbook can sound great — it was a core part of our new concept formula (change, playbook, tool) back in chapter 31 for a reason. But keep in mind what you're asking folks to do: take a risk on a new approach. Can you build an organization that doesn't just preach that message but can see it all the way through, from adoption to customer-verified success? What would that look like in your market?

Narrative for innovators vs. 9-5ers

In this positioning section of the book, we've been discussing the shape of the demand you're seeing so you can shape your offer to match, and that's true for finding a wedge opportunity, too. We've talked a lot about segment and sell vs. cast and catch because they determine the shape of the demand you see. If you're in the segment-and-sell world, for example, then you might use niche distribution to unlock niche positioning and open up a market wedge that way.

My client, however, couldn't do that. They didn't have the niche distribution that would allow them to sell to *just* the progressive folks, building enough momentum there to *then* go after the mainstream. They could, however, still pick off enough folks amenable to the progressive playbook to get the business off the ground, but it was tough going.

In part, this was a limitation of their GTM model, but it was also a limitation of the market in that the folks who were true Hardcore L&D™ were few and far between. If your narrative requires folks to take the *ideas* that drive their job very seriously, then your TAM will always be much smaller than you might otherwise hope, because plenty of your garden-variety office workers are going to be too junior, too uninterested, too unaware, and perhaps too overwhelmed to care. That's not a dig at those folks; that's just reality. Most corporate folks are just doing the 9-5, and that's it, which is fair enough. But you need to calibrate your expectations accordingly, because the shape of that demand will inevitably shape what you offer.

> **Market insights, promiscuous nerds, and product loyalty:** L&D had a long history of a handful of folks carrying the torch for the progressive playbook on how people really learn at work. That was a market insight with product *implications* without being a product *innovation* per se, but it provided enough narrative fuel to at least put something new on folks' radars and to take a position in the nerd segment of the market. The nerds — innovators and early adopters — are great in that they'll give you a go because they're always on the lookout for an edge or advantage that technology can give them. But that also can mean they're not loyal — if something newer and shinier comes along, guess what? They're going to give that a go, too. This can be confusing for founders — *I thought they were all in on our approach! Hmpf!* But the most loyal buyers often are, I imagine, the ones that just don't think too hard about the system that's already in place. It works, it does what they need, they renew. Simple. But getting there — getting to that place of being the solid default normie choice — often requires finding that wedge that the more promiscuous nerds need first.

Turning the dials

A strong, persuasive narrative can do some of the heavy lifting in the early days when your product is still more M than VP, but prospects are still going to give you the left-brain laser beam treatment as they analyze your product capabilities and seek to solve their very particular problems.

We've talked a lot about stretching your clarity strategy along one or both axes of change narrative and narrow positioning, and knowing which dial to turn — more right-brain change narrative, more left-brain narrow-focus positioning and jobs-to-be-done — can help you match the attention of your audience.

But my client faced more fundamental market strategy questions — questions that you may have considered, too — like:

- Do we sell to early adopters *only* or the more mainstream conservative folk as well?
- Do we lead with our tool and narrow-focused positioning or a broad change narrative?
- Do we target the end users or the executives?

All of this depends on the distribution, and therefore the demand, that you have access to. And when you're targeting the big, broad mid-market, you're going to get a big, broad cross-section of potential buyers. So what do you do?

Value choices

If you've fought your way into the market on the back of a very early-adopter-ish, progressive-focused narrative, then it can be tough to switch gears and figure out how you're going to serve the rest of the market, if you're going to serve them at all. You've built your reputation around a certain point of view, and now you're faced with hard choices. (This is why, incidentally, the dynamic positioning model of switching from niche to reach that we discussed in **chapter 44** isn't always as easy as it sounds.)

For example, with our LMS vendor that's strongly aligned with the progressive L&D playbook, should they support:

- **Old way:** Do they still incorporate the old, broken, engagement-and-courses approach and entrench a part of the business that's (very slowly) on the way out?
- **New way:** Do they *only* target the "progressive" approach to L&D that's clearly superior, and some folks are super into, but has had 20 years to break through and still hasn't got there yet?

In the *How Brands Grow* world that we explored in the last chapter, the answer is pretty obvious — you go for reach and try and serve all-comers. Penetration is how you grow, not loyalty from a specific segment. But if you're not a big player, you don't have the massive platform that does everything, and you've been focused on a niche, then what?

This is, in part, a matter of values — you want to take the industry forward, but you also want to sell your software. How much do you compromise on your progressive vision to support the industry laggards who may show up wanting a tool but not wanting to change anything they're doing?

That is, how much do you dilute your distinctiveness — the qualities that make you, *you* — and risk becoming yet another Frankenstein-esque product that tries to be everything to everyone? This is the challenge of maintaining distinctiveness — go too broad and build too much, too soon, and either no one gets what you're about or they just see you as a weaker knock-off of the incumbent, an incumbent that not only has much more polished features, it has the 'whole product' to match — integrations, partners, community, add ons, and whatever else constitutes its broader ecosystem.

That's not you, so what do you do?

> **Double-edged narrative:** This is what's often lost in narrative discussions — a sharp narrative, by definition, divides people. There's a new way and an old way. You're forging ahead with the new way, but are you really prepared to leave the folks clinging to the old way behind? The wedge that gets you in can also be the wedge that cuts the rest of the market out. True differentiation and going niche can mean exclusion, and that means you face efficiency challenges if your only distribution option is cast-and-catch, as we discussed in chapter 42. Something has to give, and it's perfectly acceptable to change your distribution approach instead and try and segment and sell, for example.

Concept saturation

Here's an idea. Maybe a new, killer narrative would work in the LMS market? We've been playing pretty narrow with these progressive, early-adopter types; maybe it's time to **ahem** play bigger instead?

We've talked a lot about riding waves, and in this case, LMS providers are riding the digital transformation wave for L&D. But when there are 500+ other riders on the wave you're trying to surf, you might think it's time to ride a different wave.

Typically, this is where folks reach for category creation. "Ah, you should simply break out of that crowded market and create a new category for your LMS instead!"

If so, is that microlearning? AI-driven learning? Gamified learning?

Learning in the flow of work? Maybe it's not even an LMS we're selling; maybe it's an LXP — a Learning Experience Platform. What's the difference? Who knows!

The LMS market was already drowning in concepts and playbooks. Every competitor was pushing their own shtick. Creating a category sounds great until everyone else is doing it — then it's not dozens of riders on one wave; it's dozens of tiny waves with one rider on each.

This doesn't help buyers — they want to pick the big winning wave, or at least the one that's right for them, and if they don't know what to do, they'll either default to the incumbent or fall back to no decision and stick with their status quo. That's not necessarily out of some secret love for the status quo — they're reaching out to you for a reason — but it speaks again to their fear of messing up, where simply doing nothing can be a more psychologically tolerable option. (This is where the sensemaking approach we discussed in **chapter 24** becomes important — it's the next move when buyers already have a high quantity of high-quality information.)

In this case, a new category just ain't gonna cut it.

Ok, yikes, what else can we do...

(Bear with me here, dear reader, as we go further into the pit of positioning despair. I think I see some faint light at the end of the tunnel...)

Challenging sale

Well, what if we stick with our original progressive story 1 narrative and just really challenge those mainstream prospects with it when we demo? There's a whole sales philosophy about challenging prospects, after all (*The Challenger Sale* et al.), so maybe that'd work in the cut and thrust of individual deals? Maybe our point of view will resonate then?

In this case, your pitch might involve trying to break down the old way and tell buyers their way of doing things doesn't work and they should adopt your new way instead. Alas, in my client's case, the founder was a very compelling seller and had tried that — it turns out people didn't really like being told in a sales pitch they were doing it all wrong. Ouch.[2]

Now we're back to our original conundrum. Do we stick with our focus on the progressive playbook ("solve business problems") and risk turning the traditional folks off, or do we try and sell to all comers like everyone else with the traditional playbook ("drive engagement") and sacrifice what makes us distinctive in the first place?

> **Return of the attention mismatch:** Part of the reason this challenge-the-prospect approach can fail is the attention mismatch we discussed

> at the start of this book when we looked at diagnosing narrative challenges in chapter 6. This problem rears its head when buyers are in left-brain laser-beam mode and are interested in a product's specifics, and you're trying to put something entirely new on their radar. If they haven't been primed for that, it's either not going to register or be a distraction at best. The buyer is just too far down the left-brain path, and it requires a lot of finesse to change that and get them back in right-brain radar mode, as we touched on in chapter 25 when we discussed pitching, frame-breaking insight, and nudging folks out of their left-brain laser beam.

The mid-market blues

This is the dose of realism I flagged at the start of the chapter. This stuff is *hard*.

It can be tough when you want to move an industry forward with your market insight and product, but much of the industry is quite happy to stay stuck in the past.

This is the challenge of trying to sell to the entire adoption curve at once — nerds, normies, and laggards. And that often results in vague, generic narratives and positioning — the lowest common denominator rules. This is the mid-market conundrum, and I'm not talking about the LMS market here, I'm talking about B2B SaaS in general. When faced with tough choices with real risks and tradeoffs, most folks play it safe and fall back on vague, wishy-washy narratives and positioning. (That's what got me into this work and why I wrote this book — I saw too many seemingly powerful products stuck with weak messaging, and it took me a long time to understand why.)

The 'why' is, in my view, the big gaping hole that exists where narrative and positioning strategy should be. If there's just a Post-it note with "TBD" written on it instead, then what can startups do *but* go for undifferentiated reach and opportunistic selling? (This is the temptation of *minimum* differentiation, or competing to be the same, which we'll discuss in **chapter 49**.)

TBD positioning

Again, the shape of the demand you access shapes what you offer. If you only see broad, undifferentiated, lowest-common-denominator demand, then unless you have something else you can bring to the table, that sales pressure will constantly push you to conform to the shape of your market, and your positioning will remain TBD. That's why it's often much better to find a coherent shape of demand you can address instead — a wedge — that shape that *isn't* the big messy blob of competing interests in the middle of the mid-market.

My client found some success with their narrative for L&D nerds, but then they were stuck trying to find a differentiated position — a wedge — for the rest of the market, and this brings us back to positioning by attribute.

"Owning" an attribute

In the B2C world, as we saw in the previous chapter, it's a mistake to think you can *absolutely* own an attribute. It's a mistake in much of the mainstream B2B market, too.

When researching B2B positioning, I came across one article suggesting that Sony, Dell, and IBM each "owned" one attribute: flexibility, innovation, and quality. I have no idea which was supposed to match with which. Is Sony quality or innovative? Is IBM? Who knows. (Customers certainly don't.) That approach is not what we're talking about here. We're not doing positioning by attribute in a classic B2C brand positioning sense here — we're not talking about vague stuff like being a "modern" brand or "friendly," "cutting-edge," "easy to use," or any other generic, category-level attributes. There's no radar-friendly newness in any of those attributes.

In sales-driven B2B, when targeting the mid-market, we deal with innovative products and active attention. Here, you can build your product around an attribute in a way that allows you to, if not own it, at least make it a focal point of what you're about, giving you that all-important axis of differentiated value.

Attribute as wedge

Here, we're talking about attribute-as-wedge, and that can come in a variety of forms:

- April Dunford, back in **chapter 37**, talks a lot about finding your differentiated value relative to competitors for a buyer's given job-to-be-done, particularly (but not exclusively) by finding the patterns in your existing customer base.
- David Cancel planned out his approach to differentiated value with Drift and the attribute they'd own (conversational marketing — see how he described it in his own words in **chapter 31**).
- B2B SaaS markets naturally, to a degree at least, can see competitors gravitate towards different attributes they can be 'more about' to carve out *just enough* differentiation while going for reach, as we'll see here.

Finding a wedge is, of course, business 101 in the SaaS world, but few folks do it as well as, say, Drift did. The approach I'm describing in this chapter is not nearly as sophisticated as Cancel's approach of identifying a secular trend (messaging) and bringing a *network of multiplicative choices to bear* on the market as a wedge into a broader existing category (CRM). The approach we're discussing here is what you do when other approaches have failed or aren't viable, as was true for my client. If you'll forgive another trip into the pit of despair:

- They *weren't* being carried by a unique-to-them secular change.
- Their narrative *hadn't* set the world on fire.
- They *couldn't* do segment-and-sell to find just the innovators or even, say, enterprise execs who may have been more open to right-brain change narratives.

The 'adjacent possibles' for them were limited. And that's fine. Not ideal, maybe, but not everyone is David Cancel. The secular change of *cloud software* was (and is) still in play, and in competitive markets that big, there's no shortage of customers. When you know what the routine 80% use case looks like, what you're looking for is that 20% difference that can tilt deals in your favor. That is, you want to create leverage by being 10x more about one specific thing than the competition so that customers, who pay far less attention to these things than we do, perceive maybe — *hopefully!* — 1.1x difference, with that being just enough for them to take a chance on you.

> **Doing the research:** The consistent thread that runs through B2C positioning (Mark Ritson gives a neat, bite-sized run through this process on the Everyone Hates Marketers podcast[3]) and our B2B positioning examples, from April Dunford's approach to Patreon to this case study to folks who stumble across a wedge opportunity is: market research. It's not about best guessing an attribute and YOLOing it; it's doing the research, either through formal market research or ad hoc in-market learning, where you see up close where the gaps lie. It's not rocket science, but it requires a degree of sophistication that we often lack in the B2B world where best guesses — or meaningless fluff — all too often rule the day.

1x features + 10x wedge

In light of all that, where do folks end up?

In my experience, they try and have their cake and eat it, too — a bit of niche while still trying to maintain reach at the same time. On the one hand, you're trying to maintain reach by satisfying those common, bread-and-butter 80% use cases and keeping the door open to folks in the broader market should they find their way to you. On the other hand, you're sharpening a key point of difference — your point of *distinctiveness* — in a focused way so you stand out.

This is, again, something you do because you have to. If you have a brilliant right-brain vision for your product and its *gestalt* — the whole that's greater than the sum of its parts — and you invent the next Slack, Notion, or Linear (or Salesforce, HubSpot, or Drift, for that matter) more power to you. But if you're trying to compete in a mature, competitive market as a challenger, then I imagine you're going to end up where many mid-market challenger B2B startups land with their positioning: *Yes, we do all the basics you mainstream folks want, but we also really kill it at this one thing for you cutting-edge types, too.*

It's a means of serving the mainstream market so you:

- Have no glaring reason to be ruled out when buyers make their shortlist.
- Have some distinct reason to be ruled in because of what you've built to serve that wedge you've identified.

This is especially true in the cut and thrust of sales where, again, the number one question is: why buy you? That is, what makes you different?

Pain vs. gain

The unspoken other half of the "What makes you different?" question is "in helping me do my job and helping my company succeed?" That seems obvious, but the nuance there is that the customers get to dictate what that ultimately is, not you as a vendor.

That's why the progressive narrative for my client didn't cut it for everyone — changing the way someone does their job creates risk, uncertainty, and, well, a lot of extra work. That's short-term pain for *potential* long-term gain. But what if they don't want any pain and aren't persuaded about the long-term gain? What if they just want a tool that makes their current day-to-day more pleasant?

The point of your wedge attribute, therefore, isn't to name it and claim it, declare it The One Right Way, and then just bang away at it forever — that's trying to force a wedge (or category definition) into a market that may not want it. Instead, the point is to feel out the shape of the wedge so you actually find fit, and this requires more hands-on, reality-layer contact with the real world.

And when you do find it, it can be quite a powerful thing. It's something you can then feed back into your narrative, your customer-facing messaging, your positioning, and perhaps your category-transcending playbook, too. It can unite the team and get everyone heading in the right direction.

Picking an attribute: automation

So what sort of attributes are we talking about, then? Think about the common attributes, in very broad terms, folks lean into in B2B SaaS. They might be:

- Intelligence
- Collaboration
- Automation
- Speed
- AI
- Blockchain (I kid, I kid)

Or whatever the hot trend of the moment is. I'm sure you can think of the others. This find-an-attribute approach was where my client & I landed with their LMS offering.

In my client's case, I helped them position their offer around *automation*. This wasn't due to some profound insight on my part; we was just doubled down on what they'd already learned in-market by working closely with customers and seeing where the glaring gaps were. It wasn't best guessing, either. They'd seen the need for better tools for their customers first-hand, and they could, therefore, make a bet on an area of the product that could fill that gap — that wedge they'd identified.

Having that proof is vital for both giving you a point of focus and protecting you from distractions. Plenty of startups get distracted by pointless attributes that aren't key to their *gestalt*. Sometimes, they're just tacked on to please the worst kind of startup customer, the one that doesn't even use the product: the hype-chasing investor. Try and resist those if you can. Evidence, not hype, is the guiding principle here.

> **Your insights, attributes, and brand:** You've probably discovered similarly powerful insights in dealing with your customers, but have you really doubled down on them? Have you truly tried to be more about them than the others? Once you're thoroughly sick of repeating your attribute-driven message, you're probably just starting to do it enough. Think about what we discussed in terms of brand and memory in the last couple of chapters. What name + need (or product + attribute) are you trying to drum into folks' heads? What's the recurring theme folks are going to pick up from your appearances in their inbox, along with the features you build, events you hold, and customer stories you tell?

Wait... isn't this category creation?

Earlier, I described why category creation wasn't a viable strategy in this case for the L&D market — trying to surf your own tiny wave in an ocean of folks already doing the same thing wouldn't inherently distinguish my client from other players. However, if we look at the attributes outlined above and tack on "Learning," we get:

- Learning Intelligence
- Collaborative Learning
- Learning Automation
- Fast Learning (ok, this one doesn't really work)
- AI-powered Learning
- NFTs for Learning (again I kid, I kid.)

And that looks a lot like category creation! To be fair, this is what passes for "category creation" in many cases, and one of those combinations above was indeed a *Play Bigger*-inspired attempt to create a category.[4] So... isn't this category creation by another name, like HubSpot and "marketing automation," for example?

Good question! I'd put it the other way around — these types of superficial category-creation attempts — like folks saying, "We're a Learning Intelligence tool!" — are just left-brain attempts to position around an attribute in an existing category.

So, in short, no. Neither analysts nor buyers see these attributes as new categories of software — they're still very much in the existing corporate LMS category. Instead, these are wedge attributes. And that's fine — wedge strategies are valuable in their own right, which is why we're discussing it here. But it's a different play from category creation, which we covered in excruciating detail starting at **chapter 32**. Understanding the difference between *concept* creation and this kind of positioning-by-wedge-attribute does, however, help solve the riddle of what folks think they're doing when they're "creating a category."

Wedges don't create categories

This is an important insight because it helps to set expectations and keep you grounded in reality.

If you're all about "Learning Intelligence," say, that's great, but there's probably no right-brain secular change there; no rise of a new role like Customer Success that Gainsight harnessed (**chapter 30**) or a consumer trend like social networking that Yammer harnessed for Enterprise Social Networking (**chapter 32**). You're *probably* not launching a new concept in the change + playbook + tool sense that we looked at back in **chapter 29**. (If you are, that's a different kettle of fish.)

I'm all for new concepts and story 1 narratives when you have some genuine change to ride, as we saw, but that's not what was happening here. Here, we're talking about static differentiation in the left-brain map of a very well-established category. Company A leans into "intelligence," Company B leans into "collaboration," Company C leans into "automation" (as my client did), and so on.

These wedge *opportunities* only exist insofar as incremental demand exists for them. That's the shape of demand you're trying to match to find a fit. Some companies probably *do* want greater insights and reporting ("intelligence") from their tools; other companies probably *do* wish they could deploy more collaborative tools; and others, feeling overworked and overwhelmed, probably *will* see automation as a lifeline.

The strength *and* weakness of the wedge approach is that:

- It helps you catch those prospects with an interest in the area you're exploring.
- But it doesn't help you catch everyone.

This brings us back to our niche vs. reach tradeoff. Theoretically, sure, if you could just build best-in-class versions of all these things, then great — maximum reach. But most startups can't, so they have to pick one to try and be 'more about.' Different startups pick different attributes as they look for a competitive edge, and that fills out the category map of options for a given market.

This is very much about positioning for a degree of reach in an existing category, not a new one. The terms of competition are, in this case, still very much about building the best LMS — your attribute just gives you a particular perspective on what makes a *great* LMS.

Confusing wedge for category

To create a new category would be to say, "We're not an LMS, we're an LIT" (a Learning Intelligence Tool, which I just made up). And that would quickly cut you off from the *reach* in the LMS market that you're after, relegating you to the *niche* of LIT that has zero demand because it's made up.

This is a big strategic mistake that folks make when they get carried away with the category creation playbook. They niche themselves out of the market — out of the mainstream demand they're trying to tap, and then tie themselves into knots trying to find a way back in.

The big myth of 'category creation' is that by declaring a new category definition, you somehow invent the demand that drives it and can crown yourself king of it. But if the demand stays firmly in the existing category, you're SOL.

That's why it's important not to confuse wedge opportunities for category creation. If no market is created, there's no new king to be crowned and no new demand to tap. Instead, what really happens is a battle of wedge attributes and opportunities in the existing category as vendors try and walk the line between the new and the familiar — the mildly progressive and the staunchly traditional. But this is still very much positioning for reach, not some made-up niche.[5]

Reach and the MAYA threshold

What prospects often want to buy is whatever is within their MAYA threshold (**chapter 48**) — the 'most advanced yet acceptable' option for them. Competition, in this case, is about finding the sweet spot within threshold — what's suitably advanced (along the axis of your wedge attribute) while maintaining enough familiarity that it doesn't cause pain for the typical day-to-day work folks have to get done. A little bit of gain along with absolutely no pain.

Is positioning for reach around a wedge attribute less sexy than category creation? Sure, but it's far more realistic. It has all the positive aspects of alignment, focus, and differentiation without the onerous task of trying to "educate the market" about a category that doesn't really exist. Instead, it's about your product, your insight, and your reach across a mature, very-much-existing category.

But how do you find an attribute that sits neatly in the MAYA zone? This brings us back to Jobs to be Done.

> **Do you have to stick with one attribute?** No — this is about finding the wedge opportunity to get *into* the market. That wedge opportunity might be around a single attribute you can own, but it's unlikely that any single attribute is going to be a means to dominate the market. Every company keeps building out its multifaceted *gestalt*, of course, because you want to keep building for reach. There's a point of diminishing returns for any one attribute, and that's where you may have to build your stats, RPG style, in other attributes, too. If you can do so in a way those attributes reinforce each other, then you'll create a new recognizable whole; if you don't, you'll end up in the Cheesecake Factory zone of incoherence (chapter 12).

Finding your attribute

> *"Look for the hole" in the prospect's mind is one of the best strategies in the field of marketing.*
>
> – Trout & Ries[6]

How you find a key wedge attribute to position around becomes critically important in this case, and it needs to come from your customers, otherwise what point, exactly, are you positioning around? If you pick "speed" but prospects want "intelligence" (or vice versa), then you may be backing a horse that's actually just a bunch of cats taped together.[7]

In the old-school world of *Positioning* in the 80s, the idea was, as the quote from Trout and Ries suggests, to "look for the hole" in your customer's mind. The examples those authors provide, however, are all cases of classic market segmentation — be the high price or low price offer, be the product for men or women, be the product for old people, be the product for a given time of day, or for heavy users (or drinkers — beer is the example), and so on.

In this case, however, we're not looking for a hole in the prospect's mind or at traditional market segmentation (that customers may or may not actually perceive); we're looking for a hole in our prospect's jobs-to-be-done. There are, broadly, two ways to find this:

- **Exploit what's working:** This is the April Dunford approach we discussed back in **chapter 37**, where you develop your positioning based on what's working with your best-fit customers. You're relying on customers at the edge to have found the hole in their workflow and plugged it with your product, which is something you can potentially double down on. (Loom did this, too, as we discussed back in **chapter 10**.)
- **Exploit what's missing:** Of course, that's not the only approach (and Dunford offers plenty of other insight-driven approaches, too). You can, alternatively, also study what your target market lacks, find the gap yourself, and use that as your wedge into the market.

In this case, my client discovered a unique need around what was missing while effectively consulting on-site for a customer, i.e., through services. It wasn't about what was working; it was spotting what wasn't — the hole in a user's broader JTBD that represented an underserved need.

The gap in the JTBD chain

What was the need? An LMS helps L&D teams make learning resources and courses available to company employees, but my client discovered these teams often had very few tools to automate the distribution of those resources to the right people at the right time. That is, L&D teams could create and post great resources all day long, but if no one ever heard about them or received them when they'd be useful, what was the point?

The solution, then, was building marketing-automation style tools for specific L&D needs, which helped close a gap in L&D's broader jobs-to-be-done. This was great for several reasons:

- **Nerds and normies:** Whether you were running the old-school engagement playbook *or* the new-school problem-solving playbook, this kind of functionality would improve your day-to-day.
- **Reach, not niche:** This angle gave the company reach across all the mainstream buyers they were seeing. They didn't have to stay in their progressives-only, early adopter niche (though their initial narrative was still there for those folks); they could offer a unique tool to more or less everyone.

My client wasn't the first to discover this need, and they certainly won't be the last, either. The point *isn't* that your wedge attribute has to be 100% absolutely original and never before done in any capacity; it just has to be something you can focus on more than others, give it the best UX you can, and therefore give folks a reason to consider you at all. Marry that with other secondary attributes, like simplicity relative to highly complex incumbents, and you've got a compelling offer through your unique *gestalt*.

> **Wedge first, narrative second:** There's a simplicity in this approach that can get lost when folks get too high-minded about the big-picture category, narrative, or playbook (*guilty!*), and so on. We often want to give prospects the big narrative change story or category creation pitch because it makes us sound big and important, but what if we didn't? If you're seeing normie buyers in left-brain laser beam mode comparing tools, then if you can differentiate your tool by solving an underserved link in their JTBD chain, you both (a) extend your reach and (b) open up opportunities for your narrative or playbook *after* the sale. The narrative doesn't have to come first. You can flip the approach and sell folks on a differentiated tool before you sell them on the narrative or playbook once they're a customer. Narrative doesn't stop with the sale. Once folks are bought into the tool, you've got their attention for the length of their contract to sell them on — and get feedback on — your bigger ideas, too. Come for the tool, stay for the school, as we touched on in chapter 20.

Attribute weakness

We've got our JTBD-driven wedge attribute of *automation*, now let's run it through our right vs. left-brain hierarchy of attention to understand some inherent limitations:

- **User vs. executive attention:** The insight here around automation was very much focused on the end user — their workflow, their day-to-day, their productivity, and their outcomes. That is, it was a pretty left-brain, doing-the-work, manipulating-the-world-based pitch.
- **Lack of secular change:** There was no wave driving the need for this functionality and attribute — it made for a better product, but it didn't tap a huge groundswell of energy like the secular change stories of **chapter 30**.

On the first point, the pitch the sellers developed around automation focused on the user saving time and getting their day back. That was great insofar as they were selling to end users, but it didn't fully square the circle with their broader right-brain change narrative.

The lack of secular change driving the need for this kind of tool also meant that it was something the startup would have to push uphill. It was pretty left-brained in this sense as well, given it was a wedge to differentiate in that static map of the market, not in a more right-brain, change-over-time way. But this was true for other players too, which makes competition more a battle of resources and execution than trend-riding.

Nevertheless, positioning around this attribute helped put the tool (and not the progressive narrative) first in a pitch for your average end user, giving the company both normie-friendly differentiation and an opportunity to introduce their narrative as customers became more familiar with the company. But this was only possible because the team happened to discover this need. Like many startup stories, it was a stroke of luck in some sense, but in another sense, they did the hard work to put themselves in the right place at the right time, where they could deploy their laser-beam attention to their customer's end-to-end JTBD workflow and see what was missing.

> **Getting out of the building:** It's classic startup advice for a reason, but getting out of the building helps you see things that will never show up when scrutinizing product usage stats, NPS scores, or theories about markets, for that matter (**ahem**). In this case, the insight came from an in-person understanding of the customer's entire workflow, their entire job-to-be-done. (That is, incidentally, something the Figma team excelled at, as we discussed back in chapter 27.) And that will never show up in your data or vanilla customer interviews. Customers

> can't ask for what they don't know they need; you have to find the insight and build the solution to match.

Greasing the wheels

Even with those limitations in mind, it's amazing what happens when you find that attribute, that conceptual hook, that thing to truly *be about*. It can energize the team and provide a renewed sense of direction and purpose. And even though these attributes are pretty left-brained in a sense, they can still give you something new to put on folks' radar.

Like anything, you've still got to test your assumptions. You've got to test it in person (looking for those 'eyes light up' moments we discussed in **chapter 38**) and/or in the market, and then go on the long march of finding creative and/or direct ways of using it to tell people you exist. Mental availability still needs to be solved, and that can require significant financial firepower.

The good news, however, is that if you can show your wedge translates to growth, you've got a powerful ingredient for your fundraising story. In this case:

- If the client wanted to raise more funding, they had a promising pitch about marketing automation for L&D.
- The client already had their Hemingway story-esque metaphor of choice worked out — they were going to be the 'HubSpot for L&D.'
- The potential for being the automation platform for internal communications with a TAM of just about every corporate business on earth is pretty compelling.

All because they found an attribute to be more about.

> **The market moves fast:** When there are dozens (or hundreds!) of other players sniffing around looking for opportunities, it's quite likely your competitors will see the same problems from the same customers, too. Windows of opportunity only stay open for so long. That's what HubSpot's founders realized when they came up with 'inbound' and decided to raise money and move as fast as they could, as we discussed in chapter 31. They knew many people would try to copy them and, as they tell it, many did. But in HubSpot's case, it was *multiple, successive* insights like 'inbound' that drove their success, as they went from the vertical-focused LegalSpot to the horizontal HubSpot and from marketing apps to CRM. Getting into the market

> is one thing; dominating it through speed of execution and speed of *insight* is quite another.

Attributes for everybody

This positioning-by-attribute approach wasn't unique to my client — most challengers do it, and even some incumbents too — it's just that not everyone does it very deliberately or well. In the LMS market, different players seemed to naturally gravitate to the one nook, or one attribute, they could be more about. For example:

- My client positioned around **automation**, as we've discussed.
- They were acquired by a competitor positioned around **collaboration**.
- They were all competing against an incumbent positioning themselves around **AI** (before the generative AI rush).

  And what was particularly interesting, however, was...

- They were all getting flanked by another company positioned around **content.**

(This wasn't true for everyone — there were plenty of undifferentiated, me-too offerings in the market, too. But they weren't a competitive threat.)

Ol' mate Hemingway

Remember our idea of a Hemingway story, the evocative few words that act as a micro pitch, like "use chat not email"? That last competitor, the one positioned around content, had a powerful "X for Y" micro pitch: "We're Netflix for learning." (This competitor had, I should also note, raised over $400M. Firepower can help.[8])

That simple narrative made for a diabolically good story for L&D buyers. My progressive client bemoaned that they were 'Just giving (traditional) L&D what they want' — courses and engagement, the classic old-school playbook — much to their chagrin. But this competitor had managed to dress it up as new and sexy — L&D courses on demand, just like Netflix! That gave them reach, relatability, and distinctiveness that, in narrative terms at least, was hard to match.

> **Netflix and the power of metaphor:** As we discussed in chapter 15, there's tremendous power in metaphor and analogy when it comes to your story. Netflix is so well known that borrowing their fame — and their established attributes and brand memory structures — to

> define your product works a treat. This competitor could have just been an 'on-demand content library,' which sounds super boring. But reframe it as 'the Netflix of learning,' and you're onto something far more distinctive. Soon enough, folks were writing about why 'the Netflix of learning' is bad, but that only serves to highlight the very catchy concept: *'Wait, there's a Netflix for learning? That sounds like something we should check out!'*

Play to win

This story doesn't end with a billion-dollar unicorn exit, though my client was acquired at a nice multiple. Finding a wedge-shaped opportunity made them, to a degree, acquirable — they'd excelled at a key capability that, along with their progressive narrative (and their ARR!), a bigger player wanted.

Even without that, aligning the team around an underserved JTBD, differentiating by positioning around an attribute, and continuing to leverage your wedge opportunity to build market share is just an inherently good thing to be able to do.

Ultimately, you want to keep building those multiplicative choices that help you compete to be unique and find a dominant, winning position in your market. But to try to get there, you've often got to start by being simply *more about something* than the competition, finding that wedge-shaped gap in the market. Maybe that alone will get a nice exit via acquisition, where you can play the game again, this time at a much bigger scale.

For my client, the next question was whether a similar strategy of positioning around an attribute and being more about that than the competition — combined, this time, with Scrooge McDuck levels of capital (the early '20s were a wild time!) — would work out for the company that acquired them. Will they stick with their wedge attribute or try to be all things to all people for maximum category reach? The jury's out!

Notes

1. Wensing, Matt. "Are you still adding features?," June 11, 2023. **https://mattwensing.com/posts/are-you-still-adding-features**
2. This wasn't the *Challenger* approach in a pure form by any means, but it was very much about literally challenging the prospect to adopt a new approach!
3. See Louis Grenier's interview with Ritson: "Mark Ritson: 4 Steps to Creating a Rock-Solid Marketing Strategy," *Everyone Hates Marketers*, May 22, 2018.

https://www.everyonehatesmarketers.com/podcast/63-mark-ritson-marketing-week

4. I leave identifying which one as an exercise for the reader.
5. Let me stress that if there *is* a real niche to work with — a specific vertical, for example — then great, go for that, but this is about strategies for reach when that's not possible or desirable.
6. Positioning: The Battle for Your Mind, 20th Anniversary Edition (p. 66). McGraw-Hill Education. Kindle Edition.
7. Look it up :)
8. Well it could in the ZIRP era, until that era ended, and it became a massive liability tied to an overinflated valuation. What looked like massive war chests in the heady heights of the ZIRP days eventually became massive millstones around the necks of many startups that required painful unwinding.

47

The B2B spectrum V — the enterprise chasm, 10x pricing, UiPath, & reach + niche

What we learned is that it's not just the price point. The effort that it takes to onboard an enterprise is very, very different. Enterprise operates with its own complexity. Expectations are very different from a small business. Small businesses have a very standardized experience, but big enterprises have a very different experience. So that's a lot of learning that we had to do.

- Vara Kumar Namburu, Whatfix co-founder[1]

In chapter 43, we kicked off our exploration of the B2B spectrum, which runs from (in the US) the ~8M SMBs at one end to the ~8,000 large and enterprise businesses at the other, with a couple hundred thousand (give or take) medium-sized businesses in-between.

We've looked at the consumer-style approach of positioning for reach when selling to SMBs and the classic wedge strategy for getting traction in the sales-driven mid-market. Now, let's briefly focus on positioning at the enterprise end of the spectrum.

We're not going to spend too long here as we've already explored this end of the spectrum in varying degrees in terms of:

- Story 1 pitches for the C-suite (**chapter 20** and **chapter 23**).
- Your pitch deck (**chapter 9**).
- Playing narrower before playing bigger (**chapter 35**).
- April Dunford's approach to positioning (**chapter 37**), which was also born from decades of experience positioning enterprise software.
- The traditional chasm-crossing segment-and-sell tech market strategy (**chapter 40**).
- Mid-market wedge strategies (**chapter 46**) which can also apply here, too.

In that sense, much of this book has been about strategies for selling to enterprises through narrative and positioning. There is, however, more to say about positioning for niche vs. reach at the big end of town because here, when you're at scale, you can do both.

Before we get to that, let's briefly touch on the risks and rewards of going upmarket.

Risks: Crossing the enterprise chasm

The classic startup journey is to start with a small, focused product targeted at people who have some burning need for your small, focused thing. Perhaps they're SMBs with straightforward needs; perhaps they're innovators at bigger companies who will take a chance on you for the edge you offer despite the immaturity of your product.

Either way, at some point, your product matures, your growth in your initial segment becomes predictable, and the siren song of bigger deals from companies with famous logos gets louder. These are deals with potentially thousands of seats all tied up in a single account that is, theoretically, less likely to churn. Huge deals with less churn — what's not to like?

In the eternal quest for growth, sometimes investors push founders to "go upmarket" and chase these big deals. Why put in all that work to close a deal at the SMB end, or even the mid-market end of the spectrum, competing against 500+ other players (as we saw last chapter) when you could be closing much bigger deals instead?

Let's acknowledge a few of the risks here first. Going upmarket is not as simple as hiring an "enterprise" rep and telling them to go sell bigger versions of your SMB offering. There isn't One Weird Positioning Trick that will suddenly get you into these enterprise deals. There's a lot you'll need to learn — and, perhaps, build — as the Whatfix team discovered, per the opening quote, and as we'll see below.

In fact, there's a trap waiting for startups looking to make the leap into enterprise — the so-called "enterprise chasm" (see box below) — and that's perhaps the biggest risk facing founders who want to go upmarket.

We discussed the original chasm in **chapter 40** and looked at what can happen if you fall into the chasm, as we saw Pebble found out the hard way in **chapter 41**, where the original product failed to make the leap from nerds to normies.

Normies, we learned, want very different things compared to nerds, and that's not obvious when you're getting all this great feedback from the early-adopter nerds.

Likewise, enterprises want specific things that your early adopter customers don't need, and founders who have cut their teeth on self-serve or high-velocity sales often don't know what they are. At best, perhaps they have a very foggy idea about what's needed (SSO? A dedicated rep? Security audits?), and even with a clear idea, it can still be hard to prioritize both the necessary product *and* organizational enterprise features, especially when you have a long list of requests from existing customers.

Crossing the enterprise chasm can, therefore, require enterprise-specific:

- **Awareness** of this chasm you're trying to cross.

- **Product features** and integrations that are costly to build, not to mention security reviews and compliance audits.
- **GTM and sales motions** which may be very different from the cadence you've been used to (think multi-stakeholder deals with 6-12+ month sales cycles[2]).
- **Customer success teams** and other additions to the org for post-sales support.

This is not a small undertaking! It can require preparing the whole company to cross the proverbial enterprise chasm, which can be, as you might imagine, very challenging and demand a lot of patience.

Some folks do their best to warn founders of these risks. For example, consider this warning on LinkedIn from Amy Volas, who runs a sales executive recruitment firm for startups, spent many years in sales, and therefore sees plenty of folks trying to hire their way into enterprise sales [3]:

> *Founders, DO NOT spin up an enterprise team if you don't understand how this buyer journey works. A $30K deal with a big brand is not enterprise. It's a small deal with a big company. [...] Trying to tackle this segment when you're not ready is a sobering reality for many [...] It takes AT LEAST 12 months to build a repeatable pipeline in this segment. [...] You're fooling yourselves if you think you'll get a ton of revenue from your new enterprise sellers in the first 6+ months. If you don't have enough time or money to support a 12-month ramp time, STOP.*

Or founder Sean Byrnes, CEO coach and investor, in his piece "Going Up-Market Will Not Solve Your Problems"[4]:

> *I have met with many CEOs of unicorn companies who believe going after larger contracts will solve all their problems. It has never actually worked in my experience. If you are looking for bigger contracts to solve all your problems, what you are really doing is avoiding your problems. Those problems likely include slow (or no) growth, lack of capital and negative margins. If you can't solve those problems with your existing customers, a set of larger customers will not make them better.*

That is, chasing enterprise accounts isn't a panacea for any fundamental problems your business may have. But sometimes the opportunity is very much there, as Byrnes writes:

> *Now, there are times when going upmarket is the right move! There*

> *are times when going downmarket is the right move as well. In those cases you will be pulled in those directions because larger (or smaller) customers really want to buy from you. If that's the case, then lean into the demand! In these cases you aren't looking to solve a problem, you are listening to the signals from the market.*

The market pull has to be there, and the company maturity has to be there, too, with the onboarding, legal, security, and product maturity to match. I'd argue the right-brain, exec-friendly narrative also has to be there, as we discussed back in **chapter 20**.

That doesn't mean it's impossible to go upmarket; it just isn't a shortcut or magic recipe for growth. But for those that can crack it, it's possible to 10x your prices.

> **Crossing the enterprise product chasm:** I am not the first person to coin the phrase "the enterprise chasm." For a second I thought I was, dear reader, but a quick search revealed that founder Michael Grinich built his second startup, WorkOS, entirely around the entire concept. WorkOS provides enterprise features via APIs to startups, an opportunity Grinich decided to tackle after discovering how difficult it was to build those features for his first startup, Nylas. He has a podcast discussing these issues titled, aptly enough, "Crossing the Enterprise Chasm"[5]. If you're after inspiration and tips from folks who've made the jump, that's probably a good place to start.

Rewards: 10x'ing — or 32x'ing — your prices

Pricing and positioning go hand-in-hand, and here, enterprise sales necessitates enterprise pricing, given the amount of work it takes to close a large enterprise deal.

This can be jarring for folks who aren't used to selling to this segment, where their estimates of pricing can be off by an order of magnitude or more. That's what B2B SaaS "digital adoption platform" Whatfix found when they discovered they could more than 10x their pricing when selling to enterprises. Finding the right pricing for the right segment with the right positioning can, it turns out, open a lot of doors, as we're about to see.

Now, I *love* this story, dear reader, but first, a tiny bit of background. The Whatfix founders flirted with failure several times before finding product/market fit upmarket, as Avinash Raghava describes in his account of their journey.[6] They first built an SEO tool for SMBs, but churn was high due to the complexity of their recommendations. They add a 'fix it' feature, which would guide users on 'fixing' the problems the original tool found. That 'fix it' button turns out to be a more promising opportunity — SMBs want to use it internally — but the SMB world is still a slog. Growth is slow; churn is high.

So they look upmarket and attempt to create the "digital adoption platform" (DAP) category. (This is in the true, actual-category-of-software sense, like Yammer and 'enterprise social networking' as we saw back in **chapter 32**.) For an extra degree of difficulty, they were doing all this from India, where enterprise software sales wasn't much of a thing,[7] let alone this DAP category that orbits it.

But what does a DAP — a digital adoption platform — actually do? It's software layered on top of existing enterprise software that helps with user onboarding, support, and performance. The idea is that large companies spend bucketloads of cash on a new CRM or other major piece of software to accelerate their digital transformation, perhaps with the help of a major consultancy, and then... it doesn't get used, or is only used minimally, and the result is poor ROI. Enter DAPs like Whatfix that give users much more guidance and support. Now the underlying software *does* get used and the enterprise gets the nice ROI it originally wanted.

CRM is 'low-hanging fruit' for DAPs, and on the *100x Entrepreneur* podcast, Whatfix founders Khadim Batti and Vara Kumar Namburu describe their journey with host Siddhartha Ahluwalia.[8] The key part, for our purposes here, is the bet they took as a tiny company when they decided to attend Salesforce's Dreamforce conference in the US. According to Batti:

> *We were a five, six person company, we didn't even have enough people with a visa to fill the booth of four people, actually.*

Those booths — and everything that's involved in manning one — aren't cheap, either:

> ***Batti:*** *With the booth and everything it cost $35-40,000.*
>
> ***Ahluwalia:*** *And how much did you have in the bank?*
>
> ***Batti:*** *$700k, $600k something. So it was a bet, there was a big investment.*

So off they go to Dreamforce to further validate their market. At this point, they'd been selling their product for $2,000 to $3,000 per year (i.e., $250/mo at the high end). At Dreamforce, they were aiming to pitch the product at the "expensive" price point of $8,000 to $10,000 annually instead. Makes sense, right? They're pitching enterprises, after all.

A prospect comes along, loves the concept, and asks how much it costs. The rep says "$8,000," meaning per year. The prospect does some quick mental calculations and says, '8 by 12... $96,000, that looks fine.'

That was, as you can imagine, quite the shock to the team. What they thought was an "expensive" annual price turned out to be an entirely reasonable *monthly* price for these kinds of buyers. But, as Batti noted, the prospect was spending $20 million already, so an extra $100k on top for Whatfix was fine if it helped them get ROI.

The Whatfix team didn't just 10x their prices (32x their original $250/mo pricing!) and waltz into enterprise deals, though. There's no free lunch — there was a lot of learning ahead, as the opening quote captures. But closing deals for $96k vs. $3k opens up a whole new market opportunity that can change the trajectory of a business.

Five, six, or seven-figure deals? What counts as a true enterprise deal — $50k, $100k, or even $1M in ARR? Some folks might argue that anything up to $100k is just the upper end of the mid-market, and 'real' enterprise deals start well above that. That may be true, but the cutoff is pretty arbitrary. It's more informative to look at the average ACV (Annual Contract Value) that exists across the B2B SaaS spectrum and see where you might fit. Sammy Abdullah of Blossom Street Ventures has a helpful breakdown of the ACV for 73 SaaS companies at IPO[9], with:

- Squarespace at $163 in SMB.
- Qualtrics at $49,263 with a range of mid-market and enterprise customers.
- UiPath at $76,260 in the enterprise zone.
- Cloud vendor DynaTrace at $744,615 and enterprise AI vendor C3.ai at a whopping $2,447,906 ACV at the very top end.

As we've been discussing, B2B is quite a spectrum, with a *15,000x* difference in ACV — that's not a typo — between the top and bottom ends of this sample.

UiPath & B2Enterprise

UiPath is squarely in the B2Enterprise end of the B2B spectrum, and they spent a decade toiling away in relative obscurity before cracking the enterprise market.[10] Jason Lemkin describes them as a "hyper-enterprise" company and notes that they crossed the $1B ARR mark with 10,000 customers.[11] And of those 10,000 customers, roughly 1% of them were paying more than $1 million a year for 35% of UiPath's total revenue.[12] That's the power of closing big deals.

UiPath, for their part, has traditionally described themselves as a "leading enterprise Robotic Process Automation (RPA) software company,"[13] a "Business Automation Platform,"[14] or more recently just as a "leading enterprise automation software company."[15] Whatever the label, if Whatfix helped end users get to grips with enterprise software, UiPath helps folks automate software tasks entirely, particularly with legacy systems or those outside the enterprise's control. Hence the "Robotic Process Automation" label: it could act as a robot, executing your processes, automating whatever tedious or repetitive software tasks you had.

We've been discussing the differences between B2B and B2C positioning and GTM in recent chapters, and it's worth reiterating just how different the B2Enterprise world is to the B2SMB one. Here's Daniel Dines, UiPath co-founder, explaining their sales motion off the cuff[16]:

> *We've traditionally landed within the CFO, CEO, C-suite. It's very easy to find the low-hanging fruit in that department. [...] Typically, how we progress from a small deal to a $1 million deal is kind of simple. Landing in a department 50%+ is finance. Then we show them the return on investment. It's easy to prove — you have to work with the controller to define a set of KPIs, and then you will see that the controller is willing to give you more budget, get the sponsor up in the company, and then go enterprise-wide. The key [is] connecting the dots between an executive sponsor, a business person, an IT person, and a partner. And that creates the momentum for large-scale adoption.*

We call them complex sales for a reason, and here you can see why. This is why enterprise software costs 10x+ and involves a lot of humans on the vendor's side — because it also involves a lot of humans on the buyer's side, too. (Though I suppose it must have seemed pretty "simple" for Dines, given how fast they eventually grew.) In any case, we're a world away here from Squarespace's Pavlovian mass podcast sponsorship of 'have need, think brand, buy product' customer journeys.

Instead, this is the world of the sales-led enterprise playbook: targeting the C-suite and decision makers, landing and expanding, and putting in a whole lot of leg work to close those deals — RFPs, references, case studies, webinars, roadshows, one-pagers, decks, demos, pilots, dinners, and more. And in UiPath's case, that includes the partners and integrators — i.e., big consulting firms, who Jason Lemkin says are "an important piece of the UiPath story many miss."[17]

> **The hidden services layer:** The role of these consulting firms also

highlights the massive services layer that many enterprise vendors either operate directly or rely on in a quasi-outsourced way through partners. In tech, we can get a little too myopically focused on the, well, tech, forgetting the large number of skilled humans it takes to pull off a successful implementation project, where the tool or platform is just one piece of the puzzle. If it takes a village to raise a child, in the enterprise world, it can take a swarm of service folks (for lack of a better collective noun) to deliver a successful implementation.

Going for reach *and* niche

Where does positioning fit in this world?

The idea that the "laws" of B2C brand marketing and positioning have much to do with this B2Enterprise world is kind of absurd at this point, despite our friends at the EBI insisting B2C and B2B are much the same.[18] Nevertheless, the fundamental positioning question of going *niche* or *reach* is still relevant. It turns out that when you're this big, you can do both.

Consider the reach in UiPath's homepage messaging, which says (as of writing)[19]:

> *When the AI-powered UiPath Business Automation Platform is everywhere in your business, it can be used by everyone. For everything. To benefit every business.*

Here's the messaging in context:

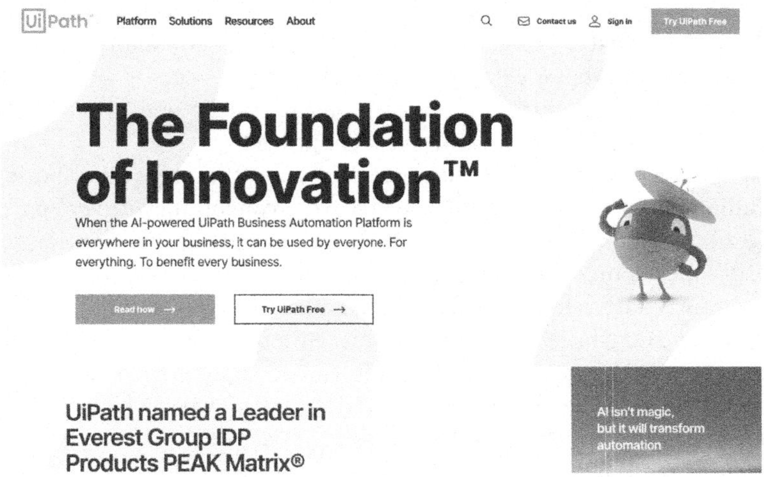

Figure 47.1: UiPath's 2023 homepage hero messaging.

Everywhere, everyone, for everything, in every business. That's pretty broad! There are absolutely no limits on their reach there. But if we explore the page further, we see:

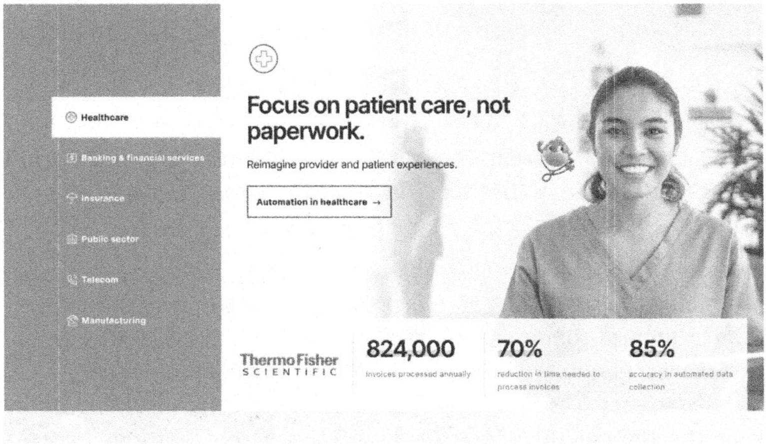

Figure 47.2: UiPath's 2023 homepage industry messaging.

Segments! That is, positioning by industry, with a focus on healthcare, banking & financial services, insurance, the public sector, telecom, and manufacturing. This is where left-brain, narrow-focus positioning comes into the picture: there's a specific niche pitch for each industry.

That is, there's both broad horizontal reach at one level *and* a deep, industry-by-industry targeting and positioning on another. This is reminiscent of both our claw game from **chapter 18** — going a long way out *and* down to win attention — and the reach + niche end of our dynamic positioning framework we discussed in **chapter 44**.

Why can they do both? Because they have the sales distribution — including an incredibly well-connected, big-consultancy-driven partners program — that allows them to do so. In this case, niche distribution in *multiple* niches unlocks niche positioning for those niches. Penetration and segmentation are, here, two sides of the same coin.

I'm not suggesting you must follow UiPath in this regard — this is description, not prescription — but you can see how it works at this scale at this particular end of the market. (And yet again, we see how novel GTM strategies — the partners program in this case — can help drive extraordinary growth.)

Narrative differentiation

Let's also note that at this end of the market, your narrative is an essential ingredient in your growth. That narrative might be driven by secular change (**chapter 30**), organizational transformation (**chapter 12**), or a new playbook (**chapter 35**).

In UiPath's case, the RPA (Robotic Process Automation) wave swept enterprises, with Deloitte, one of UiPath's partners, writing[20] (their bold):

> *In 2016, just 15 percent of respondents reported that their RPA program was part of a wider corporate initiative. In 2017, the figure had risen to* ***64 percent–organizations that have embarked on the RPA journey have RPA as either an enterprise-wide or strategic initiative.***
>
> *Not all RPA journeys start off with an enterprise-wide scope and organizations should "think big, start small." Many of our clients have started small, in a specific functional area, where there stakeholder support and clear opportunity, for Finance or HR, or in a shared services or global business services organization. Once proven, they can then showcase this to other parts of the organization and help leaders "think big" about the broader opportunity across the enterprise.*

This echoes UiPath's Dines' comments about how they land and expand. You can see the radar-and-laser-beam approach to narrative and positioning here, too:

- **Radar:** First, there's the big right-brain strategic need for the C-suite: What's your RPA strategy? Where are you on your RPA journey? You *are* going to implement RPA, right...? This is the new way to understand the world: you can robotically automate all your tedious processes!
- **Radar to laser beam:** With the big need put on an organization's radar, there's the dive into specific functional areas — Finance, HR, and so on.
- **Laser beam:** That laser beam focus needs to go all the way down to exactly what folks can do with the tool — how they can manipulate the world. (If you're curious, like everything in life, you can head over to Reddit and see what folks think and how they're evaluating it in the trenches, too.[21])

You can see how both right-brain radar attention and left-brain laser beam focus are engaged from the C-suite down to individual users and implementors. This is a different kind of reach, but it's reach nevertheless. This is where your radar-friendly narrative and positioning laser-beam work really need to work hand-in-hand. You're going to go a long way out on the x-axis *and* a long way down on the y-axis to build your unique view of the market.

Ultimately, this approach gives you compelling answers to questions like:

- **The company:** Why should we buy at all? And why should we buy you over the alternatives?
- **The executives:** What does it mean for me in the C-suite?
- **The users:** What does it mean for our employees?
- **The stakeholders:** What's in it for other folks who need to participate to make it happen?

And again, it helps to know the game you're playing, including:

- **Change:** The wave you're riding (RPA in this case).
- **Segment:** The market segment you're targeting *and* the way you're targeting it (reach across enterprises *and* niche down to specific industries).
- **Attention:** The narrative and positioning you need (both a strategic change concept and industry-specific positioning deep dives).
- **Distribution:** The distribution you're using (enterprise sales and consultant partners).

And so on. If you can weave all of these things together in a multiplicative way, you get multiplicative results, and UiPath's rocketship growth is a testament to what happens when the pieces finally click.[22]

Most Advanced, Yet Acceptable

When the stakes are high in these big enterprise and mid-market deals, with a considerable amount of time, money, and reputation on the line, an interesting dynamic comes into play — the pull and push of the new.

On the one hand, there's the pull: executives, managers, and workers are all keenly aware that there's always some new technology on offer. No one in a big corporation, surrounded by inefficiency and facing pressure to perform, is saying, "Our best bet is to simply not adopt any new technology whatsoever for the next 5 years."

On the other hand, software (and technology more broadly) isn't a magic wand that you can wave and everything magically improves. If it was, Whatfix wouldn't exist, and UiPath would be redundant.

This is the pull and push of what's *most advanced, yet acceptable*, or MAYA — an idea that's been around for decades and has its roots in consumer products. It's not easy, but if you can thread that needle, you'll make adoption that much easier. In the next chapter, we'll look at finding that MAYA sweet spot.

Notes

1. Khadim Batti and Vara Kumar Namburu were interviewed by Siddhartha Ahluwalia in "Building a Multi-Million Dollar SaaS Company: Whatfix Founders," *100x Entrepreneur,* April 17, 2023. (Lightly edited for clarity.) **https://100xentrepreneur.com/podcasts/from-scratch-to-success-whatfix-founders-kadim-batti-vara-kumar-namburu-share-struggles-and-insights-on-building-a-multi-million-dollar-saas-company/**
2. See the 2023 data for >$75k average selling price (ASP) sales from Insight Partners, for example: **https://info.insightpartners.com/2023-sales-kpi-report.html** (Note that high ASP is not a perfect proxy for customer size, as David Kellogg mentions in his analysis of this data: **https://kellblog.com/2023/06/17/interpreting-the-insight-2023-sales-kpi-report/**. It's helpful in a directional sense, though.)
3. Volas, Amy. 2023. **https://www.linkedin.com/posts/amyvolas_startups-enterprisesales-founders-activity-7030890571644706817-Femg/**
4. Byrnes, Sean. "Going Up-Market Will Not Solve Your Problems," *Breaking Point,* June 7, 2023. **https://www.breakingpoint.tech/p/going-up-market-will-not-solve-your**
5. Grinich, Michael. "Crossing the Enterprise Chasm." **https://workos.com/podcast**
6. I'm drawing in part from the podcast interview and from Avinash Raghava's piece on Whatfix's history: "Nix, Fix & Whatfix," *The Grit Stories*, February 10, 2023. **https://thegritstories.com/nix-fix-whatfix-2133af3f6f6a**
7. Ahluwalia, the podcast host, mentions this in the interview, saying, "[Whatfix] started

in 2010 [and] pivoted to enterprise software in the 2013-14 timeframe, when enterprise software was not at all a category in India. So there were the first two category creations that you did."

8. "Building a Multi-Million Dollar SaaS Company: Whatfix Founders," *100x Entrepreneur*, April 17, 2023. **https://100xentrepreneur.com/podcasts/from-scratch-to-success-whatfix-founders-kadim-batti-vara-kumar-namburu-share-struggles-and-insights-on-building-a-multi-million-dollar-saas-company/**
9. Abdullah, Sammy. "$27k is the median ACV at IPO," Blossom Street blog. **https://www.blossomstreetventures.com/post/27k-is-the-median-acv-at-ipo**
10. As Lemkin notes, they took ten years to get to 1M ARR and then hit $600M six years later. Lemkin, Jason. "Three Mega IPOs That Took A Long, Long Time To Get Big: Squarespace, Procore and UiPath," *SaaStr*, May 21, 2021. **https://www.saastr.com/three-mega-ipos-that-took-a-long-time-to-get-big-squarespace-procore-and-uipath/**
11. Ibid.
12. Lemkin, Jason. "5 Interesting Learnings from UiPath at $600,000,000 in ARR," *SaaStr*, March 31, 2021. (Lemkin notes UiPath's use of ARR is pretty loose, but you get the idea.) **https://www.saastr.com/5-interesting-learnings-from-uipath-at-600000000-in-arr/**
13. "UiPath Named Fastest Growing Technology Company in FT Americas' Fastest Growing Companies 2020," April 6, 2020. **https://www.uipath.com/newsroom/uipath-named-fastest-growing-technology-company**
14. Per their homepage, **https://www.uipath.com**.
15. "UiPath Appoints Co-CEO Rob Enslin and Veteran Technology Executive June Yang to its Board of Directors," January 10, 2024. **https://www.uipath.com/newsroom/uipath-appoints-co-ceo-enslin-and-june-yang-to-board-of-directors**
16. UiPath Inc. (PATH) CEO Daniel Dines presents at Bank of America 2022 Global Technology Conference. June 7, 2022. (Edited for clarity.) **https://seekingalpha.com/article/4516997-uipath-inc-path-ceo-daniel-dines-presents-bank-of-america-2022-global-technology-conference**
17. Lemkin, Jason. "5 Interesting Learnings from UiPath at $600,000,000 in ARR," *SaaStr*, March 31, 2021. **https://www.saastr.com/5-interesting-learnings-from-uipath-at-600000000-in-arr/**
18. To me, it's like insisting the rules of aviation apply to space flight, too, despite the vastly different nature of gravity, because of all the research you did on the ground. But I digress.
19. Homepages for companies at this stage are canvases for many competing groups and needs, including investors. I have no idea what went into UiPath's copy, but it's worth keeping in mind that internal politics can drive the output here as much as any customer-focused strategy or insight.
20. "Robotic process automation (RPA) trends and to-do list for scaling across the enterprise," Deloitte, February 16, 2018. **https://www2.deloitte.com/us/en/pages/operations/articles/global-robotic-process-automation-report.html**
21. "Skeptical Of UiPath / Trying To Understand The Options," June 3, 2021. **https://www.reddit.com/r/UiPath/comments/nqvisp/**

skeptical_of_uipath_trying_to_understand_the/

22. Until the stock market moves on, of course, with UiPath's shares being down, at one point, about 80% since IPO, leaving them valued at a 'mere' $8.88B.

48

MAYA — the positioning sweet spot & enterprise buyers

> This is the one phase of our profession that fascinates me no end. Our desire is naturally to give the buying public the most advanced product that research can develop and technology can produce. Unfortunately, it has been proved time and time again that such a product does not always sell well. There seems to be for each individual product (or service, or store, or package, etc.) a critical area at which the consumer's desire for novelty reaches what I might call the shock-zone. At that point the urge to buy reaches a plateau, and sometimes evolves into a resistance. It is a sort of tug of war between attraction to the new and fear of the unfamiliar. [...] At this point, a design has reached what I call the MAYA (Most Advanced Yet Acceptable) stage.
>
> *- Raymond Loewy*[1]

In the last couple of chapters, we've looked at the mid-market and enterprise end of the B2B spectrum, where there's a delicate balance between trying to introduce new technology, new concepts, and new playbooks for work... and not scaring folks off. We've discussed how plenty of buyers are more scared of 'messing up than missing out'[2]. The bigger deals get, the higher the stakes, the more that fear looms in the background.

I want to sneak in one more acronym that sheds light both on this phenomenon and how folks buy in general before we explore competitive differentiation in the next chapter. In that spirit of catchy, named concepts, it's MAYA — Most Advanced Yet Acceptable. Let's briefly touch on what it is, why it matters, whether you can DIY your own MAYA concept, and where you land on the MAYA spectrum.

MAYA — Most Advanced Yet Acceptable

MAYA, as a concept, dates back to the middle of last century and was coined by legendary industrial designer Raymond Loewy, the man often dubbed the father of industrial design. (His credits include everything from streamline trains to Coca-Cola bottles to Lucky Strike packets to Air Force One.) It was intended to capture the nature of consumer product acceptance as the opening quote describes (a "tug of war between attraction to the new and fear of the unfamiliar") and the sweet spot of design acceptance he aimed for.

Despite its B2C roots, MAYA still captures a good chunk of what we've been discussing here about B2B SaaS, especially at the more hands-on, big-ticket end of the spectrum. Here's Loewy again from his 1951 book *Never Leave Well Enough Alone*:

> *The adult public's taste is not necessarily ready to accept the logical solutions to their requirements if this solution implies too vast a departure from what they have been conditioned into accepting as the norm. In other words, they will go only so far.*

Normie buyers were risk-averse then, and they can be risk-averse today, too, especially in business. That's why there's a positioning sweet spot between what's *most advanced* (what makes you unique) while still being *yet acceptable* (i.e., not too great a departure, which is especially relevant when selling to cautious enterprise buyers). MAYA is the art of balancing the new and the familiar — a particular challenge in B2B where "high-tech" and "conservative buyers" meet.

What is 'Yet Acceptable'?

Loewy came up with MAYA to sum up his theory of design and the acceptance of the new, the acceptance of the *unique*. Here's Derek Thompson, writing in *The Atlantic* about Loewy (my bold)[3]:

> *Loewy had an uncanny sense of how to make things fashionable. He believed that consumers are torn between two opposing forces: neophilia, a curiosity about new things; and neophobia, a fear of anything too new. As a result, they gravitate to products that are bold, but instantly comprehensible. Loewy called his grand theory "Most Advanced Yet Acceptable"—maya.* ***He said to sell something surprising, make it familiar; and to sell something familiar, make it surprising***.

That is, that's the idea. Pretty simple. But does that not sum up B2B and enterprise SaaS? That tension that customers face — the curiosity about the new, but fear of anything *too* new? We're constantly trying to make our uniqueness understandable, and (let's be honest) what's not-so-unique seem like it is. B2B tech sales are, in some ways, a constant dance between novelty — it's new! — and reassurance — but not too new! — and the seller's attempt to calibrate for the customer's MAYA sweet spot.

> **The MAYA sweet spot and your pipeline:** If you were to analyze your

recent deals or your current pipeline, what does the dance between 'new, but not too new' look like for deals you win and deals you lose? How much time are you spending trying to engage with prospects' *neophilia* (the love of the new) vs. their *neophobia* (fear of the new)? Are prospects coming to you for your high degree of novelty (new, early-stage startups) or your high degree of acceptability (incumbents and established players)? Threading the needle between the advanced product you offer (or need to offer to stay competitive) and the acceptable baseline prospects want is the art of finding your MAYA sweet spot.

Bold, but instantly comprehensible

The MAYA line is where, as Thompson puts it, the "bold, but instantly comprehensible" lives. I admit this is vaguely reminiscent of a tagline for a new fragrance for men (look, this is the second last chapter and I'm running out of metaphors) and it sounds like the kind of impossible mission clients give creatives from time to time, to much ensuing eye-rolling. *'We want our design to be modern and cutting edge... but not too out there, we're selling to CFOs after all!'* (Again: new, but not too new.)

And yet, from a deep story point of view, doesn't 'bold, but instantly comprehensible' describe the Hemingway stories we looked at in **chapter 10**? Think about the examples we looked at in that chapter, like Salesforce (use cloud, not on-prem), Drift (use chat, not forms), Slack (ditch email, use chat), Stripe (payments for developers), and so on. Pushing what's advanced, while being catchy and easily relatable.

That's the kind of simple message some famous B2B SaaS winners have at their core — bold, but instantly comprehensible ones. It's that kind of clear concept that makes building name/need brand memories that much easier, too.

Teaching buyers where their MAYA threshold should be: MAYA was coined for consumers going off gut reaction to products for sale on a shelf or cars for sale in a lot. But in complex B2B sales, we have the opportunity to teach buyers where their MAYA sweet spot *should* be for their particular needs. Prospects get easily bamboozled by the litany of options they face, and helping them work out where their 'most advanced, yet acceptable' line is — i.e., what's realistic and plausible for their unique context — is a way sellers can add value in the sales process. Remember, you and your team might sell day in and day

> out, but B2B buyers aren't experts on buying. They don't necessarily know what's best or how the market works. That's why helpful vendors can help map the market in acceptable vs. advanced terms and show prospects where they fit in a Goldilocks spectrum of too hot, too cold, just right.

Memetic MAYA

Everyone rolls their eyes at acronyms; everyone uses and remembers acronyms. And that's what MAYA is — an acronym. That's the meta lesson here in our use of MAYA. The reason I can cite MAYA at all is that Loewy gave it a name and made it catchy.

That's why a >70-year-old consumer product adoption theory can show up in a 2020s book about B2B technology and why it wasn't forgotten as just a cursory theory of a very talented French-American dude with (apparently) an ego to match.

This is, again, what the category creation shtick we covered back in **chapter 33** tries to capture — naming a catchy concept, a playbook of sorts, that points back to you. Naming concepts is fundamental to discussing and passing on those concepts. It's hard to share what you can't name.

And, look, let's be honest about MAYA — it wasn't coined as a scientific theory at the time, which Loewy acknowledges. It's an observation from experience. He developed a loose theory around it, gave it a name, and that gave the concept a life of its own. For us, it's both a useful observation about human nature — the world may have moved on, but our brains haven't changed in 70 years — and evidence that sticky concepts stick for a reason.

> **DIY MAYA:** If you've been in the game long enough in your particular industry, with your particular customers, you've surely noticed repeated patterns of behavior that can be given their own name, their own loose theory, and can take on a life of their own. With MAYA, that was identifying these competing forces around newness that seemed intrinsic to human behavior. For Byron Sharp, Jenni Romaniuk, and the EBI back in chapter 43, it was coining the terms mental and physical availability. It's worked for plenty of other folks who have coined their own catchy concept or put their finger on an emerging wave. And like everything we've been discussing, it has to resonate as true — Loewy's observation came from many years of observation, and people accepted it because they could relate to it.

B2B and neophobia

For folks in startup land, the "Most Advanced" half of MAYA is a given. This is technology, after all. Companies get founded to explore the frontiers of modern tech, innovation, and new ways of working and getting things done. But we don't always necessarily appreciate just how conservative corporate B2B buyers can be, i.e., how much neophobia there is, especially at the enterprise end. Therefore, calibrating what we need to do to assuage the fear of an executive corporate buyer in a traditional industry is not usually high on our list. When we're tech nerds tuned into the trends of the day, it can be very hard to put ourselves in the shoes of a normie buyer who just wants to get on with their life.

That neophobia varies from country to country, too. (And industry to industry, as well.) For example, here's David Kellogg on US corporate technology buyers[4]:

> *US customers tend to buy less on perceived product superiority and on a mix of product and vendor attributes. Ultimately, for most US customers, picking the "best" product isn't about selecting the product with the best technology, but about picking a vendor who they perceive as safe and who holistically offers the best solution to their problem. That's easy to say, it's much harder to internalize.*

"*Perceive as safe.*" Corporate B2B buyers are, in that sense, usually normies, not nerds, to use our analogy from **chapter 40**, and they want a safe 'whole product,' as Geoffrey Moore popularized decades ago.[5] They want the ecosystem, the partners, the plugins, the integrations, the support... everything that helps them do closer to 100% of their JTBD on one platform.

That means that when push comes to shove — when they stop window shopping and start considering writing a fat check for a long contract — that 'neophobia' Loewy cited, the fear of anything too new, is usually going to kick in.

That fear can be entirely rational — switching tools is a pain. Getting your department or organization to adopt a new tool can be a disruptive, expensive, time-consuming slog. There are jobs-to-be-done in the buying process (**chapter 36**) that go beyond just acknowledging how advanced your product might be. For example, here's Jason Lemkin[6]:

> *There are so many products I would buy if the vendor just said "I'll do all the work for you." Instead, they all just want to tell me why they are 10x better.*
>
> *But not really help me:*

- *buy out my contract*
- *migrate my data*
- *onboard my team*

That is, vendors often want to tell a prospect why they're "most advanced," i.e., what you'll gain, but forget the "most acceptable" half of the equation, which is where all the pain is. There's still a lot of work to do to overcome the very rational fear of the new — the fear of dealing with data migration, onboarding, and existing contracts.

That is, in part, why enterprise SaaS outfits can sometimes look more like consulting shops with the emphasis on the second S — service — and not the first S. When the tool itself is just one part of the offer — unlike self-serve folks who get by with a login link and help docs — the actual offering is more like Software as a Service as a Service, if I can put it that way.

This is an interesting challenge for Product Management, too. PM is usually focused on the product, but who, in your company, is the Whole Product Manager?[7]

> **If it's hard, it's an opportunity:** Don't be put off — the fact it's hard is what creates opportunity. Most folks don't bother, and those who do can build enduring relationships with their clients. Remember, it's hard for the buyer, too, and the more effort they put in, the less likely they are to have one foot out the door when renewal time comes around. But you need to build an organization (or a partner ecosystem, as UiPath did) that reflects the shape of the demand you're serving, and that leans far more on the GTM and implementation side of your organization than most product-focused founders realize.

Narrative for the journey

This is all the more reason why your narrative and positioning matter so much — it sets the context for a long, lucrative, and high-stakes relationship. Your narrative has to hit that "instantly comprehensible" target (with your Hemingway story, for example), and it has to carry often conservative buyers through a long journey from Milan to Minsk, err, from first touch to happy, expanding, successful, rusted-on customer.

In that sense, when we talk about your broader strategy being the combination of your choices coming together as some cohesive whole, a big part of that, in enterprise at least, is all your choices coming together simply to support a very conservative buyer.

Your MAYA challenge

How does this challenge of making the 'surprising familiar, and the familiar surprising' play out for your organization with your positioning?

- Are your sales pitches and demos attempts to make people understand the surprising aspects of your product or offer?
- Or are you trying to overcome the impression that you're not surprising enough? That you're too familiar, too the same?

That is, are you working at the "most advanced" (the wedge) or "most acceptable" (the 'whole product') end of the spectrum?

Again, most founders and their GTM teams default to the "most advanced" side, and sales narratives boil down to "our product can do THIS and it can do THIS and it can do THIS and it can do..." and so on. That tendency is what Loewy identified decades ago through his own work. As he says in the opening quote, "Our desire is naturally to give the buying public the most advanced product that research can develop and technology can produce."

But that doesn't always sell. That's because there's another side to the equation — what's "most acceptable" and what conquers the *fear* of the new that buyers have, too. The fear of messing up, of making the wrong choice not just for them, but for their whole org. No one wants to bet their reputation on the wrong horse.

So, when working on your positioning, it's worth keeping MAYA in mind. Positioning isn't just driven by the new that prospects may *say* they want; it needs to factor in an (often unspoken) fear of what's too new, too. Being easy to buy also means reducing risk across all those hidden *buying* JTBD prospects have to work through so that you are the easiest to buy, even if what you offer isn't the 'most advanced.'

This echoes what B2C brand marketers try and do by making their products the most available, mentally or physically, and while complex B2B sales are very different, that same principle — being easiest to buy — still applies.

Finding *unique* ways to make those pieces click — along with the ones we've been discussing throughout this book — is the real art of positioning and the real foundation of true differentiation. If you only attempt to copy what's working for other folks, you'll be stuck competing to be the same. But if you don't match their moves, you'll leave wedge opportunities open and risk being left behind. It's this push and pull of uniqueness — competing to be unique and competing to be the same — that seems to define competition in technology markets. Next, in the final part of this book, we're going to explore uniqueness as the foundation for strategy and winning your market.

Notes

1. Loewy, Raymond. *Never Leave Well Enough Alone*, 1951. **https://babel.hathitrust.org/cgi/pt?id=uc1.b3128329&view=1up&seq=345**
2. To borrow *The JOLT Effect's* line.
3. Thompson, Derek. "The Four-Letter Code To Selling Just About Anything," *The Atlantic*, January/February 2017. **https://www.theatlantic.com/magazine/archive/2017/01/what-makes-things-cool/508772/**
4. Kellogg, David. "The Top 5 Mistakes European Startups Make in US Expansion," *Kellblog*, June 27, 2022. **https://kellblog.com/2022/06/27/the-top-5-mistakes-european-startups-make-in-us-expansion/**
5. And Theodore Levitt before that, as Moore mentioned in 2022: "The whole product, as Ted Levitt taught us a generation ago, is the complete set of products and services needed to fulfill the compelling reason to buy for the target customer." **https://www.linkedin.com/pulse/thrive-downturn-deliver-whole-product-geoffrey-moore/** There are plenty of expanding-concentric-circles models of the whole product out there, but what matters is to list out everything your specific buyers need to complete their core JTBD with your tool or platform, start to finish.
6. Lemkin, Jason. "Want To Steal a Customer from the Competition? You Gotta Do The Work," *SaaStr.* **https://www.saastr.com/want-to-steal-a-customer-from-the-competition-you-gotta-do-the-work/**
7. Another old idea of Moore's!

PART IX

CLARITY & COMPETITIVE STRATEGY

49

Competing to be unique vs. competing to be the same (Porter vs. Hotelling)

> But the most consistent theme [I hear from clients] is what I call the "more better" idea. "We do all the other things that our competitors do, only we do them better." Or, "they say they do them, but we really do them." [...] Folks, more better is not a strategy.
>
> – *David C. Baker*[1]

Throughout this book, we've explored a variety of dichotomies. Primarily, that has been:

- Right hemisphere vs. left hemisphere and the hierarchy of attention.
- Story 1 narrative vs. story 2 positioning that we derive from the hierarchy of attention.

We've also looked at various metaphors to capture the nature of this dichotomy. (The claw game, x-axis vs. y-axis, observation vs. analysis, change over time vs. static maps, and so on). While the right brain/left brain split been our primary lens, we've also explored dichotomies like:

- Nerds vs. normies in the adoption cycle.
- Cast and catch vs. segment and sell in distribution.
- Reach vs. niche in positioning.
- SMB vs. mid-market/enterprise in market segments.
- Passive vs. active attention (and distinctiveness vs. differentiation) in branding.
- Advanced vs. acceptable in MAYA.

Dichotomies are, of course, intentionally reductive, but they're great for stretching our thinking and framing common issues in stark ways. In this final chapter, I want to explore one more dichotomy, and that's:

- Competing to be unique vs. competing to be the same (or Porter vs. Hotelling).

This dichotomy addresses the underlying driving force of competition — and therefore strategy — in business in general. In the B2B SaaS world, we usually treat this as "differentiation," but uniqueness really goes to the heart of your entire business model, not just product features or choice of market niche.

The problem, however, is that when we drill down to this fundamental uniqueness (or lack thereof), we often see that what passes for differentiation is actually *undifferentiated* differentiation. Founders get confused about what really is unique and they *think* they're standing out, and they are... just like everyone else. This happens with brand, as we discussed in **chapter 45**, and it happens in competitive strategy, too.

Therefore, in this chapter, I want to give you a better sense of how uniqueness works by looking at the case for uniqueness in strategy (the Michael Porter view) as well as the case against it (the Hotelling view). Uniqueness is such a fundamental force of strategy that knowing which "uniqueness" game you're playing is critical, especially when, as we'll see, our radar for what's unique may be horribly off.

Uniqueness and "more better"

In the opening quote, David C. Baker, the positioning consultant we met back in **chapter 37**, reminds us that unique positioning — and unique strategy — isn't found in declaring yourself "more better" than the competition.

That is, *better* isn't unique, *different* is.

Play Bigger, the book on category creation we examined back in **chapter 33**, dials this idea up to 11, writing (their emphasis)[2]:

> *[The most exciting] companies don't only invent something to sell us. They are not making products or services that just incrementally improve on whatever came before. They don't sell us* better. *The most exciting companies sell us* different.

That's their springboard to launch into the world of category — you need to be *so* different that you're introducing a whole new *category* of product or service to the world. The idea that *category* is the appropriate lens to capture that uniqueness is, as we saw in that chapter, based on a lot of well-intentioned but ultimately confused thinking. Nevertheless, the idea persists for a reason. The allure of category creation is so strong because it represents extreme uniqueness. You're not just a better product/SaaS/mousetrap; you're a *completely new category* of mousetrap.

On one level, this rings true in an intuitive sense. It's now common business wisdom that the idea that you can 'build a better mousetrap, and the world will beat a path to your door' is a fallacy. Incrementalism doesn't cut it. You need to create something that is *categorically* different (pun intended); categorically unique.

But is that how B2B SaaS markets actually work?

The uniqueness disconnect

There's a night-and-day difference between what business books say and what B2B markets look like in this regard. On the one hand, there are shelves full of breezy reads exhorting leaders to stand out and be unique; on the other, there are typical B2B SaaS markets like the LMS market we saw in **chapter 46**, where 500+ competitors are battling it out. Or in martech, there's the famous Marketing Technology Landscape infographic with 11,000+ solutions, each represented by a tiny speck of pixels.[3] Then, in the B2C brand marketing world, as we saw in **chapter 45**, there's a long-running debate about whether differentiation matters at all, with some arguing brands should go for reach and 'meaningless' distinctiveness instead.

What's going on? Why is there such a disconnect?

This is the push and pull of uniqueness. Uniqueness is something everyone wants... just not too much, otherwise you're weird and incomprehensible. It's something that can give you a real edge in a market... but most competitors spend more time *denying* it (copying) than they do *creating* it (innovating).

It turns out that this push and pull of uniqueness is a well-theorized phenomenon, it's just that the two theories don't interact much.

Two ways to compete

There are two 'laws' that represent each side of our dichotomy, that is, the two broad approaches to uniqueness and competition. They are:

- **Competing to be unique**, a.k.a Porter's Law, which I'm coining right now for this chapter.
- **Competing to be the same**, a.k.a. Hotelling's Law, or the law of minimum differentiation, which has been around for a century.

First, let's explore competing to be unique.

Porter: Competing to be unique (C2BU)

> *Only by competing to be unique can an organization achieve sustained, superior performance.*
>
> – Joan Magretta, citing Porter[4]

'Competing to be unique' is a slogan of sorts from Professor Michael Porter, the OG strategy thinker of the '80s and '90s who is, perhaps, "the most famous business-school professor of all time."[5]

That fame faded from tech strategy consciousness as Clayton Christensen took over with insights like disruptive innovation in the latter half of the '90s. Porter is hardly forgotten, though — his work, especially his 'five forces,' still pervades MBA curricula around the world, for better or worse, and he more than deserves his own law, even if I'm just making it up for our purposes here.

Porter wrote dense tomes on competitive advantage[6], but here we're just going to zero in on a tiny sliver of his approach — the compete-to-be-unique side, or "C2BU," to give it its own acronym. In this slice of his philosophy, competition is a matter of uniqueness, and a leader, i.e., a CEO's job is to make *choices* that create that uniqueness:

> *Choosing the particular kind of value you will offer your customers is the core of competing to be unique.*
>
> – Joan Magretta[7]

> **Magretta and *Understanding Michael Porter:*** I'm quoting directly (and otherwise drawing from) Joan Magretta's excellent book on Porter's philosophy, the aptly titled *Understanding Michael Porter.* Magretta, a long-time colleague of Porter, has done business nerds the world over a great favor in distilling Porter's philosophy of competition and value creation into a very accessible book. Porter, for his part, appears in talks and lectures on YouTube[8], too. I know it's corny, but the internet is amazing. Want to hear a lecture from one of the greatest business strategy minds of the last 50 years? YouTube is there waiting for you.

Choosing to be unique & the left hemisphere

You can see shades of this approach in ideas around category creation — leaders should *choose* to 'create a category' as though it were that simple. But it's obviously not that simple, and that's why there are, in reality, so many different schools of strategy, focusing on people and the organization, on core competencies and resources, on emergence and complexity, and so on.[9] A domain as broad and deep as "how businesses compete" supports a multitude of explanations and points of view, but we can still bring our new hemisphere hypothesis to bear here, too.

Competitive strategy has been described as the intellectualization of business[10], and to your humble author, this often seems like the *left-brain* intellectualization of business, especially the Porter-championed "positioning" school of strategy.

Businesses, however, evolve somewhat chaotically over time — there's plenty of chance, emergence, and serendipity involved in where they end up, in what *gestalt* 'whole' emerges, and who ultimately succeeds. But strategy, as an intellectual discipline, is often about trying to analyze the factors involved in that success in a static way, focusing on the specific parts, drilling down looking for those specific variables that matter, and declaring with a degree of (often unwarranted) confidence why company X or Y succeeded while others failed.

This feels very much like the domain of the left brain. That's not inherently bad, let's remember; it's just incomplete. Strategy in practice is, at its best, far more about the right/left/right jump we looked at in **chapter 26**. This is the more integrative approach where we acknowledge there's an ambiguous, intuitive, evolving-over-time whole from a right-hemisphere point of view, pass it over to the left to take it apart for explicit analysis, before reintegrating it back into our more implicit sense of the whole.

(This is why, incidentally, when folks talk about "strategic planning," there's always a caveat that things will change as the year, quarter, or month rolls on. Or perhaps there's a jaded, or just realistic, view that the plan will be quickly forgotten once we're back in 'reality,' so what's the point of planning at all? This is the fault of disconnected left-brain analysis — static analysis of the parts — that doesn't survive reintegration with the flow of time and ongoing, emerging, implicit, right-brain 'whole.')

Don't get me wrong. We can, and at times should, stop and analyze things in a static way, breaking them down, deliberating over them, and figuring out what that means for the future. That's much of what this whole book is. But we need to balance that with a view of the emergent and implicit that will inevitably hit when we reengage with the flow of time, i.e., back to reality and out of a planning session.

In this very left-brain world, however, the latter is often neglected or forgotten, and so we get stuck with a mismatch of strategic static "plans" and emergent events over time and wonder why the two don't match up. They don't match up because, while we use our whole brain for everything all the time, when we're doing "strategy" by analysis only, we're focusing on one mode of attention over the other, and that neglects the final L → R jump in the R → L → R pattern.

Hopefully, this chapter helps you sharpen your intuition for strategy that's both a product of analysis *and* emergence, of the parts *and* the whole, and of flowing time, not just arbitrary goals.

> **Porter, Mintzberg, and the Honda effect:** There's a famous story that

> highlights the dangers of strategy as analysis over emergence, which Daniel Good neatly summarizes on Medium[11], and which I'll draw from here. Honda entered the US motorcycle market in 1959, and by 1966 had captured 63% of the market. This was a blow to the British, whose market share fell from 49% to 9%. The British commissioned Boston Consulting Group (BCG) to find out why. BCG delivers a lengthy report full of strategy jargon, Good says, on "how the Japanese had positioned its bikes and won a beachhead with new customer segments using low-cost production." Honda's success then becomes the basis of case studies taught in business schools, including by Porter. A guy called Richard T. Pascale, however, goes to Japan and interviews the relevant managers from Honda. Did their internal strategy match what was being described in the West? Nope, they had "no strategy, other than the idea of seeing if we could sell something in the United States." They arrived with poor timing, little money, limited English, and poor quality flagship bikes. They reluctantly decided to sell their small 50cc bikes instead, largely out of necessity. But that worked! That's what got them into the market. Pascale would go on to tell the *real* story of Honda's emergent success at a strategy conference, with Porter presenting the now-known-to-be-artificial BCG view at the same conference the next day. Professor Henry Mintzberg (champion of the emergent strategy school and Porter critic) said it "was kind of a high point in the conference because everybody was sitting there thinking, 'Mike, you really should have been here yesterday.'"

Emergence + analysis

The Honda story highlights the challenge of analysis vs. emergence in strategy and positioning. I see it as a case of attention preferences — the left-brain-favoring academics lapped up the analysis, and in a sense the analysis, insofar as it was a snapshot of what happened, *was* accurate. Honda *did* win a beachhead with a new customer segment with low-cost production of their small 50cc Super Cub scooters. But the way they got there wasn't *through* analysis and deliberation in the office; it was getting out there and learning in the field.[12]

The truth is that analysis and emergence can both work. That's a theme I've tried to weave through this book, with examples of strategy-by-analysis and strategy-by-emergence. We saw that analysis worked for Drift (as we discussed back in **chapter 31**), and we've also looked at how emergence worked for Segment and Loom (**chapter 8** and **chapter 10**), too. HubSpot (**chapter 31** as well) was a mix of both — analysis of the market to focus on 'inbound' and emergence in that LegalSpot wasn't a viable path forward.

We've also discussed Figma (**chapter 27**), as they've excelled at both, as we've seen. On the analysis side, Figma used the Blue Ocean Strategy's simple framework for differentiation, with Figma founder Dylan Field saying in an interview with the *Distributed Podcast*[13] (edited for clarity):

> *There was a point where we had to say we're going to focus on interface design [...] We used a cool framework — omit, raise, reduce, create — from* Blue Ocean Strategy *to figure out, okay, what are we doing that's different than the competition? What are we not doing that the condition is doing? What are we doing worse and where are we going to make this brand new?*

Both can work! You need to engage with the emergent reality layer (as discussed in **chapter 28**) *and* what could be true by way of analysis, i.e., to understand where you can zig (or omit, raise, reduce, or create) where other players are zagging and truly differentiate.

Porter and positioning

For Michael Porter, choices and analysis were the name of the game, and it's easy to see why (even if we accept the blind spot around emergence). In very simple terms, leaders have to make choices, and the question is often: do we do something new, or do we match the competition? That is, do we compete to be unique, or do we compete to be the same?

This is the positioning school of strategy where you deliberately choose the position your company will take in the market. That's not quite "positioning" as we've been using it; it's a level of abstraction higher, looking at your place in your industry as a fundamental set of strategic choices. Porter famously described three of these fundamental strategies, as Walter Kiechel describes in *Lords of Strategy*[14]:

> *There were essentially three strategies a company could choose, he posited: low-cost leadership (beloved of fans of the experience curve), product differentiation (making your offering so distinctive that you could charge more for it), or market specialization (pick a niche and dominate it).*

These generic strategies have been expanded and heavily discussed over the years, and for our part, while not exactly the same thing, we've discussed similar themes in terms of niche approaches (**chapter 37**), product differentiation in terms of attributes (**chapter 45**) and wedge strategies (**chapter 46**).

These really are just "generic" strategies — your own strategy will obviously be something much more tailored to your specific context. But regardless of the particulars of such strategies, the underlying theme is the same: uniqueness. You need to find a way to be unique.

Standard uniqueness

This core insight has been around for decades and has popped up over and over again in various guises in the years since. Many influential business books parrot a similar line that can be traced back to Porter:

- *Zero to One*, and the 'monopolies are good; competition is bad' shtick? Competing to be unique (i.e., so unique you're a "monopoly" free of competitors — it's Porter with edgy metaphors).
- Category creation? Competing to be unique.
- Finding a blue ocean? Competing to be unique.
- Fighting for a position in the buyer's mind? Competing to be unique.

These folks all have their own spin on *how* you compete to be unique, to be sure, but the gist is that you need to make unique, compounding, hard-to-copy choices so you stand on your own, playing your game — be it your *monopoly*, your *category*, your *blue ocean*, or your *position*.

It's timeless advice. So why doesn't everyone do it?

You think you're different

Quick quiz: how unique is your strategy and your product?

In more objective terms, how unique would your customers rate it?

We've been talking about uniqueness in the abstract, but if we focus on your product for a moment, and you were asked in what quadrant your features sit on an axis of value and customer willingness to pay (WTP), I'm guessing you'd probably answer that you're building stuff that's high value *and* that customers have a strong WTP for.

That is, highly differentiated stuff folks want.

The reason I'd make that guess is that ProfitWell (acquired by Paddle) ran just such a survey with 2,500 product leaders, and that's how folks responded[15]:

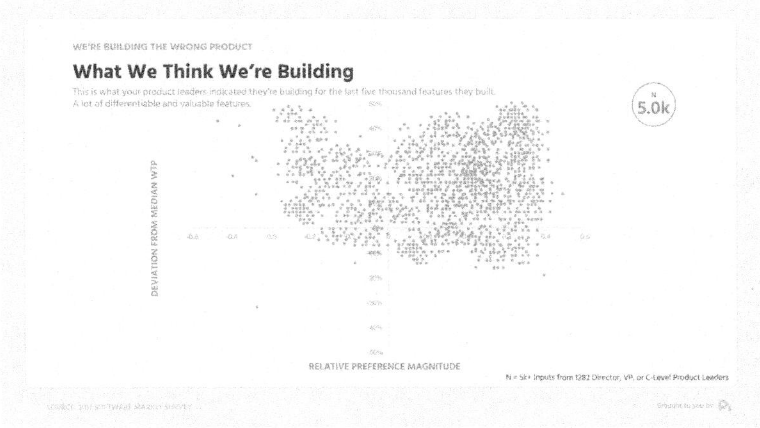

Figure 49.1: ProfitWell's product leader survey results.

That's great — everything is up in the high-value quadrant. Except ProfitWell also asked 1.2 million customers the same thing, and the results were strikingly different:

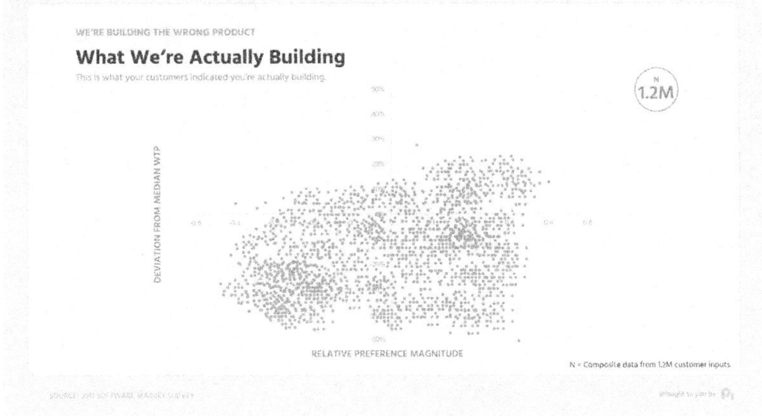

Figure 49.2: ProfitWell's customer survey results.

Uh oh. That's... not so great. That's a lot of low value, low WTP product. As ProfitWell founder Patrick Campbell says:

> *Essentially your product teams are extremely confident they're building value when in reality they aren't and they don't even know they're doing this, because you're not talking to your customers.*

This has shades of the EBI's research in the B2C brand world — brand marketers, using active attention, thought they were imbuing their brands with unique, differentiated qualities, but when customers were surveyed, the degrees of difference were vastly smaller than imagined. (We discussed this phenomenon of active vs. passive attention back in **chapter 43**.)

Overexposure risk

This is the trap of overexposure rearing its ugly head again. We see the intricacies of our product and strategy; customers don't. We apply a lot of active attention; customers don't.

Again, I don't want to throw out the baby with the bathwater here. It wasn't that customers in the survey above didn't detect *any* high-value/reasonable WTP feature differentiation. It's just that their perceptions were vastly out of sync with the folks building those features.

Maybe that doesn't apply to you, and you truly are different. If so, you're my favorite kind of business (reach out and say hi!).

But this also represents opportunity. When the rest of the market is deluded about what their customers perceive, the opportunity is there to be the one that is different in a way customers *do* perceive.

Let's also be clear about how hard it is to truly be different *in the eyes of the customer*. This is what taking competitive uniqueness seriously means and what much of this book has been about — finding genuine points of competitive uniqueness. That may be by finding that secular trend and stretching your narrative out or stretching your positioning down to find that unique value that only you can deliver. Either way, it's about doing everything you possibly can to get out of that undifferentiated quadrant of sameness.

Competitive uniqueness

> *Competing to be the best [i.e., "more better"] feeds on imitation. Competing to be unique thrives on innovation.*
>
> – Joan Magretta[16]

And that is *really* hard.

Part of the challenge is that as the SaaS world matures, it gets harder and harder to meaningfully differentiate. The low-hanging fruit is gone, and while it's easy enough to pick an attribute to position around, finding a true wedge opportunity that customers see and appreciate in a high-value/high-WTP way gets harder and harder.

Contrast this with the way uniqueness gets celebrated in business books as though it's normal and obvious. It *is* obvious — in the very rare mega-winners everyone loves to cite (Apple, IKEA, Southwest, etc.). These companies, built over decades, are what connoisseurs of business strategy like to savor, and in an evolutionary sense, it *is* interesting to observe the kinds of businesses that survive and thrive. If competition is the great leveler in markets, then by definition, it's almost an evolutionary necessity to compete in a way that others can't match, that is, to make compounding choices unique to you.

That's very easy to say, but from a psychological point of view, "competing to be unique" seems to be punishingly hard for us to accept as humans because we almost always default to competing to be the same as other players — we imitate first, perhaps with a little tiny twist of differentiation.

That little twist of differentiation is usually that we're "more better," as David C. Baker mentions in the opening quote. But being "more better" isn't being unique; it's being the same. That's because your narrative is functionally the same as your competitors — they're all "more better" too. It's a "more better" bake-off. There's no meaningful secular change to drive new behavior or some powerful insight you've found by drilling right down to some tears of gratitude-inducing value that only you can provide.

The irony is that everyone in a crowded category all think they're standing out to some degree. They've all got a 'unique' brand. They might pick up a 'unique' attribute. They've all got a 'unique' set of features. And they'd all put themselves in that top differentiated quadrant.

But so does everyone else. This is *undifferentiated* differentiation.

Customers see undifferentiated differentiation for what it is and mentally place you in the undifferentiated quadrant instead. *You're a bit more Y, the competition a bit more X... all looks the same to us. We'll flip a coin and choose one.*

> **Competing on GTM vs. product:** Being in a "more better" bake-off is not an inherently insurmountable problem, so long as you know what game you're playing. You just have to pull other levers. You can try and go for reach by maximizing market penetration, for example, perhaps while fast following any major features competitors roll out. You'll need a lot of capital to pull it off — you're substituting innovation for CAC warfare, but it is possible for some players to outspend the competition. (Making that money back is another thing, of course.) You can compete on who is easier to buy — the most memorable brand, the friendliest features, the most accessible product, the easiest to get started with, etc. But I'm always wary of trying to cordon off competition to just one axis. In reality, you need to consider the whole — if you can stay ahead of the competition on GTM execution, then

> there's no reason you can't catch up and pull ahead on product roadmap, too, ideally carving your own path that makes your whole model more robust. But it requires being clear-eyed about the exact strategy you're pursuing — where you'll be the same and where you'll be genuinely different in a customer-perceived way.

Competitive denial of uniqueness

> *Deny differentiation. Without ever giving them credit, blunt every innovation competitors use to try and differentiate from you, the market leader.*
>
> – David Kellogg on the 'market leader' play in enterprise software[17]

In the world of startups, technology, and innovation, the emphasis is — quite rightly — on the new. That's why we're doing this, after all. We're trying to play positive-sum games and create new things. The world only gets better (or, at least, more productive in B2B) if people get together to try and do new things to *make* it better.

But there's a more elemental game of survival, too. If uniqueness *is* such a powerful force in competitive markets, then the corollary is that denying uniqueness is powerful, too. And that's what incumbent market leaders do — just look at Microsoft or Meta, for example, and ask Slack or Snapchat. Incumbents know the disruption playbook, and they're more than happy to snuff out anyone trying to exploit a wedge in a niche. They'll imitate, adapt, and exploit the innovation with their incredible reach instead. They want to keep their spot at the top of the adoption curve and are more than happy to stomp on any competitors rushing up it.

And to be fair, sometimes competing to be the same is just "keeping up" with competitors — par for the business course — where everyone evolves together. Sometimes ideas are just in the air, or there's a spectrum of JTBD coverage that everyone is slowly biting off. It may be less of a one-on-one boxing slugfest (like Yammer and Jive in **chapter 32**) and more of a Tour de France endurance test with a pack of competitors. But it's a race with no end, and the question is who can iterate — and learn — fast enough to break away.

Hotelling: The anti-Porter

This phenomenon of competing to be the same is not new. It was described over a century ago, as mentioned. But it's not very popular. It doesn't make for catchy pop business books. No one's shifting units with buzzy books titled "Copy Their Why" or "Play the Same." And yet, isn't that how markets often play out?

Perhaps it doesn't get much attention because it is such a default move for us as humans — to see growth or profit anywhere as a signal to rush in and grab some while the going's good. And maybe that's why 'compete to be unique' is a mantra that *needs* to be repeated over and over — because it is so counterintuitive to our default reaction of: 'I don't know, just copy whatever they're doing!'

Imitation is, after all, fundamental to how we learn and who we are. As McGilchrist says:

> *[W]e become who we are by imitating the models of people we admire or respect. [Imitation] is also how we acquire most of our skills, even though at times we may resort to copying by rote.*
>
> – Dr Iain McGilchrist[18]

And that seems to be true from giant, world-spanning enterprises down to each of us as individuals. This isn't just a function of human nature, either, but a *rational response to competition.*

That was the observation of Harold Hotelling, the anti-Porter, back in 1929 [19]:

> *Hotelling's law is an observation in economics that in many markets it is rational for producers to make their products as similar as possible. This is also referred to as the principle of minimum differentiation.*

The *principle of minimum differentiation* describes how very similar businesses end up clustering together, like fast food joints that all exist within a short walk of each other at one end of town or, say, CRM or project management tools that all do the same thing (and are all oddly "number one").

This is the denial of uniqueness — the very opposite of competing to be unique.

> **Hotelling, Sharp, and market nihilism:** Brand strategist Nick Liddell has an insightful article both illustrating Hotelling's law and examining how Byron Sharp and the EBI's ideas drive brands right to the dead center of the market in an almost perfect demonstration of what Hotelling observed.[20] This is the gravitational pull of the market that draws everyone to much the same point — same features, same

> marketing strategies, and same positioning, no matter how much folks wish they could 'stand out.' Why? Because everyone is serving the same cross-section of undifferentiated market demand, and so all arrive at the same point. But if you take this too far, you get a form of market nihilism. Differences don't matter; just be the same with distinct brand assets. This is the mistake of confusing what often *is* for what *ought* to be. But unique companies exist! It *is* possible to build a unique strategy! You *can* get closer to your customers, innovate in a way they care about, break away from the competition, *and* reach more of them through 'creative publicity' (and/or an abundance of capital). That's what makes tech special — unique innovation that's hard to match *is* the ultimate creative publicity.

'Fair' competition

> *Managers often think about competition as a form of warfare, a zero-sum battle for dominance in which only the alphas prevail. This [...] is a deeply flawed and destructive way of thinking. The key to competitive success—for businesses and nonprofits alike—lies in an organization's ability to create unique value. Porter's prescription: aim to be unique, not best. Creating value, not beating rivals, is at the heart of competition.*
>
> – Joan Magretta[21]

Fast food joints *could*, in theory, all be placed equidistant across town and, theoretically, all do completely different things. The even distribution and 'unique value' gives them equal access to customers (and vice versa), and everyone is — in theory — happy.

Creating value is great, but contra Magretta and Porter, sometimes markets *are* zero-sum, and you do need to beat your rivals, too, or at least cut off what makes them unique, or they'll drive you out of business.

That's what Hotelling's law is about. In a nutshell, it works like this: CRMs could, for example, all serve distinct, mutually exclusive verticals — placed equidistant across industries if you like, and everyone gets an equal slice of the pie. But because we're a greedy little lot, we end up trying to grab more and more of the same pie until an equilibrium of sorts is reached, where players have moved so close to each other that there's no real advantage to be gained by moving *away* from the group.

And this is what we see in practice — folks piling into the same categories over and over, trying their best to pick up market share by mimicking (i.e., moving closer to) the leader. That might work, for a while at least. But if the incumbent is competently run, they'll respond, mimicking (i.e., moving closer to) whatever made the challenger unique — deny and blocking their differentiation — while still preserving their reach.

And thus, this game of edging closer and closer continues until everyone settles on a local maxima. In mature markets, this results in fast food joints (or gas stations, or convenience stores) quite literally moving as close as possible to each other, selling the same stuff at the same price with the slimmest of differentiation.

The same is true for mature technology markets, too — Windows and macOS, iOS and Android — similar products serving similar needs in a similar way. Convergence and a dominant design emerge.

> **The SaaS 'sea of sameness' & the shape of B2B demand:** We're exploring uniqueness from the point of view of the business here, but they're not the only ones with agency in a market. We can infer certain things about the shape of demand by seeing how B2B SaaS markets shake out. If your market is indeed a 'sea of sameness,' then that suggests (a) it's a big market — a sea — that can support many players, and (b) the needs of the market are either very similar and/or the buyers are very conservative. If there was a market for wildly different solutions, then that's what we'd see instead. Furthermore, if a winner hasn't taken all, and the market is mature, that suggests two things. One, buyers aren't loyal and are happy to churn to you — hooray, new customers for you, and two, that lack of loyalty (and the resulting game of vendor musical chairs) suggests significant churn away from you is inevitable too — boo, no winner-takes-all opportunity for anyone. This is why understanding the narrow MAYA threshold in your market, as we saw previously in chapter 48, is so important. You need to be advanced enough — unique enough — to be worth churning to while acceptable enough — similar enough — to not be too painful to adopt when there are plenty of other worthwhile alternatives. (Of course, if you *can* completely disrupt the market instead, go for it, but you need to be cognizant of the market forces other players have already discovered, too.)

Eras of ferment vs. dominant designs

> *The early days of an industry are an "era of ferment:" many solutions are possible & entering is cheap, breaking through is hard. Then a dominant design emerges, the focus goes to price & quality. R&D*

> *costs go up, competition is fierce. Discovering what consumers want & what can be produced is an evolutionary process. Once companies figure that out, they drive down prices and fight hard to be the winner.*
>
> – Ethan Mollick[22]

Here's where founders can get confused.

Investors tell them to, essentially, compete to be unique in the most extreme form possible: create a category! That's where all the returns are! (Remember that is, as we saw in **chapter 34**, not particularly true.)

Founders declare that's what they're doing and tend to believe their own hype. *Look out world, we're creating a new category here!*

Maybe that's true, but your competitive strategy needs to take into account the broader era you're operating in and whether it's an 'era of ferment' (where uniqueness and novelty are highly prized as teams try and iterate towards an evolutionary winner), or whether a 'dominant design' has already emerged, and you're more likely to be fighting it out to be the *ideal form* of the 'dominant design' where customer expectations are relatively mature. There, it's more a question of execution excellence than raw innovation.[23]

Who learns fastest, wins

How you compete — to be unique, to be the same — is, in part, a function of how fast you learn. That can be learning through analysis or learning through doing, i.e., emergence. If all you "learn" about the market comes from competitors, then you'll never develop the uniqueness to out-compete them. But if you *never* learn from competitors, then you'll get left behind with a product with clear vulnerabilities.

What happens if we map this to our hierarchy of attention?

This isn't a perfect analogy (hey, it hasn't stopped me thus far), but in my view, it sheds important light on the way folks compete.

Competing to be the same is often the product of narrow, laser-beam attention applied to competitors. It's about that static, side-by-side competitive snapshot and how you compare. In sales terms, it's about battle cards and 'versus' marketing pages.

Competing to be unique, on the other hand, is a balance of both right-brain observation and left-brain analysis — looking *out* at what's new, what's possible, and where opportunities for innovation lie on the right-brain radar, while also looking *down* with left-brain analysis at both your customers and competitors to cover the core JTBD on your platform. Here, opportunity comes from looking beyond the fishbowl you and your competitors are swimming in and out into the broader tech environment or deeper into your customers' work lives.

What happens when you do both?

Give 'em the ol' radar and laser beam

Truly well-run businesses end up both competing to be unique *and* competing to be the same. They can play both offense and defense. The strongest players can really drive both at once, charting their own unique path and loudly touting their unique approach while furiously matching the incumbent feature for feature (or job-to-be-done for job-to-be-done).

And it was ever thus. Here's Ben Horowitz (the Horowitz in Andreessen Horowitz, a.k.a. a16z) writing in 2010 about his experience leading Loudcloud/Opsware in the late '90s and early '00s[24]:

> *But in a bust (like the one we were in), having a lot of cash can be a huge competitive advantage because you can use that cash to put enormous pressure on your underfunded competitors. And that's what we did. We spent aggressively to match our best competitor's product, feature for feature. And we used our public currency to acquire important adjacent functionality (network, process and storage management) that our competitors did not have and couldn't acquire because they didn't have the cash (or the equity).*

This is the laser beam and radar in action. The competition gets the laser beam treatment — matched feature for feature. Strong defense. The market gets the radar treatment — acquiring functionality to build outwards. Strong offense.

And the cash meant they could simply do it faster — and do it at all — when their competitors could not. That's significant strategic differentiation! I'm sure this pattern will repeat as downturns come and go, decades on from Horowitz's experience.

The point here is that being a strong competitor is not just dabbling in uniqueness. It's not about having a little nibble at both — *we have this thing that's kind of neat, and we do this other thing that's kind of what everyone else does*. We're a long way from "more better" here.

People talk about having an *aggressive* strategy, but really it's just about having a proper strategy at all, and taking competitive uniqueness, in both forms, more seriously than the competition.

That's easy to say, but how do you actually do it?

Finding your uniqueness

The 'competing to be the same' side is obvious — go toe-to-toe with true competitors feature for feature (we're imitators at heart, after all) — so let's focus on uniqueness. In this book, we've discussed narrative, differentiation, and positioning — i.e., finding your uniqueness — from a variety of angles, including:

- **Secular change & narrative:** Developing a unique narrative and strategy by riding a wave of secular change and developing your unique playbook to match (**chapter 31** and Qualtrics in **chapter 35**).
- **Positioning & JTBD:** Developing your unique insight by looking at the JTBDs your best-fit customers perform and positioning around them (**chapter 36** and **chapter 37**).
- **Category creation:** We unbundled this as generic category (i.e., literally a new category of software), owned concepts, and Aggregators (**chapter 34**), and this may well be the lens you're using to define your uniqueness, too.
- **Brand and distinctiveness:** Unique *distinctiveness* in a visual identity sense and the unique, attribute-driven *gestalt* that defines your brand on a more fundamental level (**chapter 45**).
- **Wedge positioning:** Finding a unique wedge attribute, as we saw competitors naturally try and do to carve out their own unique space in a crowded market (**chapter 46**).
- **MAYA:** Finding that unique sweet spot around what's "most advanced" but still "yet acceptable" and therefore easy to buy (**chapter 48**).

There are plenty of angles to find uniqueness in your narrative and positioning. But how do you take that further? How do you take it beyond a simple strategy and turn it into something more unique and defensible? How do you make it into a moat, as they say?

This is the world of *compounding* strategy and the *complexification* that emerges over time.

Compounding choices vs. a pile of features

> *It's easy to claim that "our peculiar set of features makes us unique," but that's not strategic unless it forms a mutually-reinforcing network, including weaknesses. It's not durable leverage unless a written strategy explains this network, so that the company can continue to reinforce the entire network with its actions and investments, including accepting the weaknesses, not treating them as "bugs to be fixed."*
>
> *Otherwise, it's just a pile of features.*
>
> – Jason Cohen, channeling his inner Michael Porter[25]

A "mutually-reinforcing network" of choices is the key idea here, and this is where we jump into the Michael Porter world of strategy with both feet. (Or the Jason Cohen world, as the case may be, thanks to his fantastic long-form writing on strategy.[26])

Let me stress that, for a lot of folks, this is not where you start. This is not how you get into the market. It's how you build a durable position in the market once you've got the flywheel going. But it's worth understanding so you don't fall victim to the 'pile of features' trap.

One of the most straightforward but far-reaching ideas of Porter's is the simple concept that different outcomes require different activities, and different activities that compound build durable competitive advantage (i.e., a competitive moat). That's an idea I've tried to sprinkle in across the book (e.g., in our discussion of positioning in **chapter 37**), and it becomes particularly necessary the bigger companies get.

This is where you build a *gestalt* not just on fuzzy brand attributes but hard-nosed strategic choices. To do that in a way that creates leverage requires compounding choices, and it's compounding choices that separate the pros from the amateurs.

That's easy to say but brutally hard to do, as Dropbox has found.

Dropbox's failure to compound

Dropbox, who went public in 2018, rode to fame and fortune as a B2C-ish file-syncing solution, attracting hundreds of millions of users. But file syncing has become both commoditized and increasingly irrelevant in a cloud-first world, a fact Dropbox's leadership (and the market) are acutely aware of.

With secular trends working against their long-term viability, Dropbox has spent years looking for its next act, acquiring a variety of companies in the process. That has included Dropbox trying its hand at email and collaboration (Mailbox[27]), chat (Zulip), photo management (Loom, Snapjoy turned Carousel), web documents (Hackpad, which became Dropbox Paper), web project management (Paper again), and more recently document sending and signing (DocSend and HelloSign).

But none of these, perhaps save the last ones, meaningfully compounded on Dropbox's strength in behind-the-scenes file syncing, which is, I'd suggest, why these quasi pivots haven't stuck. I can see the logic in *trying* to capitalize on Dropbox's enormous installed base of users, but evidently, these additions never added up to a meaningful new strategic position. Dropbox's share price has (excluding the bubble of the early 2020s) mostly traded below its initial share price in its 2018 IPO, and all this 'strategy thrashing' is a symptom of bad or non-existent strategy.

Perhaps they'll unlock a new phase of growth with document signing —

at least it compounds on their file-centric background — but it remains to be seen if they can make strategic choices that really do build on their (very impressive!) initial base.

> **Making lightning strike twice:** Dropbox's litany of failed acquisitions goes to show how non-determinative success is, especially in the B2C world. Dropbox has all the money, all the talent, all the users, a great culture, access (one would hope) to the best advice, and all the experience in driving viral growth loops, but so far, they haven't been able to repeat their initial success. Perhaps they will in the future, and in the meantime, I'm sure they'll remain being great at *being Dropbox*, but so far those lessons evidently don't generalize. Part of the trap is the difficulty of moving brand memories — Dropbox has a very strong brand in terms of memory association with file sync and storage, but creating new brand associations such that when people hear "Dropbox" they think "document signing" (or whatever comes next) *without* killing the golden goose is one devilishly hard problem.

Lead and stack

> *Good strategies depend on many choices, not one, and on the connections among them. A core competence alone will rarely produce a sustainable competitive advantage.*
>
> – Joan Magretta[28]

It's compounding choices, however, that build tight strategies. If you want to build a different kind of company, you need to stack your choices in a different kind of way. Again, different activities result in different outcomes. What can you stack that drives compounding advantage? For example, consider how these approaches might combine and multiply:

- **Product and insight:** The unique tech, market, or UX insight or 'secret' that's driving what you do.
- **GTM:** Sales-led that's far faster than the competition, product-led that's more collaborative and viral, or unique hybrid approaches that make the most of both worlds.
- **Concept:** The secular change you ride, the playbook you name, the concept you coin.
- **Positioning:** The market you choose, the niche you focus on, the category you define, the JTBD you solve for, and/or the distinctiveness of the brand you build.

- **Team:** There's unique advantage in having a team of folks with more subject matter expertise, more experience, or that just care more.
- **Success:** Ultimately, how can you deliver value to your customers in a way that makes them uniquely successful? How can you go beyond software alone to deliver that success?

What are your unique advantages in this mix of factors?

IKEA's compounding choices and activities: IKEA, while B2C, is the classic business case study for these kinds of compounding choices. In *Understanding Michael Porter*, Magretta describes the 'fit' of IKEA's multiplicative choices, including huge stores with warehouses attached, loads of free parking, everything on display for you to see and try, and a cafe so you can spend the day. All these choices — along with the cheap prices, which are a product of an additional set of choices — add up to an experience that makes it incredibly easy for customers to buy and incredibly hard for competitors to copy. IKEA's products are therefore cheap, available, accessible, exhaustive in range, low pressure, and easy to commit decision-making time to (it's a big store with a cafe), which adds up to a brand that is both highly differentiated and highly distinct. I fear in B2B SaaS that even that list above (product, GTM, people, success, etc.) is too typical, too much within the realm of competing like everyone else. As SaaS gets more competitive, I look forward to more creative big bets on building truly unique companies that transcend the traditional playbook, both in terms of unique product offerings and a unique, incredibly buyer-centric buying experience. ProfitWell, for example, bootstrapped for 10 years as a tech-enabled service with excellent margins and sold for $200M.[29] Vertical SaaS has also emerged where money is made on transaction fees and the software is given away. Unique models get overlooked when everyone tries to de-risk by following "proven" playbooks, but by definition, that's competing to be the same by doing what everyone else is doing.

Compounding and expertise

Let's revisit some examples of compounding choices in some of the startup stories we've looked at so far:

- **David Sacks** talked about Yammer's unique product, category positioning, architecture (SaaS, not on-prem), and go-to-market model (bottoms up, a.k.a. product-led), as we saw in **chapter 32**. That's a lot of compounding.

- **David Cancel** likewise talked about Drift's market choice (CRM), the trend it would ride (shift in buyer power), the market and technology shift (messaging and live chat), and the named playbook (conversational marketing), as we saw in **chapter 31**. That's compounding through astute observation, strategy and planning.
- **Nathan Barry** bootstrapped through ultra-niche positioning, founder-led sales, founder onboarding (by hand!), and eventually references and affiliates to get the flywheel going, as we saw in **chapter 39**. That's compounding by sheer force of will.
- **April Dunford's** discovery and position work with a database product (as we discussed in **chapter 37**) turned up a niche use case that necessitated a completely different way of selling the product, where compounding choices were made on the back of a unique insight about customer success.

These are just a few great examples where compounding choices added up to a unique *gestalt* on a business model level that delivered unique value in very competitive markets.

> **Seasoned operators & richer strategy:** What I find particularly interesting, dear reader, is the 'seasoned operator' phenomenon where folks who have done their time on the startup merry-go-round just tend to have a more sophisticated, more intuitive, and deeper sense of compounding choices both in the company they're building *and* the macro environment they're operating in. Their strategy is richer because their experience has been richer, and that sustains them and their company for a very long time.

Activities → outcomes

Let's look at a few more examples from earlier chapters where folks have managed to make compounding choices in a smart way, picking a unique evolutionary path that was fundamentally different from everyone else. Again, different activities = different outcomes:

- **Figma** spent *four years* building their still-to-be-replicated tech (using WebGL and WebAssembly), used a generous product-led approach to drive usage, figured out their competitive differentiation with a *Blue Ocean Strategy* framework as mentioned, and leveraged the rise of design systems, as discussed in **chapter 27**. Different activities, different outcomes.

- **Qualtrics'** first act was to go super niche selling to marketing professors, but later success came from repositioning around their XM concept and playbook, changing their GTM, focusing on the C-suite, and more (**chapter 35**). Different activities, different outcomes.
- **Squarespace**, focusing on SMBs, didn't just innovate with their design-oriented product, they innovated with a novel podcast advertising program riding the wave of renewed interest in podcasts (**chapter 44**). Different activities, different outcomes.

This approach, where differentiated results require differentiated choices and activities by definition, may be driven by choice and planning, emergence and good timing, or a combination of both. Either way, it's the opposite of silver bullet thinking. Instead, it's stacked thinking: *and, and, and*, or & + & + & as I like to describe it.

These stacked choices include your narrative and positioning. The secular change *and* the technology wedge. The catchy name *and* the playbook. You've got to start small and find fit — the simple system has to come first — but then it's a question of stacking advantages and doing what other competitors can't or won't.

Complexification and clarity

> *A complex system that works is invariably found to have evolved from a simple system that worked. A complex system designed from scratch never works and cannot be patched up to make it work. You have to start over with a working simple system.*
>
> – Gall's law[30]

This is all starting to sound a bit complex, and it is, but it needs to start from something simple that works.

The idea behind compounding choices is that they add up to something unique and recognizable. We've also talked a lot about the need for simplicity in your narrative, messaging, and positioning (e.g., your Hemingway story in **chapter 10**), and unique choices can create a distinct whole that, paradoxically enough, makes you an easier, simpler choice to buy. When you're unique, buyers aren't trying to parse minute differences between dozens of competitors that look just like you, they can just buy *you*.

But the operative word there is *can*. We need to beware of the 'Honda effect' we touched on earlier, where we derive strategy concepts from analysis of the survivors of a long, emergent, evolutionary process for any given company. That kind of analysis is great for knowing what a successful end state looks like, but the evolutionary process to get there, though, was probably driven by a simple system that worked in the first place, as Gall's Law reminds us. In the tech world, that simple system is your initial product/market fit, and without that, no amount of strategy complexity will help.

It took IKEA, the pinnacle of compounding choices, for example, more than a decade for its basic model to emerge, which came about at times out of sheer necessity.[31] Building a deep strategy means playing the long game, not rushing for an exit.

In any case, once you've found that simple system that works, what happens?

Inevitably, things get more complex. Growth usually always begets complexity. But complexity isn't accidental; it's something we actively do. We *complexify*. Complexification seems to be intrinsic to evolution and the universe itself — here we are, as complex beings, doing complex things in a universe that we know wasn't always like this.

This is the journey of every growing company — complexifying with additional features, products, use cases, infrastructure, documentation, marketing programs, market segments, sales approaches, and so on. Every function in a growing company starts as someone else's part-time job before blossoming — and complexifying — into a role, team, department, and organization in its own right.

Even startup websites usually start simple, and then the content grows and the design matures, and what's the result? Something more complex. When folks want their landing page design to communicate something impressive about their company, what do they do? They complexify, with more pages, more links, more design details, more value props, more logos, case studies, and statistics. Complexification itself seems to signal — or at least folks want to believe it signals — a complex, evolved, mature organization.

The task, then, is to find clarity in that complexity and not succumb to complicated approaches for their own sake. You can be clear in message with a complex understanding of a user's exact needs. You can be clear in design because of the technical complexity under the hood that ultimately lets users do more.

When it comes to strategy, the choices you make are fundamentally about creating clarity in how you operate and who you serve *while* your product and organization complexify. You'll inevitably choose to do more as you grow; it's just a question of whether you can choose to do that in a way that continues to bring clarity to how you help your customers succeed, clarity to how you derive returns from that value, and clarity to how you operate that's different from your peers. IKEA might be complex, but it's very clearly *IKEA*.

The question, then, is whether you complexify with clarity as you balance competing to be unique and competing to be the same. That becomes incredibly hard to do if it's not baked in from the very start of your journey. In the ZIRP era, this degree of clarity wasn't that important, as there was always the next funding round around the corner. But in a world that's returned to efficiency and fundamentals, it's never been more important to have strategic clarity at every stage of your journey so you can complexify intelligently and evolve in a coherent, compounding way that's hard for competitors to match.

Competing in a post-ZIRP world

The best way to do that is to build on fundamentals that we know work. Not work in a 'proven playbook' sense, but in terms of first principles — how you compete, how you direct your attention, and what traditional business discipline and sound strategy looks like — the stuff that's been known for decades but has been forgotten in the drunk-on-capital, ZIRP-fueled euphoria of the late 2010s and early 2020s.

In times of euphoria, bets go all-in on emergence, assuming market forces are strong enough to take everyone to the moon, and it's all right-brain, ride-the-new-wave narrative. But when markets fall back to earth, analysis comes back into fashion, and it's all spreadsheets, efficiency, and left-brain number crunching.

For high-risk bets like venture-funded technology companies, there always needs to be a balance of both — ambitious right-brain narrative and sober left-brain analysis — and my hope is that you, as a founder or operator, can find a powerful way to combine those modes of attention.

And if the choice ultimately comes down to competing to be the same and competing to be unique, I hope you'll choose the latter, because there really are so many different ways to win.

Attention as a strategy

> *Competition focuses more on meeting customer needs than on demolishing rivals. Just look around. Because there are so many needs to serve, there are many ways to win.*
>
> – Joan Magretta[32]

I love this quote from Magretta. "Just look around." *Just look around*. We let our attention get so fixated on rivals and peers that we forget to zoom out and *look around*.

Your differentiation and uniqueness is, like everything else, only as strong as your discovery. The inputs matter. Otherwise, it's garbage in, garbage out. And that discovery is, fundamentally, a product of how you direct your attention — what else could it be? That's why we've spent all this time discussing ways to stretch your attention out and down, to get the most out of you and your organization's right-brain radar and left-brain laser beam. To discover, by analysis or emergence, what can make you unique and competitive in the incredibly crowded evolutionary environment that is the world of B2B SaaS.

Back in **chapter 21**, we discussed Peta the pigeon, with its attention on the seeds on the ground *and* its attention looking out for changes in the environment. As leaders caught up in the day-to-day, it's very easy to stay focused on the seeds on the ground. But sometimes, we need to look up and look around. Consider the many ways you could win, not just one. What seems most plausible? What seems most worth trying? What are customers actually trying to do? And what's the simplest system that could work that you can evolve into a complex one?

It's that kind of discovery that will drive your success.

And while I can't tell you how that will work, I can tell you what we know does work, so you can focus on the discovery that matters and not on rediscovering the things we already know. And the good news is that we know quite a bit about what can work — that's what this book has been about, after all. In the next and final chapter, we'll recap those things that we know and finish on how essential it is to get started *writing*.

Notes

1. Baker, David C.. ""More Better" Is Not a Strategy," November 16, 2017. **https://www.davidcbaker.com/more-better-is-not-a-strategy**
2. Ramadan, Al; Peterson, Dave; Lochhead, Christopher; Maney, Kevin. *Play Bigger: How Rebels and Innovators Create New Categories and Dominate Markets*. Little, Brown Book Group. Kindle Edition.
3. Brinker, Scott. "2023 Marketing Technology Landscape Supergraphic: 11,038 solutions searchable on martechmap.com," May 2, 2023. **https://chiefmartec.com/2023/05/2023-marketing-technology-landscape-supergraphic-11038-solutions-searchable-on-martechmap-com/**
4. Magretta, Joan. *Understanding Michael Porter* (p. 12). Harvard Business Review Press. Kindle Edition.
5. Kiechel, Walter. *Lords of Strategy* (p. 117). Harvard Business Review Press. Kindle Edition.
6. Like 1985's *Competitive Advantage: Creating and Sustaining Superior Performance.*
7. Magretta, Joan. *Understanding Michael Porter* (p. 75). Harvard Business Review Press. Kindle Edition.

8. "Keynote on Strategy By Michael Porter, Professor, Harvard Business School," Institute for Competitiveness, India, July 6, 2017. **https://www.youtube.com/watch?v=EvvnoNAUPS0**
9. If you really want to nerd out on management strategy, JP Castlin's paid *Strategy in Praxis* newsletter (and archives) is a fantastic resource for getting up to speed quickly: **https://strategyinpraxis.substack.com/archive?sort=new**
10. Kiechel, Walter. *Lords of Strategy.* Harvard Business Review Press. Kindle Edition.
11. Good, Daniel. "The Emperor's New Strategy," *Make Work Better,* November 30, 2018. **https://medium.com/make-work-better/the-emperors-new-strategy-37b17afe420e** (*Lords* also covers this in some detail.)
12. Ultimately, as *Lords* mentions, Honda would go on to become a client of BCG, too, so it wasn't like they never saw the value in BCG's approach.
13. Connie Yang interviewed Field in "Episode 29: Dylan Field, Figma Co-founder, Talks Design, Digital Economy, and Remote Culture with Host Connie Yang," *Distributed Podcast,* November 18, 2021. **https://distributed.blog/2021/11/18/dylan-field-figma-talks-design-with-connie-yang/**
14. Kiechel, Walter. *Lords of Strategy* (p. 132). Harvard Business Review Press. Kindle Edition.
15. Campbell, Patrick. "Customer Research: Benchmarks and Common Mistakes," May 29, 2018. **https://web.archive.org/web/20231004103406/https://www.profitwell.com/recur/all/customer-research-benchmarks** Charts used with permission.
16. Magretta, Joan. *Understanding Michael Porter* (p. 26). Harvard Business Review Press. Kindle Edition.
17. Kellogg, David. "3. Deny differentiation. Without ever giving them credit, blunt every innovation competitors use to try and differentiate from you, the market leader." March 21, 2022. **https://twitter.com/Kellblog/status/1505735756257525760**
18. McGilchrist, Iain. *The Master and His Emissary* (p. 121). Yale University Press. Kindle Edition.
19. See "Hotelling's law" on Wikipedia: **https://en.wikipedia.org/wiki/Hotelling%27s_law**
20. Liddell, Nick. "Byron Sharp – Brand Purpose and the Tyranny of the Majority," *Brandingmag,* June 30, 2021. **https://www.brandingmag.com/2021/06/30/byron-sharp-brand-purpose-and-the-tyranny-of-the-majority/**
21. Magretta, Joan. *Understanding Michael Porter* (p. 23). Harvard Business Review Press. Kindle Edition.
22. Mollick, Ethan. June 14, 2021. **https://twitter.com/emollick/status/1404283103284281347**
23. Though with the generative AI era here, perhaps we are back to an era of ferment.
24. Horowitz, Ben. "The Case for the Fat Startup," March 17, 2010. **https://a16z.com/the-case-for-the-fat-startup/**
25. Cohen, Jason. "The three kinds of leverage that anchor effective strategies," June 18, 2023. **https://longform.asmartbear.com/leverage/**

26. Jason's essays are some of the best on modern strategy for startups: **https://longform.asmartbear.com/**

27. Casey Newton covered the $100M Mailbox acquisition (announced a month after the app launched!) and the product's axing for *The Verge* in 2015: **https://www.theverge.com/2015/12/8/9873268/why-dropbox-mailbox-shutdown**

28. Magretta, Joan. *Understanding Michael Porter* (p. 198). Harvard Business Review Press. Kindle Edition.

29. ProfitWell founder Patrick Campbell discussed the acquisition in "#37: Bootstrapped for 10 years and Sold ProfitWell for $200 million" on the *Practical Founders* podcast in 2023: **https://practicalfounders.com/podcast/37-bootstrapped-for-10-years-and-sold-profitwell-for-200-million-patrick-campbell/**

30. Per Wikipedia: "Gall's Law is a rule of thumb for systems design from Gall's book *Systemantics: How Systems Really Work and How They Fail.*" **https://en.wikipedia.org/wiki/John_Gall_(author)#Gall's_law**

31. IKEA's origin story is discussed by Amy Lamare in her August, 2020 piece "A brief history of IKEA: From cheap bookcases to Swedish meatballs," which includes the emergence of flat packing by taking the legs off a coffee table, and the need for in-house design because their original manufacturers boycotted due to low prices! **https://www.businessofbusiness.com/articles/ikea-history-founder-data-hiring/**

32. Magretta, Joan. *Understanding Michael Porter* (p. 20). Harvard Business Review Press. Kindle Edition.

PART X

CLARITY IN **CONCLUSION**

50

We know how things work

> A business is talked into existence... Entrepreneurs, through communicating, provide order and familiarity to previously unordered things... **The business does not precede the narrative, but vice versa.**
>
> *– Pollak, Rutherford, & Nagy (my bold), via Ethan Mollick*[1]

Writing this book has been quite the journey for your humble author. If you've made it this far, I'm sure it's been quite the journey for you, too. I started writing *Positioning Playbook* while the good times in venture-backed SaaS were still rolling and finished when they were very much not. But the times are still fundamentally good — there was no market meltdown, no dot-com era tech wreck. There was, instead, a reversion to the mean, a sobering up, and, I hope, a return to some degree of sanity in the market. With efficiency back in fashion, it's time to return to the fundamentals, to what we know works, at least on the market side.

When it comes to B2B startups and technology, there's so much that's emergent and unpredictable. Ideas are often stumbled upon — like Slack when working on a MMORPG, or Loom when working on usability testing, or even HubSpot when trying to build a tool for lawyers — and fragile, emergent ideas need as much clarity as possible to help them succeed, both in narrative, and in how markets work.

My personal takeaway from this project is that we know quite a bit about how product marketing, including positioning, narratives, and markets themselves work; that knowledge is just not evenly distributed. My hope is this book makes some small difference in that regard, **and I very much appreciate any and all help in getting the word out.** Any help you can provide — really, any — means the world to me.

Attention <> discovery

The core of this book has been about the science and application of attention — attention through analysis and observation, planning and emergence, looking both down and out, or as I put it way back in **chapter 4**, story 1 and story 2, narrative and positioning, radar and laser beams.

The point of that attention is to help you ultimately discover what works in your specific context by applying your attention in a considered way, especially to the world of your customers and *their* attention.

Your customers are trying to discover what might work for them, too, which is why it's so important to meet them where they're at and avoid the attention mismatches we discussed back in **chapter 6**.

When attention and mutual discovery meet, good things happen — growth accelerates, learning accelerates, and a compounding flywheel kicks in where more growth drives more discovery, all of which helps you build a strong, unique, defensible position in your market.

The fuel for that flywheel is market insight, be that a wave of secular change to drive your story 1 or untapped needs around jobs-to-be-done that power your story 2. Those insights may not always be easy to find, but you'll never find them if you don't look or can't see them. I hope the tools of attention and ways of seeing that we've covered in this book:

- Radically improve your chances of finding, understanding, and capitalizing on those kinds of insights.
- Help drive your organization further along the adoption curve and find true product/market fit, where 'market' isn't an afterthought but an integral part of your strategy.
- And (if I may be so bold) provide, through Iain McGilchrist's work, a deeper sense of how we as humans exist in and 'attend to' the world around us.

To wrap up, let's briefly recap what we know works.

> **Finding my secular trend:** Speaking of secular trends and narratives, I didn't start this book with any secular trend in mind; I just thought the rough sketch of ideas I had could be valuable if they were fleshed out a little more. I ended up fleshing them out a *lot* more, and meanwhile, the market went through quite a shift. Hey presto: an emergent secular trend to ride (or die trying). You better believe I'm going to dine out on that trend, especially the end of the ZIRP era and the return to fundamentals. That's what this book is about, after all — fundamentals and, fittingly enough, the emergent nature of opportunity, too. It's funny how narratives — introductions, really — sometimes only appear at the end once we've done the work and built (or written) the thing. Now the real work begins. Wish me luck riding the wave — I'm just one dude on a rickety surfboard out there, so I'm going to need it!

We know how the fundamentals work

We know how attention works

We know how right-brain 'story 1' narrative works with a focus on change over time and new ways of *understanding* the world.

We know that some prospects engage primarily with that 'radar' attention, e.g., people keeping an eye out for what's new and executives wondering where they should take their company.

We know how left-brain 'story 2' positioning works with a focus on jobs-to-be-done and specific wedge attributes.

We know that in-the-trenches prospects, in particular, engage with narrow-focus laser-beam attention, especially late in the sales process.

We know what makes messaging stick.

We know that metaphors and memetics can help create those 'information packets with attitudes' that help our messages spread.

We know how adoption works

We know there's a tech adoption curve and that your positioning has to match your stage.

We know sometimes you've got to play narrower before you play bigger, and go niche before you go for reach.

We know sometimes you've got to work with the nerds before you cross the chasm to the normies, and that what appealed to the nerds might turn off the normies.

We know 'category creation' is a bundle of ideas from new concepts to new categories of software to unique exemplars and Aggregators. We know that today, the vast majority of companies IPO in an existing category.

We know that different segments behave very differently across the B2B spectrum, especially B2SMB vs. B2Enterprise, and addressing each end is like running two separate businesses.

We know how memory and branding works

We know that branding is about building memory associations in the minds of buyers over a long period of time.

We know that branding isn't just a visual identity; it's a name/need association on the one hand and a broader *gestalt* on the other.

We know the first challenge of branding is to get prospects to think of you at all.

We know how positioning and strategy works

We know that sometimes momentum is found in a niche, and sometimes it's found in going for reach.

We know that finding a wedge is important, but building a network of hard-to-copy compounding choices is better.

We know how competition plays out around uniqueness — competing to be the same and competing to be unique.

We know all these things

But knowing isn't enough.

It's the application that matters, and getting the application right is a matter of writing and endless, bloody, brutal iteration.

We know business is built on writing

Before the in-person pitch, there's a deck. Before the product, there's a plan. Before there's a company, there are scraps of notes and bullet points and scribbled ideas.

Growing your business comes from iterating on your ideas, and the absolute fastest, cheapest, and most efficient way to iterate is to get words on a page and start thinking them through.

The act of writing, of making your ideas explicit, lets your subconscious go to work on them in the background until the task finishes and gets bubbled back up as inspiration while you do the dishes, have a shower, or wake up in the middle of the night and reach for the Notes app.

That, at least, has been my experience in writing this book. Writing fuels thinking, so long as you give yourself space to think. That's a virtuous cycle that, in my experience, is well worth unlocking.

I front-loaded this book with as many product marketing tools to get you started writing as I could, and they're worth revisiting now. Those tools include:

- Your narrative deck (**chapter 9**).
- A simple Hemingway story (**chapter 10**).
- A 5W1H snapshot of where you're at (**chapter 11**).
- A checklist for messaging (**chapter 14**).
- A cheatsheet on naming things (**chapter 17**).

There are also plenty of resources to draw from in terms of how you write a memo or pitch, including:

- Stewart Butterfield's famous Slack memo (**chapter 12**).
- Three ways to pitch (**chapter 23**).
- The Thinking, Aligning, Executing framework (**chapter 13**), which crucially starts with a diagnosis, as all good strategy should.

Any of those chapters make for an excellent place to start. Or, perhaps your best option might just be bullet points in your notes app or notebook or scribbles on a whiteboard app or canvas. It doesn't matter where it happens; it just matters that it happens.

I hope those resources provide ample inspiration to get started. You can do it. I believe in you.

Now go sell some software :)

> **Thank you so much for reading!** I'm honored you picked up *Positioning Playbook,* whether you read every page cover to cover or jumped straight here (it's a choose-your-own-adventure book, after all). Either way, I truly appreciate you making it this far. This chapter was about my takeaways, but I'd love to know what yours were, too. What stood out to you? What was helpful? What was confusing? What are you planning to run with? What worked? What didn't? **I want to hear everything. Don't be a stranger:** luke@lukestevens.co

Notes

1. From Mollick's excellent Twitter account: "In the beginning, a business is just words; deep truths on startup pitches: A business is talked into existence... Entrepreneurs, through communicating, provide order and familiarity to previously unordered things... The business does not precede the narrative, but vice versa."" January 30, 2022. **https://twitter.com/emollick/status/1487609125756452864** Original paper and source of the quote: **https://www.researchgate.net/publication/233733878_Preparedness_and_Cognitive_Legitimacy_as_Antecedents_of_New_Venture_Funding_in_Televised_Business_Pitches**

Acknowledgements

I'd like to thank my wife, Nancy, for her love, support and endless patience as this small, exploratory "writing project" turned into a massive, self-published mega-book.

I'd also like to thank:

Dr Iain McGilchrist for his theory of attention and the profound insight into the human experience that has provided.

April Dunford for almost single-handedly putting B2B positioning back on the map and providing a gateway to this world and beyond.

All the writers, founders, creators, doers, and thinkers whose work I was able to draw on, quote from, and otherwise riff on. Thanks for putting something out in the world worth talking about.

And thank you, dear reader, for reading any or all of this book. I sincerely appreciate it.

— Luke Stevens

Consulting & writing at: **lukestevens.co**

Say hello: **luke@lukestevens.co**

Made in the USA
Monee, IL
04 November 2026

67378326R00351